Bioenergy and Biofuels: Advances and Applications

Bioenergy and Biofuels: Advances and Applications

Edited by **Robbie Larkin**

$\mathcal{C}$LANRYE
INTERNATIONAL

New Jersey

Published by Clanrye International,
55 Van Reypen Street,
Jersey City, NJ 07306, USA
www.clanryeinternational.com

Bioenergy and Biofuels: Advances and Applications
Edited by Robbie Larkin

International Standard Book Number: 978-1-63240-529-6 (Hardback)

The publisher's policy is to use permanent paper from mills that operate a sustainable forestry policy. Furthermore, the publisher ensures that the text paper and cover boards used have met acceptable environmental accreditation standards.

Trademark Notice: Registered trademark of products or corporate names are used only for explanation and identification without intent to infringe.

Printed in the United States of America.

Contents

Preface

This book has been an outcome of determined endeavour from a group of educationists in the field. The primary objective was to involve a broad spectrum of professionals from diverse cultural background involved in the field for developing new researches. The book not only targets students but also scholars pursuing higher research for further enhancement of the theoretical and practical applications of the subject.

This book primarily deals with bioenergy and biofuels. It also discusses the applications and advancements in these fields. Bioenergy is a form of renewable energy which is produced by the combustion of biofuels. Ethanol, syngas, biogas, charcoal, wood, etc. are some examples of biofuels. These are the potential alternatives of fossil fuels. Scientists and researchers across the globe are constantly pushing the boundaries in this field to make biofuels as useful and productive as possible for commercial use. This book is meant for students who are looking for an elaborate reference text on the evolution of bioenergy. The readers would gain knowledge that would broaden their perspective about these renewable resources.

It was an honour to edit such a profound book and also a challenging task to compile and examine all the relevant data for accuracy and originality. I wish to acknowledge the efforts of the contributors for submitting such brilliant and diverse chapters in the field and for endlessly working for the completion of the book. Last, but not the least; I thank my family for being a constant source of support in all my research endeavours.

Editor

Identification and molecular characterization of the switchgrass AP2/ERF transcription factor superfamily, and overexpression of *PvERF001* for improvement of biomass characteristics for biofuel

Wegi A. Wuddineh [1,2], Mitra Mazarei [1,2], Geoffrey B. Turner [2,3], Robert W. Sykes [2,3], Stephen R. Decker [2,3], Mark F. Davis [2,3] and C. Neal Stewart Jr. [1,2]*

[1] Department of Plant Sciences, University of Tennessee, Knoxville, TN, USA, [2] Bioenergy Science Center, Oak Ridge National Laboratory, Oak Ridge, TN, USA, [3] National Renewable Energy Laboratory, Golden, CO, USA

Edited by:
Robert Henry,
The University of Queensland,
Australia

Reviewed by:
Jaime Puna,
Instituto Superior Engenharia Lisboa,
Portugal
Arumugam Muthu,
Council of Scientific and Industrial
Research, India

*Correspondence:
C. Neal Stewart Jr.,
Department of Plant Sciences,
University of Tennessee, 2431 Joe
Johnson Drive, 252 Ellington Plant
Sciences, Knoxville,
TN 37996-4561, USA
nealstewart@utk.edu

The APETALA2/ethylene response factor (AP2/ERF) superfamily of transcription factors (TFs) plays essential roles in the regulation of various growth and developmental programs including stress responses. Members of these TFs in other plant species have been implicated to play a role in the regulation of cell wall biosynthesis. Here, we identified a total of 207 AP2/ERF TF genes in the switchgrass genome and grouped into four gene families comprised of 25 AP2-, 121 ERF-, 55 DREB (dehydration responsive element binding)-, and 5 RAV (related to API3/VP) genes, as well as a singleton gene not fitting any of the above families. The ERF and DREB subfamilies comprised seven and four distinct groups, respectively. Analysis of exon/intron structures of switchgrass AP2/ERF genes showed high diversity in the distribution of introns in AP2 genes versus a single or no intron in most genes in the ERF and RAV families. The majority of the subfamilies or groups within it were characterized by the presence of one or more specific conserved protein motifs. *In silico* functional analysis revealed that many genes in these families might be associated with the regulation of responses to environmental stimuli via transcriptional regulation of the response genes. Moreover, these genes had diverse endogenous expression patterns in switchgrass during seed germination, vegetative growth, flower development, and seed formation. Interestingly, several members of the ERF and DREB families were found to be highly expressed in plant tissues where active lignification occurs. These results provide vital resources to select candidate genes to potentially impart tolerance to environmental stress as well as reduced recalcitrance. Overexpression of one of the ERF genes (*PvERF001*) in switchgrass was associated with increased biomass yield and sugar release efficiency in transgenic lines, exemplifying the potential of these TFs in the development of lignocellulosic feedstocks with improved biomass characteristics for biofuels.

Keywords: AP2, ethylene response factors, stress response, transcription factors, biofuel, PvERF001, overexpression, sugar release

Introduction

Switchgrass (*Panicum virgatum*) is an outcrossing perennial C4 grass known for its vigorous growth and wide adaptability and, hence, is being developed as a candidate lignocellulosic biofuel feedstock (Yuan et al., 2008). The feasibility of commercial production of liquid transportation biofuel from switchgrass biomass is hampered by biomass recalcitrance (the resistance of cell wall to enzymatic breakdown into simple sugars). Lignin is considered to be a primary contributor to biomass recalcitrance as it hinders the accessibility of cell wall carbohydrates to hydrolytic enzymes. Substantial progress has been made in engineering the switchgrass lignin biosynthesis pathway to reduce lignin content and/or modify its composition (Fu et al., 2011a,b; Shen et al., 2012, 2013a,b; Baxter et al., 2014, 2015). The downregulation of individual genes in the lignin biosynthesis pathway has been effective to reduce lignin, but can result in the production of metabolites that can impede downstream fermentation processes (Tschaplinski et al., 2012). Alternatively, overexpression of transcription factors (TFs), such as switchgrass *MYB4*, has been shown to circumvent this inhibitory effect while leading to significantly reduced biomass recalcitrance and improved ethanol production (Shen et al., 2012, 2013b; Baxter et al., 2015).

The master regulators of gene cluster TFs with altered expression could, in turn, endow such traits as increased biomass yield, tiller number, improved germination/plant establishment, or root growth as well as tolerance to environmental stresses (Xu et al., 2011; Licausi et al., 2013; Ambavaram et al., 2014). Therefore, identification of TFs with such putative roles would provide a dynamic approach to developing better biofuel feedstocks that could thrive under adverse environmental conditions. The availability of switchgrass ESTs (Zhang et al., 2013) and draft genome sequences produced by Joint Genome Institute (JGI), Department of Energy, USA, provides a vital resource for the discovery of relevant target genes that could be utilized in the genetic improvement of perennial grasses, which could be used as dedicated bioenergy feedstocks. However, compared to dicots such as *Arabidopsis*, relatively little is known about the key regulatory mechanisms in monocots that control lignification and cell wall formation; this is especially true of switchgrass. Likewise, we also have depauperate knowledge about stress responses and defense against pests in these species.

APETALA2/ethylene responsive factor (AP2/ERF) is a large group of regulatory protein families in plants that are characterized by the presence of one or two conserved AP2 DNA binding domains. AP2/ERF TFs are involved in the transcriptional regulation of various growth and developmental processes and responses to environmental stressors. The AP2 domain is a stretch of 60–70 conserved amino acid sequences that is essential for the activity of AP2/ERF TFs (Jofuku et al., 2005). It has been demonstrated that the AP2 domain binds the *cis*-acting elements including the GCC box motif (Ohme-Takagi and Shinshi, 1995), the dehydration responsive element (DRE)/C-repeat element (CRT) (Sun et al., 2008), and/or TTG motif (Wang et al., 2015) present in the promoter regions of target genes thereby regulating their expression. The AP2/ERF superfamily can be divided into three major families, namely ERF, AP2, and RAV (related to API3/VP)

(Licausi et al., 2013). The ERF family is further subdivided into two subfamilies, ERF and dehydration responsive element binding proteins (DREB) based on similarities in amino acid residues in the AP2 domain. The DREB subfamily in *Arabidopsis* and rice has been further classified into 4 distinct groups while ERF subfamily was clustered into 8 groups in *Arabidopsis* and 11 groups in rice based on analysis of gene structure and conserved motifs (Nakano et al., 2006). The AP2 family comprises two groups of proteins differing in the number of AP2 domain in their amino acid sequences. The majority of proteins in this group are characterized by the presence of two AP2 domains, but a few members of this group have only a single AP2 domain that is more similar to the AP2 domains in the double domain groups. RAV proteins, on the other hand, are a small family TFs characterized by the presence of B3 DNA binding domain besides a single AP2 domain. Genome-wide analysis of AP2/ERF TFs has been extensively studied in many dicots including *Arabidopsis* (Nakano et al., 2006), *Populus* (Zhuang et al., 2008; Vahala et al., 2013), Chinese cabbage (Liu et al., 2013), grapevine (Licausi et al., 2010), peach (Zhang et al., 2012), and castor bean (Xu et al., 2013). However, with the exception of rice (Nakano et al., 2006; Rashid et al., 2012), and foxtail millet (Lata et al., 2014), little information is available on the AP2/ERF TF families in monocots such as switchgrass.

Numerous genes coding for AP2/ERF superfamily TFs have been identified and functionally characterized in various plant species (Xu et al., 2011; Licausi et al., 2013). The DREB subfamily proteins have been extensively studied with regard to tolerance to abiotic stress such as freezing (Jaglo-Ottosen et al., 1998; Ito et al., 2006; Fang et al., 2015), drought (Hong and Kim, 2005; Oh et al., 2009; Fang et al., 2015), heat (Qin et al., 2007), and salinity (Hong and Kim, 2005; Bouaziz et al., 2013). Moreover, it has been reported that DREB genes play roles in the regulation of ABA-mediated gene expression in response to osmotic stress during germination and early vegetative growth stage (Fujita et al., 2011). ERF TFs, on the other hand, have been shown to participate in the regulation of defense responses against various biotic stresses (Guo et al., 2004; Dong et al., 2015) and/or tolerance to environmental stressors, such as drought (Aharoni et al., 2004; Zhang et al., 2010b), osmotic stress (Zhang et al., 2010a), salinity (Guo et al., 2004), hypoxia (Hattori et al., 2009), and freezing (Zhang and Huang, 2010). Moreover, AP2/ERF TFs in aspen (PtaERF1) and *Arabidopsis* (AtERF004 and AtERF038) have been suggested to be associated with the regulation of cell wall biosynthesis in some tissues (Van Raemdonck et al., 2005; Lasserre et al., 2008; Ambavaram et al., 2011). The functions of AP2 family TFs, on the other hand, have been associated with plant organ-specific regulation of growth and developmental programs (Elliott et al., 1996; Jofuku et al., 2005; Horstman et al., 2014). Genes in the RAV TF family have been shown to play a role in the regulation of gene expression in response to phytohormones such as ethylene and brassinosteroid as well as in response to biotic and abiotic stresses (Mittal et al., 2014). Therefore, AP2/ERF TF superfamily may hold tremendous potential for the improvement of bioenergy feedstocks, such as switchgrass, that is intended to be grown on marginal lands that could impose undue environmental stress.

In this study, we report the identification of 207 AP2/ERF TF genes in the switchgrass genome. Cluster analysis of the identified proteins, distribution of conserved motifs, analysis of their gene structure, and expression profiling were presented. We highlight the potential application of these data to identify putative target genes that might be exploited to improve bioenergy feedstocks. To that end, we cloned one of the ERF subfamily genes, which was subsequently overexpressed in switchgrass to improve biomass productivity and sugar release efficiency.

Materials and Methods

Identification of AP2/ERF Gene Families in Switchgrass Genome

We used representative genes from appropriate rice gene families as the basis to search for orthologs in switchgrass. The amino acid sequences of AP2 domain-containing rice genes represented three families: AP2 (Os02g40070), ERF (Os06g40150), and RAV (Os01g04800). These proteins were used to query the derived amino acid sequences of all switchgrass AP2/ERF TFs using tblastn against the switchgrass EST database (Zhang et al., 2013) or blastp against the *P. virgatum* draft genome (Phytozome v1.1 DOE-JGI)[1]. The sequences were retrieved and evaluated for the presence of AP2 domains by searching against the conserved domain database (CDD) at NCBI. The AP2-containing switchgrass sequences were further evaluated for any redundant and missing sequences by blastp searches using the previously identified homologous counterparts of the foxtail millet (Lata et al., 2014) and rice (Nakano et al., 2006; Rashid et al., 2012). The presence of multiple gene copies from the tetraploid switchgrass genome was addressed by the identification of only a single gene copy with the highest similarity to the corresponding homologs in foxtail millet or rice. Genes with additional domains besides the AP2 domain with no corresponding homologs in foxtail millet, rice, and *Arabidopsis* AP2/ERF TFs were excluded from our subsequent analysis.

Cluster and Protein Sequence Analysis of AP2/ERF TFs

The amino acid sequences of the AP2/ERF TFs were imported into the MEGA6 program and multiple sequence alignment analysis was conducted using MUSCLE with default parameters (Edgar, 2004). Construction of cluster trees was performed using the neighbor-joining (NJ) method by the MEGA6 program using a bootstrap value of 1000, Poisson correction and pairwise deletion (Tamura et al., 2013). Conserved motifs in switchgrass AP2/ERF TFs were identified with the online tool, MEME version 4.10.0[2] using the following parameters: optimum width, 6–200 amino acids; with any number of repetitions and maximum number of motifs set at 25 (Bailey and Elkan, 1994).

Analysis of Gene Structure and Gene Ontology Annotation

The genomic and coding DNA sequences of the identified AP2/ERF TFs were retrieved from the Phytozome (*P. virgatum* v1.1 DOE-JGI). The exon–intron organizations in these genes

were visualized by the gene structure display server[3] (Guo et al., 2007). To evaluate the gene ontology (GO) annotation of the identified AP2/ERF TFs, their amino acid sequences were imported into the Blast2GO suite (Conesa and Gotz, 2008). Blastp search was performed against rice protein sequences at NCBI. The resulting hits were mapped to obtain the GO terms, which were annotated to assign functional terms to the query sequences. Plant GOslim was used to filter the annotation to plant-related terms. The protein subcellular localization prediction tool WOLF PSORT[4] was used to complement the results of the cellular localization predicted by blast2GO.

Analysis of Transcript Data from the Switchgrass Gene Expression Atlas

The transcript data for the AP2/ERF superfamily TFs were extracted from the publicly available switchgrass gene expression atlas (PviGEA)[5] (Zhang et al., 2013), which was obtained by Affymetrix microarray analysis. The probe set IDs of 108 matching genes representing the switchgrass unitranscripts (PviUT) were identified by tblastn query search using the amino acid sequences of the AP2/ERF TFs. The transcript data for each tissues and stage of development were retrieved using the probe set IDs. The expression values of the genes were log2 transformed and a heatmap was created using an online graphing tool, Plotly[6]. Tissues used for the extraction of RNA to determine the level of expression included the following: whole seeds for seed germination at 24, 48, 72 and 96 h intervals post-imbibition, whole shoots and roots at vegetative stages, V1–V5, pooled leaf sheath (LSH), leaf blade (LB) and nodes, whole crown, the bottom, middle, and top portions of the fourth internode, vascular bundle tissues, and middle portion of the third internode all at E4 (stem elongation stage 4) developmental stage. For analysis of the expression level during reproductive developmental stages, inflorescence tissues and whole seeds along with floral tissues such as lemma and palea were used.

Vector Construction and Plant Transformation

Cloning and tissue culture was performed as previously described (Wuddineh et al., 2015). Briefly, the putative homolog of *Arabidopsis* AtSHN2 (At5g11190) and rice OsSHN (Os06g40150) was identified by tblastn or blastp against the switchgrass EST database or draft genome (Phytozome v1.1 DOE-JGI) followed by cluster and multiple sequence alignment analysis to discriminate the most closely related gene for cloning. For construction of overexpression cassette, the open reading frame (ORF) of *PvERF001* was isolated from cDNA obtained from ST1 clonal genotype of 'Alamo' switchgrass using gene-specific primers flanking the ORF of the gene and cloned into pANIC-10A expression vector by GATEWAY recombination (Mann et al., 2012). The primer pairs used for cloning are shown in Table S1 in Supplementary Material. Embryogenic callus derived from SA1 clonal genotype of 'Alamo' switchgrass (King et al., 2014) was transformed with the expression vector construct through *Agrobacterium*-mediated

---

[1] http://phytozome.jgi.doe.gov/pz/portal.html
[2] http://meme-suite.org/

[3] http://gsds.cbi.pku.edu.cn/
[4] http://www.genscript.com
[5] http://switchgrassgenomics.noble.org/
[6] https://plot.ly/plot

transformation (Burris et al., 2009). Antibiotic selection was carried out for about 2 months on 30–50 mg/L hygromycin followed by regeneration of orange fluorescent protein reporter-positive callus sections on regeneration medium (Li and Qu, 2011) containing 400 mg/L timentin. Regenerated plants were rooted on MSO medium (Murashige and Skoog, 1962) with 250 mg/L cefotaxime to assure elimination of *Agrobacterium* from the tissues as well as promote shoot regeneration from transgenic callus (Grewal et al., 2006), and the transgenic lines were screened based on the presence of the insert and expression of the transgene. Simultaneously a non-transgenic control line was also generated from callus.

Plants and Growth Conditions

T0 transgenic and non-transgenic control plants were grown in growth chambers under standard conditions (16 h·day/8 h·night light at 24°C, 390 μE·m^{-2} s^{-1}) and watered three times per week, including weekly nutrient supplements with 100 mg/L Peter's 20-20-20 fertilizer. Transgenic and non-transgenic control lines were propagated from a single tiller to produce three clonal replicates for measuring growth parameters (Hardin et al., 2013). The plants were grown in 12-L pots in Fafard 3B soil mix (Conrad Fafard, Inc., Agawam, MA, USA) and grown for 4 months to the R1 stage, in which shoot samples were collected to assay the transgene transcript abundance (Moore et al., 1991; Shen et al., 2009). Each sample was snap frozen in liquid nitrogen and macerated with mortar and pestle. The macerated samples were used for RNA extraction as described below.

RNA Extraction and Quantitative Reverse Transcription Polymerase Chain Reaction

RNA extraction and analysis of transgene transcripts were performed as previously described (Wuddineh et al., 2015). Briefly, total RNA was extracted from shoot tip samples of transgenic and non-transgenic control lines using Tri-Reagent (Molecular Research Center, Cincinnati, OH, USA), and 3 μg of the RNA was treated with DNase-I (Promega, Madison, WI, USA). High-Capacity cDNA Reverse Transcription kit (Applied Biosystems, Foster City, CA, USA) was used for the synthesis of first-strand cDNA. Power SYBR Green PCR master mix (Applied Biosystems) was utilized to conduct quantitative reverse transcription polymerase chain reaction (qRT-PCR) analysis according to the manufacturer's protocol. All the experiments were conducted in triplicates. The list of all primer pairs used for qRT-PCR is shown in Table S1 in Supplementary Material. Analysis of the relative expression was done as previously described (Wuddineh et al., 2015). There was no amplification products observed with all the primer pairs when using only the RNA samples or the water instead of cDNA.

Determination of Leaf Water Loss

The rate of water loss via leaf epidermal layer was determined as previously described (Zhou et al., 2014). The second fully expanded leaves of both transgenic and non-transgenic plants were excised and soaked in 50 mL distilled water for 2 h in the dark to saturate the leaves. Subsequently, the excess water was removed and initial leaf weight was measured and water loss determined by weighing the leaves every 30 min for at least 3 h. Subsequently, the detached leaves were dried for 24 h at 80°C to determine the final dry weight. The rate of water loss was calculated as the weight of water lost divided by the initial leaf weight.

Analysis of Lignin Content and Composition

Both qualitative (phloroglucinol–HCl staining) and quantitative [pyrolysis molecular beam mass spectrometry (py-MBMS)] analysis of lignin content was performed as previously described (Wuddineh et al., 2015). Briefly, leaf samples collected at the R1 developmental stage and cleared in a 2:1 solution of ethanol and glacial acetic acid for 5 days were used for staining analysis. The cleared leaf samples were immersed in 1% phloroglucinol (in 2:1 ethanol/HCl) overnight for staining and the pictures were taken at 2× magnification. For the quantification of lignin content and S:G lignin monomer ratio by NREL high-throughput py-MBMS method, tillers were collected at R1 developmental stage, air-dried for 3 weeks at room temperature and milled to 1 mm (20 mesh) particle size. Lignin content and composition were determined on extractives- and starch-free samples (Sykes et al., 2009).

Determination of Sugar Release

For analysis of sugar release efficiency, tiller samples at R1 developmental stage were collected and air-dried for 3 weeks at room temperature. The dry samples were pulverized to 1 mm (20 mesh) particle size and sugar release efficiency was determined via NREL high-throughput sugar release assays on extractives- and starch-free samples (Decker et al., 2012). Glucose release and xylose release were measured by colorimetric assays and summed for total sugar release.

Statistical Analysis

To analyze the differences between treatment means, analysis of variance (ANOVA) with least significant difference (LSD) procedure was used while PROC TTEST procedure was used to examine the statistical difference between the expression of target genes in transgenic vs non-transgenic lines using SAS version 9.3 (SAS Institute Inc., Cary, NC, USA). Pearson's correlation coefficient to determine the relationship between relative transcript levels and growth parameters was calculated by SAS.

Results

Identification of AP2/ERF TFs in Switchgrass Genome

A total of 207 unique switchgrass genes containing one or two AP2 DNA binding domain were identified from the currently available switchgrass EST and genome databases. Amino acid sequence similarities within the conserved AP2 domain between these proteins and previously characterized AP2/ERF TFs from rice and *Arabidopsis* along with the presence of conserved B3 domain suggest that these proteins might be categorized as putative AP2/ERF TFs. The characteristic features of these genes are summarized in Table S2 in Supplementary Material. The amino acid sequences of AP2/ERF TFs showed wide variation in size (ranging from 119 to 666 amino acids) and sequence composition. Twenty-two of these TFs contained two AP2 DNA-binding domains and hence

TABLE 1 | Summary of the AP2/ERF superfamily gene members found in various plant species.

Family	Subfamily	Group	*Panicum virgatum*	*Oryza sativa*[a]	*Arabidopsis thaliana*[a]	*Populus trichocarpa*[b]
AP2			25	29	18	26
ERF	DREB	I	12	9	10	5
		II	11	15	15	20
		III	27	26	23	35
		IV	5	6	9	6
		Total	55	56	57	66
	ERF	V	10	8	5	10
		VI	9	6	8	11
		VI-L	7	3	4	4
		VII	17	15	5	6
		VIII	25	13	15	17
		IX	37	18	17	42
		X	12	13	8	9
		Xb-L	–	–	3	4
		XI	4	7	–	–
		Total	121	76	65	103
RAV			5	5	6	6
Singleton			1	1	1	1
Total			207	174	147	202

The switchgrass (P. virgatum) data are from this study. Note that switchgrass is the only polyploidy species listed above.
[a]*Nakano et al. (2006).*
[b]*Zhuang et al. (2008).*

were classified under AP2 family. Five of the AP2/ERF proteins had a B3 conserved domain at the C-terminus in addition to the common AP2 domain, and these genes were grouped into the RAV family. Three of the remaining 180 proteins, namely PvERF049, PvERF160, and PvERF177 with a single AP2 domain, which is more similar to the amino acid sequences of AP2 domains in the AP2 family TFs, were also grouped under the AP2 family. Moreover, one AP2/ERF protein showed a distinct AP2 domain different from all other switchgrass AP2/ERF proteins but with higher shared sequence similarity with the previously identified genes in rice and *Arabidopsis*. The remaining 176 proteins were grouped into ERF family, which was further subdivided into either one of two subfamilies (ERF and DREB) based on sequence similarity in the AP2 domain. The ERF subfamily members included 121 proteins while DREB had only 55 proteins (**Table 1**).

The distribution of the identified switchgrass AP2/ERF genes across the nine chromosomes was also evaluated. Thus far, only about half of the switchgrass genomic sequences are mapped into their chromosomal locations based on the draft genome assembly by JGI-DOE available at Phytozome. Accordingly, 166 of the 207 genes could be assigned a chromosomal location. The genes were non-evenly distributed across the nine switchgrass chromosomes wherein the highest number of genes was localized on chromosomes 9, 2, and 1, with the fewest number of genes being assigned to chromosome 8 (Table S3 in Supplementary Material).

Cluster Analysis of Switchgrass AP2/ERF Proteins

To confirm the classification and evaluate the sequence similarities between the switchgrass AP2/ERF TFs, a dendrogram was constructed by NJ method using the whole amino acid sequences of the proteins. The analysis showed distinct clustering

of the proteins into specific groups and families as previously described in other species (**Figure 1**). Specifically, these clusters highlighted the distinction between the switchgrass AP2, ERF, and RAV families as well as between the ERF and DREB subfamilies. The ERF and DREB subfamilies were further subdivided into seven (groups V–XI) and four (I–IV) distinct groups, respectively. The cluster analysis also resolved the RAV protein family and the singleton into separate clusters, which was in accordance with the sequence similarities in the conserved domains as well as the presence of additional domains in the families/clusters.

Characterization of AP2/ERF Gene Structures and Conserved Motifs

To complement the cluster analysis-based classification, the exon–intron structures of AP2/ERF genes were evaluated. The schematic representations of protein and gene structures of switchgrass AP2/ERF superfamily are presented in **Figure 2** (ERF), **Figure 3** (DREB), and **Figure 4** (AP2, RAV, and Singleton). The ORF lengths of these genes vary from 394 bp for the shortest gene to 5409 bp for the longest gene. Analysis of their gene structure showed highly diverse distribution of intron regions within the ORF of the different gene groups or families. The majority of genes belonging to ERF and DREB subfamilies and all but one of the RAV genes appeared to be intronless. Only nine DREB genes (16%) belonging to group I, III, and VI had a single intron in their gene structures. Among ERF genes, 45 (37%) had a single intron in their ORF while eight genes had two and three of them with three introns in its ORF. On the other hand, genes in the AP2 family contained a higher number of introns; ranging from 1 to 10. Only one gene in the AP2 family had a single intron while majority of the genes had more than five introns. The position and state of the introns in the ORF

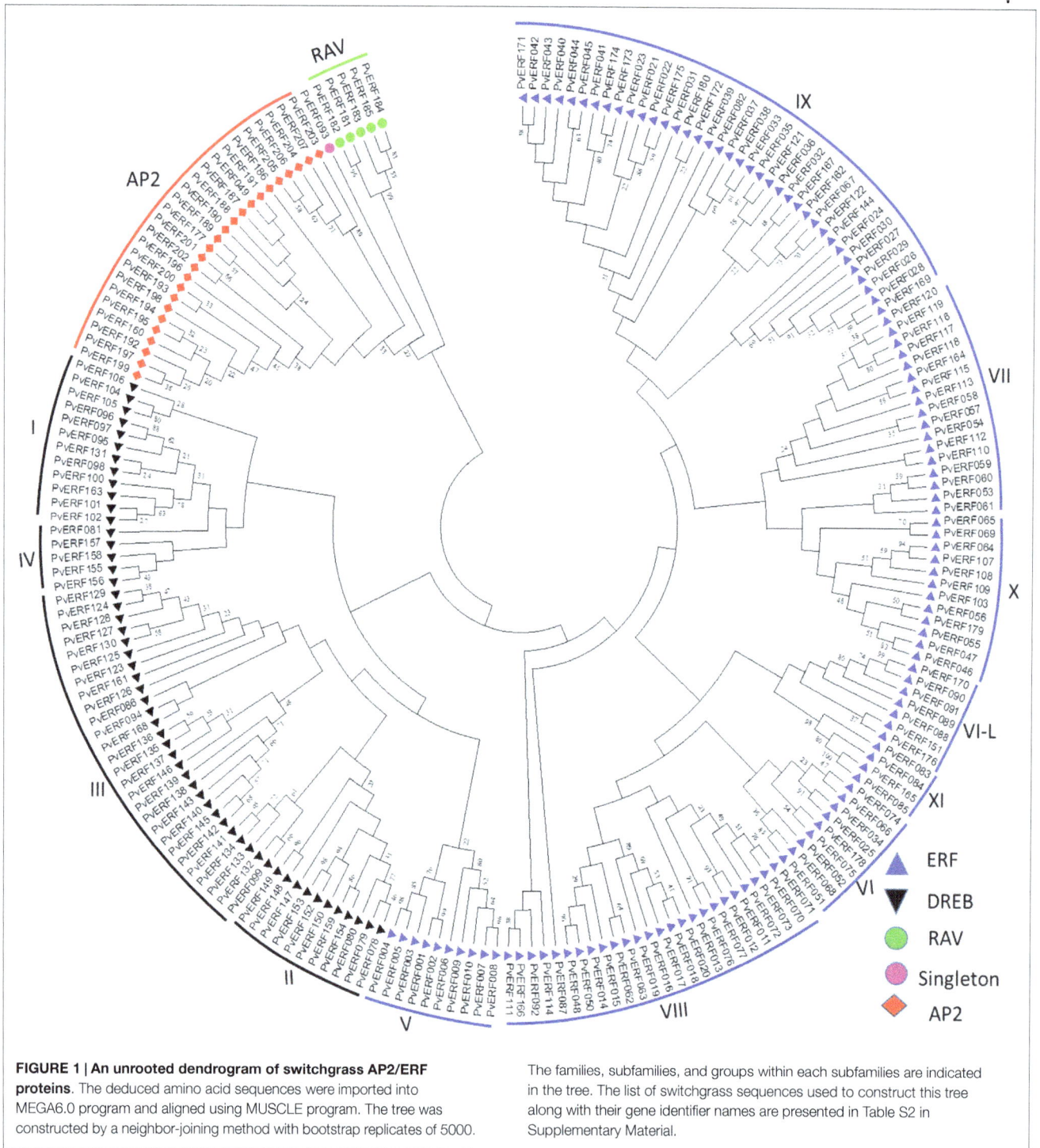

FIGURE 1 | An unrooted dendrogram of switchgrass AP2/ERF proteins. The deduced amino acid sequences were imported into MEGA6.0 program and aligned using MUSCLE program. The tree was constructed by a neighbor-joining method with bootstrap replicates of 5000.

The families, subfamilies, and groups within each subfamilies are indicated in the tree. The list of switchgrass sequences used to construct this tree along with their gene identifier names are presented in Table S2 in Supplementary Material.

of ERF family genes belonging to groups V, VII, and X show high functional conservation. For instance, about half of the genes belonging to phylogenetic group V in the ERF family showed highly conserved intron positions with an intron phase of two, meaning the location of the intron is found between the second and third nucleotides in the codon. Similarly, the intron positions and splicing phases seems conserved in group VII of the ERF subfamily (**Figures 2–4**).

Analysis of amino acid sequence conservation in the whole proteins of AP2/ERF superfamily showed the presence of unique conserved motifs shared between proteins within families, subfamilies, or groups (**Figures 2–4**). Moreover, shared conserved motifs across families, subfamilies, or between groups within subfamilies were also detected, signifying the conservation of the proteins in the AP2/ERF superfamily. In general, a total of 25 conserved motifs (M1–M25) were identified in the superfamily

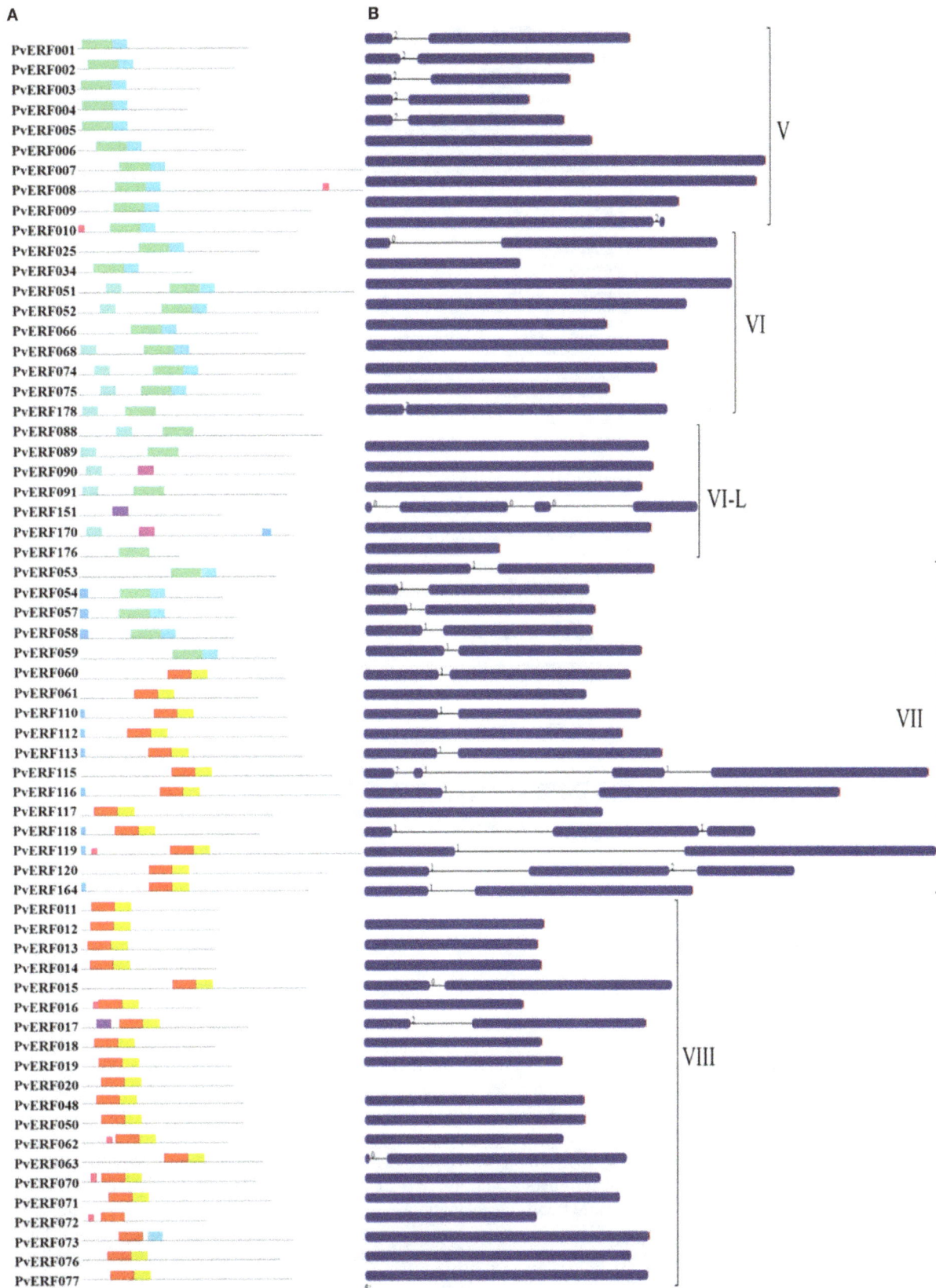

FIGURE 2 | The schematic representation of protein and gene structures of switchgrass ERF subfamily.

(Continued)

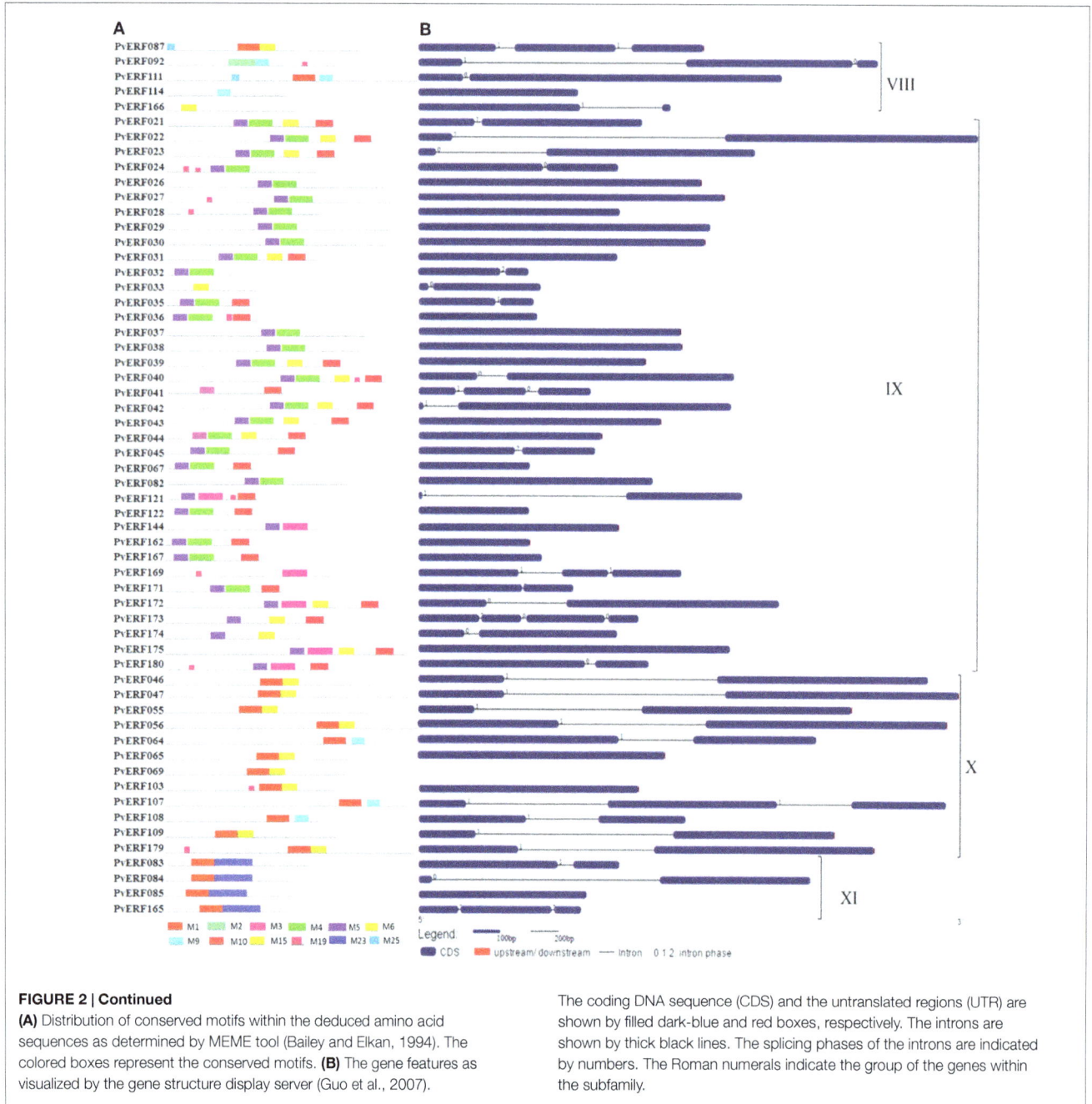

FIGURE 2 | Continued

(A) Distribution of conserved motifs within the deduced amino acid sequences as determined by MEME tool (Bailey and Elkan, 1994). The colored boxes represent the conserved motifs. **(B)** The gene features as visualized by the gene structure display server (Guo et al., 2007).

The coding DNA sequence (CDS) and the untranslated regions (UTR) are shown by filled dark-blue and red boxes, respectively. The introns are shown by thick black lines. The splicing phases of the introns are indicated by numbers. The Roman numerals indicate the group of the genes within the subfamily.

of which 14 motifs, M1–M7, M9, M11, M12, M16, M20, M22, and M23, were related to the AP2 domain (Table S4 in Supplementary Material). The conserved motifs from the non-AP2 domain region appear to specify individual groups within the subfamilies. Among the ERF subfamily, proteins in groups VII and IX have the most diverse set of motifs compared to others while proteins in group XI harbors merely two motifs, M1 and M23 with the last motif being unique to the group (**Figure 2**). Moreover, shared unique motifs were found in the ERF subfamily proteins belonging to group VII (M25), IX (M10 and M15), VI (M18), and VI-L (M18). Most of the DREB genes belonging to group II have only one specific motif (M12) while a few others

have additional motifs such as M5 (**Figure 3**). The pattern of conserved motif distribution within the largest group in the DREB subfamily (group III) showed the presence of two unique subgroups sharing a set of three conserved motifs, (M2, M9, and M16) and (M4, M11, and M21), respectively. Three of these motifs (11, 16, and 21) were specific to proteins in group III DREB subfamily. DREB subfamily proteins in group I were distinguished by conserved motif-M13 and motif-M24, while group IV DREB genes have unique motif-M2 (**Figure 3**). Proteins of AP2 family genes harbor four family-specific motifs, namely M7, M8, M20, and M22 (**Figure 4**). In addition, the majority of AP2 family proteins share M3 with ERF proteins belonging to group IX.

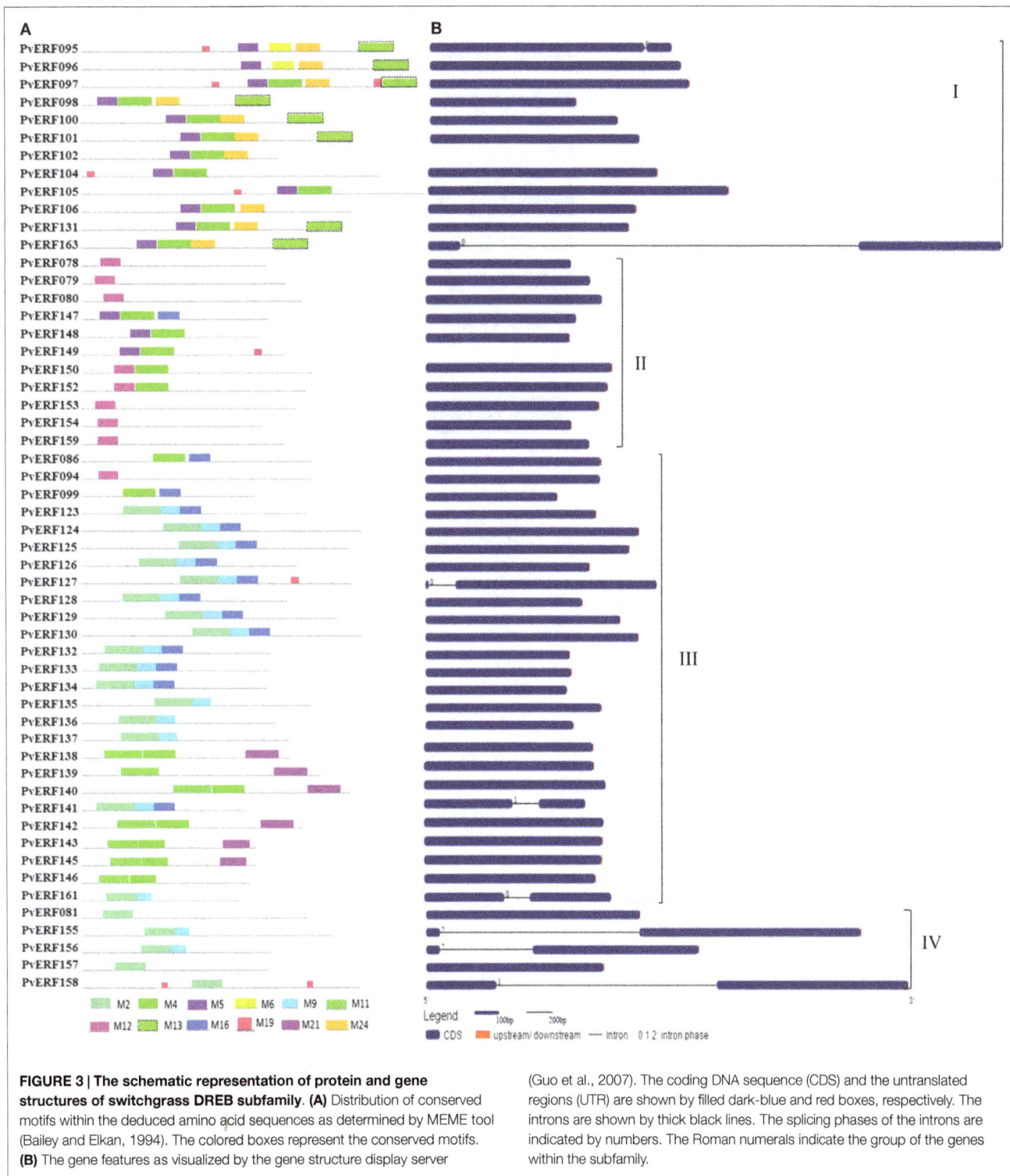

FIGURE 3 | The schematic representation of protein and gene structures of switchgrass DREB subfamily. (A) Distribution of conserved motifs within the deduced amino acid sequences as determined by MEME tool (Bailey and Elkan, 1994). The colored boxes represent the conserved motifs. **(B)** The gene features as visualized by the gene structure display server

(Guo et al., 2007). The coding DNA sequence (CDS) and the untranslated regions (UTR) are shown by filled dark-blue and red boxes, respectively. The introns are shown by thick black lines. The splicing phases of the introns are indicated by numbers. The Roman numerals indicate the group of the genes within the subfamily.

Similarly, RAV proteins also possess two unique motifs, M14 and M17 spanning the B3 DNA binding domain, in addition to M6 and M12 spanning the AP2 domain (**Figure 4**). M6 and M12 motifs are also present in most proteins in the ERF and DREB (group II) subfamilies (**Figures 2** and **3**; Table S4 in Supplementary Material).

Gene Ontology Annotation

Gene ontology analysis of switchgrass AP2/ERF TFs, based on rice reference sequences, predicted candidate genes' molecular functions, putative roles in the regulation of diverse biological processes, and their cellular localization (**Figure 5**; Table S5 in Supplementary Material). According to blast2GO outputs, over 95% of

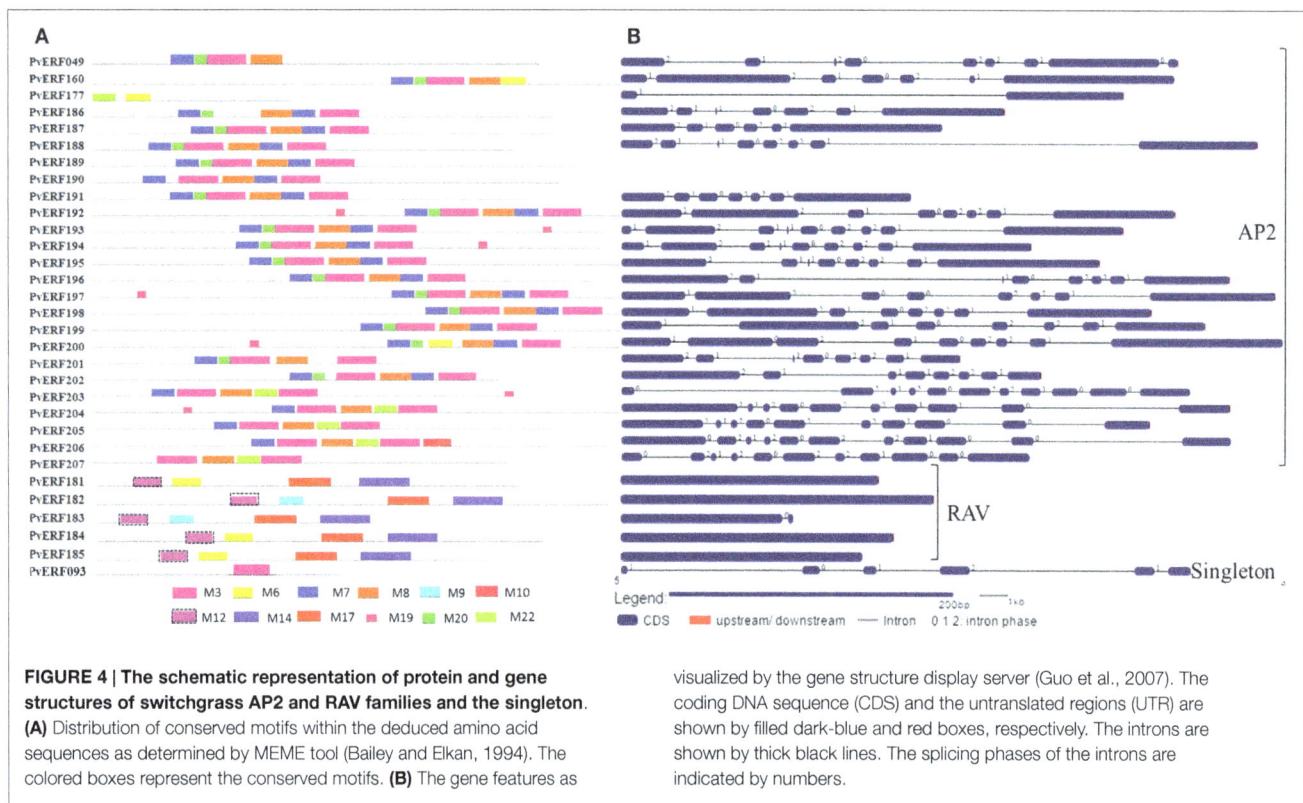

FIGURE 4 | The schematic representation of protein and gene structures of switchgrass AP2 and RAV families and the singleton. **(A)** Distribution of conserved motifs within the deduced amino acid sequences as determined by MEME tool (Bailey and Elkan, 1994). The colored boxes represent the conserved motifs. **(B)** The gene features as visualized by the gene structure display server (Guo et al., 2007). The coding DNA sequence (CDS) and the untranslated regions (UTR) are shown by filled dark-blue and red boxes, respectively. The introns are shown by thick black lines. The splicing phases of the introns are indicated by numbers.

the switchgrass genes in the AP2/ERF superfamily were predicted to have sequence-specific DNA binding activities (**Figure 5A**; Table S5 in Supplementary Material). Furthermore, these genes were anticipated to be involved in the regulation of various biosynthetic processes, which could include the biosynthesis of cuticle, waxes, hormones, and other organic compounds. Importantly, many of these genes were also predicted to participate in the regulation of responses to various environmental stresses caused either by biotic factors such as pathogens and insect pests or abiotic factors such as flooding, water deprivation, wounding, and osmotic stress (**Figure 5B**; Table S5 in Supplementary Material). Cellular localization of the AP2/ERF TFs was predicted by Blast2GO analysis complemented with subcellular localization prediction tool, WoLF PSORT for proteins with heretofore ambiguous results. The results showed that majority of switchgrass AP2/ERF proteins (>80%) were at least dual targeted, i.e., localized to nucleus, plastid, and/or mitochondrion (**Figure 5C**; Table S5 in Supplementary Material). Only 39 gene products (20%) were predicted to be localized solely to the nucleus (Table S5 in Supplementary Material).

Expression Pattern of Switchgrass *AP2/ERF* Genes

A switchgrass gene expression atlas (PviGEA) containing expression data for about 78,000 unique transcripts in various tissues was recently developed (Zhang et al., 2013) and is publicly available at web server[7]. To investigate whether the identified switchgrass AP2/ERF genes may have any association with various biological

processes that occur during seed germination, vegetative, and reproductive development as well as lignification or cell wall development, transcript data were pooled from the PviGEA web server to assess their expression profile.

During seed germination (**Figure 6**; Table S6 in Supplementary Material), some genes in the DREB subfamily showed high expression at early stages of germination (radicle emergence) (48 h after imbibition) while others showed increased expression at later stages of germination (mainly coleoptile emergence) (**Figure 6**; Table S6 in Supplementary Material). Similarly, the expression of many ERF genes showed dramatic increase during early germination stage while numerous others had peak expression at later stages (coleoptile emergence (72 h) and mesocotyl elongation (96 h) stages. Four of the AP2 family genes (*PvERF193*, *PvERF194*, *PvERF195*, and *PvERF201*) displayed increased expression level at radicle emergence whereas the other two (*PvERF049* and *PvERF203*) showed increased expression at coleoptile emergence. The expression of the RAV genes and the singleton gene were apparently relatively less variable throughout the seed germination process (**Figure 6**; Table S6 in Supplementary Material).

Comparison of the expression pattern of AP2/ERF genes in roots and shoots at three vegetative phases of development (first, third, or fifth fully collared leaf stages) revealed apparent differential expression pattern between the organs and different stages of vegetative development (Figure S1 and Table S6 in Supplementary Material). Moreover, the expression pattern of AP2/ERF genes during reproductive development also showed differential expression between the reproductive tissues from the initiation of inflorescence meristem to the maturation of the seeds (Figure S2 and Table S6 in Supplementary Material).

---

[7]http://switchgrassgenomics.noble.org/

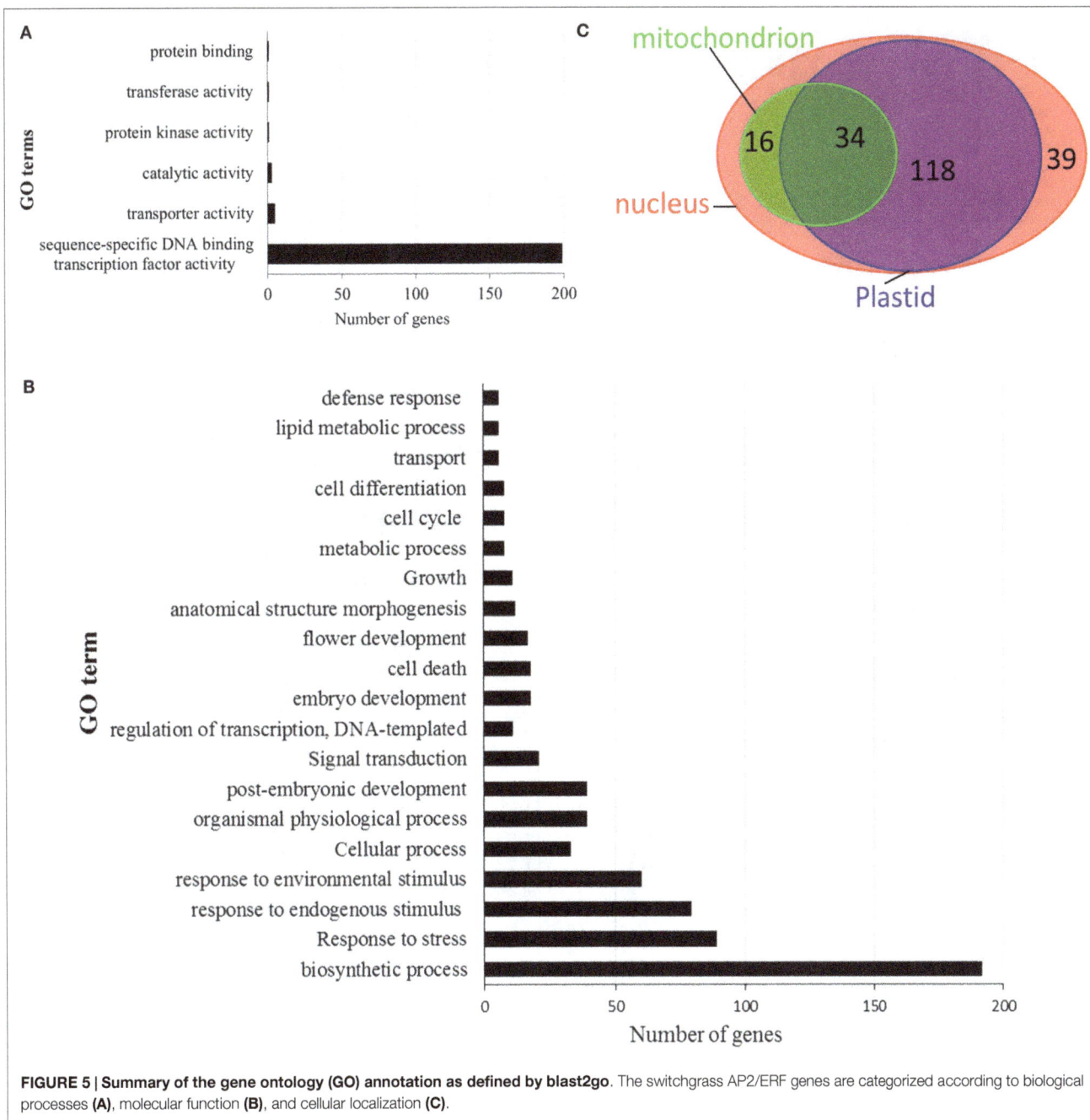

FIGURE 5 | Summary of the gene ontology (GO) annotation as defined by blast2go. The switchgrass AP2/ERF genes are categorized according to biological processes **(A)**, molecular function **(B)**, and cellular localization **(C)**.

Expression Profiles of Switchgrass AP2/ERF Genes in Lignified Tissues

To evaluate whether the identified switchgrass genes coding for AP2/ERF TFs are associated with the regulation of the cell wall biosynthetic genes during cell wall formation or lignification, the transcripts of the genes extracted from the PviGEA web server were used to compare the level of expression in the lignified tissues of vascular bundles and internode fragments against the expression level in less lignified plant tissues such as LBs and sheath. Four genes in group I (*PvERF95*, *PvERF98*, *PvERF101*, and *PvERF102*) and one gene in group II (*PvERF148*) of the DREB subfamily showed highest expression in vascular bundles

and internode tissues followed by internode portions where active lignification is expected (**Figure 7**; Table S6 in Supplementary Material). The majority of DREB genes belonging to group III were highly expressed mainly in the vascular bundles. Similarly, many genes in the ERF subfamily group VIII (*PvERF013*, *PvERF015*, *PvERF016*, *PvERF018*, *PvERF019*, and *PvERF020*) and X (*PvERF047*, *PvERF065*, and *PvERF103*) showed the highest expression in the vascular bundles followed by youngest internode sections (**Figure 7**; Table S6 in Supplementary Material). In comparison, only two genes in group IX (*PvERF037* and *PvERF038*), one gene in group VI-L (*PvERF088*), and three genes in group VII (*PvERF111*, *PvERF112*, and *PvERF116*) had high

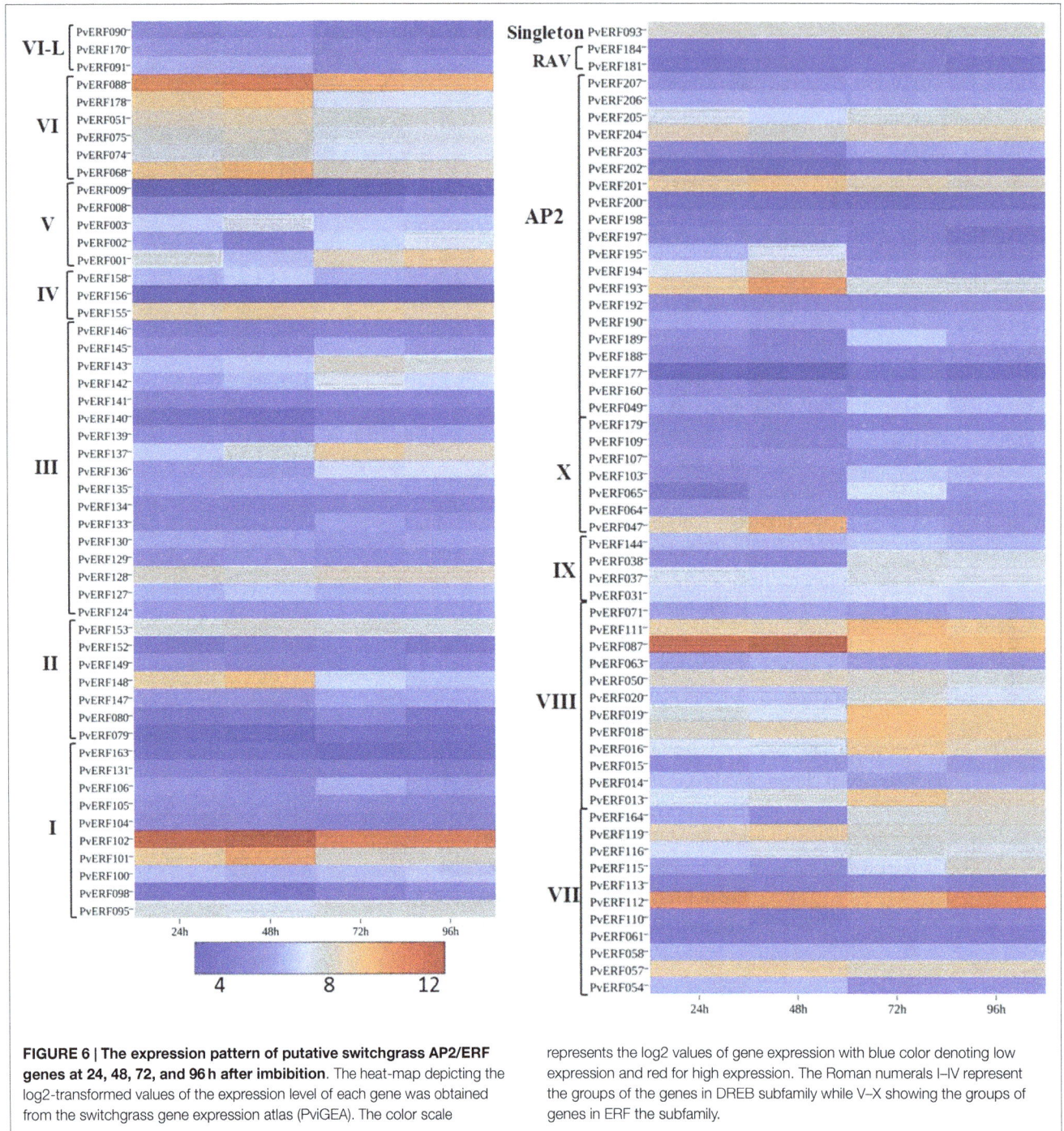

FIGURE 6 | The expression pattern of putative switchgrass AP2/ERF genes at 24, 48, 72, and 96 h after imbibition. The heat-map depicting the log2-transformed values of the expression level of each gene was obtained from the switchgrass gene expression atlas (PviGEA). The color scale represents the log2 values of gene expression with blue color denoting low expression and red for high expression. The Roman numerals I–IV represent the groups of the genes in DREB subfamily while V–X showing the groups of genes in ERF the subfamily.

expression in vascular bundles. Contrastingly, some genes in the ERF subfamily belonging to group V (*PvERF001* and *PvERF002*) and VI (*PvERF068*) showed the highest expression in the basal fragments of the fourth internodes (E4) that is under less active lignification. Other genes including *PvERF178* (VI); *PvERF110* (VII), *PvERF115* (VII), and *PvERF164* (VII); and *PvERF038* (IX) had notably high relative expression in roots than in other tissues. Compared to the ERF family genes, the expression of AP2 genes was highly diverse with some genes having high specificity to roots and vascular bundles. The expression of the two RAV genes

analyzed was uniformly low throughout whereas the singleton gene was highly expressed in the LBs, LSH as well as the vascular bundles, and young internode sections (**Figure 7**; Table S6 in Supplementary Material).

Overexpression of *PvERF001* in Switchgrass Have Enhanced Plant Growth and Sugar Release Efficiency

Transgenic switchgrass is desired for less recalcitrance biomass for biofuels. To that end, we selected PvERF001, a putative

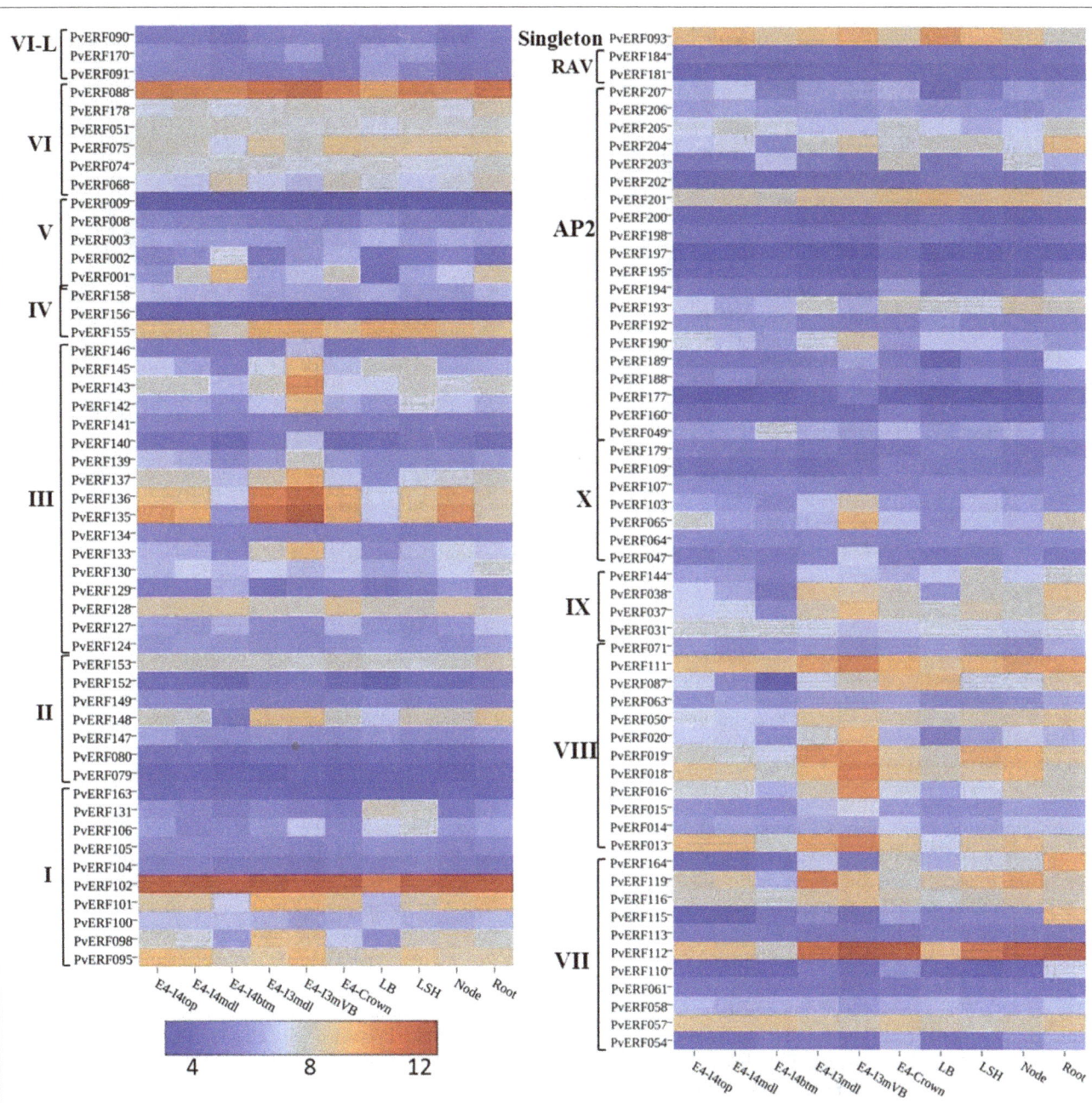

FIGURE 7 | The expression pattern of putative switchgrass AP2/ERF genes in roots and shoot parts including portions of developing internodes and vascular bundles at stem elongation stage 4 (E4). The heat-map depicting the log2-transformed values of the expression level of each gene was obtained from the switchgrass gene expression atlas (PviGEA). The color scale represents the log2 values of gene expression with blue color denoting low expression and red for high expression. The level of expression was reported for roots, nodes, leaf sheath (LSH), leaf blade (LB), whole crown (E4-crown), vascular bundle isolated from fragments of the third internode (E4-I3mVB), middle fragments of the third internode (E4-I3mdl) and from the bottom (E4-I4btm), middle (E4-I4mdl), and top (E4-I4top) fragments of the fourth internode. The Roman numerals I–IV represent the groups of the genes in DREB subfamily while V–X showing the groups of genes in ERF the subfamily.

switchgrass homolog of *Arabidopsis* AtERF004 (AtSHN2) and rice OsERF057 (OsSHN) in ERF subfamily group V, for overexpression analysis in switchgrass. This gene was selected since the expression of its *Arabidopsis* homolog in transgenic rice resulted in modified cell wall composition (Ambavaram et al., 2011). Sequence grouping/cluster and sequence alignment analysis suggested that PvERF001 is closely related with its rice and *Arabidopsis* homologs, sharing two highly conserved motifs: the middle motif (mm) and the C-terminal motif (cm) specific to the *Arabidopsis* SHINE clade of TFs (AtERF001, AtERF004, and AtERF005) and OsERF012 and OsERF057 (**Figures 8A,B**). Thus, the ORF of *PvERF001* was cloned and overexpressed in

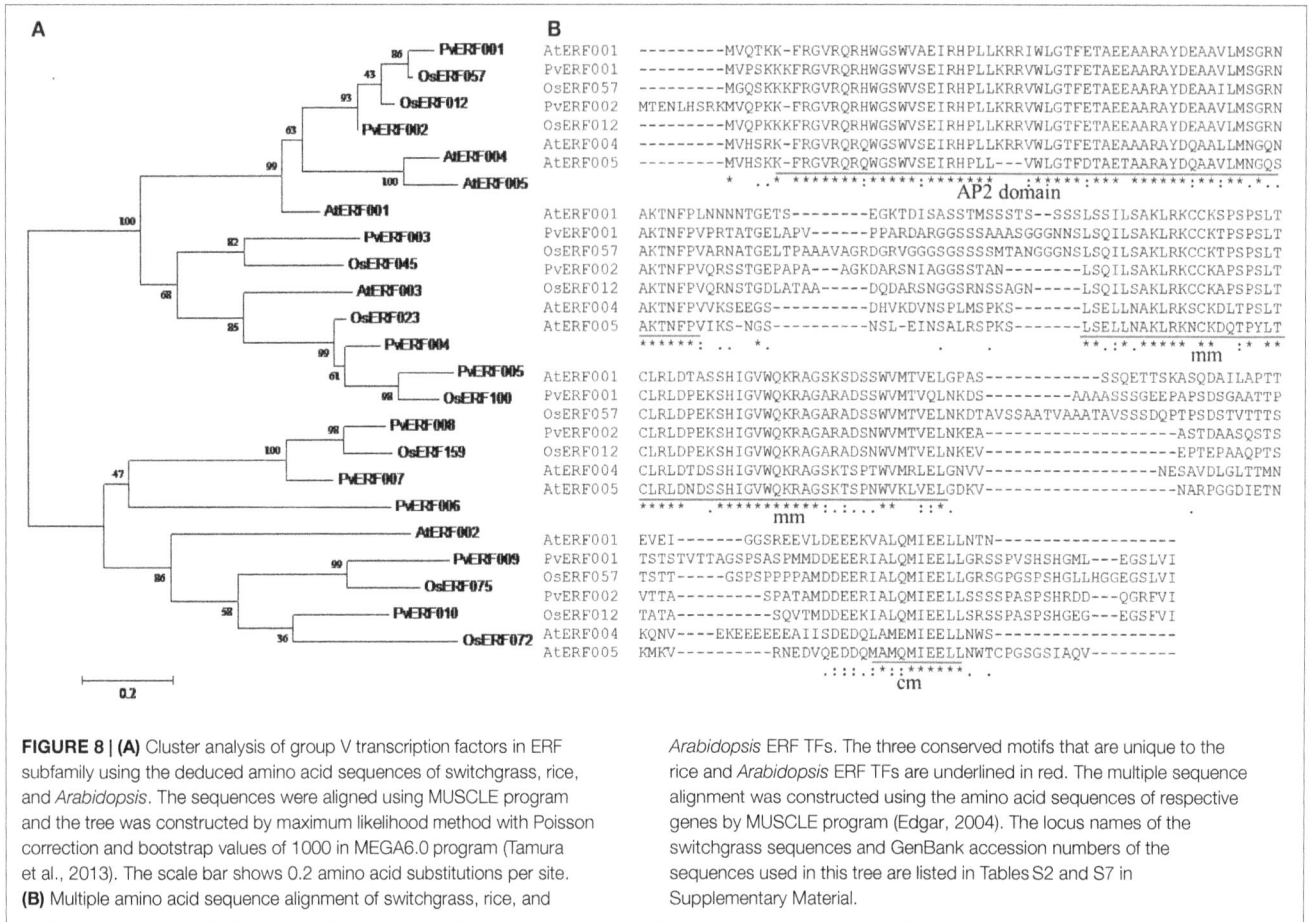

FIGURE 8 | (A) Cluster analysis of group V transcription factors in ERF subfamily using the deduced amino acid sequences of switchgrass, rice, and *Arabidopsis*. The sequences were aligned using MUSCLE program and the tree was constructed by maximum likelihood method with Poisson correction and bootstrap values of 1000 in MEGA6.0 program (Tamura et al., 2013). The scale bar shows 0.2 amino acid substitutions per site. **(B)** Multiple amino acid sequence alignment of switchgrass, rice, and *Arabidopsis* ERF TFs. The three conserved motifs that are unique to the rice and *Arabidopsis* ERF TFs are underlined in red. The multiple sequence alignment was constructed using the amino acid sequences of respective genes by MUSCLE program (Edgar, 2004). The locus names of the switchgrass sequences and GenBank accession numbers of the sequences used in this tree are listed in Tables S2 and S7 in Supplementary Material.

switchgrass producing more than six independent transgenic lines, which were confirmed based on genomic PCR for the insertion of the transgene and the hygromycin-resistance gene, as well as visualization of OFP in transgenic plants compared to the non-transgenic control lines (**Figure 9A**; Figures S3A–C in Supplementary Material). Analysis of the transgene expression level by qRT-PCR showed 1–12-fold overexpression in transgenic lines (**Figure 9B**). The expression of the endogenous gene in transgenic lines was not affected compared to the non-transgenic control line (**Figure 9C**). All transgenic lines had equivalent or improved vegetative growth metrics relative to the non-transgenic control lines under greenhouse conditions, which was congruent with the relative transcript levels of the transgene [Pearson's correlation for biomass weights ($R = 0.77$ at $P < 0.05$) and tiller height ($R = 0.73$ at $P = 0.06$)] (**Figure 9B**; **Table 2**; Figure S4 in Supplementary Material). Three transgenic lines (3, 7, and 9) had increased biomass. Line 3 had statistically significant increases in four of the six growth traits and approximately twice the dry biomass of the control line (**Table 2**).

To investigate whether *PvERF001* overexpression could affect the leaf cuticular permeability, the water retention capacity in transgenic and non-transgenic control lines was analyzed in detached leaves measured in the dark to minimize transpirational water loss through stomata. Transgenic lines showed relative reduction in rate of water loss compared with the control lines (Figure S5 in Supplementary Material). However, no tangible difference was observed in the rate of leaf chlorophyll leaching between transgenic and the control lines (data not shown). Subsequently, we analyzed whether the changes in leaf morphology might be accompanied by changes in the expression level of genes in the cutin and wax biosynthesis pathway, in which none were observed (Figure S6 in Supplementary Material). Moreover, overexpression of *PvERF001* in transgenic switchgrass showed relatively reduced expression of some lignin (*PvC4H* and *PvPAL*), hemicellulose (*PvCSLS2*), and cellulose (*PvCESA4*) biosynthetic genes, as well as some of the transcriptional regulators (*PvMYB48/59* and *PvNST1*) of cell wall biosynthesis (Figures S7A–C in Supplementary Material). The total lignin content in R1 tillers determined by Py-MBMS of cell wall residues and in leaves determined by phloroglucinol–HCl staining did not show sizeable difference between the transgenic and non-transgenic control lines (Figures S8 and S9A in Supplementary Material). Similarly, analysis of the S/G lignin monomer ratio in transgenic lines did not significantly change as compared to that of the non-transgenic control line (Figure S9B in Supplementary Material). However, significant improvement in glucose release efficiency was observed in lines 7 (10%) and 8 (16%) relative to the non-transgenic control line (**Table 3**). In contrast, none of the transgenic lines released significantly more xylose than the control. The total sugar release, however, was significantly increased in transgenic line 8 by 11% relative to the non-transgenic control (**Table 3**).

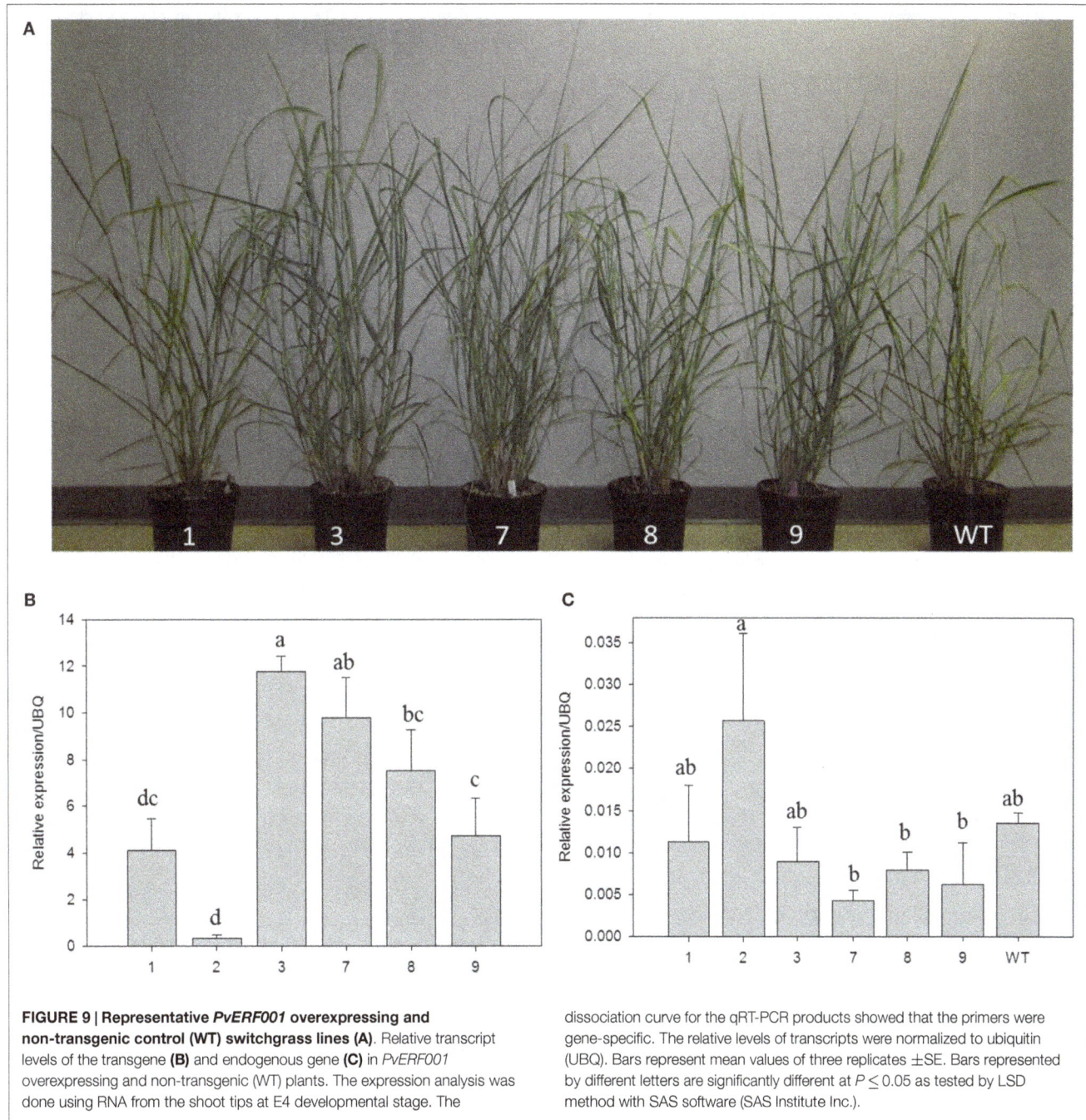

FIGURE 9 | Representative *PvERF001* overexpressing and non-transgenic control (WT) switchgrass lines (A). Relative transcript levels of the transgene (B) and endogenous gene (C) in *PvERF001* overexpressing and non-transgenic (WT) plants. The expression analysis was done using RNA from the shoot tips at E4 developmental stage. The dissociation curve for the qRT-PCR products showed that the primers were gene-specific. The relative levels of transcripts were normalized to ubiquitin (UBQ). Bars represent mean values of three replicates ±SE. Bars represented by different letters are significantly different at $P \leq 0.05$ as tested by LSD method with SAS software (SAS Institute Inc.).

TABLE 2 | Morphology and biomass yields of transgenic switchgrass lines overexpressing *PvERF001* and non-transgenic control (WT) plants.

Lines	Tiller height (cm)	Tiller number	Fresh weight (g)	Dry weight (g)	Plant diameter (cm)	Fresh/dry weight ratio
1	98.9 ± 2.0^b	13.3 ± 1.5^{bc}	40.5 ± 1.7^{cd}	12.8 ± 0.3^c	1.38 ± 0.06^b	3.15 ± 0.07^a
2	105.8 ± 2.9^{ab}	12.3 ± 2.6^c	45.8 ± 11.3^{bcd}	15.1 ± 3.9^{bc}	1.36 ± 0.06^{bc}	3.05 ± 0.04^a
3	115.3 ± 1.2^a	17.7 ± 1.9^{ab}	70.2 ± 9.9^a	21.9 ± 3.3^a	1.54 ± 0.03^a	3.21 ± 0.08^a
7	115.7 ± 3.6^a	21.0 ± 1.6^a	66.7 ± 3.6^{ab}	17.7 ± 5.5^{ab}	1.23 ± 0.06^{cd}	3.03 ± 0.17^a
8	101.3 ± 2.1^b	15.3 ± 0.9^{abc}	42.9 ± 1.8^{cd}	13.3 ± 0.9^c	1.20 ± 0.02^d	3.23 ± 0.14^a
9	109.7 ± 3.1^{ab}	17.0 ± 0.8^{abc}	61.4 ± 3.6^{abc}	16.1 ± 5.2^{ab}	1.44 ± 0.05^{ab}	3.00 ± 0.11^a
WT	85.8 ± 2.8^c	15.3 ± 1.5^{abc}	34.6 ± 4.7^d	10.7 ± 1.6^c	1.05 ± 0.03^e	3.27 ± 0.08^a

Tiller height estimates were determined for each plant by taking the mean of the five tallest tillers within each biological replicate. The fresh and dry biomass measurements were obtained from aboveground plant biomass harvested at similar growth stages. Values are means of three biological replicates ±SEs (n = 3). Values represented by different letters are significantly different at $P \leq 0.05$ as tested by LSD method with SAS software (SAS Institute Inc.).

TABLE 3 | Sugar release by enzymatic hydrolysis in transgenic and non-transgenic control (WT) lines.

Lines	Glucose release (g/g CWR)	Xylose release (g/g CWR)	Total sugar release (g/g CWR)
1	0.214 ± 0.009^d	0.175 ± 0.003^c	0.389 ± 0.011^c
2	0.238 ± 0.008^{bc}	0.182 ± 0.014^{abc}	0.420 ± 0.018^b
3	0.234 ± 0.003^{bcd}	0.192 ± 0.008^a	0.427 ± 0.006^{ab}
7	0.247 ± 0.003^{ab}	0.176 ± 0.009^{bc}	0.423 ± 0.007^b
8	0.261 ± 0.003^a	0.188 ± 0.005^{ab}	0.448 ± 0.012^a
9	0.227 ± 0.020^{bcd}	0.188 ± 0.007^{ab}	0.415 ± 0.024^b
WT	0.225 ± 0.007^{cd}	0.181 ± 0.007^{abc}	0.405 ± 0.003^{bc}

All data are means $\pm$ SE (n = 3). CWR, cell wall residues. Values represented by different letters are significantly different at $P \leq 0.05$ as tested by LSD method with SAS software (SAS Institute Inc.).

Discussion

Significance of AP2/ERF TFs for Improvement of Bioenergy Crops

AP2/ERF TFs constitute one of the largest protein superfamilies in plants. These TFs play a role in regulating a wide array of developmental and growth processes. Thus, they are interesting targets for crop genetic engineering and breeding (Licausi et al., 2013; Bhatia and Bosch, 2014). Numerous TFs belonging to this superfamily have been characterized in various plant species and their potential biotechnological applications in crop improvement has focused primarily on biotic and abiotic stress tolerance (Xu et al., 2011; Licausi et al., 2013; Hoang et al., 2014). However, less effort has been made to utilize this potential for genetic improvement of bioenergy feedstocks such as switchgrass (Bhatia and Bosch, 2014). We found this lack of development to be somewhat anachronistic since these TFs are variably associated with plant growth and cell wall biosynthesis, which are directly related to two most important traits to a bioenergy crops, such as switchgrass: biomass and cell wall recalcitrance.

Sequence-Based Classification of Putative AP2/ERF TFs in Switchgrass

With this in mind, we conducted a whole genome search for putative switchgrass AP2/ERF superfamily of TFs and found 207 members (**Figure 1**; **Table 1**; Table S2 in Supplementary Material). Based on comparative genome analysis with the published results in rice, foxtail millet, and *Arabidopsis*, the identified proteins were classified into three families, namely AP2, RAV, and ERF with the later further divided into two subfamilies (ERF and DREB) (Nakano et al., 2006; Lata et al., 2014). The number of genes in the DREB subfamily found in switchgrass (55) was comparable with that of rice (56), *Arabidopsis* (57), and *Populus* (66). All three species along with switchgrass have a singleton in their genome. Consistent with the previous report in rice (Nakano et al., 2006), the switchgrass DREB and ERF subfamilies comprise four and seven groups, respectively. Moreover, based on comparative analysis of the AP2/ERF TFs between different plant species, it seems that group XI of ERF subfamily is specific to monocots as the Xb-L was reported only in dicots (Nakano et al., 2006; Liu et al., 2013). In general, the relative distribution of genes within the different groups in each subfamily appears to be conserved between the three plant species (**Table 1**). Classification

of the switchgrass AP2/ERF TFs into distinct groups was clearly supported by the amino acid sequence-based dendrogram of the identified proteins suggesting robust evolutionary conservation between the superfamily among plant species.

In Silico Predicted Gene Functions and Subcellular Localization of AP2/ERF TFs in Switchgrass

Consistent with the purported role of AP2/ERF proteins as transcriptional regulators of target genes (Magnani et al., 2004), GO analysis predicted that the majority of the switchgrass AP2/ERF genes appear to have DNA-binding activity consistent with the previous observation in foxtail millet (Lata et al., 2014). Therefore, these genes might be associated with the regulation of various biosynthetic processes as well as responses to environmental stimuli as previously demonstrated for numerous genes in other plant species (Xu et al., 2011; Mizoi et al., 2012; Licausi et al., 2013) (**Figures 5A,B**). The predicted subcellular localization pattern of AP2/ERF superfamily genes in switchgrass, which was mainly to the nucleus as would be expected for transcriptional regulators but also to the plastids and/or mitochondria in addition to the nucleus, was comparable to that reported in foxtail millet (**Figure 5C**) (Lata et al., 2014). Such multi-localization of the proteins could be attributed to post-translational modifications, protein folding, or interactions with other proteins (Karniely and Pines, 2005), and might serve to facilitate the coordinated regulation of the expression of nuclear and organellar genomes (Duchene and Giege, 2012).

Gene and Protein Sequence Diversity of Switchgrass AP2/ERF TFs

The exon/intron structures of switchgrass *AP2/ERF* genes were analogous with that of foxtail millet (Lata et al., 2014), castor bean (Xu et al., 2013), rice, and *Arabidopsis* (Nakano et al., 2006). Consistent with these species, we observed a high diversity in the distribution of the intron regions of AP2 genes versus a single or no intron in most genes in the ERF and RAV families (**Figures 2 and 4**). The pattern of intron distribution within the ORF and their splicing phases was highly conserved in genes within specific groups as reported in castor bean (Xu et al., 2013). Consistent with the observation in rice, the majority of proteins in the groups or subfamilies of switchgrass AP2/ERF superfamily could be distinguished by the presence of one or more diagnostic motifs located outside the AP2 domain region (Table S4 in Supplementary Material) (Rashid et al., 2012). These groups or subfamily-specific conservation in gene structures and protein motifs supported the accuracy of the predicted cluster relationships between the switchgrass AP2/ERF TFs.

AP2/ERF TFs that function as repressors or activators of specific target genes are distinguished by the presence of conserved motifs called repression domains (RD) that are highly conserved, or by the presence of activation domains which are generally less conserved (Licausi et al., 2013). One of the characteristic motif in AP2/ERF transcriptional activators is the activation domain, EDLL motif (Tiwari et al., 2012), while repressors have unique RD namely the ERF-associated amphiphilic repression (EAR) motif (LxLxL or DLNxxP)

(Kagale and Rozwadowski, 2011) and B3 repression domain (BRD: R/KLFGV) motif (Ikeda and Ohme-Takagi, 2009). Analysis of the switchgrass AP2/ERF TF sequences also indicated the presence of these motifs in many proteins (Table S4 in Supplementary Material). For instance, many genes in group IX of ERF subfamily appear to be transcriptional activators due to the presence of motif M10, which is an EDLL-like motif. Moreover, this motif is rich in acidic amino residues which has been suggested as the characteristics of transcriptional activators (Licausi et al., 2013). Majority of the ERF family TFs in group VIII and DREB family TFs in group I displayed a DLNxxP-like motifs. Four TFs belonging to the AP2 family (PvERF204, PvERF205, PvERF206, and PvERF207) also displayed similar EAR motif while PvERF203 and PvERF207 harbors DLELSL and NLDLS-like RD, respectively. Similarly, switchgrass TFs in RAV family also displayed unique repression domain, RLFGV (Ikeda and Ohme-Takagi, 2009). ERF subfamily TFs in groups VI and VI-L share a characteristic motif at the N-terminus (M18), also known as the cytokinin responsive factor (CRF) domain in *Arabidopsis* that is also shared by rice ERF genes belonging to same group in rice ERF subfamily (Nakano et al., 2006). Genes containing the CRF domain (VI and VI-L) were shown to be responsive to cytokinin (Rashotte et al., 2006). The distinguishing N-terminal motif in group VII ERF subfamily proteins, M25 was conserved in both *Arabidopsis* and rice as described previously (Nakano et al., 2006). This motif was shown to dictate the stability of proteins based on the level of oxygen via N-end rule pathway (Dubouzet et al., 2003; Licausi et al., 2011). DREB genes in rice with characteristic LWSY motif have been shown to function in regulation of drought, cold, and salinity responsive gene expression (Dubouzet et al., 2003). Switchgrass genes belonging to group III in DREB subfamily (*PvERF133, PvERF134, PvERF135, PvERF136, PvERF137, PvERF139, PvERF140, PvERF141, PvERF142, PvERF143, PvERF145,* and *PvERF146*) displayed LWSY conserved motif (M21) at the C-terminal and thus may play similar roles. No information is available in the literature on some of the conserved motifs identified here including M8, M13, M14, M15, M17, and M24 (Table S4 in Supplementary Material), which might potentially be specific to switchgrass.

Diverse Expression Profiles of Switchgrass AP2/ERF TFs and Functional Implications

Differential expression of genes according to developmental stages and tissue or organ types may provide an insight into the specialized biological processes that are taking place in the specific plant parts (Cassan-Wang et al., 2013; Zhang et al., 2013). The observed pattern of expression for the majority of switchgrass AP2/ERF genes at different stages of plant development as well as in different tissues/organ types highlight the significance of these genes in the regulation of various plant growth and developmental processes at the specific stages (**Figures 6** and 7; Figures S1 and S2 in Supplementary Material). One of the engrossing observations in this study is the association of the expression of numerous genes with tissues/organs undergoing lignification or secondary cell wall development/modification, suggesting that these genes may have intrinsic association with the regulatory

machinery of cell wall formation/lignification, which is not as well characterized compared to their roles in stress response (Licausi et al., 2013; Bhatia and Bosch, 2014). Activation of genes responsible for cell wall modification has already been reported to be key during the initiation of seed germination in barley (Sreenivasulu et al., 2008; An and Lin, 2011). In agreement with this, we reported here the transcriptional upregulation of ERF (*PvERF057, PvERF068, PvERF088,* and *PvERF119*), DREB (*PvERF101, PvERF102,* and *PvERF148*), and AP2 genes (*PvERF193, PvERF201,* and *PvERF204*) during the initiation of seed germination as well as in vascular bundles and internode sections. Moreover, the observed robust expression of 14 DREB, 17 ERF, and 3 AP2 genes in tissues or organs undergoing active lignification (vascular bundles, top or middle internode sections as well as roots) but less robust expression in less lignified tissues (leaves) also supports this assertion (**Figure 7**). It should also be noted that the transcript levels of several of these genes showed a relative increase with the developmental stage of the plants (**Figure 7**; Figure S1 in Supplementary Material) while exhibiting only marginal expression in less lignified tissues such as inflorescence meristem and germinating seedlings (**Figure 6**; Figure S2 in Supplementary Material). Differential gene expression profiling between elongating and non-elongating internodes in maize was used to identify a total of seven AP2/ERF TFs that are highly expressed in non-elongating internodes undergoing secondary wall development suggesting that these genes may involve in the regulation of secondary cell wall formation (Bosch et al., 2011). Moreover, recent study in *Arabidopsis* and rice identified several putative secondary cell wall-related AP2/ERF TFs based on preferential expression in secondary cell wall-related tissues and coexpression analysis (Cassan-Wang et al., 2013; Hirano et al., 2013a; Bhatia and Bosch, 2014). Some of the switchgrass genes identified in this study (PvERF037, PvERF115, PvERF116, PvERF143, PvERF148, and PvERF164) appear to be putative homologs of maize, rice, and *Arabidopsis* genes identified in the aforementioned studies. Overexpression of *Populus* ERF genes in wood-forming tissues of hybrid aspen was recently shown to result in modified stem growth (including increased stem diameter following the overexpression of five different ERF genes), reduced lignification, and enhanced carbohydrate content (cellulose) in the wood of transgenic lines hinting that these TFs may indeed interact with the transcriptional machinery regulating cell wall biosynthesis (Vahala et al., 2013). Another evidence supporting this is a recent study suggesting that an ERF TF from loquat fruit (*Eriobotrya japonica*) (EjAP2-1) is an indirect transcriptional repressor of lignin biosynthesis via interaction with EjMYB1 TFs (Zeng et al., 2015).

Overexpression of *PvERF001* Improved Biomass Productivity and Sugar Release Efficiency in Switchgrass

Based on global gene coexpression analysis, the rice homolog of AtSHN2, OsSHN (OsERF057) was proposed to have a native association with cell wall regulatory and biosynthetic pathways, yet this was not experimentally verified (Ambavaram et al., 2011). In this study, we investigated whether PvERF001, the closest putative switchgrass homolog of these genes based on clustering, sequence alignment analysis, and the sharing of conserved motifs (mm and cm) specific to *Arabidopsis* SHN clade of TFs and the rice SHN, may participate in the regulation of cell wall biosynthesis (**Figure 8**). Our results suggest that PvERF001 may not be directly involved in the regulation cell wall biosynthesis though its transgenic overexpression resulted in increased sugar release efficiency (Figure S7 in Supplementary Material; **Table 3**). Despite the observed reduction in relative expression of some lignin biosynthetic genes and their transcriptional regulators in switchgrass that seem to relate with the results in rice overexpressing *AtSHN2*, no significant changes in the lignin content and composition was detected in transgenic switchgrass in contrast to the reduced lignin content observed in rice overexpressing *AtSHN2* (Ambavaram et al., 2011) (Figures S7, S8, and S9A in Supplementary Material). The increased sugar release might be attributed to altered storage carbohydrates such as starches as recently reported in *Arabidopsis* where ectopic expression of rice ERF TF (SUB1A-1) gene resulted in improved enzymatic saccharification efficiency via increased level of starch (Nunez-Lopez et al., 2015). Similar results were obtained from overexpression of maize *corngrass1* microRNA in switchgrass (Chuck et al., 2011). However, whether PvERF001 is associated with starch biosynthesis remains to be determined. Moreover, in contrast to the previous reports where heterologous expression of *AtSHN2* in rice did not significantly affect the growth characteristics of transgenic lines (Ambavaram et al., 2011), overexpression of *PvERF001* resulted in increased plant growth including plant height, stem diameter and aboveground biomass weight in transgenic lines (**Table 2**). The discrepancy in lignin content and biomass productivity traits between the AtSHN2 and PvERF001 may indicate the differences in functional specialization between the two genes in monocots and dicots even though sequence analysis seems to suggest that they might be homologs. The fact that overexpression of *AtSHN* genes in *Arabidopsis* rather showed association with the regulation of wax, cutin, and pectin biosynthesis supports this assertion (Aharoni et al., 2004; Shi et al., 2011). Moreover, recent study showed that the homolog of *Arabidopsis SHN* genes in tomato (*SlERF52*) was expressed mainly in the abscission zone and functionally associated with the regulation of the pedicel abscission zone-specific transcription of genes including cell wall-hydrolytic enzymes (polygalacturonase and Cellulase) required for abscission (Nakano et al., 2014). These differences in the expression pattern and function may suggest functional divergence between *SlERF52* and its *Arabidopsis* homologs. Functional divergence between homologous TFs in monocots and dicots has also been reported in previous studies involving the homologs of AtMYB58/63, which is a known activator of lignin biosynthesis that did not appear to play similar roles in rice (Hirano et al., 2013b).

A recent study involving overexpression of rice homolog of *AtSHN2*, *OsSHN*, in rice showed enhanced tolerance of transgenic plants to water deprivation and association of the gene with the regulation of wax and cutin biosynthesis and hence named rice wax synthesis regulatory gene (OsWR2) (Zhou et al., 2014). The closest homolog of this gene, OsERF012 (OsWR1), was also shown to be induced by drought stress and involved in the regulation of wax synthesis (Wang et al., 2012). Therefore, we examined whether PvERF001 might be involved in the regulation of wax and cutin biosynthesis. Consistent with previous studies in rice, relative increase in leaf water retention capacity was detected in transgenic plants though the effect on the expression of wax and cutin biosynthetic genes was minimal (Figures S5 and S6 in Supplementary Material). Possible explanation for the observed differences between overexpression of rice and switchgrass homologs might be an indication of the functional divergence in the switchgrass genes due to gene duplication. This may explain the discrepancy between transgenic rice overexpressing rice *SHN* (*OsWR2*) exhibiting reduction in plant height but increase in the number of tillers (Zhou et al., 2014) and transgenic switchgrass overexpressing *PvERF001* showing increased plant height but no difference in number of tillers. This suggests that ERF genes might functionally be highly diversified and PvERF001 may be part of a different pathway than we anticipated such as regulation of responses to biotic stress or other abiotic stress or regulation of cell elongation or division in coordination with the cytokinin pathway, with the latter perhaps explaining the observed increase in biomass and vegetative growth in transgenic lines.

In summary, the expression profiling of the switchgrass AP2/ERF genes provides baseline information as to the putative roles of these genes and thus a useful resource for future reverse genetic studies to characterize genes for economically important bioenergy crops. With the current advancements in switchgrass research and establishment of efficient transformation system, this inventory of genes along with the information provided here could facilitate our understanding regarding the functional roles of AP2/ERF TFs in plant growth and development. Furthermore, it would aid in the identification of potential target genes that may be used to improve stress adaptation, plant productivity, and sugar release efficiency in bioenergy feedstocks such as switchgrass. The increased biomass yield and sugar release efficiency from overexpressing *PvERF001* highlight the potential of these TFs for improvement of bioenergy feedstocks.

Author Contributions

WW designed and performed the experiments, analyzed the data, and wrote the manuscript. MM participated in experimental design and data analysis, assisted with revisions to the manuscript and coordination of the study. GT, RS, SD, and MD assisted with performing lignin and sugar release assays and contributed in revision of the manuscript. CS conceived the study and its design and coordination, and assisted with revisions to the manuscript. All authors read and consented to the final version of the manuscript.

Acknowledgments

We thank Angela Ziebell, Erica Gjersing, Crissa Doeppke, and Melvin Tucker for their assistance with the cell wall characterization and Susan Holladay for her assistance with data entry into LIMS. We thank Wayne Parrot for providing the switchgrass SA1 clone. This work was supported by funding from the BioEnergy Science Center (DE-PS02-06ER64304). The BioEnergy Science Center is a U.S. Department of Energy Bioenergy Research

Center supported by the Office of Biological and Environmental Research in the DOE Office of Science. We also thank Tennessee Agricultural Experiment Station for providing partial financial support for WW.

References

Aharoni, A., Dixit, S., Jetter, R., Thoenes, E., Van Arkel, G., and Pereira, A. (2004). The SHINE clade of AP2 domain transcription factors activates wax biosynthesis, alters cuticle properties, and confers drought tolerance when overexpressed in *Arabidopsis*. *Plant Cell* 16, 2463–2480. doi:10.1105/tpc.104.022897

Ambavaram, M. M., Krishnan, A., Trijatmiko, K. R., and Pereira, A. (2011). Coordinated activation of cellulose and repression of lignin biosynthesis pathways in rice. *Plant Physiol.* 155, 916–931. doi:10.1104/pp.110.168641

Ambavaram, M. M. R., Basu, S., Krishnan, A., Ramegowda, V., Batlang, U., Rahman, L., et al. (2014). Coordinated regulation of photosynthesis in rice increases yield and tolerance to environmental stress. *Nat. Commun.* 5, 5302. doi:10.1038/ncomms6302

An, Y.-Q., and Lin, L. (2011). Transcriptional regulatory programs underlying barley germination and regulatory functions of gibberellin and abscisic acid. *BMC Plant Biol.* 11:105. doi:10.1186/1471-2229-11-105

Bailey, T. L., and Elkan, C. (1994). Fitting a mixture model by expectation maximization to discover motifs in biopolymers. *Proc. Int. Conf. Intell. Syst. Mol. Biol.* 2, 28–36.

Baxter, H., Poovaiah, C., Yee, K., Mazarei, M., Rodriguez, M. Jr., Thompson, O., et al. (2015). Field evaluation of transgenic switchgrass plants overexpressing *PvMYB4* for reduced biomass recalcitrance. *Bioenerg. Res.* 1–12. doi:10.1007/s12155-014-9570-1

Baxter, H. L., Mazarei, M., Labbe, N., Kline, L. M., Cheng, Q., Windham, M. T., et al. (2014). Two-year field analysis of reduced recalcitrance transgenic switchgrass. *Plant Biotechnol. J.* 12, 914–924. doi:10.1111/pbi.12195

Bhatia, R., and Bosch, M. (2014). Transcriptional regulators of *Arabidopsis* secondary cell wall formation: tools to re-program and improve cell wall traits. *Front. Plant Sci.* 5:192. doi:10.3389/fpls.2014.00192

Bosch, M., Mayer, C.-D., Cookson, A., and Donnison, I. S. (2011). Identification of genes involved in cell wall biogenesis in grasses by differential gene expression profiling of elongating and non-elongating maize internodes. *J. Exp. Bot.* 62, 3545–3561. doi:10.1093/jxb/err045

Bouaziz, D., Pirrello, J., Charfeddine, M., Hammami, A., Jbir, R., Dhieb, A., et al. (2013). Overexpression of *StDREB1* transcription factor increases tolerance to salt in transgenic potato plants. *Mol. Biotechnol.* 54, 803–817. doi:10.1007/s12033-012-9628-2

Burris, J., Mann, D. J., Joyce, B., and Stewart, C. N. Jr. (2009). An improved tissue culture system for embryogenic callus production and plant regeneration in switchgrass (*Panicum virgatum* L.). *Bioenerg. Res.* 2, 267–274. doi:10.1007/s12155-009-9048-8

Cassan-Wang, H., Goué, N., Saidi, M. N., Legay, S., Sivadon, P., Goffner, D., et al. (2013). Identification of novel transcription factors regulating secondary cell wall formation in *Arabidopsis*. *Front. Plant Sci.* 4:189. doi:10.3389/fpls.2013.00189

Chuck, G. S., Tobias, C., Sun, L., Kraemer, F., Li, C., Dibble, D., et al. (2011). Overexpression of the maize Corngrass1 microRNA prevents flowering, improves digestibility, and increases starch content of switchgrass. *Proc. Natl. Acad. Sci. U.S.A.* 108, 17550–17555. doi:10.1073/pnas.1113971108

Conesa, A., and Gotz, S. (2008). Blast2GO: a comprehensive suite for functional analysis in plant genomics. *Int. J. Plant Genomics* 2008, 619832. doi:10.1155/2008/619832

Decker, S. R., Carlile, M., Selig, M. J., Doeppke, C., Davis, M., Sykes, R., et al. (2012). Reducing the effect of variable starch levels in biomass recalcitrance screening. *Methods Mol. Biol.* 908, 181–195. doi:10.1007/978-1-61779-956-3_17

Dong, L., Cheng, Y., Wu, J., Cheng, Q., Li, W., Fan, S., et al. (2015). Overexpression of *GmERF5*, a new member of the soybean EAR motif-containing ERF transcription factor, enhances resistance to *Phytophthora sojae* in soybean. *J. Exp. Bot.* 66, 2635–2647. doi:10.1093/jxb/erv078

Dubouzet, J. G., Sakuma, Y., Ito, Y., Kasuga, M., Dubouzet, E. G., Miura, S., et al. (2003). *OsDREB* genes in rice, *Oryza sativa* L., encode transcription activators that function in drought-, high-salt- and cold-responsive gene expression. *Plant J.* 33, 751–763. doi:10.1046/j.1365-313X.2003.01661.x

Duchene, A. M., and Giege, P. (2012). Dual localized mitochondrial and nuclear proteins as gene expression regulators in plants? *Front. Plant Sci.* 3:221. doi:10.3389/fpls.2012.00221

Edgar, R. C. (2004). MUSCLE: multiple sequence alignment with high accuracy and high throughput. *Nucleic Acids Res.* 32, 1792–1797. doi:10.1093/nar/gkh340

Elliott, R. C., Betzner, A. S., Huttner, E., Oakes, M. P., Tucker, W. Q., Gerentes, D., et al. (1996). *AINTEGUMENTA*, an *APETALA2*-like gene of *Arabidopsis* with pleiotropic roles in ovule development and floral organ growth. *Plant Cell* 8, 155–168. doi:10.1105/tpc.8.2.155

Fang, Z., Zhang, X., Gao, J., Wang, P., Xu, X., Liu, Z., et al. (2015). A buckwheat (*Fagopyrum esculentum*) DRE-Binding transcription factor gene, *FeDREB1*, enhances freezing and drought tolerance of transgenic *Arabidopsis*. *Plant Mol. Biol. Rep.* 1–16. doi:10.1007/s11105-015-0851-4

Fu, C., Mielenz, J. R., Xiao, X., Ge, Y., Hamilton, C. Y., Rodriguez, M., et al. (2011a). Genetic manipulation of lignin reduces recalcitrance and improves ethanol production from switchgrass. *Proc. Natl. Acad. Sci. U.S.A.* 108, 3803–3808. doi:10.1073/pnas.1100310108

Fu, C., Xiao, X., Xi, Y., Ge, Y., Chen, F., Bouton, J., et al. (2011b). Downregulation of *Cinnamyl Alcohol Dehydrogenase* (*CAD*) leads to improved saccharification efficiency in switchgrass. *Bioenerg. Res.* 4, 153–164. doi:10.1007/s12155-010-9109-z

Fujita, Y., Fujita, M., Shinozaki, K., and Yamaguchi-Shinozaki, K. (2011). ABA-mediated transcriptional regulation in response to osmotic stress in plants. *J. Plant Res.* 124, 509–525. doi:10.1007/s10265-011-0412-3

Grewal, D., Gill, R., and Gosal, S. S. (2006). Influence of antibiotic cefotaxime on somatic embryogenesis and plant regeneration in indica rice. *Biotechnol. J.* 1, 1158–1162. doi:10.1002/biot.200600139

Guo, A. Y., Zhu, Q. H., Chen, X., and Luo, J. C. (2007). [GSDS: a gene structure display server]. *Yi Chuan* 29, 1023–1026. doi:10.1360/yc-007-1023

Guo, Z. J., Chen, X. J., Wu, X. L., Ling, J. Q., and Xu, P. (2004). Overexpression of the AP2/EREBP transcription factor *OPBP1* enhances disease resistance and salt tolerance in tobacco. *Plant Mol. Biol.* 55, 607–618. doi:10.1007/s11103-004-1521-3

Hardin, C. F., Fu, C., Hisano, H., Xiao, X., Shen, H., Stewart, C. N. Jr., et al. (2013). Standardization of switchgrass sample collection for cell wall and biomass trait analysis. *Bioenerg. Res.* 6, 755–762. doi:10.1007/s12155-012-9292-1

Hattori, Y., Nagai, K., Furukawa, S., Song, X. J., Kawano, R., Sakakibara, H., et al. (2009). The ethylene response factors SNORKEL1 and SNORKEL2 allow rice to adapt to deep water. *Nature* 460, 1026–1030. doi:10.1038/nature08258

Hirano, K., Aya, K., Morinaka, Y., Nagamatsu, S., Sato, Y., Antonio, B. A., et al. (2013a). Survey of genes involved in rice secondary cell wall formation through a co-expression network. *Plant Cell Physiol.* 54, 1803–1821. doi:10.1093/pcp/pct121

Hirano, K., Kondo, M., Aya, K., Miyao, A., Sato, Y., Antonio, B. A., et al. (2013b). Identification of transcription factors involved in rice secondary cell wall formation. *Plant Cell Physiol.* 54, 1791–1802. doi:10.1093/pcp/pct122

Hoang, X. L. T., Thu, N. B. A., Thao, N. P., and Tran, L.-S. P. (2014). "Transcription factors in abiotic Stress responses: their potentials in crop improvement," in *Improvement of Crops in the Era of Climatic Changes*, eds P. Ahmad, M. R. Wani, M. M. Azooz and L.-S. Phan Tran (New York, NY: Springer), 337–366.

Hong, J. P., and Kim, W. T. (2005). Isolation and functional characterization of the *Ca-DREBLP1* gene encoding a dehydration-responsive element binding-factor-like protein 1 in hot pepper (*Capsicum annuum* L. cv. Pukang). *Planta* 220, 875–888. doi:10.1007/s00425-004-1412-5

Horstman, A., Willemsen, V., Boutilier, K., and Heidstra, R. (2014). AINTEGUMENTA-LIKE proteins: hubs in a plethora of networks. *Trends Plant Sci.* 19, 146–157. doi:10.1016/j.tplants.2013.10.010

Ikeda, M., and Ohme-Takagi, M. (2009). A novel group of transcriptional repressors in *Arabidopsis*. *Plant Cell Physiol.* 50, 970–975. doi:10.1093/pcp/pcp048

Ito, Y., Katsura, K., Maruyama, K., Taji, T., Kobayashi, M., Seki, M., et al. (2006). Functional analysis of rice DREB1/CBF-type transcription factors involved in cold-responsive gene expression in transgenic rice. *Plant Cell Physiol.* 47, 141–153. doi:10.1093/pcp/pci230

Jaglo-Ottosen, K. R., Gilmour, S. J., Zarka, D. G., Schabenberger, O., and Thomashow, M. F. (1998). *Arabidopsis CBF1* overexpression induces *COR* genes and enhances freezing tolerance. *Science* 280, 104–106. doi:10.1126/science.280.5360.104

Jofuku, K. D., Omidyar, P. K., Gee, Z., and Okamuro, J. K. (2005). Control of seed mass and seed yield by the floral homeotic gene *APETALA2*. *Proc. Natl. Acad. Sci. U.S.A.* 102, 3117–3122. doi:10.1073/pnas.0409893102

Kagale, S., and Rozwadowski, K. (2011). EAR motif-mediated transcriptional repression in plants: an underlying mechanism for epigenetic regulation of gene expression. *Epigenetics* 6, 141–146. doi:10.4161/epi.6.2.13627

Karniely, S., and Pines, O. (2005). Single translation – dual destination: mechanisms of dual protein targeting in eukaryotes. *EMBO Rep.* 6, 420–425. doi:10.1038/sj.embor.7400394

King, Z. R., Bray, A. L., Lafayette, P. R., and Parrott, W. A. (2014). Biolistic transformation of elite genotypes of switchgrass (*Panicum virgatum* L.). *Plant Cell Rep.* 33, 313–322. doi:10.1007/s00299-013-1531-1

Lasserre, E., Jobet, E., Llauro, C., and Delseny, M. (2008). AtERF38 (At2g35700), an AP2/ERF family transcription factor gene from *Arabidopsis thaliana*, is expressed in specific cell types of roots, stems and seeds that undergo suberization. *Plant Physiol. Biochem.* 46, 1051–1061. doi:10.1016/j.plaphy.2008.07.003

Lata, C., Mishra, A. K., Muthamilarasan, M., Bonthala, V. S., Khan, Y., and Prasad, M. (2014). Genome-wide investigation and expression profiling of AP2/ERF transcription factor superfamily in foxtail millet (*Setaria italica* L.). *PLoS ONE* 9:e113092. doi:10.1371/journal.pone.0113092

Li, R., and Qu, R. (2011). High throughput *Agrobacterium*-mediated switchgrass transformation. *Biomass Bioenergy* 35, 1046–1054. doi:10.1016/j.biombioe.2010.11.025

Licausi, F., Giorgi, F. M., Zenoni, S., Osti, F., Pezzotti, M., and Perata, P. (2010). Genomic and transcriptomic analysis of the AP2/ERF superfamily in *Vitis vinifera*. *BMC Genomics* 11:719. doi:10.1186/1471-2164-11-719

Licausi, F., Kosmacz, M., Weits, D. A., Giuntoli, B., Giorgi, F. M., Voesenek, L. A. C. J., et al. (2011). Oxygen sensing in plants is mediated by an N-end rule pathway for protein destabilization. *Nature* 479, 419–422. doi:10.1038/nature10536

Licausi, F., Ohme-Takagi, M., and Perata, P. (2013). APETALA2/Ethylene Responsive Factor (AP2/ERF) transcription factors: mediators of stress responses and developmental programs. *New Phytol.* 199, 639–649. doi:10.1111/nph.12291

Liu, Z., Kong, L., Zhang, M., Lv, Y., Liu, Y., Zou, M., et al. (2013). Genome-wide identification, phylogeny, evolution and expression patterns of AP2/ERF genes and cytokinin response factors in *Brassica rapa* ssp. *pekinensis*. *PLoS ONE* 8:e83444. doi:10.1371/journal.pone.0083444

Magnani, E., Sjolander, K., and Hake, S. (2004). From endonucleases to transcription factors: evolution of the AP2 DNA binding domain in plants. *Plant Cell* 16, 2265–2277. doi:10.1105/tpc.104.023135

Mann, D. G., Lafayette, P. R., Abercrombie, L. L., King, Z. R., Mazarei, M., Halter, M. C., et al. (2012). Gateway-compatible vectors for high-throughput gene functional analysis in switchgrass (*Panicum virgatum* L.) and other monocot species. *Plant Biotechnol. J.* 10, 226–236. doi:10.1111/j.1467-7652.2011.00658.x

Mittal, A., Gampala, S. S., Ritchie, G. L., Payton, P., Burke, J. J., and Rock, C. D. (2014). Related to ABA-Insensitive3(ABI3)/Viviparous1 and AtABI5 transcription factor coexpression in cotton enhances drought stress adaptation. *Plant Biotechnol. J.* 12, 578–589. doi:10.1111/pbi.12162

Mizoi, J., Shinozaki, K., and Yamaguchi-Shinozaki, K. (2012). AP2/ERF family transcription factors in plant abiotic stress responses. *Biochim Biophys Acta* 1819, 86–96. doi:10.1016/j.bbagrm.2011.08.004

Moore, K. J., Moser, L. E., Vogel, K. P., Waller, S. S., Johnson, B. E., and Pedersen, J. F. (1991). Describing and quantifying growth stages of perennial forage grasses. *Agron. J.* 83, 1073–1077. doi:10.2134/agronj1991.00021962008300060027x

Murashige, T., and Skoog, F. (1962). A revised medium for rapid growth and bio assays with tobacco tissue cultures. *Physiol. Plant.* 15, 473–497. doi:10.1111/j.1399-3054.1962.tb08052.x

Nakano, T., Fujisawa, M., Shima, Y., and Ito, Y. (2014). The AP2/ERF transcription factor SlERF52 functions in flower pedicel abscission in tomato. *J. Exp. Bot.* 65, 3111–3119. doi:10.1093/jxb/eru154

Nakano, T., Suzuki, K., Fujimura, T., and Shinshi, H. (2006). Genome-wide analysis of the ERF gene family in *Arabidopsis* and rice. *Plant Physiol.* 140, 411–432. doi:10.1104/pp.105.073783

Nunez-Lopez, L., Aguirre-Cruz, A., Barrera-Figueroa, B. E., and Pena-Castro, J. M. (2015). Improvement of enzymatic saccharification yield in *Arabidopsis thaliana* by ectopic expression of the rice *SUB1A-1* transcription factor. *PeerJ* 3, e817. doi:10.7717/peerj.817

Oh, S. J., Kim, Y. S., Kwon, C. W., Park, H. K., Jeong, J. S., and Kim, J. K. (2009). Overexpression of the transcription factor AP37 in rice improves grain yield under drought conditions. *Plant Physiol.* 150, 1368–1379. doi:10.1104/pp.109.137554

Ohme-Takagi, M., and Shinshi, H. (1995). Ethylene-inducible DNA binding proteins that interact with an ethylene-responsive element. *Plant Cell* 7, 173–182. doi:10.1105/tpc.7.2.173

Qin, F., Kakimoto, M., Sakuma, Y., Maruyama, K., Osakabe, Y., Tran, L. S., et al. (2007). Regulation and functional analysis of ZmDREB2A in response to drought and heat stresses in *Zea mays* L. *Plant J.* 50, 54–69. doi:10.1111/j.1365-313X.2007.03034.x

Rashid, M., Guangyuan, H., Guangxiao, Y., Hussain, J., and Xu, Y. (2012). AP2/ERF transcription factor in rice: genome-wide canvas and syntenic relationships between monocots and eudicots. *Evol. Bioinform. Online* 8, 321–355. doi:10.4137/EBO.S9369

Rashotte, A. M., Mason, M. G., Hutchison, C. E., Ferreira, F. J., Schaller, G. E., and Kieber, J. J. (2006). A subset of *Arabidopsis* AP2 transcription factors mediates cytokinin responses in concert with a two-component pathway. *Proc. Natl. Acad. Sci. U.S.A.* 103, 11081–11085. doi:10.1073/pnas.0602038103

Shen, H., Fu, C., Xiao, X., Ray, T., Tang, Y., Wang, Z., et al. (2009). Developmental control of lignification in stems of lowland switchgrass variety Alamo and the effects on saccharification efficiency. *Bioenerg. Res.* 2, 233–245. doi:10.1007/s12155-009-9058-6

Shen, H., He, X., Poovaiah, C. R., Wuddineh, W. A., Ma, J., Mann, D. G., et al. (2012). Functional characterization of the switchgrass (*Panicum virgatum*) R2R3-MYB transcription factor PvMYB4 for improvement of lignocellulosic feedstocks. *New Phytol.* 193, 121–136. doi:10.1111/j.1469-8137.2011.03922.x

Shen, H., Mazarei, M., Hisano, H., Escamilla-Trevino, L., Fu, C., Pu, Y., et al. (2013a). A genomics approach to deciphering lignin biosynthesis in switchgrass. *Plant Cell* 25, 4342–4361. doi:10.1105/tpc.113.118828

Shen, H., Poovaiah, C. R., Ziebell, A., Tschaplinski, T. J., Pattathil, S., Gjersing, E., et al. (2013b). Enhanced characteristics of genetically modified switchgrass (*Panicum virgatum* L.) for high biofuel production. *Biotechnol. Biofuels* 6, 71. doi:10.1186/1754-6834-6-71

Shi, J. X., Malitsky, S., De Oliveira, S., Branigan, C., Franke, R. B., Schreiber, L., et al. (2011). SHINE transcription factors act redundantly to pattern the archetypal surface of *Arabidopsis* flower organs. *PLoS Genet.* 7:e1001388. doi:10.1371/journal.pgen.1001388

Sreenivasulu, N., Usadel, B., Winter, A., Radchuk, V., Scholz, U., Stein, N., et al. (2008). Barley grain maturation and germination: metabolic pathway and regulatory network commonalities and differences highlighted by new MapMan/PageMan profiling tools. *Plant Physiol.* 146, 1738–1758. doi:10.1104/pp.107.111781

Sun, S., Yu, J. P., Chen, F., Zhao, T. J., Fang, X. H., Li, Y. Q., et al. (2008). TINY, a dehydration-responsive element (DRE)-binding protein-like transcription factor connecting the DRE- and ethylene-responsive element-mediated signaling pathways in *Arabidopsis*. *J. Biol. Chem.* 283, 6261–6271. doi:10.1074/jbc.M706800200

Sykes, R., Yung, M., Novaes, E., Kirst, M., Peter, G., and Davis, M. (2009). High-throughput screening of plant cell-wall composition using pyrolysis molecular beam mass spectroscopy. *Methods Mol. Biol.* 581, 169–183. doi:10.1007/978-1-60761-214-8_12

Tamura, K., Stecher, G., Peterson, D., Filipski, A., and Kumar, S. (2013). MEGA6: molecular evolutionary genetics analysis version 6.0. *Mol. Biol. Evol.* 30, 2725–2729. doi:10.1093/molbev/mst197

Tiwari, S. B., Belachew, A., Ma, S. F., Young, M., Ade, J., Shen, Y., et al. (2012). The EDLL motif: a potent plant transcriptional activation domain from AP2/ERF transcription factors. *Plant J.* 70, 855–865. doi:10.1111/j.1365-313X.2012.04935.x

Tschaplinski, T. J., Standaert, R. F., Engle, N. L., Martin, M. Z., Sangha, A. K., Parks, J. M., et al. (2012). Down-regulation of the *caffeic acid O-methyltransferase* gene in switchgrass reveals a novel monolignol analog. *Biotechnol. Biofuels* 5, 71. doi:10.1186/1754-6834-5-71

Vahala, J., Felten, J., Love, J., Gorzsas, A., Gerber, L., Lamminmaki, A., et al. (2013). A genome-wide screen for ethylene-induced ethylene response factors (ERFs)

in hybrid aspen stem identifies ERF genes that modify stem growth and wood properties. *New Phytol.* 200, 511–522. doi:10.1111/nph.12386

Van Raemdonck, D., Pesquet, E., Cloquet, S., Beeckman, H., Boerjan, W., Goffner, D., et al. (2005). Molecular changes associated with the setting up of secondary growth in aspen. *J. Exp. Bot.* 56, 2211–2227. doi:10.1093/jxb/eri221

Wang, L., Wang, C., Qin, L., Liu, W., and Wang, Y. (2015). ThERF1 regulates its target genes via binding to a novel cis-acting element in response to salt stress. *J. Integr. Plant Biol.* doi:10.1111/jipb.12335

Wang, Y., Wan, L., Zhang, L., Zhang, Z., Zhang, H., Quan, R., et al. (2012). An ethylene response factor OsWR1 responsive to drought stress transcriptionally activates wax synthesis related genes and increases wax production in rice. *Plant Mol. Biol.* 78, 275–288. doi:10.1007/s11103-011-9861-2

Wuddineh, W. A., Mazarei, M., Zhang, J., Poovaiah, C. R., Mann, D. G., Ziebell, A., et al. (2015). Identification and overexpression of *gibberellin 2-oxidase (GA2ox)* in switchgrass (*Panicum virgatum* L.) for improved plant architecture and reduced biomass recalcitrance. *Plant Biotechnol. J.* 13, 636–647. doi:10.1111/pbi.12287

Xu, W., Li, F., Ling, L., and Liu, A. (2013). Genome-wide survey and expression profiles of the AP2/ERF family in castor bean (*Ricinus communis* L.). *BMC Genomics* 14:785. doi:10.1186/1471-2164-14-785

Xu, Z. S., Chen, M., Li, L. C., and Ma, Y. Z. (2011). Functions and application of the AP2/ERF transcription factor family in crop improvement. *J. Integr. Plant Biol.* 53, 570–585. doi:10.1111/j.1744-7909.2011.01062.x

Yuan, J. S., Tiller, K. H., Al-Ahmad, H., Stewart, N. R., and Stewart, C. N. Jr. (2008). Plants to power: bioenergy to fuel the future. *Trends Plant Sci.* 13, 421–429. doi:10.1016/j.tplants.2008.06.001

Zeng, J. K., Li, X., Xu, Q., Chen, J. Y., Yin, X. R., Ferguson, I. B., et al. (2015). *EjAP2-1*, an AP2/ERF gene, is a novel regulator of fruit lignification induced by chilling injury, via interaction with EjMYB transcription factors. *Plant Biotechnol. J.* doi:10.1111/pbi.12351

Zhang, C. H., Shangguan, L. F., Ma, R. J., Sun, X., Tao, R., Guo, L., et al. (2012). Genome-wide analysis of the AP2/ERF superfamily in peach (*Prunus persica*). *Genet. Mol. Res.* 11, 4789–4809. doi:10.4238/2012.October.17.6

Zhang, H., Liu, W., Wan, L., Li, F., Dai, L., Li, D., et al. (2010a). Functional analyses of ethylene response factor JERF3 with the aim of improving tolerance to drought and osmotic stress in transgenic rice. *Transgenic Res.* 19, 809–818. doi:10.1007/s11248-009-9357-x

Zhang, Z., Li, F., Li, D., Zhang, H., and Huang, R. (2010b). Expression of ethylene response factor JERF1 in rice improves tolerance to drought. *Planta* 232, 765–774. doi:10.1007/s00425-010-1208-8

Zhang, J. Y., Lee, Y. C., Torres-Jerez, I., Wang, M., Yin, Y., Chou, W. C., et al. (2013). Development of an integrated transcript sequence database and a gene expression atlas for gene discovery and analysis in switchgrass (*Panicum virgatum* L.). *Plant J.* 74, 160–173. doi:10.1111/tpj.12104

Zhang, Z., and Huang, R. (2010). Enhanced tolerance to freezing in tobacco and tomato overexpressing transcription factor TERF2/LeERF2 is modulated by ethylene biosynthesis. *Plant Mol. Biol.* 73, 241–249. doi:10.1007/s11103-010-9609-4

Zhou, X., Jenks, M., Liu, J., Liu, A., Zhang, X., Xiang, J., et al. (2014). Overexpression of transcription factor OsWR2 regulates wax and cutin biosynthesis in rice and enhances its tolerance to water deficit. *Plant Mol. Biol. Rep.* 32, 719–731. doi:10.1007/s11105-013-0687-8

Zhuang, J., Cai, B., Peng, R. H., Zhu, B., Jin, X. F., Xue, Y., et al. (2008). Genome-wide analysis of the AP2/ERF gene family in *Populus trichocarpa. Biochem. Biophys. Res. Commun.* 371, 468–474. doi:10.1016/j.bbrc.2008.04.087

Conflict of Interest Statement: The authors declare that the research was conducted in the absence of any commercial or financial relationships that could be construed as a potential conflict of interest.

2

Use of nanostructure-initiator mass spectrometry to deduce selectivity of reaction in glycoside hydrolases

Kai Deng[1,2], Taichi E. Takasuka[3†], Christopher M. Bianchetti[3,4], Lai F. Bergeman[3],
Paul D. Adams[1,5,6], Trent R. Northen[1,5] and Brian G. Fox[3,7]*

[1] US Department of Energy Joint BioEnergy Institute, Emeryville, CA, USA, [2] Sandia National Laboratories, Livermore, CA, USA, [3] US Department of Energy Great Lakes Bioenergy Research Center, Madison, WI, USA, [4] Department of Chemistry, University of Wisconsin-Oshkosh, Oshkosh, WI, USA, [5] Lawrence Berkeley National Laboratory, Berkeley, CA, USA, [6] Department of Bioengineering, University of California Berkeley, Berkeley, CA, USA, [7] Department of Biochemistry, University of Wisconsin-Madison, Madison, WI, USA

Edited by:
Robert Henry,
The University of Queensland,
Australia

Reviewed by:
Lixin Cheng,
Aarhus University, Denmark
Chiranjeevi Thulluri,
Jawaharlal Nehru Technological
University Hyderabad, India

*Correspondence:
Brian G. Fox
bgfox@biochem.wisc.edu

†Present address:
Taichi E. Takasuka,
Research Faculty of Agriculture,
Hokkaido University, Sapporo, Japan

Chemically synthesized nanostructure-initiator mass spectrometry (NIMS) probes derivatized with tetrasaccharides were used to study the reactivity of representative *Clostridium thermocellum* β-glucosidase, endoglucanases, and cellobiohydrolase. Diagnostic patterns for reactions of these different classes of enzymes were observed. Results show sequential removal of glucose by the β-glucosidase and a progressive increase in specificity of reaction from endoglucanases to cellobiohydrolase. Time-dependent reactions of these polysaccharide-selective enzymes were modeled by numerical integration, which provides a quantitative basis to make functional distinctions among a continuum of naturally evolved catalytic properties. Consequently, our method, which combines automated protein translation with high-sensitivity and time-dependent detection of multiple products, provides a new approach to annotate glycoside hydrolase phylogenetic trees with functional measurements.

Keywords: cellulase, assay, kinetics, Nimzyme, mass spectrometry, protein engineering, biofuels

INTRODUCTION

The enzymatic hydrolysis of plant cell wall material is a formidable task because of the complexity of the plant cell wall (Himmel et al., 2007). In most currently deployed cellulosic ethanol plants, enzyme cocktails containing multiple classes of polysaccharide-degrading enzymes are used to hydrolyze plant biomass into fermentable sugars. Understanding the function, synergy, and stability of enzymes is thus of paramount importance in biofuels production.

Polysaccharide-degrading enzymes are classified into families in the carbohydrate active enzyme (CAZy) database (Henrissat and Davies, 1997; Cantarel et al., 2009; Levasseur et al., 2013), including glycoside hydrolases (GHs), pectic lyases (PLs), carbohydrate esterases (CEs), and others. Only a small fraction of the enzymes included in CAZy have a function assigned by biochemical analyses. One root of this limitation arises from difficulties in succeeding with heterologous expression of enzymes after selection from phylogenetic trees (Watson et al., 2007; Fox et al., 2008; Markley

Abbreviations: AFEX-SG, ammonia fiber expansion pretreated switchgrass; CBM, carbohydrate-binding module; CelE, broad specificity GH family 5 (GH5) domain from *C. thermocellum* Cthe_0797; GH, glycoside hydrolase; IL-SG, ionic liquid pretreated switchgrass; NIMS, nanostructure-initiator mass spectrometry.

et al., 2009; Nair et al., 2009; Pieper et al., 2013). As an option to address this limitation, we (Takasuka et al., 2014; Bianchetti et al., 2015) and others (Beebe et al., 2011, 2014; Madono et al., 2011; Hirano et al., 2013, 2015; Makino et al., 2014) have found that wheat germ cell-free protein translation can be used as an effective expression platform to make functional assignments of enzyme function.

Another limitation arises from experimental complications of carrying out high-throughput multisubstrate assays to screen for enzyme function (Gerlt et al., 2011). A breadth of assay methods have been developed for GHs, including use of soluble and insoluble chromogenic and/or fluorogenic substrates, HPLC, and others (Sharrock, 1988; Decker et al., 2003; Chundawat et al., 2008; Bansal et al., 2009; Dowe, 2009; Dashtban et al., 2010; Selig et al., 2011; Eklof et al., 2012; Horn et al., 2012; Kosik et al., 2012; McCleary et al., 2012; Pena et al., 2012; Whitehead et al., 2012; Wischmann et al., 2012). Each of these approaches has intrinsic advantages, but can suffer in sensitivity, complexity of analysis, throughput time, and volumes of reagents and enzyme needed. In comparison, nanostructure-initiator mass spectrometry (NIMS) offers high sensitivity, simplicity of detection of products derived from biomass hydrolysis, microliters or smaller volumes for reaction, and options for automation (Northen et al., 2008; Deng et al., 2012; de Rond et al., 2013; Heins et al., 2014). Recently, we used oxime-NIMS and numerical integration methods to provide time-dependent, quantitative characterization of reducing sugars released by individual enzymes in reactions with pretreated biomass (Deng et al., 2014).

Here, we report a new use of NIMS to provide quantitative analysis of time-dependent reactions of cellulases. The enzymes selected for this study were from *Clostridium thermocellum*, a Gram-positive anaerobe with high cellulolytic capacity (Ding et al., 2008; Fontes and Gilbert, 2010; Smith and Bayer, 2013). The *C. thermocellum* genome encodes ~130 CAZyme domains and ~90 carbohydrate-binding module (CBM) domains (Feinberg et al., 2011). The majority of CAZyme domains also possess dockerin domains, which serve to recruit these enzymes into the cellulosome via dockerin–cohesin interactions (Ding et al., 2008; Smith and Bayer, 2013). The specific gene regulatory and protein secretory patterns of this model consolidated bioprocessing organism have also been well described (Brown et al., 2007; Gold and Martin, 2007; Roberts et al., 2010; Feinberg et al., 2011; Raman et al., 2011; Riederer et al., 2011), and many of the enzymes have been characterized. Given this state of knowledge, individual enzymes from *C. thermocellum* have proven useful for the development and testing of new approaches for assignment of GH function.

In this work, we have used chemically synthesized tetrasaccharide-NIMS probes to study the reactivity of some cellulases from *C. thermocellum*. Patterns of reactivity identified by using the tetrasaccharide-NIMS probes provide a diagnostic approach to assess reaction specificity and also provide comparative apparent rate information. Our results show diagnostic patterns for reactions of a β-glucosidase, relaxed but varied specificity of several endoglucanases and high specificity of a cellobiohydrolase with the model substrate. Time-dependent reactions of these polysaccharide-selective enzymes were modeled by numerical integration, which provides a quantitative basis to make functional distinctions among a continuum of naturally evolved reactive properties. Consequently, this method, which combines high-sensitivity detection of multiple products with quantitative numerical analysis of their time-dependent formation, provides a new approach to enhance the annotation of GH phylogenetic trees with functional measurements.

MATERIALS AND METHODS

Enzyme Preparation

Methods for cloning, cell-free translation, and purification of the enzymes studied have been reported elsewhere (Takasuka et al., 2014). Briefly, enzymes were cloned by PCR amplification of catalytic domains as indicated by the first and last codons indicated in **Table 1**. Cloned genes were transferred into an optimized wheat germ cell-free translation plasmid pEU-HSCB (Beebe et al., 2011; Takasuka et al., 2014), which is also available from the NIH Protein Structure Initiative Materials Repository (http://psimr. asu.edu/). Enzymes were prepared by cell-free translation using either bilayer or dialysis methods (Beebe et al., 2011, 2014; Makino et al., 2014), and active enzymes were identified (Takasuka et al., 2014). The enzymes listed in **Table 1** were also cloned by PCR into the *Escherichia coli* expression vector pEC_CBM3a to create enzyme_CBM3a fusion proteins, CelAcc_CBM3a. The vector pEC_CBM3a is a derivative of pEU_HSBC_CBM3a (Takasuka et al., 2014) that yields fusion proteins having an N-terminal enzyme catalytic domain fused by an ~40 aa linker sequence to the CBM3a domain from Cthe_3077. A stop codon was added to the PCR primer used to amplify the 3′ end of the BglA gene so that no fusion to CBM3a was produced from pEU_HSBC_CBM3a. As needed, protein coding sequences were transferred between pEU and pEC vectors by use of FlexiVector cloning (Blommel et al., 2009). Methods for PCR amplification, capture and sequence verification of protein coding sequences, and transformation into *E. coli* 10G competent cells (Lucigen, Middleton, WI, USA) for DNA manipulations and *E. coli* B834 for protein expression were as previously reported (Takasuka et al., 2014). Additional details of the properties and methods for the use of pEU and pVP are described elsewhere (Aceti et al., 2015).

Synthesis of Cellotetraose-NIMS Substrate

The cellotetraose-NIMS substrate (**Figure 1A**) is an amphiphilic molecule that has a sugar head group coupled to a perfluorinated (F17) tag. The detailed synthetic procedure has been reported previously (Deng et al., 2012).

Enzyme Reactions

An enzyme reaction consisted of 10 μL of 50 mM phosphate, pH 6.0, supplemented with 1 μL of 1 mM cellotetraose-F17 dissolved in water. An aliquot of each enzyme preparation (containing 1–10 ng of enzyme) was added to initiate the reaction and the resulting mixture was incubated at 37°C. At times of 5, 10, 20, 40, 80, and 120 min, 0.2 μL of the reaction mixture was withdrawn for analysis.

TABLE 1 | *Clostridium thermocellum* enzymes studied in this work.

Gene locus	Name	CAZy family	First codon[a]	Last codon[b]	Function[c]	CMX class[d]	Reference
Cthe_0212	BglA	GH1	1	448	Exo-β-glucosidase		Grabnitz et al. (1991)
Cthe_0269	CelA	GH8	34	368	Endo-β-1,4-glucanase	CX	1CEM; Alzari et al. (1996)
Cthe_0040	CelI	GH9	28	887	Endo-β-1,4-glucanase	C	2XFG
Cthe_0797	CelE	GH5	36	388	Cellulase, xylanase, mannanase	CMX	Deng et al. (2014, Takasuka)
Cthe_0578	CelR	GH9	27	640	Endo-β-1,4-glucanase; cellotetraohydrolase	C	Zverlov et al. (2005)
Cthe_0405	CelL	GH5	31	430	–	CX	Deng et al. (2014)
Cthe_0412	CelK	GH9	28	809	Cellobiohydrolase	C	Kataeva et al. (1999)
Cthe_3077[e]	CipA	CBM3a	323	523	Cellulose-binding module		Kataeva et al. (1999)

[a]*First codon of the indicated gene locus that was included in the PCR primer design (Takasuka et al., 2014).*
[b]*Last codon of the indicated gene locus that was included in the PCR primer design.*
[c]*Function assigned from annotation as defined in CAZy (Cantarel et al., 2009), from experimental evidence cited in the table, or a combination of both.*
[d]*Representation of the breadth of substrate specificity for each enzyme (Deng et al., 2014). The CMX classification indicates that CelE can hydrolyze cellulose, xylan, or mannan; CX indicates that CelA and CelL can hydrolyze cellulose and xylan, while CelI, CelR, and CelK can only hydrolyze cellulose. This classification derives from reactions with pure polysaccharides and pretreated biomass (Deng et al., 2014; Takasuka et al., 2014).*
[e]*CBM3a was subcloned from the scaffoldin CipA gene.*

Mass Spectrometry

In each case, 0.2 µL per reaction sample was spotted onto the NIMS surface and removed after an incubation of ~30 s. A grid drawn manually on the NIMS chip using a diamond-tip scribe helped with spotting and identification of sample spots in the spectrometer. Chips were loaded using a modified standard MALDI plate. NIMS was performed on a 4800 MALDI TOF/TOF mass analyzer from AB Sciex (Foster City, CA, USA). In each case, signal intensities were identified for the ions of the cellotetraose substrate and, when present, each product shown in **Figure 1**. For each assay, ~1000 laser shots were collected. Enzyme activities were determined by measuring the intensity ratios of each product over the intensity total of ions of for the cellotetraose-, cellotriose-, cellobiose-, glucose-, and aglycone-NIMS (**Figure 2**).

Kinetic Analyses

The time dependence of hydrolysis of the tetrasaccharide-NIMS was analyzed by non-linear global optimization of differential equations accounting for the appearance and decay of products (Deng et al., 2014) using Mathematica routine NDSolve and the Nelder-Mead simplex method for constrained minimization (Nelder and Mead, 1965). The differential equations corresponding to the kinetic scheme of **Figure 3** are as follows:

$$y[1] = \text{cellotetraose-NIMS} \quad (1)$$

$$y[2] = \text{cellotriose-NIMS} \quad (2)$$

$$y[3] = \text{cellobiose-NIMS} \quad (3)$$

$$y[4] = \text{glucose-NIMS} \quad (4)$$

$$y[5] = \text{aglycone-NIMS} \quad (5)$$

$$dy[1]/d[t] = -\left(k1 + k9 + k11 + k13\right) y[1][t] \quad (6)$$

$$dy[2]/d[t] = \left(k1\right) y[1][t] - \left(k3 + k15 + k17\right) y[2][t] \quad (7)$$

$$dy[3]/d[t] = \left(k9\right) y[1][t] + \left(k3\right) y[2][t] \\ -\left(k5 + k19\right) y[3][t] \quad (8)$$

$$dy[4]/d[t] = \left(k11\right) y[1][t] + \left(k15\right) y[2][t] \\ +\left(k5\right) y[3][t] - \left(k7\right) y[4][t] \quad (9)$$

$$dy[5]/d[t] = \left(k13\right) y[1][t] + \left(k17\right) y[2][t] \\ +\left(k19\right) y[3][t] + \left(k7\right) y[4][t] \quad (10)$$

Initial guesses for apparent rate constants were made by visual inspection of the match between the results of single NDSolve calculations and the experimental data. This process was continued in an iterative way until a set of initial apparent rates that adequately matched the experimental data was obtained. Successive rounds of least squares parameter optimization with adjustment of parameter constraints were carried out until the sum of the squares difference between calculated and experimental values reached a minimum and no parameter was artificially constrained.

RESULTS AND DISCUSSION

Enzymes Chosen for Study

Clostridium thermocellum enzymes were chosen for this study based on previous transcriptomic and proteomic results (Gold and Martin, 2007; Raman et al., 2011; Riederer et al., 2011) and other biochemical and structural results (**Table 1**). Genes encoding these enzymes were expressed using wheat germ cell-free protein synthesis and the translated proteins were assayed using fluorogenic substrates (Takasuka et al., 2014); among the synthesized enzymes, 13 reacted with MUG or MUC, 11 reacted with MUX or MUX2, and 5 reacted with other diagnostic fluorogenic substrates. Reactions of these enzymes with ionic liquid pretreated switchgrass (IL-SG) have been published (Deng et al., 2014). Enzymes from cell-free translation reactions that showed promising characteristics were produced by expression in *E. coli* and purified for use in the studies described here.

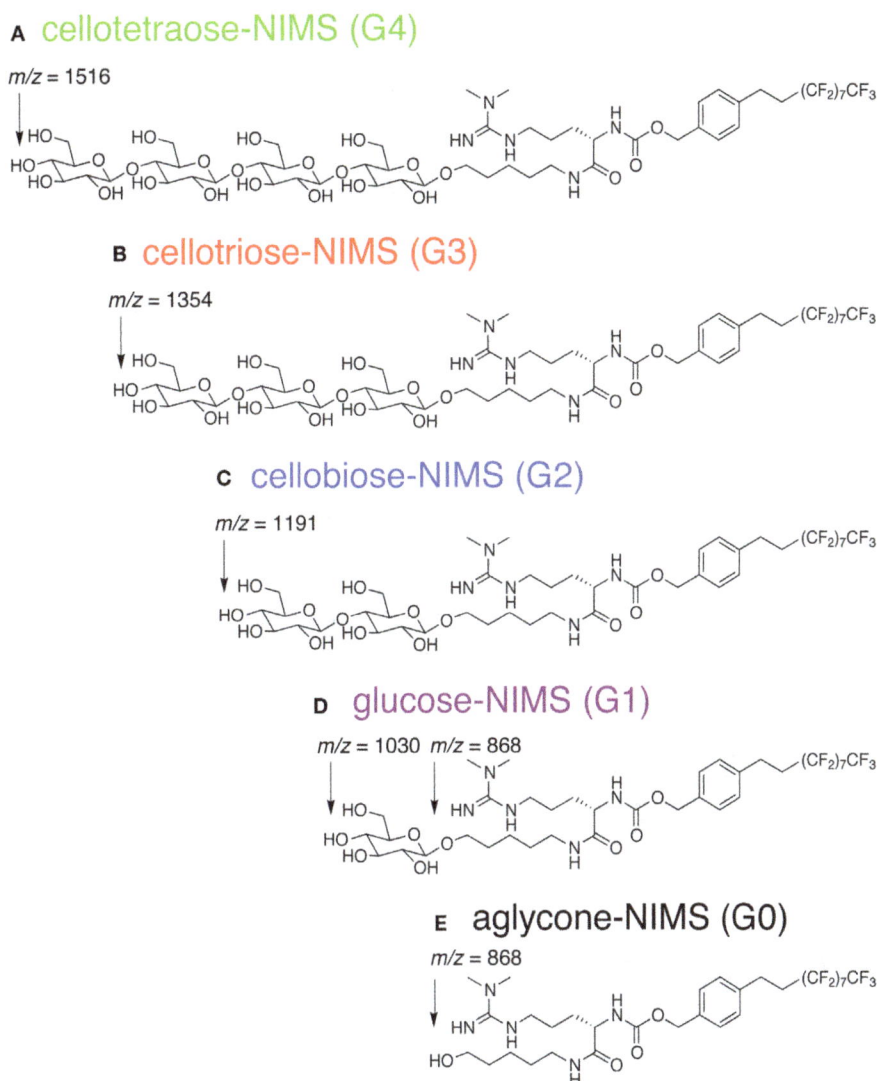

FIGURE 1 | Structure of cellotetraose-NIMS and *m/z* values for products obtained from hydrolysis at the indicated anomeric position. (A) cellotetraose-NIMS; **(B)** cellotriose-NIMS; **(C)** cellobiose-NIMS; **(D)** glucose-NIMS; **(E)** aglycone-NIMS.

Cellotetraose-NIMS Substrate

Figure 1 shows the structure of cellotetraose-NIMS and the products that can be formed by various GH reactions. In the synthesized probe, the tetra-saccharide is linked to the NIMS probe by a potentially hydrolyzable anomeric linkage. Synthesis of the NIMS probe and the tetra-saccharide derivatives are summarized in Materials and Methods (Deng et al., 2012; de Rond et al., 2013). The guanidium group on the NIMS probe provides improved ionization properties in the mass spectrometry experiment, while the perfluorinated portion of the NIMS probe provides hydrophobic anchoring of the molecule into the NIMS surface. Enzyme-catalyzed hydrolysis of the anomeric linkages give rise to a cascade of potential products retained on the NIMS surface. Reactions of GHs can progressively remove single glucose units or carry out other reactions that remove cellobiose, cellotriose, or cellotetraose.

Kinetic Scheme

Figure 2 shows a representative mass spectrum obtained after partial reaction with BglA (Cthe_0212), a β-glucosidase. At the selected time point in the reaction (120 min), the cellotetraose-NIMS probe (G4, green) has been partially converted into a mixture of cellotriose (G3, red), cellobiose (G2, blue), glucose (G1, purple), and aglycone (G0, black) derivatives of the NIMS probe. **Figure 3** shows a kinetic scheme that accounts for the potential products shown in **Figure 1**. The scheme accounts for release of one or more glucose units from the cellotetraose-NIMS probe (G4) and its successive products. Time course profiles provide the fundamental data used in this work for numerical analysis of enzyme hydrolysis reactions.

β-Glucosidase BglA Reaction

The nucleotide sequence of BglA (Grabnitz et al., 1991) was published before the genome sequence and annotated to be

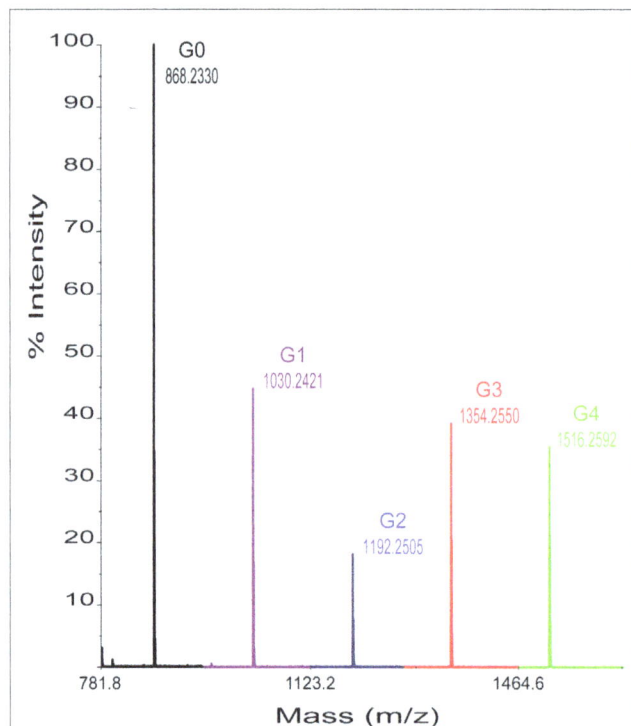

FIGURE 2 | Representative mass spectrum obtained from enzyme hydrolysis of cellotetraose-NIMS. Mass peaks corresponding to cellotetraose-NIMS (green), cellotriose-NIMS (red), cellobiose-NIMS (blue), glucose-NIMS (purple), and aglycone-NIMS (black) are indicated. The products shown are from reaction of BglA.

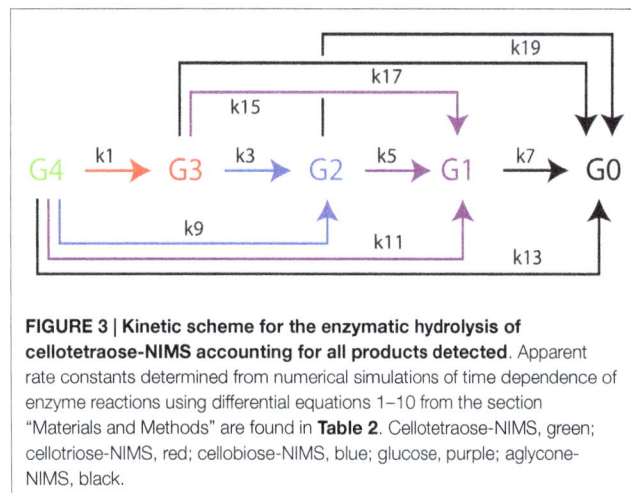

FIGURE 3 | Kinetic scheme for the enzymatic hydrolysis of cellotetraose-NIMS accounting for all products detected. Apparent rate constants determined from numerical simulations of time dependence of enzyme reactions using differential equations 1–10 from the section "Materials and Methods" are found in Table 2. Cellotetraose-NIMS, green; cellotriose-NIMS, red; cellobiose-NIMS, blue; glucose, purple; aglycone-NIMS, black.

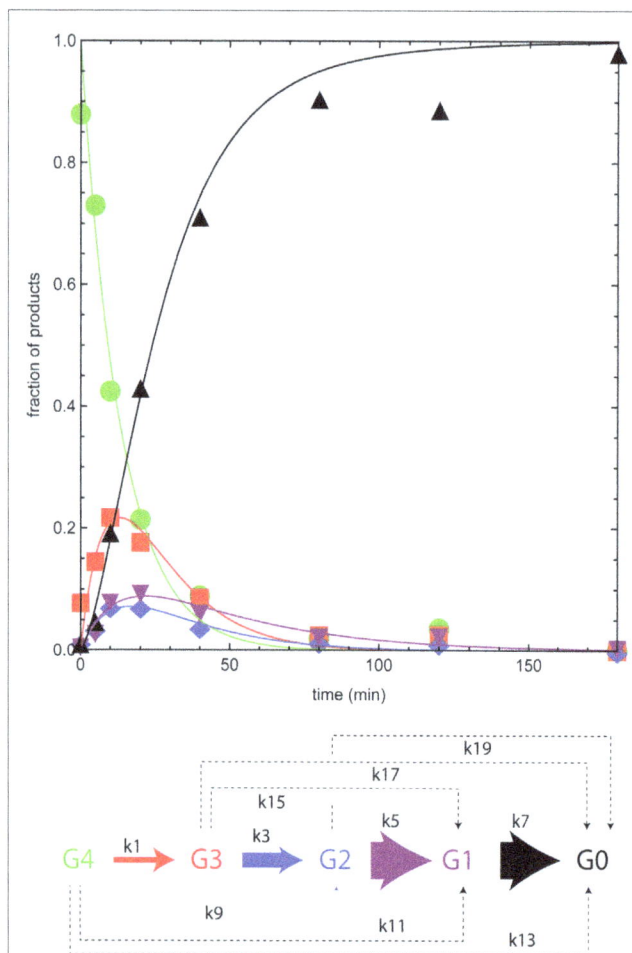

FIGURE 4 | Numerical analysis of the time course for reaction of BglA with cellotetraose-NIMS. Products are cellotetraose-NIMS (green), cellotriose-NIMS (red), cellobiose-NIMS (blue), glucose- NIMS (purple), and aglycone-NIMS (black). Relative magnitude of the apparent rates shown in Table 2 indicated by width of arrows in the modified kinetic scheme. A dashed line indicates that the apparent rate was zero.

a β-glucosidase from the GH1 family (Cantarel et al., 2009). The Cthe_02012 gene does not encode a signal peptide, so the entire gene was cloned for the studies described here. Beyond our characterization of the reaction of BglA with IL-SG (Deng et al., 2014), no other functional studies have been reported for this enzyme.

Figure 4 shows the time course for reaction of BglA with cellotetraose-NIMS. The plotted proportions of the different products come from time series of mass spectra like those shown in **Figure 2**. The solid colored lines are results of simulations of the concentration of individual products based on the kinetic scheme of **Figure 3** and the differential equations shown in the section "Materials and Methods." The apparent rate constants provided by the numerical simulation are given in **Table 2**, and a pictorial representation of the relative magnitudes of the apparent rate constants is also given in **Figure 4**. In the time course of the BglA reaction, cellotetraose-NIMS (green circles) was converted to a succession of intermediates by hydrolysis of a single glucose from the position most distal to the NIMS probe. This pattern of reactivity is as expected for the reaction of an exo-β-glucosidase with an oligosaccharide. Thus, cellotriose-NIMS (red squares) accumulated was subsequently converted to cellobiose-NIMS (purple down triangles), to glucose-NIMS (blue diamonds), and ultimately to aglycone-NIMS (black up triangles).

TABLE 2 | Apparent rate constants (min^{-1}) from numerical integration of time course reactions with cellotetraose-NIMS.

Rate	BglA	CelL	CelR	CelE	CelA	CelI	CelK
k1	0.05320	0.000	0.001	0.001	0.000001	0.0002	0.0000
k3	0.14805	0.0002	0.0000	0.0000	0.0000	0.0002	0.0000
k5	0.36803	0.0000	0.0000	0.0000	0.0000	0.0000	0.0000
k7	0.36803	0.0000	0.0000	0.0000	0.0000	0.0000	0.0000
k9	0.00047	0.0490	0.0170	0.0092	0.0004	0.0020	0.0439
k11	0.00006	0.0038	0.0453	0.0390	0.0116	0.0479	0.0000
k13	0.00006	0.0000	0.0001	0.0000	0.0000	0.0000	0.0000
k15	0.00006	0.0000	0.0000	0.0000	0.0010	0.0000	0.0004
k17	0.00006	0.0000	0.0000	0.0000	0.0000	0.0000	0.0000
k19	0.00006	0.0000	0.0000	0.0000	0.0000	0.0000	0.0001

Rates for individual enzymes are color coded with smallest values in blue and larger values in red.

There are several features of the BglA reaction and simulation that warrant attention. The apparent rates k1, k3, k5, and k7, which correspond to successive removal of single glucose groups, dominate the numerical solution (**Table 2**; **Figure 4**). Under the reaction conditions used, BglA was able to completely convert cellotetraose-NIMS to aglycone-NIMS. It is also noteworthy that shortening the oligosaccharide chain led to an enhancement in the rate of hydrolysis, with reactions k5 (converting cellobiose-NIMS to glucose-NIMS) and k7 (converting glucose-NIMS to aglycone-NIMS) being fastest. Other apparent rates corresponding to side reactions for removal of cellobiose or larger oligosaccharides (e.g., k9 for removal of cellobiose from cellotetraose-NIMS) were less than 1/100th of the value observed for k1, the smallest of the central reactions. These simulation results are consistent with the assigned function of BglA as a β-glucosidase. Indeed, prior oxime-NIMS studies of the reaction of BglA with IL-SG revealed that glucose was the only product released from the biomass substrate (Deng et al., 2014). In the following paragraphs, these diagnostic behaviors of a beta-glucosidase are contrasted with two other classes of GHs, including five phylogenetically diverse endoglucanases and one cellobiohydrolase.

Endoglucanase and Cellobiohydrolase Reactions

Figure 5 shows time courses for reactions of endoglucanases CelA, CelI, CelE, CelR, CelL, and cellobiohydrolase CelK with cellotetraose-NIMS. The reactions of the individual enzymes were carried out and evaluated as described above for **Figure 4**. The appearance of the reaction time courses and the relative rates observed are markedly different than observed for BglA. Unlike the β-glucosidase reaction, no intermediates were observed to form and decay, and the central reactions corresponding to release of glucose units were negligible. This seemingly corresponds with the requirement of endoglucanases for a longer oligosaccharide chain to occupy the active site as a determinant of productive binding and catalysis. In effect, the endoglucanases and cellobiohydrolase primarily reacted only once with the cellotetraose-NIMS probe, leading to a markedly simpler cascade of products than observed for the beta-glucosidase. None of the enzymes characterized in **Figure 5** was able to carry out reactions that yielded the aglycone-NIMS product

(black up triangles), suggesting unproductive binding or blocking steric interactions of the NIMS product with adjacent features of the active site. In contrast, the β-glucosidase BglA (**Figure 4**) was able to successively remove all glucose groups from cellotetraose-NIMS to yield aglycone-NIMS.

Endoglucanase CelA Reactions

CelA (Cthe_0269) is a GH8 endoglucanase. It is one of the most abundantly transcribed and secreted proteins in *C. thermocellum* during growth on cellulosic substrates (Brown et al., 2007; Gold and Martin, 2007; Raman et al., 2011; Riederer et al., 2011). Analysis of the crystal structure of the enzyme suggested that the substrate binding channel was optimally configured to bind a cellopentaose molecule (Alzari et al., 1996).

The functional characterizations of **Figure 5** demonstrate a progression in reaction selectivity among the enzymes studied. This is a unique power arising from the combination of time-dependent NIMS with numerical analysis. For CelA (**Figure 5A**), k11 governed removal of cellotriose from cellotetraose-NIMS, leading to the predominant accumulation of glucose-NIMS (88%, purple down triangles). The alternative removal of cellotriose via the two step pathway of k1 (removal of glucose) and k15 (removal of cellobiose) contributed ~9% to the overall product yields, while reaction via k9 (removal of cellobiose) added only ~3% of total products as cellobiose-NIMS (blue diamonds). It is worth noting that CelA gave the slowest hydrolysis of cellotetraose-NIMS of all enzymes tested, which is reflected in the values of apparent rates reported in **Table 2** and also in the shape of the plots in **Figure 5**. This may also reflect a partial rate diminution caused by a mismatch between cellotetraose-NIMS and a preferred cellopentaose occupying the active site channel.

In our earlier reactions of CelA with IL-SG (Deng et al., 2014), a mixture of glucose, cellobiose, triose, and tetraose was observed. Other than cellotetraose, whose release from cellotetraose-NIMS was probably prevented by improper binding of the NIMS moiety in the active site channel, the suite of products given by CelA reaction with cellotetraose-NIMS was comparable to that observed from reactions with the pretreated biomass (Deng et al., 2014).

Endoglucanase CelI, CelE, and CelR Reactions

For the reactions of CelI (**Figure 5B**), CelE (**Figure 5C**), and CelR (**Figure 5D**), the dominant pattern of preferred removal of cellotriose units to yield glucose-NIMS (purple down triangles) was retained. However, functional differences of these three enzymes were identified as the removal of cellobiose leading to cellobiose-NIMS (blue diamonds) assumed an increasing contribution to the total product distribution. For example, the observed change corresponds to an approximately eightfold increase in k9 between CelI and CelR. In the middle of these boundary enzymes, CelE was unique among the endoglucanases tested as it was also able to release a glucose unit from cellotetraose-NIMS in ~2% yield. In reactions with IL-SG and ammonia fiber expansion pretreated switchgrass (AFEX-SG) (Deng et al., 2014), these three enzymes released a mixture of glucose, cellobiose, and cellotriose, with the distribution of products in the biomass reaction shifted toward cellobiose and glucose. However, this shift is, in part, due

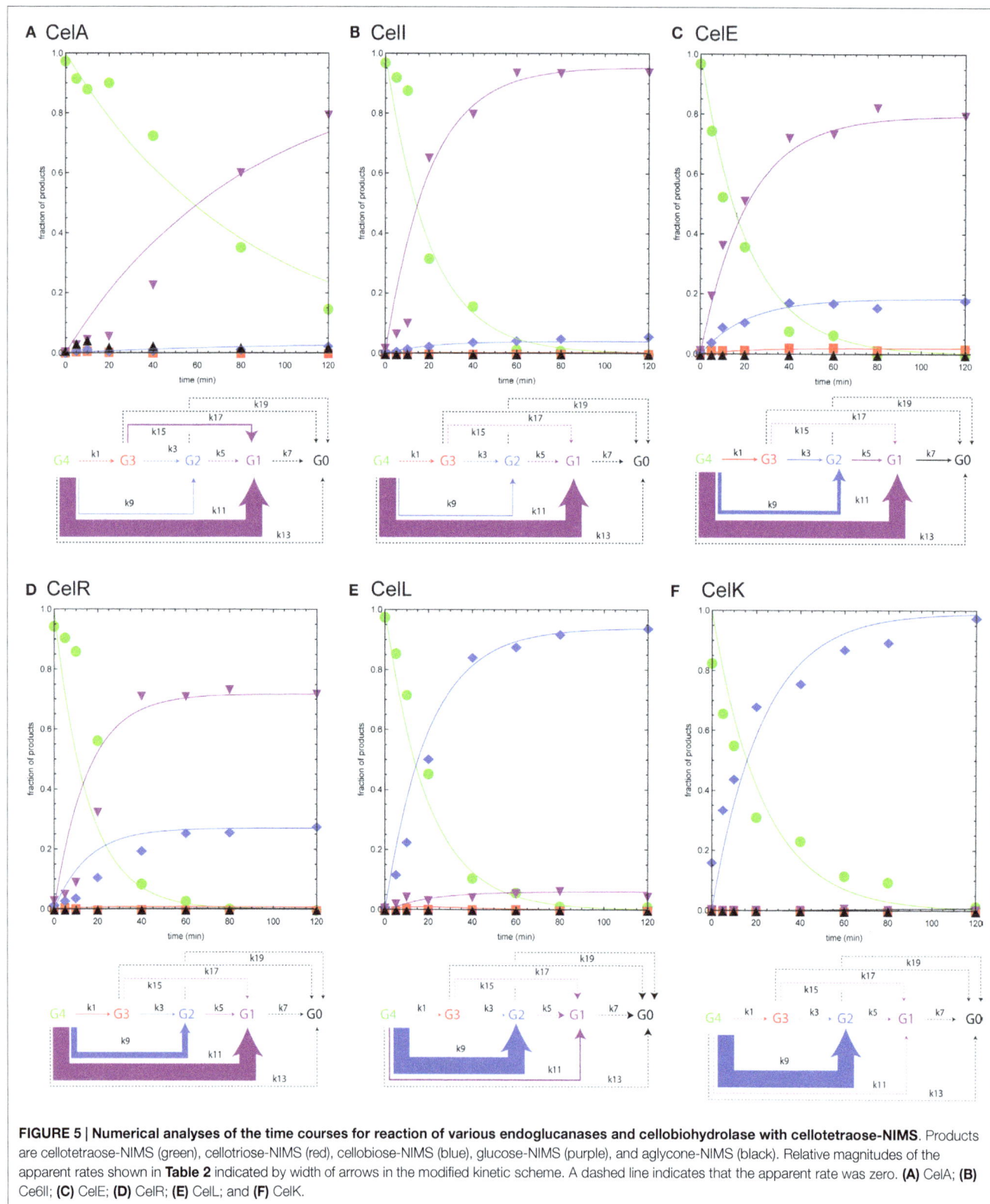

FIGURE 5 | Numerical analyses of the time courses for reaction of various endoglucanases and cellobiohydrolase with cellotetraose-NIMS. Products are cellotetraose-NIMS (green), cellotriose-NIMS (red), cellobiose-NIMS (blue), glucose-NIMS (purple), and aglycone-NIMS (black). Relative magnitudes of the apparent rates shown in **Table 2** indicated by width of arrows in the modified kinetic scheme. A dashed line indicates that the apparent rate was zero. **(A)** CelA; **(B)** Ce6II; **(C)** CelE; **(D)** CelR; **(E)** CelL; and **(F)** CelK.

to the ability of these enzymes to cleave solubilized cellotriose into cellobiose and glucose. Subsequent hydrolysis of released oligosaccharides could not be detected when cellotetraose-NIMS was the substrate.

CelI (Cthe_0040) has a structure consisting of GH9 and two CBM3 domains (Hazlewood et al., 1993). It catalyzes the hydrolysis of 1,4-β-glucosidic linkages in cellulose and other glucans. The structure suggests the position of a tunnel that can permit the release of either cellotriose or cellobiose from cellotetraose-NIMS (PDB 2XFG, no associated publication).

CelE (Cthe_0797) is a multidomain enzyme consisting of GH5, dockerin, and GSDL-lipase domains. Our work has shown that the GH5 domain has broad specificity for reaction with cellulose, xylan, mannan, xyloglucan, and other polysaccharides (Deng et al., 2014; Takasuka et al., 2014). The active site channel of this enzyme is open and tolerates the placement of each of these different linear and branched polysaccharides in a way that a glycosidic bond can be placed in the appropriate position for hydrolysis (Bianchetti et al., 2015). The release of cellotriose, cellobiose, and glucose from cellotetraose-NIMS is compatible with this broad specificity active site. Nevertheless, the active site is not sufficiently tolerant to remove cellotetraose, leading to the formation of aglycone-NIMS.

Previous studies have reported that CelR (Cthe_0578) is a β-glucanase with preference for release of cellotetraose in reactions with amorphous cellulose (Zverlov et al., 2005). Subsequently, CelR was able to convert the longer solubilized oligosaccharide to shorter oligosaccharides. The present studies provide support for this conclusion, as k11 for release of cellotriose was the predominant reaction with cellotetraose-NIMS. Our studies of CelR in reactions with IL-SG and AFEX-SG gave glucose and cellobiose as the dominant hydrolysis products (Deng et al., 2014), suggesting a kinetically rapid conversion of longer oligosaccharides to shorter during the duration of the reaction. Removal of cellotetraose was not observed from cellotetraose-NIMS, which as proposed above likely represents ineffective binding of the NIMS probe in the active site adjacent to the active site.

Endoglucanase CelL and Cellobiohydrolase CelK Reactions

We tested the cellotetraose-NIMS reactions with an additional endoglucanase, CelL (Cthe_0405, **Figure 5E**), and a reducing end cellobiohydrolase, CelK (Cthe_0212, **Figure 5F**). These enzymes show a shift in reaction specificity so that removal of cellobiose to produce cellobiose-NIMS (blue diamonds) became the dominant pattern of reaction. Notably, CelL had an approximately threefold enhanced ability to remove cellobiose relative to CelR because of a higher k9 value and also an ~10-fold decrease in the ability to remove cellotriose associated with a lower k11 value (**Table 2**). CelL reacted with IL-SG also showed preference for release of cellobiose (Deng et al., 2014). Furthermore, although CelK also had an approximately threefold enhanced ability to remove

cellobiose relative to CelR because of a higher k9 value, it showed no ability to produce either cellotriose or glucose (e.g., k1 and k11 = 0; **Table 2**).

The high specificity for release of cellobiose by a cellobiohydrolase is a characteristic reactivity (Amano et al., 1996; Barr et al., 1996; Divne et al., 1998), including CelK (Kataeva et al., 1999) and also CelK reacted with IL-SG (Deng et al., 2014). Thus, cellotetraose-NIMS clearly reports on this catalytic function of CelK. There are no previously published reactivity studies or crystal structures of CelL, beyond our studies of reaction with IL-SG, where CelL showed strong preference for release of cellobiose and xylobiose from the pretreated biomass (Deng et al., 2014).

CONCLUSION

This work establishes the utility of a chemically synthesized mass spectral probe for characterization of GHs. We have shown remarkable correspondence between the products obtained from enzyme reactions with the synthetic cellotetraose-NIMS probe and IL- and AFEX-pretreated switchgrass (Deng et al., 2014). Because of the emerging success of robotic cell-free translation to provide active enzyme samples from synthesized genes (Takasuka et al., 2014; Bianchetti et al., 2015), the substantial advantages of automation and miniaturization afforded by the Nimzyme platform (Deng et al., 2012, 2014; de Rond et al., 2013; Heins et al., 2014), and the predictive power inherent in numerical analysis of enzyme reaction time courses (Cleland, 1975; Orsi and Tipton, 1979; Duggleby, 1995; Marangoni, 2003), our combination offers a powerful new approach for functional annotation of bioenergy phylogenetic space.

AUTHOR CONTRIBUTIONS

KD, TT, CB, LB, PA, TN, and BF designed experiments, carried out experimental work, analyzed results, and prepared the manuscript. All authors read and approved the final manuscript.

FUNDING

The DOE Great Lakes Bioenergy Research Center and DOE Joint BioEnergy Institute are supported by the US Department of Energy, Office of Science, Office of Biological and Environmental Research, through contract DE-FC02-07ER64494 and through contract DE-AC02-05CH11231, respectively.

REFERENCES

Aceti, D. J., Bingman, C. A., Wrobel, R. L., Frederick, R. O., Makino, S., Nichols, K. W., et al. (2015). Expression platforms for producing eukaryotic proteins: a comparison of E. coli cell-based and wheat germ cell-free synthesis, affinity and solubility tags, and cloning strategies. J. Struct. Funct. Genomics 16, 67–80. doi:10.1007/s10969-015-9198-1

Alzari, P. M., Souchon, H., and Dominguez, R. (1996). The crystal structure of endoglucanase CelA, a family 8 glycosyl hydrolase from Clostridium thermocellum. Structure 4, 265–275. doi:10.1016/S0969-2126(96)00031-7

Amano, Y., Shiroishi, M., Nisizawa, K., Hoshino, E., and Kanda, T. (1996). Fine substrate specificities of four exo-type cellulases produced by Aspergillus niger, Trichoderma reesei, and Irpex lacteus on (1 − >3), (1 − >4)-beta-D-glucans and

xyloglucan. J. Biochem. 120, 1123–1129. doi:10.1093/oxfordjournals.jbchem.a021531

Bansal, P., Hall, M., Realff, M. J., Lee, J. H., and Bommarius, A. S. (2009). Modeling cellulase kinetics on lignocellulosic substrates. Biotechnol. Adv. 27, 833–848. doi:10.1016/j.biotechadv.2009.06.005

Barr, B. K., Hsieh, Y. L., Ganem, B., and Wilson, D. B. (1996). Identification of two functionally different classes of exocellulases. Biochemistry 35, 586–592. doi:10.1021/bi9520388

Beebe, E. T., Makino, S., Markley, J. L., and Fox, B. G. (2014). Automated cell-free protein production methods for structural studies. Methods Mol. Biol. 1140, 117–135. doi:10.1007/978-1-4939-0354-2_9

Beebe, E. T., Makino, S., Nozawa, A., Matsubara, Y., Frederick, R. O., Primm, J. G., et al. (2011). Robotic large-scale application of wheat cell-free translation to structural studies including membrane proteins. *N Biotechnol.* 28, 239–249. doi:10.1016/j.nbt.2010.07.003

Bianchetti, C. M., Takasuka, T. E., Deutsch, S., Udell, H. S., Yik, E. J., Bergeman, L. F., et al. (2015). Active site and laminarin binding in glycoside hydrolase family 55. *J. Biol. Chem.* 290, 11819–11832. doi:10.1074/jbc.M114.623579

Blommel, P. G., Martin, P. A., Seder, K. D., Wrobel, R. L., and Fox, B. G. (2009). Flexi vector cloning. *Methods Mol. Biol.* 498, 55–73. doi:10.1007/978-1-59745-196-3_4

Brown, S. D., Raman, B., Mckeown, C. K., Kale, S. P., He, Z., and Mielenz, J. R. (2007). Construction and evaluation of a *Clostridium thermocellum* ATCC 27405 whole-genome oligonucleotide microarray. *Appl. Biochem. Biotechnol.* 13, 663–674. doi:10.1007/s12010-007-9087-6

Cantarel, B. L., Coutinho, P. M., Rancurel, C., Bernard, T., Lombard, V., and Henrissat, B. (2009). The carbohydrate-active enzymes database (CAZy): an expert resource for glycogenomics. *Nucleic Acids Res.* 37, D233–D238. doi:10.1093/nar/gkn663

Chundawat, S. P., Balan, V., and Dale, B. E. (2008). High-throughput microplate technique for enzymatic hydrolysis of lignocellulosic biomass. *Biotechnol. Bioeng.* 99, 1281–1294. doi:10.1002/bit.21805

Cleland, W. W. (1975). Partition analysis and the concept of net rate constants as tools in enzyme kinetics. *Biochemistry* 14, 3220–3224. doi:10.1021/bi00685a029

Dashtban, M., Maki, M., Leung, K. T., Mao, C., and Qin, W. (2010). Cellulase activities in biomass conversion: measurement methods and comparison. *Crit. Rev. Biotechnol.* 30, 302–309. doi:10.3109/07388551.2010.490938

de Rond, T., Peralta-Yahya, P., Cheng, X., Northen, T. R., and Keasling, J. D. (2013). Versatile synthesis of probes for high-throughput enzyme activity screening. *Anal. Bioanal. Chem.* 405, 4969–4973. doi:10.1007/s00216-013-6888-z

Decker, S. R., Adney, W. S., Jennings, E., Vinzant, T. B., and Himmel, M. E. (2003). Automated filter paper assay for determination of cellulase activity. *Appl. Biochem. Biotechnol.* 105–108, 689–703. doi:10.1385/ABAB:107:1-3:689

Deng, K., George, K. W., Reindl, W., Keasling, J. D., Adams, P. D., Lee, T. S., et al. (2012). Encoding substrates with mass tags to resolve stereospecific reactions using Nimzyme. *Rapid Commun. Mass Spectrom.* 26, 611–615. doi:10.1002/rcm.6134

Deng, K., Takasuka, T. E., Heins, R., Cheng, X., Bergeman, L. F., Shi, J., et al. (2014). Rapid kinetic characterization of glycosyl hydrolases based on oxime derivatization and nanostructure-initiator mass spectrometry (NIMS). *ACS Chem. Biol.* 9, 1470–1479. doi:10.1021/cb5000289

Ding, S. Y., Xu, Q., Crowley, M., Zeng, Y., Nimlos, M., Lamed, R., et al. (2008). A biophysical perspective on the cellulosome: new opportunities for biomass conversion. *Curr. Opin. Biotechnol.* 19, 218–227. doi:10.1016/j.copbio.2008.04.008

Divne, C., Stahlberg, J., Teeri, T. T., and Jones, T. A. (1998). High-resolution crystal structures reveal how a cellulose chain is bound in the 50 A long tunnel of cellobiohydrolase I from *Trichoderma reesei. J. Mol. Biol.* 275, 309–325. doi:10.1006/jmbi.1997.1437

Dowe, N. (2009). Assessing cellulase performance on pretreated lignocellulosic biomass using saccharification and fermentation-based protocols. *Methods Mol. Biol.* 581, 233–245. doi:10.1007/978-1-60761-214-8_15

Duggleby, R. G. (1995). Analysis of enzyme progress curves by nonlinear regression. *Meth. Enzymol.* 249, 61–90. doi:10.1016/0076-6879(95)49031-0

Eklof, J. M., Ruda, M. C., and Brumer, H. (2012). Distinguishing xyloglucanase activity in endo-beta(1 – >4)glucanases. *Meth. Enzymol.* 510, 97–120. doi:10.1016/B978-0-12-415931-0.00006-9

Feinberg, L., Foden, J., Barrett, T., Davenport, K. W., Bruce, D., Detter, C., et al. (2011). Complete genome sequence of the cellulolytic thermophile *Clostridium thermocellum* DSM1313. *J. Bacteriol.* 193, 2906–2907. doi:10.1128/JB.00322-11

Fontes, C. M., and Gilbert, H. J. (2010). Cellulosomes: highly efficient nanomachines designed to deconstruct plant cell wall complex carbohydrates. *Annu. Rev. Biochem.* 79, 655–681. doi:10.1146/annurev-biochem-091208-085603

Fox, B. G., Goulding, C., Malkowski, M. G., Stewart, L., and Deacon, A. (2008). Structural genomics: from genes to structures with valuable materials and many questions in between. *Nat. Methods* 5, 129–132. doi:10.1038/nmeth0208-129

Gerlt, J. A., Allen, K. N., Almo, S. C., Armstrong, R. N., Babbitt, P. C., Cronan, J. E., et al. (2011). The enzyme function initiative. *Biochemistry* 50, 9950–9962. doi:10.1021/bi201312u

Gold, N. D., and Martin, V. J. (2007). Global view of the *Clostridium thermocellum* cellulosome revealed by quantitative proteomic analysis. *J. Bacteriol.* 189, 6787–6795. doi:10.1128/JB.00882-07

Grabnitz, F., Seiss, M., Rucknagel, K. P., and Staudenbauer, W. L. (1991). Structure of the beta-glucosidase gene bglA of *Clostridium thermocellum*. Sequence analysis reveals a superfamily of cellulases and beta-glycosidases including human lactase/phlorizin hydrolase. *Eur. J. Biochem.* 200, 301–309. doi:10.1111/j.1432-1033.1991.tb16186.x

Hazlewood, G. P., Davidson, K., Laurie, J. I., Huskisson, N. S., and Gilbert, H. J. (1993). Gene sequence and properties of CelI, a family E endoglucanase from *Clostridium thermocellum. J. Gen. Microbiol.* 139, 307–316. doi:10.1099/00221287-139-2-307

Heins, R. A., Cheng, X., Nath, S., Deng, K., Bowen, B. P., Chivian, D. C., et al. (2014). Phylogenomically guided identification of industrially relevant GH1 beta-glucosidases through DNA synthesis and nanostructure-initiator mass spectrometry. *ACS Chem. Biol.* 9, 2082–2091. doi:10.1021/cb500244v

Henrissat, B., and Davies, G. (1997). Structural and sequence-based classification of glycoside hydrolases. *Curr. Opin. Struct. Biol.* 7, 637–644. doi:10.1016/S0959-440X(97)80072-3

Himmel, M. E., Ding, S. Y., Johnson, D. K., Adney, W. S., Nimlos, M. R., Brady, J. W., et al. (2007). Biomass recalcitrance: engineering plants and enzymes for biofuels production. *Science* 315, 804–807. doi:10.1126/science.1137016

Hirano, K., Nihei, S., Hasegawa, H., Haruki, M., and Hirano, N. (2015). Stoichiometric assembly of the cellulosome generates maximum synergy for the degradation of crystalline cellulose, as revealed by in vitro reconstitution of the *Clostridium thermocellum* cellulosome. *Appl. Environ. Microbiol.* 81, 4756–4766. doi:10.1128/AEM.00772-15

Hirano, N., Hasegawa, H., Nihei, S., and Haruki, M. (2013). Cell-free protein synthesis and substrate specificity of full-length endoglucanase CelJ (Cel9D-Cel44A), the largest multi-enzyme subunit of the *Clostridium thermocellum* cellulosome. *FEMS Microbiol. Lett.* 344, 25–30. doi:10.1111/1574-6968.12149

Horn, S. J., Sorlie, M., Varum, K. M., Valjamae, P., and Eijsink, V. G. (2012). Measuring processivity. *Meth. Enzymol.* 510, 69–95. doi:10.1016/B978-0-12-415931-0.00005-7

Kataeva, I., Li, X. L., Chen, H., Choi, S. K., and Ljungdahl, L. G. (1999). Cloning and sequence analysis of a new cellulase gene encoding CelK, a major cellulosome component of *Clostridium thermocellum*: evidence for gene duplication and recombination. *J. Bacteriol.* 181, 5288–5295.

Kosik, O., Bromley, J. R., Busse-Wicher, M., Zhang, Z., and Dupree, P. (2012). Studies of enzymatic cleavage of cellulose using polysaccharide analysis by carbohydrate gel electrophoresis (PACE). *Meth. Enzymol.* 510, 51–67. doi:10.1016/B978-0-12-415931-0.00004-5

Levasseur, A., Drula, E., Lombard, V., Coutinho, P. M., and Henrissat, B. (2013). Expansion of the enzymatic repertoire of the CAZy database to integrate auxiliary redox enzymes. *Biotechnol. Biofuels* 6, 41. doi:10.1186/1754-6834-6-41

Madono, M., Sawasaki, T., Morishita, R., and Endo, Y. (2011). Wheat germ cell-free protein production system for post-genomic research. *N Biotechnol.* 28, 211–217. doi:10.1016/j.nbt.2010.08.009

Makino, S., Beebe, E. T., Markley, J. L., and Fox, B. G. (2014). Cell-free protein synthesis for functional and structural studies. *Methods Mol. Biol.* 1091, 161–178. doi:10.1007/978-1-62703-691-7_11

Marangoni, A. G. (2003). *Enzyme Kinetics*. Hoboken, NJ: John Wiley & Sons, Inc.

Markley, J. L., Aceti, D. J., Bingman, C. A., Fox, B. G., Frederick, R. O., Makino, S., et al. (2009). The center for eukaryotic structural genomics. *J. Struct. Funct. Genomics* 10, 165–179. doi:10.1007/s10969-008-9057-4

McCleary, B. V., Mckie, V., and Draga, A. (2012). Measurement of endo-1,4-beta-glucanase. *Meth. Enzymol.* 510, 1–17. doi:10.1016/B978-0-12-415931-0.00001-X

Nair, R., Liu, J., Soong, T. T., Acton, T. B., Everett, J. K., Kouranov, A., et al. (2009). Structural genomics is the largest contributor of novel structural leverage. *J. Struct. Funct. Genomics* 10, 181–191. doi:10.1007/s10969-008-9055-6

Nelder, J. A., and Mead, R. (1965). A simplex-method for function minimization. *Comput. J.* 7, 308–313. doi:10.1093/comjnl/7.4.308

Northen, T. R., Lee, J. C., Hoang, L., Raymond, J., Hwang, D. R., Yannone, S. M., et al. (2008). A nanostructure-initiator mass spectrometry-based enzyme activity assay. *Proc. Natl. Acad. Sci. U.S.A.* 105, 3678–3683. doi:10.1073/pnas.0712332105

Orsi, B. A., and Tipton, K. F. (1979). Kinetic analysis of progress curves. *Meth. Enzymol.* 63, 159–183. doi:10.1016/0076-6879(79)63010-0

Pena, M. J., Tuomivaara, S. T., Urbanowicz, B. R., O'neill, M. A., and York, W. S. (2012). Methods for structural characterization of the products of cellulose- and xyloglucan-hydrolyzing enzymes. *Meth. Enzymol.* 510, 121–139. doi:10.1016/B978-0-12-415931-0.00007-0

Pieper, U., Schlessinger, A., Kloppmann, E., Chang, G. A., Chou, J. J., Dumont, M. E., et al. (2013). Coordinating the impact of structural genomics on the human alpha-helical transmembrane proteome. *Nat. Struct. Mol. Biol.* 20, 135–138. doi:10.1038/nsmb.2508

Raman, B., Mckeown, C. K., Rodriguez, M. Jr., Brown, S. D., and Mielenz, J. R. (2011). Transcriptomic analysis of *Clostridium thermocellum* ATCC 27405 cellulose fermentation. *BMC Microbiol.* 11:134. doi:10.1186/1471-2180-11-134

Riederer, A., Takasuka, T. E., Makino, S., Stevenson, D. M., Bukhman, Y. V., Elsen, N. L., et al. (2011). Global gene expression patterns in *Clostridium thermocellum* as determined by microarray analysis of chemostat cultures on cellulose or cellobiose. *Appl. Environ. Microbiol.* 77, 1243–1253. doi:10.1128/AEM.02008-10

Roberts, S. B., Gowen, C. M., Brooks, J. P., and Fong, S. S. (2010). Genome-scale metabolic analysis of *Clostridium thermocellum* for bioethanol production. *BMC Syst. Biol.* 4:31. doi:10.1186/1752-0509-4-31

Selig, M. J., Tucker, M. P., Law, C., Doeppke, C., Himmel, M. E., and Decker, S. R. (2011). High throughput determination of glucan and xylan fractions in lignocelluloses. *Biotechnol. Lett.* 33, 961–967. doi:10.1007/s10529-011-0526-7

Sharrock, K. R. (1988). Cellulase assay methods: a review. *J. Biochem. Biophys. Methods* 17, 81–105. doi:10.1016/0165-022X(88)90040-1

Smith, S. P., and Bayer, E. A. (2013). Insights into cellulosome assembly and dynamics: from dissection to reconstruction of the supramolecular enzyme complex. *Curr. Opin. Struct. Biol.* 23, 686–694. doi:10.1016/j.sbi.2013.09.002

Takasuka, T. E., Walker, J. A., Bergeman, L. F., Vander Meulen, K. A., Makino, S., Elsen, N. L., et al. (2014). Cell-free translation of biofuel enzymes. *Methods Mol. Biol.* 1118, 71–95. doi:10.1007/978-1-62703-782-2_5

Watson, J. D., Sanderson, S., Ezersky, A., Savchenko, A., Edwards, A., Orengo, C., et al. (2007). Towards fully automated structure-based function prediction in structural genomics: a case study. *J. Mol. Biol.* 367, 1511–1522. doi:10.1016/j.jmb.2007.01.063

Whitehead, C., Gomez, L. D., and Mcqueen-Mason, S. J. (2012). The analysis of saccharification in biomass using an automated high-throughput method. *Meth. Enzymol.* 510, 37–50. doi:10.1016/B978-0-12-415931-0.00003-3

Wischmann, B., Toft, M., Malten, M., and Mcfarland, K. C. (2012). Biomass conversion determined via fluorescent cellulose decay assay. *Meth. Enzymol.* 510, 19–36. doi:10.1016/B978-0-12-415931-0.00002-1

Zverlov, V. V., Schantz, N., and Schwarz, W. H. (2005). A major new component in the cellulosome of *Clostridium thermocellum* is a processive endo-beta-1,4-glucanase producing cellotetraose. *FEMS Microbiol. Lett.* 249, 353–358. doi:10.1016/j.femsle.2005.06.037

Conflict of Interest Statement: Kai Deng and Trent R. Northen are coinventors on a patent application that covers the oxime-NIMS assay. Taichi E. Takasuka, Christopher M. Bianchetti, and Brian G. Fox are coinventors on a patent application that covers use of multifunctional enzymes. Lai F. Bergeman and Paul D. Adams have no conflict of interest to declare.

3

Potential for Genetic Improvement of Sugarcane as a Source of Biomass for Biofuels

Nam V. Hoang[1,2]*, Agnelo Furtado*[1]*, Frederik C. Botha*[1,3]*, Blake A. Simmons*[1,4] *and Robert J. Henry*[1]

[1] *Queensland Alliance for Agriculture and Food Innovation, The University of Queensland, St. Lucia, QLD, Australia,* [2] *College of Agriculture and Forestry, Hue University, Hue, Vietnam,* [3] *Sugar Research Australia, Indooroopilly, QLD, Australia,* [4] *Joint BioEnergy Institute, Emeryville, CA, USA*

Edited by:
P. C. Abhilash,
Banaras Hindu University, India

Reviewed by:
Yu-Shen Cheng,
National Yunlin University of Science
and Technology, Taiwan
Tianju Chen,
Chinese Academy of Sciences, China

*Correspondence:
Nam V. Hoang
hoang.nam@uq.net.au

Sugarcane (*Saccharum* spp. hybrids) has great potential as a major feedstock for biofuel production worldwide. It is considered among the best options for producing biofuels today due to an exceptional biomass production capacity, high carbohydrate (sugar + fiber) content, and a favorable energy input/output ratio. To maximize the conversion of sugarcane biomass into biofuels, it is imperative to generate improved sugarcane varieties with better biomass degradability. However, unlike many diploid plants, where genetic tools are well developed, biotechnological improvement is hindered in sugarcane by our current limited understanding of the large and complex genome. Therefore, understanding the genetics of the key biofuel traits in sugarcane and optimization of sugarcane biomass composition will advance efficient conversion of sugarcane biomass into fermentable sugars for biofuel production. The large existing phenotypic variation in *Saccharum* germplasm and the availability of the current genomics technologies will allow biofuel traits to be characterized, the genetic basis of critical differences in biomass composition to be determined, and targets for improvement of sugarcane for biofuels to be established. Emerging options for genetic improvement of sugarcane for the use as a bioenergy crop are reviewed. This will better define the targets for potential genetic manipulation of sugarcane biomass composition for biofuels.

Keywords: sugarcane, biofuels, biomass for biofuels, biofuel traits, association studies

INTRODUCTION

Plant biomass from grasses such as sugarcane or woody species contains mostly cellulose, hemicellulose, and lignin (also referred to as lignocellulosic biomass), which can be converted to biofuels as a source of renewable energy. At the moment, plant biomass-derived biofuels have great potential in countries that have limited oil resources because they reduce the dependence on fossil fuel, mitigate air pollution by cutting down greenhouse gas emissions, and can be produced from a wide range

Abbreviations: AFLP, amplified fragment length polymorphism; BAC, bacterial artificial chromosome; CAD, cinnamyl alcohol dehydrogenase (EC 1.1.1.195); cDNA, complementary DNA; COMT, caffeic acid *O*-methyltransferase (EC 2.1.1.68); DArT, diversity array technology; EST, expressed sequence tag; Gb/Mb, gigabase/megabase; LD, linkage disequilibrium; Lignin G, lignin guaiacyl; Lignin H, lignin hydroxyphenyl; Lignin S, lignin syringyl; NGS, next-generation sequencing; QTL, quantitative trait locus; RFLP, restricted fragment length polymorphism; RNAi, RNA interference; S/G ratio, syringyl/guaiacyl ratio; SNP, single nucleotide polymorphism; SUCEST, sugarcane EST database; TF, transcription factor; TIGR, the institute for genome research.

of abundant sources (Matsuoka et al., 2009). Biofuels generated from plant lignocellulosic biomass (also known as the second generation of biofuels) have been shown to be advantageous over the first generation (from plant starches, sugar, and oil) in terms of net energy and CO_2 balance and, more importantly, they do not compete with food industries for supplies (Yuan et al., 2008). To date, producing bioethanol from the sugar in sugarcane has been one of the world's most commercially successful biofuel production systems, with the potential to deliver second-generation fuels with a high positive energy balance and at a relatively low production cost (Yuan et al., 2008; Botha, 2009; Matsuoka et al., 2009). The rapid growth and high yield of sugarcane compared to other grasses and woody plants makes it a good candidate for ethanol processing platform and the second generation of biofuels in general (Pandey et al., 2000). Sugarcane has an exceptional ability to produce biomass as a C4 plant with the potential of a perennial grass crop allowing harvest four to five times by using ratoons without requiring replanting (Verheye, 2010), resulting in a lower cost of energy production from sugarcane than for most of the other potential sources of biomass (Botha, 2009). Brazil is the world's first country to launch a national fuel alcohol program (ProAlcooL). This program is based on the use of sugarcane and substitutes the usage of gasoline by ethanol (Dias De Oliveira et al., 2005). Approximately, 23.4 billion liters (6.19 billion U.S. liquid gallons) of ethanol was produced in Brazil in the year 2014 (Renewable Fuels Association, 2015). As of 2009, sugarcane bagasse contributes to about 15% of the total electricity consumed in Brazil, and it is predicted that energy generated from sugarcane stalks could supply more than 30% of the country energy needs by 2020 and will be equal to or more than the electricity produced from hydropower (Matsuoka et al., 2009).

Conventionally, sugarcane bagasse is usually burned to produce fertilizer or steam and electricity to fuel the boilers in sugar mills (Pandey et al., 2000). Recently, it has been used for biofuel production; however, the production cost of biofuels from lignocellulosic biomass is still considered to be relatively high, which makes it difficult to be price-competitive and commercialized on a large scale (Halling and Simms-Borre, 2008). At the moment, the cost of bagasse pretreatment (to remove or separate its recalcitrant components before converting to biofuels) and microbial enzymes contributes mostly to the total production cost, resulting in reducing the incentive to transition from first generation to second generation of biofuels in sugarcane (Yuan et al., 2008). To maximize the efficiency of conversion of sugarcane biomass into biofuels, it is imperative to generate improved sugarcane cultivars with not only high biomass yield and fiber content but also better biomass degradability for conversion to biofuels in addition to improving the pretreatment and enzyme digestion technologies.

This review focuses on the potential for the genetic improvement of sugarcane as a source of biomass for biofuels, exploring the beneficial characteristics of sugarcane, the available genetic resources and germplasm, the potential of cell wall modification by breeding and biotechnology, and the potential of whole genome/transcriptome sequencing applications in dissecting important biofuel traits to improve sugarcane biomass composition. This will define the targets for potential genetic manipulation and better exploitation of sugarcane biomass for biofuels.

SUGARCANE AT A QUICK GLANCE

Biology and Genetics

Taxonomically, sugarcane belongs to the genus *Saccharum* (established by Carl Linnaeus in 1753), in the grass family *Poaceae* (or *Gramineae*), subfamily *Panicoideae*, tribe *Andropogoneae*, subtribe *Sacharinae*, under the group *Saccharastrae* and has a very close genetic relationship to sorghum and other grass family members such as *Erianthus* and *Miscanthus* (Amalraj and Balasundaram, 2006). Typically, the genus is divided into six different species namely *Saccharum barberi*, *Saccharum edule*, *Saccharum officinarum*, *Saccharum robustum*, *Saccharum sinense*, and *Saccharum spontaneum* (Daniels and Roach, 1987; Amalraj and Balasundaram, 2006), in which *S. spontaneum* and *S. robustum* are wild species; *S. officinarum*, *S. barberi*, and *S. sinense* are early cultivars while *S. edule* is a marginal specialty cultivar. All genotypes of the *Saccharum* genus are reported to be polyploid with the ploidy level ranging from $5\times$ to $16\times$ and are considered as among the most complex plant genomes (Manners et al., 2004). The cytotype ($2n$, the number of chromosomes in the cell) was reported to be different in each species as follows: *S. officinarum* ($2n = 80$), *S. spontaneum* ($2n = 40-128$), *S. barberi* ($2n = 111-120$), *S. sinense* ($2n = 81-124$), *S. edule* ($2n = 60-80$), and *S. robustum* ($2n = 60, 80$); hence, the basic chromosome number (x, the monoploid set of chromosomes in the cell) ranges from 5, 6, 8, 10 to 12 (Sreenivasan et al., 1987). The basic chromosome number of *S. spontaneum* is 8 (even though a number of very variable cytotype is observed) and of *S. officinarum* and *S. robustum* is 10 [Panje and Babu, 1960, D'Hont et al. (1998), and Piperidis et al. (2010)]. For the other three species, *S. sinense*, *S. barberi*, and *S. edule*, due to the fact that these are early interspecific hybrid cultivars, there have not been a consensus reported, but a study by Ming et al. (1998) suggested that the basic chromosome number for these three species could also be 10.

Hybrid sugarcane was derived from crosses between a female *S. officinarum* ($2n = 80$) and a male *S. spontaneum* ($2n = 40-128$). Due to the female restitution phenomenon, at first, the F1 hybrid conserves the whole *S. officinarum* chromosome set and half of the *S. spontaneum* which was $2n + n$, then a few backcrosses later, this hybrid breaks down to $n + n$, establishing the hybrid chromosome set of modern sugarcane hybrid (Bremer, 1961). For this reason, current sugarcane cultivars (*Saccharum* spp. hybrids) have a combination of a highly aneuploid and interspecific set of chromosomes. By using genomic *in situ* hybridization (GISH) and fluorescent *in situ* hybridization (FISH), it is revealed that among chromosomes in the nucleus of modern hybrid sugarcane, approximately 80% are contributed by *S. officinarum*, 10–20% from *S. spontaneum*, and less than 5–17% from recombination of chromosomes of the two species (D'Hont et al., 1996; Piperidis et al., 2001; Cuadrado et al., 2004). Modern sugarcane hybrids are normally crosses between varieties/clones, which makes the combination of the chromosomes in each offspring unique and unpredictable due to the random sorting of the chromosomes in the genome (Grivet and Arruda, 2002). The first sugarcane breeding program, which started more than one century ago, generated a few interspecific hybrids and constitutes the basic germplasm used by sugarcane breeding programs (Ming et al., 2010). Modern sugarcane cultivars are derived from the basic germplasm, but there has been only a few generations for chromosome recombination opportunities (the number of meiosis that chromosomes have undergone is mainly about 2–7) as the sugarcane breeding processes normally take between 10 and 15 years (Raboin et al., 2008; Ming et al., 2010). As a result, the modern sugarcane population has a narrow genetic basis and high linkage disequilibrium (Roach, 1989; Lima et al., 2002; Raboin et al., 2008).

The Nature of a Complex, Polyploid, and Repetitive Genome

The complex and polyploid genome of sugarcane makes the process of analyzing and understanding difficult by normal methods applied to diploid plants. The size of the sugarcane genome is about 10 Gb while its genome complexity is due to the mixture of euploid and aneuploid chromosome sets with homologous genes present in from 8 to 12 copies (Souza et al., 2011). The estimated monoploid genome size is about 750–930 Mb (the monoploid genome size of the two parental species, S. officinarum and S. spontaneum, are 930 Mb and 750 Mb, respectively), not much larger than the sorghum genome (~730 Mb) and about twice the size of the rice genome (~380 Mb) (D'Hont and Glaszmann, 2001). On the other hand, studies revealed that despite this complex and polyploid genome, sugarcane showed synteny with other grasses, especially sorghum (collinear, due to the limited divergence time) and maize (orthologous but altered loci collinearity) [reviewed in Grivet and Arruda (2002)]. It was thought that the sugarcane genome contains roughly the same amount of repetitive DNA as in the sorghum genome (Jannoo et al., 2007); however, studies on BAC-end sequences by Wang et al. (2010), Figueira et al. (2012), and Kim et al. (2013) suggested that there is less repetitive content in the sugarcane genome (e.g., 45.2% and 42.8% repetitive sequences observed in large BAC collections in comparison to 61% in the sorghum genome). More recently, using the k-mer approach, Berkman et al. (2014) found that the repetitive proportion in three sugarcane hybrid cultivars ranges from 63.74 to 78.37% and higher than that in the sorghum genome (55.5%) using the same approach. The authors postulated that the increased proportion could be attributed to ploidy level rather than repetitive content in the sugarcane genome. A high gene-copy number, the integration of two chromosome sets from two different species, and a significant repeat content hinder the understanding of how the genome functions and obtaining a genuine assembled monoploid genome (Souza et al., 2011; Figueira et al., 2012).

Candidate Crop for Future Biofuels

To date, sugarcane is among the most efficient crops in the world together with other C4 grasses such as switch grass (*Panicum virgatum*), *Miscanthus* species (*Miscanthus x giganteus*), and *Erianthus* species (*Erianthus arundinaceus* Retz.) in terms of converting solar energy into stored chemical energy and biomass accumulation (Tew and Cobill, 2008; Furtado et al., 2014). In general, C4 plants outperform C3 plants in biomass yield, including grain, stem, and leaf yield (Jakob et al., 2009; Wang and Paterson, 2013). Sugarcane and other C4 grasses are the highest yield potential feedstocks (**Table 1**), and for sugarcane, the potential yield can exceed 100 tons dry matter per hectare per year (Jakob

TABLE 1 | Average lignocellulosic biomass yield (dry matter) from sugarcane compared to other sources.

Plant name	Yield (tons/ha/year)	Reference
Sugarcane	22.9[a]	Van Der Weijde et al. (2013)
Switch grass	7–35	Reviewed in Hattori and Morita (2010)
Miscanthus	12–40	Reviewed in Hattori and Morita (2010)
Erianthus	40–60	Reviewed in Hattori and Morita (2010)
Eucalyptus	15–40	Reviewed in Johansson and Burnham (1993)

[a] Average total cane biomass dry matter is 39 tons/ha/year (Moore, 2009).

et al., 2009; Moore, 2009; Henry, 2010a). At present, the most suitable energy crop is probably sugarcane because of its high biomass yield and the potential for production on other than prime agricultural land avoiding competing with the land used for food industries (Waclawovsky et al., 2010). Globally, sugarcane is the most important crop in about 100 countries with a production area of 26.9 million hectares, total production of ~1.9 billion tons, and yield of 70.9 tons of fresh cane per hectare (FAOSTAT, 2015). At present, Brazil is the world's largest sugarcane producer followed by India, China, Thailand, Pakistan, Mexico, Colombia, Indonesia, Philippines, U.S., and Australia (FAOSTAT, 2015). In sugarcane internodal tissue, sucrose concentration ranges from 14 to 42% of the dry weight (Whittaker and Botha, 1997), while the rest of dry biomass comes from the cell wall lignocellulose, mostly containing cellulose, hemicellulose, lignin, and ash (Pereira et al., 2015). Biofuels from sugarcane can be produced extensively not only from its soluble sugar but also from main residues in sugarcane production, bagasse and trash, on the same production area (Seabra et al., 2010; Alonso Pippo et al., 2011a,b; Macrelli et al., 2012). The total estimated available lignocellulosic biomass from sugarcane worldwide was 584 million dry tons per year, with an average lignocellulosic biomass yield of 22.9 dry tons per hectare per year (Van Der Weijde et al., 2013). Sugarcane bioethanol yield from bagasse is estimated at about 3,000 L per hectare in a total yield of 9,950 L per hectare from sugar and bagasse (Somerville et al., 2010).

AVAILABLE SUGARCANE GENETIC RESOURCES FOR BIOFUELS

Existing Variations within *Saccharum* Germplasm

Genetically diverse sugarcane germplasm may play a key role in improving sugarcane for biofuels through breeding and biotechnological approaches. Genetic variation may be found in biomass yield, fiber content, and sugar composition in the *Saccharum* germplasm. This includes the diversity among the cultivars within one species and also diversity among species within the genus. A relatively high genetic variability within sugarcane hybrid cultivars was reported thanks to their heterozygosity and high polyploidy despite their originating from a few clones of a narrow genetic base (Aitken and McNeil, 2010). There is also great genetic and morphological diversity within *Saccharum* species, *Miscanthus* species, and *Erianthus* species to be potentially exploited and incorporated to broaden the genetic base in breeding programs (Harvey et al., 1994; Aitken and McNeil, 2010). To date, the genetic diversity of *S. officinarum* has been exploited in breeding programs; however, the diversity of *S. spontaneum* and other species have not been used much (Aitken and McNeil, 2010). *Saccharum* species have also been shown to have varied genome size, *S. officinarum* genome is about 7.50–8.55 Gb, *S. robustum* ranging from 7.56 to 11.78 Gb, and *S. spontaneum* ranging from 3.36 to 12.64 Gb, whereas the other three species – *S. sinense*, *S. barberi*, *S. edule* – and modern sugarcane are interspecific hybrids whose genome size depends upon each cross (Zhang et al., 2012).

There are two world largest collections of germplasm of *Saccharum* species, one is located in Florida (USA) while the other is in Kerala (India), containing approximately 1,200 accessions collected from 45 countries (Tai and Miller, 2001; Todd et al., 2014). These collections could be potentially selected and utilized for breeding purpose to improve sugarcane germplasm for new biofuel traits (Todd et al., 2014). The wild sugarcane species show wider variability in comparison to the domesticated species. In the *Saccharum* genus, *S. spontaneum* has the widest range of morphological variability, ratoon yielding, as well as biotic and abiotic stress tolerance (Tai and Miller, 2001; Aitken and McNeil, 2010; Govindaraj et al., 2014). The coefficient of variation (CV%) for some of the traits such as internode length, midrib width, leaf width, plant height, and stalk height studied by Govindaraj et al. (2014) were reported to be between 15 and 30%, which indicates a very high variability within the collection. It has been shown that the diversity within modern sugarcane hybrids was mostly contributed by the introgression from *S. spontaneum* (D'Hont et al., 1996). On the other hand, *S. robustum* also possesses a large amount of phenotypic variations in many traits studied (Aitken and McNeil, 2010). Sugarcane parental species (*S. officinarum*, *S. spontaneum*, and *S. robustum*), *Miscanthus* species, *Erianthus* species, and sorghum species with their diversity in genome content, structure, and tremendous allelic variation are a valuable and significant genetic reservoir which could be exploited for improving sugarcane biomass.

Genetic Markers and Maps

To support the effort of understanding the sugarcane genome, many physical maps, molecular markers, and resources such as RFLP, RAPD, AFLP, SSR, and ESTs have been developed over time. These common markers have been applied for genetic studies such as diversity, mapping, quantitative trait loci (QTL), and synteny definition; however, these systems have been developed mostly for well-established diploid species and are less effective for polyploidy plants (Garcia et al., 2013). Markers like AFLP, SSR, and RFLP are unable to estimate the number of allelic copies and level of polyploidy in complicated genomes such as potato, strawberry, and sugarcane (Garcia et al., 2013). More recently, the use of SNPs markers, which are distributed at high density across the genome, for complex genomes can allow estimation of the number of allelic copies and the ploidy level of genomes (Zhu et al., 2008; Hall et al., 2010). The currently available genetic maps and markers have been generated for sugarcane by using low-throughput methods, providing limited information on genome organization due to the low density of markers and coverage (most of them have less than 1,000 markers) (Aitken et al., 2014).

Therefore, it is difficult to allocate these markers into linkage groups or cosegregation groups or sugarcane expected chromosome number (Souza et al., 2011). More detailed linkage maps of *S. officinarum* cultivar IJ76-545 (534 markers in 123 linkage groups) and cultivar Green German (615 markers in 72 linkage groups); *S. spontaneum* cultivar IND (536 markers in 69 linkage groups); and the hybrid cultivars R570 and Q165 (with 2,000 markers placed in more than 100 linkage groups) have been constructed (Souza et al., 2011; Aitken et al., 2014). Most recently, using Diversity Array Technology (DArT), Aitken et al. (2014) integrated DArT markers, RFLPs, AFLPs, SSRs, and SNPs into the largest marker collection for sugarcane, which contains 2,467 single-dose markers for the cross between Q165 and IJ76-514 (a *S. officinarum* accession) and 2,267 markers from the cultivar Q165. These were placed into 160 linkage groups and eight homology groups, with some uncategorized linkage groups indicating that more markers are required. There is still a need to develop high-throughput marker arrays for sugarcane association studies, to generate more markers, and also to make use of the available markers. These markers will be a valuable resource in facilitating and unraveling the complex genome structure of sugarcane. It is worth considering that information on DNA-based molecular markers of progenitor plants can potentially reveal available genetic polymorphism for the analysis of their progenies (Henry et al., 2012). This could be a useful strategy in the case of sugarcane, where the genomes of the parental species are less complex than that of the hybrids.

Transcriptome Sequences and Transcription Factors

Expressed sequence tags (ESTs) and complementary DNA (cDNA) sequences provide direct evidence of the genes present in the samples, and this sequence information is very useful for genome exploration, gene prediction/discovery, genome structure identification, SNP characterization, and transcriptome and proteome analysis (Nagaraj et al., 2007). As of May 2015, the GenBank EST database (dbEST) was composed of 75,906,308 ESTs from different organisms of which 284,818 hits were detected under the search term sugarcane ("*S. officinarum*" or "*Saccharum* hybrid cultivar" or sugarcane). In the last 20 years, sugarcane ESTs have been used for gene discovery, BAC clone selection, and dissecting the coding regions of the genome, involving many projects in South Africa, France, U.S., Australia, and Brazil (Carson and Botha, 2000, 2002; Vettore et al., 2001; Casu et al., 2003, 2004; Grivet et al., 2003; Pinto et al., 2004; Bower et al., 2005). The largest collection of sugarcane ESTs was generated by SUCEST, which is composed of approximate 238,208 ESTs from 26 diverse cDNA libraries of different tissues of sugarcane cultivars, e.g., SP80-3280, SP70-1143, RB845205, RB845298, and RB805028 (Vettore et al., 2001, 2003; Souza et al., 2011). These sequences were assembled into 42,982 sugarcane assembled sequences representing more than 30,000 unique genes (~90% of the estimated genes, about 43,141, of *S. officinarum*) (Vettore et al., 2003; Hotta et al., 2010; Grassius: Grass Regulatory Information Server, 2015). There are other sugarcane EST collections containing less EST entries generated by Casu et al. (2003, 2004) (8,342 ESTs), Ma et al. (2004) (7,993 ESTs), Gupta et al. (2010) (~35,000 ESTs) and small number of ESTs by Carson and Botha (2000, 2002).

Due to the homology between genomes, genome-wide mapping of ESTs of one species provides an important framework for the genome structure of other related species (Sato et al., 2011). However, it is noteworthy that the discovery of the ESTs may be restricted to specific cultivars, as within sugarcane germplasm each cultivar has been shown to have different gene expression level [reviewed in Hotta et al. (2010)]. Moreover, for biofuel trait analysis, the TFs regulating monolignol biosynthesis in lignin pathway have received attention as understanding this allows reducing and modifying lignin content and composition which are essential in addressing the recalcitrant problem in biomass conversion (Santos Brito et al., 2015). It is shown that the lignin regulation can be species specific and information on TFs obtained from model plants such as *Arabidopsis* may require to be validated in other species (Santos Brito et al., 2015). A limited number of TFs in grass and sugarcane have been preliminarily characterized recently including those involved in monolignol biosynthesis, for example, in grass (Handakumbura and Hazen, 2012), rice (Yoshida et al., 2013), sorghum (Yan et al., 2013), and sugarcane (Santos Brito et al., 2015). Gene discovery of sugarcane has progressed to some extent despite the complexity of the genome. The valuable information of ESTs, TFs, full-length cDNAs, and BACs provides an understanding of allelic variations in the genome while a full-genome sequence is not available.

BAC Libraries to Construct a Reference Genome for Sugarcane

Sugarcane cultivar R570 and other cultivars including ones from the parental species *S. officinarum* and *S. spontaneum* have been used for constructing bacterial artificial chromosome (BAC) libraries (Hotta et al., 2010). BAC libraries from the sugarcane include hybrid cultivar R570 (103,296 clones, average insert size of 130 kb and two other libraries of 100,000 clones) (Tomkins et al., 1999; Grivet and Arruda, 2002), *S. spontaneum* cultivar SES208 (38,400 clones, average insert size of 120 kb), and *S. officinarum* cultivar LA Purple (74,880 clones, average insert size of 150 kb) generated from different restriction enzymes, e.g., *Hind*III and *Bam*H1 [reviewed in Souza et al. (2011)]. BAC sequencing in sugarcane is currently based on the sequencing of BAC clones anchored to an available physical map. Even though it requires a higher cost compared to the whole-genome shotgun sequencing (using high-throughput platforms, Illumina, for example), it is a reliable approach for reference construction, especially, for highly repetitive genomes which cannot be sequenced and resolved by a short-read method (Eversole et al., 2009; Steuernagel et al., 2009). This BAC sequencing approach has been used successfully in sequencing of *Arabidopsis*, rice, and maize genomes and producing the barley reference genome [reviewed in Steuernagel et al. (2009)]. The ongoing Sugarcane Genome Sequencing Initiative (SUGESI) has selected 5,000 BAC clones for sequencing from a library by Tomkins et al. (1999) of cultivar R570, the most intensively characterized cultivar to date, to help assembly of the monoploid coverage (monoploid tiling path) of the sugarcane genome using the sorghum sequence as the guide (Souza et al., 2011; Sugesi, 2015).

Sorghum bicolor Genome as the Closest Related Reference Genome

Sorghum is the most closely related species to sugarcane (Grivet et al., 1994; Dillon et al., 2007). The sorghum genome sequencing project was initiated and completed in 2007 with the total genome size of ~730 Mb, and 34,496 protein-coding loci, at the coverage of 8.5× using whole-genome shotgun sequencing by standard Sanger methodologies (Paterson et al., 2009). The sequenced genome is composed of 10 pairs of chromosomes and 3,294 supercontigs (most of these have been placed into chunks on 10 chromosomes), covering 90% of the genome and 99% of protein-coding regions (including the majority of available non-repetitive markers, known sorghum protein-coding genes, and the majority of ESTs) (Paterson et al., 2009). The sorghum genome has approximately 61% repetitive DNA, a low level of gene duplication compared to other C4 grasses, and a high degree of gene parallelism with sugarcane, even though the sugarcane genome is much more polyploid (Paterson et al., 2009, 2010). Microcollinearity between sugarcane and sorghum genomes indicated that sorghum is suitable as the template for sugarcane genome assembly (Ming et al., 1998; Wang et al., 2010; Figueira et al., 2012). It has been suggested that the sugarcane genome could be 20–30% smaller than that of sorghum despite the estimated monoploid genome size of sugarcane being about 760–930 Mb, at approximately the size of the sorghum genome (Figueira et al., 2012).

BIOMASS-DERIVED BIOFUELS AND THE CHALLENGING ISSUES IN BIOMASS CONVERSION TO BIOFUELS

The Second Generation of Biofuels – Cell Walls for Fuels

Due to the depletion of fossil fuel sources, the potential for oil to become more expensive, and the raising awareness of the negative impact of fossil fuels on the environment, biomass-derived biofuels have been investigated and developed recently as an alternative source of renewable, sufficient, and clean energy (Botha, 2009). The demand for renewable biofuels is predicted to be increasing (Fedenko et al., 2013). The first generation of biofuels from plant biomass involved the process of conversion of stored polysaccharides, non-structural carbohydrates, and oils from plants (starchy, sugary, and oily parts of plants such as corn starch, sugarcane molasses, soybeans, canola seeds, and palm oil) into fuels like ethanol and diesel (Schubert, 2006; Yuan et al., 2008). However, these sources are also used as food supplies and are limited due to the increasing demand from the growing world's population (Schubert, 2006). The second generation of biofuels can be generated by using the non-food parts of plants such as cell walls, composed of structural polysaccharides, such as cellulose and hemicellulose (Schubert, 2006; Yuan et al., 2008; Henry, 2010a). This is considered to be advantageous over the first generation of biofuels as it has a higher energy production potential, lower cost, sustainable CO_2 balance, no competition with the food production, and a wide range of plant biomass sources are available at costs affordable to a biorefinery (Yuan et al., 2008; Henry, 2010a). As of 2009, sugarcane biomass as sucrose accounted for about 40% of biofuels feedstock worldwide for first-generation biofuel production (Lam et al., 2009). Using sugarcane bagasse as a feedstock for second-generation biofuels would lead to doubling the current output of biofuel production from sugarcane (Halling and Simms-Borre, 2008).

Sugarcane Cell Wall and Biomass Composition

Physically, sugarcane biomass can be divided into four major fractions, whose content depends on the industrial process: fiber (heterogeneous organic solid fraction), non-soluble solids (inorganic substances), soluble solids (sucrose, waxes, and other chemicals), and water (Canilha et al., 2012; Shi et al., 2013). Second generation of biofuels focuses on using the fiber fraction especially the cell wall constituents of the plant to produce biofuels (Schubert, 2006; Henry, 2010a). This approach may be made more efficient by optimizing the composition of the biomass source for biofuel production. This could be achieved by advances in pretreatment methods or biotechnological modification of cell wall synthesis pathways to create a biomass that can be more efficiently processed (Sims et al., 2006; Yuan et al., 2008; Simpson, 2009; Viikari et al., 2012). Three major components make up the fiber fraction of sugarcane, namely, cellulose, hemicellulose (or non-cellulosic polysaccharide components), and lignin. Cellulose constitutes around 50% of the dry weight sugarcane bagasse while hemicellulose and lignin each account for about 25% (Loureiro et al., 2011). These three components are biosynthesized through different complex pathways (Higuchi, 1981; Whetten and Ron, 1995; Saxena and Brown, 2000; Mutwil et al., 2008; Harris and DeBolt, 2010; Pauly et al., 2013). Cellulose and hemicellulose molecules form the cell walls which act as the skeleton of plants and are strengthened by lignin and phenolic cross-linkages (Carpita, 1996; Henry, 2010b). The complex interlinking between cell wall components plays an important role in grass defense and yet challenges the biofuel production by requiring the pretreatment to separate them (De O. Buanafina, 2009).

The sugarcane and grass cell wall are categorized as type II cell wall, which differs from the type I and type III cell walls of other plants [reviewed in Souza et al. (2013)]. In general, there is little pectin, less lignin, and less structural proteins in grass cell walls than that in the non-grasses (Carpita, 1996; Henry, 2010b; Saathoff et al., 2011). There is similar cellulose content between grass and non-grass primary and secondary cell walls; however, hemicellulose composition is different between two groups. Grass cell walls have four to eight times more xylans, higher mixed linkage glucans, and lower levels of xyloglucans, mannans, glucomannans, and pectin in primary cell wall, but higher phenolics and lignin in the secondary cell wall (Loureiro et al., 2011). Grassy lignin is composed of three monolignols (lignin syringyl – S, lignin guaiacyl – G and lignin hydroxyphenyl – H subunits) forming various ratios of them and normally has more H subunit (more coumaryl derivatives) than in non-grasses (Vogel, 2008). A recent study by Bottcher et al. (2013) showed that sugarcane lignin content and composition are varied depending on tissue types and stem positions on the plant. Within one plant, the bottom internode has higher lignin accumulation than the top internode,

and the inner part of stem has higher syringyl/guaiacyl (S/G) ratio than the outer part. Polysaccharides found in sugarcane leaf and culm walls were similar but different in the proportions of xyloglucan and arabinoxylan (Souza et al., 2013). The major monosaccharides released from sugarcane cell walls were glucose, xylose, and arabinose (Loureiro et al., 2011; Rabemanolontsoa and Saka, 2013; Souza et al., 2013). Understanding the fine structure and detailed composition of sugarcane cell wall will assist in optimizing the tissue pretreatment and cell wall hydrolysis protocol. At present, converting sugarcane lignocellulosic biomass to ethanol includes (1) pretreatment to remove the lignin and other recalcitrant cellular constituents (or hemicellulose) to free cellulose, (2) enzyme-mediated action to depolymerize carbohydrates to simple sugars, and (3) fermentation of sugars and distillation of ethanol as the end product (Canilha et al., 2012).

Dealing with the Conversion Issues

Even though sugarcane biomass is less resistant to enzymatic digestion compared to that from woody plants, it is reported that biomass recalcitrant components impede the efficiency of the conversion to ethanol (Jung, 1989; Anterola and Lewis, 2002; Chen and Dixon, 2007; Himmel et al., 2007; Balat et al., 2008; Li et al., 2013). Biomass recalcitrance is caused by many factors such as the presence of epidermal and sclerenchyma tissues, vascular bundle density and arrangement, degree of lignification, heterogeneity and complexity of cell wall constituents, insoluble matter, natural inhibitors, and cellulose crystallinity (Himmel et al., 2007). Most approaches for producing biofuels from biomass at the moment rely on the disruption of the biomass, to separate lignocellulose and remove lignin in the biomass, and then conversion using microbial enzymes (Sticklen, 2006). In general, overcoming the recalcitrant issue can be addressed by physical, chemical, and genetic approaches. Physical and chemical strategies deal mainly with the pretreatment and involve loosening the cell wall structure, lowering the biomass heterogeneity, providing the enzymes access to the cellulose, cleaving the crossing linkages, and removing enzymatic inhibitors (Balat et al., 2008; Saathoff et al., 2011). To make the physical and chemical changes in plant biomass, pretreatment processing conditions must be tailored to the specific chemical and structural composition of the various and variable sources of lignocellulosic biomass (Mosier et al., 2005). Currently available physical and chemical pretreatment methods are varied and can be listed as uncatalyzed steam explosion, flow-through acid, liquid hot water, pH-controlled hot water, dilute acid, ammonia, lime and, more recently, the method using ionic liquids (Mosier et al., 2005; Shi et al., 2013; Sun et al., 2013). Genetic approaches involve genetic enhancement, molecular biology, and plant breeding efforts to improve biomass sources by having crops with less lignin, modified lignin, crops that self-produced enzymes, and crops with increased cellulose and biomass overall [reviewed in Sticklen (2006)]. The costs of the enzymatic pretreatment of cellulosic biomass (which accounts for about 25% of total processing expenses), biomass conversion, and microbial tanks limit the price-competitiveness of biofuel from lignocellulosic biomass in comparison to fossil fuel (Gnansounou and Dauriat, 2010; Macrelli et al., 2012, 2014; Van Der Weijde et al., 2013). This emphasizes the value of genetic improvement of biomass composition to reduce processing costs.

POTENTIAL IMPROVEMENT OF SUGARCANE BY BREEDING FOR BIOFUELS

The complex and highly polyploid genome of sugarcane poses a great challenge in unraveling and studying its functions. Each cross of modern sugarcane cultivar has a unique set of chromosomes due to the random sorting of chromosomes and recombination of alleles from two progenitor species (Grivet and Arruda, 2002). There are several distinct alleles at each locus in sugarcane chromosomes, making the characteristics of the offspring unpredictable and requiring evaluation of thousands of lines from many parents to gather sufficient information in breeding programs (Matsuoka et al., 2009). In conventional breeding, after crossing and obtaining the F1 generation, hundreds of thousands of F1 seedlings are used for screening for the desired traits such as disease resistance, sugar content, agronomic characteristics, and adaptability (Matsuoka et al., 2009). The process is normally repeated for some vegetatively propagated generations to obtain the required stability of the traits. For industrial purpose, after a long process of selection, from hundreds of thousands seedlings at the beginning, breeders normally end up at a limited number of clones for release as commercial lines or cultivars.

To facilitate the second generation of biofuels, sugarcane breeding programs need to be focusing not only on important traits such as total biomass yield, sugar yield adaptability to local environment, and resistance to major pathogens but also on biofuel traits (e.g., less lignin, improve biomass composition for conversion) as a whole (Matsuoka et al., 2009; Waclawovsky et al., 2010). In sugarcane breeding, to maximize heterosis, the parents are usually selected from divergent genotypes of genetic background (Tabasum et al., 2010). Increasing sugarcane biomass yield and productivity is getting more and more difficult to achieve by conventional methods; hence, broadening the sugarcane genetic basis by introgression of its ancestors or closely related species such as *Miscanthus* and *Erianthus* is being explored in sugarcane improvement [reviewed in Dal-Bianco et al. (2012) and De Siqueira Ferreira et al. (2013)]. This is normally done by crossing *S. officinarum* and *Erianthus*, *Miscanthus*, or backcrossing the hybrids to *S. spontaneum* (Matsuoka et al., 2009). Dual-purpose cane and energy cane, sugarcane lines for lignocellulosic biomass production, have been derived from two sugarcane species, *S. spontaneum* and *S. robustum*, by crossing to develop lines with a high ability to accumulate fiber and high biomass content in addition to accumulating soluble sugars (De Siqueira Ferreira et al., 2013). Another case is *Miscane*, which was the result of crossing between *Saccharum* x *Miscanthus*. This produces cane varieties with more biomass (lignocellulose and total fermentable sugars), disease resistance, and cold tolerance. This effectively adapts *Miscanthus* to a tropical climate and expands sugarcane production to temperate, dry, and cold conditions (Alexander, 1985; Burner et al., 2009; Lam et al., 2009). Recently, using molecular markers in sugarcane breeding program (marker-assisted selection) allows the direct comparison of DNA genetic diversity and provides a precise tool in assessing the genetic diversity of the germplasm (Tabasum et al., 2010; Berkman et al., 2012). The use of markers associated with the desired traits in combination with the advances in next-generation sequencing (NGS) technology, bioinformatics tools, and high-throughput phenotyping methods will significantly improve the sugarcane breeding programs (Lam et al., 2009). NGS will allow a great number of markers such as SNPs to be generated, which could be used to obtain a high density of marker at high coverage across the genome, to dissect the important traits they associate with. These sources of markers will be essential in breeding programs for screening of the parental plants from germplasm collection and of progenies derived from the crosses, selecting traits where the phenotypic methods are not practical (Berkman et al., 2012). High-throughput phenotyping methods will collect data from a large number of samples to overcome the small effects of genes, especially the QTL, controlling the traits (Lam et al., 2009).

POTENTIAL IMPROVEMENT BY MOLECULAR GENETICS FOR BIOFUELS

The competitiveness of biofuels over other options relies on biotechnology advancement. Efficient conversion of plant biomass to biofuels requires the supply of appropriate feedstocks that can be sustainably produced in large quantities at high yields. The efficient conversion of the biomass in these feedstocks will be facilitated by having a composition that is optimized for efficient processing to deliver high yields of the desired end products. Manipulating of the carbohydrates of the cell walls is the key of improving the biomass composition for biofuels (Harris and DeBolt, 2010). Powerful tools of biotechnology could aim to produce genetically modified sugarcane plants with a favorable ratio of cellulose to non-cellulose content; with *in planta* enzymes that can digest the biomass or degrade the lignin prior to its conversion to ethanol; with pest and disease resistance, flower inhibition, abiotic resistance; or incorporate them into elite sugarcane cultivars for better agronomic performance (Sticklen, 2006; Yuan et al., 2008; Matsuoka et al., 2009; Arruda, 2012).

Among the grasses potentially used for biofuel production such as sugarcane, switch grass, *Miscanthus*, and *Erianthus*, sugarcane has been used more for gene transformation studies (Falco et al., 2000; Manickavasagam et al., 2004; Basnayake et al., 2011) and the first transgenic sugarcane was established by Bower and Birch (1992). The current status of improving sugarcane biomass by using the genetic tools is hindered by its genome complexity, low transformation efficiency, transgene inactivation (gene silencing and regulation), somaclonal variation, and difficulty in backcrossing (Ingelbrecht et al., 1999; Hotta et al., 2010; Arruda, 2012; Dal-Bianco et al., 2012). Targets tackled so far on sugarcane include sucrose and biomass yield increase [i.e., in Ma et al. (2000) and Botha et al. (2001)], downregulation of lignin content or monolignol changes in lignin to lower biomass recalcitrance (described later), expression and accumulation of microbial cellulosic enzymes in leaf [i.e., in Harrison et al. (2011)], herbicide tolerance [i.e., in Gallo-Meagher and Irvine (1996) and Enríquez-Obregón et al. (1998)], disease or pest resistance [i.e., in Joyce et al. (1998), Arencibia et al. (1999), and Zhang et al.

(1999)], flowering inhibition [reviewed in Matsuoka et al. (2009) and Hotta et al. (2010)], and drought tolerance [i.e., in Zhang et al. (2006)]. Genetically modified sugarcane has great potential to contribute to biofuel production, with new varieties incorporating these characteristics (Arruda, 2012). Unexploited genes not only from the *Saccharum* germplasm but also in other related species, such as cold-tolerant genes in *S. spontaneum* and *Miscanthus* or drought-tolerant genes in sorghum, once identified would allow their integration into the sugarcane genome, facilitating the production of more sugarcane biomass in temperate areas or under dry conditions (Lam et al., 2009).

Increasing plant cellulose and total biomass content may be achieved by using approaches such as manipulation of growth regulators or key nutrients, increasing the ability of the plant to fix carbon by increasing atmospheric CO_2 and also manipulating some key metabolic enzymes in biomass synthesis pathways [reviewed in Sticklen (2006)]. Reduction of the cross-links of the maize cell walls (including ferulate and diferulate cross-links; benzyl ether and ester cross-links) has been shown to increase the initial hydrolysis of its cell wall polysaccharides by up to 46% (Grabber, 2005). In general, selection of grasses with less ferulate cross-linking or potent microbial xylanases by breeding or engineering tools are more attractive than pretreatment of the cell wall with a feruloyl esterase (Grabber, 2005).

Lignin content accounts for about 25% of sugarcane total lignocellulosic biomass and is probably the main obstacle affecting the efficiency of saccharification during conversion to ethanol (Canilha et al., 2012, 2013). Lignin and other recalcitrant components in cell walls prevent cellulase accessing the cellulose molecules and need to be removed before further processing (Sticklen, 2006). Lignin biosynthesis pathways are complicated and at least 10 different enzymes have been found involved in the lignin pathway in sugarcane (Higuchi, 1981; Whetten and Ron, 1995) and a total of 28 unigenes associated with monolignol biosynthesis were identified in sugarcane using SUCEST database and annotated genes from closely related species such as sorghum, maize, and rice (Bottcher et al., 2013). Tailoring sugarcane biomass composition for biofuels can be achieved by manipulating some of the key genes in lignin pathway (downregulation of some key enzymes), mostly targeting genes which encode the terminal enzymes such as caffeic acid O-methyltransferase (COMT) and cinnamyl alcohol dehydrogenase (CAD), to minimize the impact of the modifications on growth and development of the plant [as reviewed in Sticklen (2006), Jung et al. (2012), and Furtado et al. (2014)]. Not only lignin content but also the lignin S/G ratio is a very important aspect to consider in terms of modifying the lignin content because these two are both associated with biomass recalcitrance (Chen and Dixon, 2007; Li et al., 2010). Sugarcane lignin content was reduced by 3.9–13.7% using RNA interference (RNAi) suppression to downregulate the *COMT* gene [which has at least 31 different ESTs involved (Ramos et al., 2001)] by 67–97% and at the same time, the lignin S/G ratio was reduced from 1.47 to 1.27–0.79 (Jung et al., 2012). This resulted in an increase of up to 29% in total sugar yield without pretreatment (34% with dilute acid pretreatment). This study suggests that RNAi-mediated gene suppression is a promising method for suppression of target genes not only in lignin pathway but also for cell wall constituent biosynthesis (Jung et al., 2012; Bottcher et al., 2013).

Producing enzymes *in planta* is another way to cut the cost of biofuel production as it reduces the expense of enzymes and enzyme treatment. *Cellulase* has been produced within the plant (in the apoplast) of *Arabidopsis*, rice, and maize without effects on the growth and development of the host plants [reviewed in Sticklen (2006)]. *In planta* enzyme expression in sugarcane is still in its infancy; however, a high-yield biofuel plant such as sugarcane must be a target for the production of enzymes within the biomass. Recombinant protein enzymes have been targeted to organelles such as chloroplasts, vacuoles, and the endoplasmic reticulum to separate the enzymes produced and their substrates (Harrison et al., 2011). In sugarcane, thanks to its well-established transformation methods via *Agrobacterium*, the expression of enzymes in leaves and other tissues is feasible (Manickavasagam et al., 2004; Taylor et al., 2008). Endoglucanases and exoglucanases have been overexpressed in sugarcane leaves by using the maize *PepC* promoter achieving an accumulation level of 0.05% of total soluble proteins (endoglucanase, in chloroplast) and less of exoglucanases without altering the phenotype (Harrison et al., 2011). In the future, enzymes might be synthesized in specific energy cane plants that could be coprocessed with other biomass sources from sugarcane for sugar and biomass production (e.g., bagasse from sugar mills) (Arruda, 2012).

POTENTIAL OF SUGARCANE WHOLE GENOME AND TRANSCIPTOME SEQUENCING FOR BIOFUELS

The advent of NGS technology and a sharp reduction in per-base cost in the past decade [as reviewed in Van Dijk et al. (2014)] allows us to sequence the whole genome of a species, even a complex genome such as sugarcane, at a relatively low price within a relatively short time. At present, the cost of sequencing of a human genome at 30× coverage using the latest Illumina's Hiseq X is around US $1,000. Since the first plant genome was completely sequenced (*Arabidopsis thaliana* in 2000) using the traditional Sanger sequencing platform, the sequencing strategies have moved to high-throughput and cost-effective approaches (Henry et al., 2012). High-throughput genome sequencing platforms have recently advanced and facilitated improved genotyping, allowing huge data output to be generated for polymorphism detection (especially SNPs) and marker discovery.

Potential Strategies in Dissection of Biofuel Traits in Sugarcane

At present, a whole-genome sequence of sugarcane is not available to support its biofuel trait analysis. However, a strategy to overcome this using the currently available resources, for dissecting biofuel traits, for example, in sugarcane biomass, is to carry on the association studies, in which a population of genetic variability is selected, phenotyped, and genotyped. Association studies use the molecular markers from the genetic variability to detect the association between markers and traits of interest in order to validate the location of the genes, especially for quantitative traits (Huang et al., 2010). This strategy has been used for human and animal genetic studies since it was first established and more recently also for plants. To date, association studies have been applied successfully to many different plants including *Arabidopsis*, wheat, barley, rice, cotton, maize, potato, soybean, sugar beet, *Pinus*, *Eucalyptus*, ryegrass [also Zhu et al. (2008); for a review, see Hall et al. (2010)], and sugarcane (Aitken et al., 2005; Wei et al., 2006) for important traits like pathogen resistance, flowering time, grain composition, and quality. Association studies differ from traditional QTL studies, where in QTL analysis the linkage disequilibrium between markers and QLTs from a segregating population is established in a cross of different genotypes, whereas in association studies a non-structured population is used (Neale and Savolainen, 2004; Ingvarsson and Street, 2011). Therefore, association studies investigate variations of the whole population not just variations between parents. Association studies analyze the direct linkage disequilibrium between genetic markers and traits to overcome the limitations of the traditional QTL in sample size, low variation, and recombination in the population (Ingvarsson and Street, 2011). In sugarcane, association studies are a powerful method for understanding the complex traits which are controlled by many loci and dosage effects (i.e., Ming et al., 2001; Wei et al., 2006; Banerjee et al., 2015). In general, association studies involve population selection, phenotyping, genotyping, population structure, and statistical testing for the association. For these, there is a requirement to have a population with genetic variability and high linkage disequilibrium; and for sugarcane, the most important aspect of doing association studies is having marker data and a breeding population of elite varieties (Huang et al., 2010). Due to the limited number of generations, low recombination rate between chromosomes, and strong founder effect, it is expected that sugarcane has an extensive linkage disequilibrium despite the large number of chromosomes and being an outcrossing species (Huang et al., 2010). In fact, attaining a F2 population (such as inbred backcrosses or recombinant inbred lines and double haploid lines) in sugarcane is not practical due to its clonal propagation, high heterozygosity, and inbreeding depression (Aitken and McNeil, 2010; Sreedhar and Collins, 2010). Therefore, more commonly, a segregating F1 population from biparental crosses or self-pollinated progenies from heterozygous parents (as the *pseudo* F2 population) are used, and hence, most of sugarcane linkage maps (as AFLP, RAPD, isozyme, and SSR) were developed on this type of F1 population (Sreedhar and Collins, 2010). To date, most of these maps have low coverage and a limited number of markers because of the genome complexity and high cost of marker generation (Aitken et al., 2014). The high redundancy of the chromosomes in the sugarcane genome implies that with conventional approaches only the single-dose markers (present on only one of the homologous/homoeologous haplotype) can be used to obtain a high-resolution mapping (Hoarau et al., 2002; Le Cunff et al., 2008).

The potential applications of the current genotyping technologies to sugarcane association studies employ both whole-genome sequencing and whole transcriptome sequencing technologies. Genotyping is normally either by analysis candidate genes or genome-wide approaches, in which the candidate gene approach is restricted to genes which are likely thought to be associated with traits of interest based on prior knowledge (Hirschhorn and Daly, 2005; Ingvarsson and Street, 2011). At present, whole-genome sequencing based on the random sequencing of fragments of whole genomic DNA has been successfully applied to medium-size genomes with limited amount of repetitive elements, genome resequencing with the guide of a reference sequence, or *de novo* assembly of small genomes (Steuernagel et al., 2009; Henry et al., 2012; Xu et al., 2012; Edwards et al., 2013). The large genome size of sugarcane is partially attributable to sugarcane being a polyploid and the genome having a significant amount of repetitive sequences (Berkman et al., 2014). As a result, the current short reads generated from NGS technologies cannot resolve completely the challenges in the sugarcane genomes. For highly repetitive genomes, the genomic complexity will be lost or reduced by using the *de novo* assembly approaches of NGS-derived short reads as the identical repeat sequences in the genome will be collapsed (Green, 2002). Therefore, it is required to develop efficient genotyping strategies using whole-genome sequencing data for sugarcane system to overcome the challenges. Moreover, whole transcriptome sequencing gives details of the entire transcript expressed in the samples across the whole genome and could be applicable to the sugarcane genome in identifying biological significant variations (SNPs) between different developmental stages, between varieties, or for transcripts *de novo* assembly and gene discovery (Henry et al., 2012).

For large and polyploid genomes, there are still requirements to enrich the genomic DNA to capture the coding regions to ensure the depth of coverage, resolve the variable short reads, and lessen the effect of repetitive sequences in the genome on discovery of polymorphisms (Bundock et al., 2012; Henry et al., 2012). Selective sequencing of genomic loci of interest (genes or exomes) can reduce the cost compared to whole-genome sequencing and therefore simplify the data interpretation since non-coding regions are not abundant in the data. The enrichment techniques can be hybrid capture (e.g., Agilent SureSelect, NimbleGen, FlexGen) or selective circularization (e.g., Selector probes) or PCR amplification (e.g., Raindance). Hybrid capture supported by a microarray platform has been applied to sugarcane and other complex genomes due to its high capacity to enrich large regions of interest (1–50 Mb), the possibility of multiplexing, the availability of kits, and a the small amount of input DNA required (<1–3 μg) (Mertes et al., 2011). This approach uses a selection library of fragmented DNA or RNA representing the targets (normally oligonucleotides from 80 to 180 bases produced from known information such as gene indices, ESTs) to capture the cDNA fragments from a shotgun DNA library based on the hybridization and then sequence the captured fragments (Mertes et al., 2011; Bundock et al., 2012). Bundock et al. (2012) conducted the solution-based hybridization (Agilent SureSelect) to capture the exome regions of sugarcane using sorghum and sugarcane coding probes, enriched the genome 10–11 folds, and detected 270,000–280,000 SNPs in each genotype of the material tested. At the moment, a great number of SNPs from a genome or haplotype can be generated by using high-capacity genome sequencing instruments or high-density oligonucleotide arrays (Zhu et al., 2008). The continuous advancement in genotyping technology allows generation of up to 1 million SNPs spanning across the entire genome in one reaction (e.g., using SNP chip), and the newest SNP chip can measure the copy number as well as the allelic variation. Examples of available platforms are Affymetrix (e.g., Affymetrix Genome-Wide Human SNP Array 6.0) and Illumina (e.g., Illumina's WGGT Infinium BeadChips). Due to the multiple chromosomes in the homologous groups of sugarcane genome and the number of alleles at each locus (and the SNPs numbers consequently), an allele would likely be defined by a combination of SNPs, not just a single SNP (McIntyre et al., 2006, 2015). SNP genotyping including SNP calling and statistical methods to estimate the ploidy level and allele dosage within homologous groups have been developed for sugarcane by Garcia et al. (2013) to allow in-depth association analysis of the genome. In this study, SNPs were developed by SEQUENOM iPLEX MassARRAY and capture primers and then discovered by QualitySNP software, mass-based procedures, and the SuperMASSA software. For whole transcriptome sequencing, Cardoso-Silva et al. (2014) identified 5,106 SSRs and 708,125 SNPs from the unigenes assembled from RNA-seq data of contrasting sugarcane varieties. These advances in sugarcane genotyping technology, together with well-developed high-throughput phenotyping methods for biofuel traits [reviewed in Lupoi et al. (2013) and Lupoi et al. (2015)] and bioinformatics tools, could accelerate sugarcane analysis while a reference genome is not available.

Some of the association studies have been carried out on sugarcane recently such as those for QTLs which control the *Pachymetra* root rot and brown rust resistance on 154 genotypes (McIntyre et al., 2005); genetics of root rot, leaf scald, Fiji leaf gall, cane sugar, and yield using 1,068 AFLP, 141 SRR (on 154 genotypes), and 1,531 DArT markers (on 480 genotypes) (Wei et al., 2006, 2010); smut and *eldana* stalk borer using 275 RFLP and 1,056 AFLP markers on 77 genotypes (Butterfield, 2007); resistance to sugarcane yellow leaf virus using 3,949 polymorphic markers (DArT and AFLP) on 189 genotypes (Debibakas et al., 2014); markers agro-morphological traits, sugar yield disease resistance, and bagasse content using 3,327 DArT, AFLP, and SSR markers on 183 genotypes (Gouy et al., 2015); and sucrose and yield contributing traits using 989 SSR markers on 108 genotypes (Banerjee et al., 2015). Using the Affymetrix GeneChip Sugarcane Genome Array, Casu et al. (2007) identified 119 transcripts associated with enzymes of cell wall metabolism and development on sugarcane variety Q177. These promising preliminary studies were carried out on small sample sizes and limited numbers of markers (even though a small number of significant associations have been identified) while the polyploid sugarcane genome and small effect of quantitative traits requires larger sample sizes and more markers (e.g., genome-wide markers) so that significant association can be detected (Huang et al., 2010; Gouy et al., 2015).

The Reference Sequence Matters

As mentioned earlier, construction of a sugarcane nuclear genome reference sequence is an important objective, even though it might take some time to finish. However, in the meantime, sugarcane genome analysis still can exploit the currently available genetic

resources such as the sorghum gene indices (sorghum gene models), sugarcane gene indices (DFCI Sugarcane Gene Index version 3.0, an integrated collection of sugarcane ESTs, complete cDNA sequences, non-redundant data of all sugarcane genes and their related information), transcription factors (TFs), and sugarcane tentative consensus/assembled sequences. For example, in the study mentioned earlier, Bundock et al. (2012), based on the gene sequences in the sorghum genome and sugarcane gene indices, captured the exomic regions of two sugarcane genotypes Q165 and IJ76-514, detected SNPs present in 13,000–16,000 targeted genes from Illumina short read data of these samples, and 87–91% of SNPs were validated and confirmed by 454 sequencing. For transcript profiling, the reference transcriptome sequence can be constructed for specific tissues using *de novo* assembly such as in Vargas et al. (2014) and Cardoso-Silva et al. (2014) and validated to find suitable reference gene sets to be used for gene expression normalization as in Guo et al. (2014). The currently available resources, on the other hand, are also utilized. Park et al. (2015) used the Sugarcane Assembled Sequences from SUCEST-FUN database as reference sequences in a study on cold-responsive gene expression profiling of sugarcane hybrids and S. *spontaneum* and found that more than 600 genes are differentially expressed in each genotype after applying stress.

CONCLUSION

Sugarcane has been shown to be a good candidate for use as a lignocellulosic biomass feedstock for second-generation biofuel production. However, its genome complexity still remains a great bottleneck restricting the dissection of biofuel traits. The most significant achievements in improving sugarcane biomass for biofuels so far have been the establishment of the high fiber cane varieties to generate more lignocellulosic biomass, and preliminary results in modifying biomass with more cellulose, less lignin content, a preferable lignin S/G ratio, and enzyme expressed *in planta* (in leaves) for easy conversion to biofuels. The improvement of sugarcane biomass has been by traditional breeding, molecular genetics approaches and, more recently, accelerated with the use of NGS technology. The future of second-generation biofuel production using sugarcane lignocellulosic biomass will depend greatly on advances in understanding of the key biofuel traits required to deliver more efficient and price-competitive biofuels. This objective will be facilitated once the whole genome of sugarcane is fully sequenced. Optimizing sugarcane lignocellulosic bagasse composition may result in biomass with better digestibility, modified carbohydrates, and reduction of cross-linking or self-produced enzymes (in planta). Currently available sugarcane genetic resources include diverse germplasm in the genus *Saccharum*, genetic markers and maps, ESTs, and the sequence of a closely related species genome. However, novel strategies need to be developed to overcome the challenges posed by the complex genetics. Traditional approaches using breeding and molecular genetics have potential for wider use improving sugarcane while the advent of NGS technology and high-throughput phenotyping technologies will accelerate the process of dissection of biofuel traits, genome-wide. By using these approaches, the loci of interest will be defined for use to improve sugarcane biomass. Once a better understanding of the genes controlling cell wall biosynthesis is achieved, breeding programs will be able to accelerate the selection and development of varieties with optimized biomass composition to generate better sugarcane biomass sources to meet the demand of biofuel production.

AUTHOR CONTRIBUTIONS

NVH wrote the paper. AF, FCB, BAS, and RJH discussed and edited the manuscript. All authors read and approved the final manuscript.

REFERENCES

Aitken, K., Jackson, P., and McIntyre, C. (2005). A combination of AFLP and SSR markers provides extensive map coverage and identification of homo(eo)logous linkage groups in a sugarcane cultivar. *Theor. Appl. Genet.* 110, 789–801. doi:10.1007/s00122-004-1813-7

Aitken, K., and McNeil, M. (2010). "Diversity analysis," in *Genetics, Genomics and Breeding of Sugarcane*, eds R. Henry and C. Kole (Enfield, NH: Science Publishers), 19–42.

Aitken, K. S., McNeil, M. D., Hermann, S., Bundock, P. C., Kilian, A., Heller-Uszynska, K., et al. (2014). A comprehensive genetic map of sugarcane that provides enhanced map coverage and integrates high-throughput diversity array technology (DArT) markers. *BMC Genomics* 15:152. doi:10.1186/1471-2164-15-152

Alexander, A. G. (1985). *The Energy Cane Alternative.* Amsterdam: Elsevier Science Publishers BV.

Alonso Pippo, W., Luengo, C. A., Alonsoamador Morales Alberteris, L., Garzone, P., and Cornacchia, G. (2011a). Energy recovery from sugarcane-trash in the light of 2nd generation biofuel. Part 2: socio-economic aspects and techno-economic analysis. *Waste Biomass Valorization* 2, 257–266. doi:10.1007/s12649-011-9069-3

Alonso Pippo, W., Luengo, C. A., Alonsoamador Morales Alberteris, L., Garzone, P., and Cornacchia, G. (2011b). Energy recovery from sugarcane-trash in the light of 2nd generation biofuels. Part 1: current situation and environmental aspects. *Waste Biomass Valorization* 2, 1–16. doi:10.1007/s12649-010-9048-0

Amalraj, V. A., and Balasundaram, N. (2006). On the taxonomy of the members of 'Saccharum complex'. *Genet. Resour. Crop Evol.* 53, 35–41. doi:10.1007/s10722-004-0581-1

Anterola, A. M., and Lewis, N. G. (2002). Trends in lignin modification: a comprehensive analysis of the effects of genetic manipulations/mutations on lignification and vascular integrity. *Phytochemistry* 61, 221–294. doi:10.1016/S0031-9422(02)00211-X

Arencibia, A., Carmona, E., Cornide, M., Castiglione, S., O'relly, J., Chinea, A., et al. (1999). Somaclonal variation in insect-resistant transgenic sugarcane (*Saccharum* hybrid) plants produced by cell electroporation. *Transgenic Res.* 8, 349–360. doi:10.1023/A:1008900230144

Arruda, P. (2012). Genetically modified sugarcane for bioenergy generation. *Curr. Opin. Biotechnol.* 23, 315–322. doi:10.1016/j.copbio.2011.10.012

Balat, M., Balat, H., and Öz, C. (2008). Progress in bioethanol processing. *Prog. Energy Combust. Sci.* 34, 551–573. doi:10.1016/j.pecs.2007.11.001

Banerjee, N., Siraree, A., Yadav, S., Kumar, S., Singh, J., Kumar, S., et al. (2015). Marker-trait association study for sucrose and yield contributing traits in sugarcane (Saccharum spp. hybrid). *Euphytica* 205, 185–201. doi:10.1007/s10681-015-1422-3

Basnayake, S. W. V., Moyle, R., and Birch, R. G. (2011). Embryogenic callus proliferation and regeneration conditions for genetic transformation of diverse sugarcane cultivars. *Plant Cell Rep.* 30, 439–448. doi:10.1007/s00299-010-0927-4

Berkman, P. J., Bundock, P. C., Casu, R. E., Henry, R. J., Rae, A. L., and Aitken, K. S. (2014). A survey sequence comparison of *Saccharum* genotypes reveals allelic diversity differences. *Trop. Plant Biol.* 7, 71–83. doi:10.1007/s12042-014-9139-3

Berkman, P. J., Lai, K., Lorenc, M. T., and Edwards, D. (2012). Next-generation sequencing applications for wheat crop improvement. *Am. J. Bot.* 99, 365–371. doi:10.3732/ajb.1100309

ACKNOWLEDGMENTS

We are grateful to the Australian Agency for International Development (AusAID) for financial support through an Australian Development Scholarship (ADS) to NVH and to the Queensland Government for funding this research. This work was part of the DOE Joint BioEnergy Institute (http://www.jbei.org) supported by the U. S. Department of Energy, Office of Science, Office of Biological and Environmental Research, through contract DE-AC02-05CH11231 between Lawrence Berkeley National Laboratory and the U. S. Department of Energy.

Botha, F. (2009). "Energy yield and cost in a sugarcane biomass system," in: *Proceedings Australian Society Sugarcane Technologists.* 1–9.

Botha, F., Sawyer, B., and Birch, R. (2001). "Sucrose metabolism in the culm of transgenic sugarcane with reduced soluble acid invertase activity," in *Proceedings of the International Society of Sugar Cane Technologists*, 588–591.

Bottcher, A., Cesarino, I., Santos, A. B. D., Vicentini, R., Mayer, J. L. S., Vanholme, R., et al. (2013). Lignification in sugarcane: biochemical characterization, gene discovery, and expression analysis in two genotypes contrasting for lignin content. *Plant Physiol.* 163, 1539–1557. doi:10.1104/pp.113.225250

Bower, N. I., Casu, R. E., Maclean, D. J., Reverter, A., Chapman, S. C., and Manners, J. M. (2005). Transcriptional response of sugarcane roots to methyl jasmonate. *Plant Sci.* 168, 761–772. doi:10.1016/j.plantsci.2004.10.006

Bower, R., and Birch, R. G. (1992). Transgenic sugarcane plants via microprojectile bombardment. *Plant J.* 2, 409–416. doi:10.1111/j.1365-313X.1992.00409.x

Bremer, G. (1961). Problems in breeding and cytology of sugar cane. *Euphytica* 10, 59–78. doi:10.1007/BF00037206

Bundock, P. C., Casu, R. E., and Henry, R. J. (2012). Enrichment of genomic DNA for polymorphism detection in a non-model highly polyploid crop plant. *Plant Biotechnol. J.* 10, 657–667. doi:10.1111/j.1467-7652.2012.00707.x

Burner, D. M., Tew, T. L., Harvey, J. J., and Belesky, D. P. (2009). Dry matter partitioning and quality of *Miscanthus, Panicum,* and *Saccharum* genotypes in Arkansas, USA. *Biomass Bioenergy* 33, 610–619. doi:10.1016/j.biombioe.2008.10.002

Butterfield, M. K. (2007). *Marker Assisted Breeding in Sugarcane: A Complex Polyploidy.* PhD Thesis, Univ Stellenbosch, Matieland, Stellenbosch, South Africa.

Canilha, L., Kumar Chandel, A., Dos Santos Milessi, T. S., Fernandes Antunes, F. A., Da Costa Freitas, W. L., Das Gracas Almeida Felipe, M., et al. (2012). Bioconversion of sugarcane biomass into ethanol: an overview about composition, pretreatment methods, detoxification of hydrolysates, enzymatic saccharification, and ethanol fermentation. *J. Biomed. Biotechnol.* 2012, 989572. doi:10.1155/2012/989572

Canilha, L., Rodrigues, R., Antunes, F. A. F., Chandel, A. K., Milessi, T., Felipe, M. D. G. A., et al. (2013). "Bioconversion of hemicellulose from sugarcane biomass into sustainable products," in *Sustainable Degradation of Lignocellulosic Biomass-Techniques, Applications and Commercialization*, eds A. K. Chandel and S. S. da Silva (Rijeka: InTech), 15–45. doi:10.5772/53832

Cardoso-Silva, C. B., Costa, E. A., Mancini, M. C., Balsalobre, T. W. A., Canesin, L. E. C., Pinto, L. R., et al. (2014). *De novo* assembly and transcriptome analysis of contrasting sugarcane varieties. *PLoS ONE* 9:e88462. doi:10.1371/journal.pone.0088462

Carpita, N. C. (1996). Structure and biogenesis of the cell walls of grasses. *Annu. Rev. Plant Physiol. Plant Mol. Biol.* 47, 445–476. doi:10.1146/annurev.arplant.47.1.445

Carson, D., and Botha, F. (2002). Genes expressed in sugarcane maturing internodal tissue. *Plant Cell Rep.* 20, 1075–1081. doi:10.1007/s00299-002-0444-1

Carson, D. L., and Botha, F. C. (2000). Preliminary analysis of expressed sequence tags for sugarcane. *Crop Sci.* 40, 1769–1779. doi:10.2135/cropsci2000.4061769x

Casu, R., Dimmock, C., Chapman, S., Grof, C. L., McIntyre, C. L., Bonnett, G., et al. (2004). Identification of differentially expressed transcripts from maturing stem of sugarcane by in silico analysis of stem expressed sequence tags and gene expression profiling. *Plant Mol. Biol.* 54, 503–517. doi:10.1023/B:PLAN.0000038255.96128.41

Casu, R., Grof, C. L., Rae, A., McIntyre, C. L., Dimmock, C., and Manners, J. (2003). Identification of a novel sugar transporter homologue strongly expressed in maturing stem vascular tissues of sugarcane by expressed sequence tag and microarray analysis. *Plant Mol. Biol.* 52, 371–386. doi:10.1023/A:1023957214644

Casu, R. E., Jarmey, J. M., Bonnett, G. D., and Manners, J. M. (2007). Identification of transcripts associated with cell wall metabolism and development in the stem of sugarcane by Affymetrix GeneChip Sugarcane Genome Array expression profiling. *Funct. Integr. Genomics* 7, 153–167. doi:10.1007/s10142-006-0038-z

Chen, F., and Dixon, R. A. (2007). Lignin modification improves fermentable sugar yields for biofuel production. *Nat. Biotechnol.* 25, 759–761. doi:10.1038/nbt1316

Cuadrado, A., Acevedo, R., Moreno Diaz De La Espina, S., Jouve, N., and De La Torre, C. (2004). Genome remodelling in three modern *S. officinarumx S. spontaneum* sugarcane cultivars. *J. Exp. Bot.* 55, 847–854. doi:10.1093/jxb/erh093

Dal-Bianco, M., Carneiro, M. S., Hotta, C. T., Chapola, R. G., Hoffmann, H. P., Garcia, A. A. F., et al. (2012). Sugarcane improvement: how far can we go? *Curr. Opin. Biotechnol.* 23, 265–270. doi:10.1016/j.copbio.2011.09.002

Daniels, J., and Roach, B. T. (1987). "Chapter 2 – taxonomy and evolution," in *Developments in Crop Science*, ed. J. H. Don (Amsterdam: Elsevier), 7–84.

De O. Buanafina, M. M. (2009). Feruloylation in grasses: current and future perspectives. *Mol. Plant.* 2, 861–872. doi:10.1093/mp/ssp067

De Siqueira Ferreira, S., Nishiyama, M. Y. Jr., Paterson, A. H., and Souza, G. M. (2013). Biofuel and energy crops: high-yield Saccharinae take center stage in the post-genomics era. *Genome Biol.* 14, 210. doi:10.1186/gb-2013-14-6-210

Debibakas, S., Rocher, S., Garsmeur, O., Toubi, L., Roques, D., D'Hont, A., et al. (2014). Prospecting sugarcane resistance to sugarcane yellow leaf virus by genome-wide association. *Theor. Appl. Genet.* 127, 1719–1732. doi:10.1007/s00122-014-2334-7

D'Hont, A., and Glaszmann, J. C. (2001). "Sugarcane genome analysis with molecular markers, a first decade of research," in *Proceedings of International Society of Sugar Cane Technologists*, 556–559.

D'Hont, A., Grivet, L., Feldmann, P., Rao, S., Berding, N., and Glaszmann, J. C. (1996). Characterisation of the double genome structure of modern sugarcane cultivars (*Saccharum* spp.) by molecular cytogenetics. *Mol. Gen. Genet.* 250, 405–413. doi:10.1007/s004380050092

D'Hont, A., Ison, D., Alix, K., Roux, C., and Glaszmann, J. C. (1998). Determination of basic chromosome numbers in the genus *Saccharum* by physical mapping of ribosomal RNA genes. *Genome* 41, 221–225. doi:10.1139/g98-023

Dias De Oliveira, M. E., Vaughan, B. E., and Rykiel, E. J. (2005). Ethanol as fuel: energy, carbon dioxide balances, and ecological footprint. *Bioscience* 55, 593–602. doi:10.1641/0006-3568(2005)055[0593:EAFECD]2.0.CO;2

Dillon, S. L., Shapter, F. M., Henry, R. J., Cordeiro, G., Izquierdo, L., and Lee, L. S. (2007). Domestication to crop improvement: genetic resources for *Sorghum* and *Saccharum* (*Andropogoneae*). *Ann. Bot.* 100, 975–989. doi:10.1093/aob/mcm192

Edwards, D., Batley, J., and Snowdon, R. J. (2013). Accessing complex crop genomes with next-generation sequencing. *Theor. Appl. Genet.* 126, 1–11. doi:10.1007/s00122-012-1964-x

Enríquez-Obregón, G. A., Vázquez-Padrón, R. I., Prieto-Samsonov, D. L., De La Riva, G. A., and Selman-Housein, G. (1998). Herbicide-resistant sugarcane (*Saccharum officinarum* L.) plants by *Agrobacterium*-mediated transformation. *Planta* 206, 20–27. doi:10.1007/s004250050369

Eversole, K., Graner, A., and Stein, N. (2009). "Wheat and barley genome sequencing," in *Genetics and Genomics of the Triticeae*, eds G. J. Muehlbauer and C. Feuillet (New York, NY: Springer), 713–742.

Falco, M. C., Tulmann Neto, A., and Ulian, E. C. (2000). Transformation and expression of a gene for herbicide resistance in a Brazilian sugarcane. *Plant Cell Rep.* 19, 1188–1194. doi:10.1007/s002990000253

FAOSTAT. (2015). Available at: http://faostat3.fao.org/home/E [accessed May 25, 2015].

Fedenko, J., Erickson, J., Woodard, K., Sollenberger, L., Vendramini, J. B., Gilbert, R. A., et al. (2013). Biomass production and composition of perennial grasses grown for bioenergy in a subtropical climate across Florida, USA. *BioEnergy Res.* 6, 1082–1093. doi:10.1007/s12155-013-9342-3

Figueira, T., Okura, V., Rodrigues Da Silva, F., Jose Da Silva, M., Kudrna, D., Ammiraju, J. S. et al. (2012). A BAC library of the SP80-3280 sugarcane variety (*Saccharum* sp.) and its inferred microsynteny with the *Sorghum* genome. *BMC Res. Notes* 5:185. doi:10.1186/1756-0500-5-185

Furtado, A., Lupoi, J. S., Hoang, N. V., Healey, A., Singh, S., Simmons, B. A., et al. (2014). Modifying plants for biofuel and biomaterial production. *Plant Biotechnol. J.* 12, 1246–1258. doi:10.1111/pbi.12300

Gallo-Meagher, M., and Irvine, J. E. (1996). Herbicide resistant transgenic sugarcane plants containing the bar gene. *Crop Sci.* 36, 1367–1374. doi:10.2135/cropsci1996.0011183X003600050047x

Garcia, A. A. F., Mollinari, M., Marconi, T. G., Serang, O. R., Silva, R. R., Vieira, M. L. C., et al. (2013). SNP genotyping allows an in-depth characterisation of the genome of sugarcane and other complex autopolyploids. *Sci. Rep.* 3, 3399. doi:10.1038/srep03399

Gnansounou, E., and Dauriat, A. (2010). Techno-economic analysis of lignocellulosic ethanol: a review. *Bioresour. Technol.* 101, 4980–4991. doi:10.1016/j.biortech.2010.02.009

Gouy, M., Rousselle, Y., Thong Chane, A., Anglade, A., Royaert, S., Nibouche, S., et al. (2015). Genome wide association mapping of agro-morphological and disease resistance traits in sugarcane. *Euphytica* 202, 269–284. doi:10.1007/s10681-014-1294-y

Govindaraj, P., Amalraj, V. A., Mohanraj, K., and Nair, N. V. (2014). Collection, characterization and phenotypic diversity of *Saccharum spontaneum* L. from arid and semi arid zones of northwestern India. *Sugar Tech* 16, 36–43. doi:10.1007/s12355-013-0255-4

Grabber, J. H. (2005). How do lignin composition, structure, and cross-linking affect degradability? A review of cell wall model studies. *Crop Sci.* 45, 820. doi:10.2135/cropsci2004.0191

Grassius: Grass Regulatory Information Server. (2015). Available at: http://grassius.org/ [accessed May 25, 2015].

Green, P. (2002). Whole-genome disassembly. *Proc. Natl. Acad. Sci. U.S.A.* 99, 4143–4144. doi:10.1073/pnas.082095999

Grivet, L., and Arruda, P. (2002). Sugarcane genomics: depicting the complex genome of an important tropical crop. *Curr. Opin. Plant Biol.* 5, 122–127. doi:10.1016/S1369-5266(02)00234-0

Grivet, L., Dhont, A., Dufour, P., Hamon, P., Roques, D., and Glaszmann, J. C. (1994). Comparative genome mapping of sugar-cane with other species within the *Andropogoneae* tribe. *Heredity* 73, 500–508. doi:10.1038/hdy.1994.148

Grivet, L., Glaszmann, J. C., Vincentz, M., Da Silva, F., and Arruda, P. (2003). ESTs as a source for sequence polymorphism discovery in sugarcane: example of the Adh genes. *Theor. Appl. Genet.* 106, 190–197. doi:10.1007/s00122-002-1075-1

Guo, J., Ling, H., Wu, Q., Xu, L., and Que, Y. (2014). The choice of reference genes for assessing gene expression in sugarcane under salinity and drought stresses. *Sci. Rep.* 4, 7042. doi:10.1038/srep07042

Gupta, V., Raghuvanshi, S., Gupta, A., Saini, N., Gaur, A., Khan, M. S., et al. (2010). The water-deficit stress- and red-rot-related genes in sugarcane. *Funct. Integr. Genomics* 10, 207–214. doi:10.1007/s10142-009-0144-9

Hall, D., Tegstrom, C., and Ingvarsson, P. K. (2010). Using association mapping to dissect the genetic basis of complex traits in plants. *Brief. Funct. Genomics* 9, 157–165. doi:10.1093/bfgp/elp048

Halling, P., and Simms-Borre, P. (2008). Overview of lignocellulosic feedstock conversion into ethanol-focus on sugarcane bagasse. *Int. Sugar J.* 110, 191.

Handakumbura, P. P., and Hazen, S. P. (2012). Transcriptional regulation of grass secondary cell wall biosynthesis: playing catch-up with *Arabidopsis thaliana*. *Front. Plant Sci.* 3:74. doi:10.3389/fpls.2012.00074

Harris, D., and DeBolt, S. (2010). Synthesis, regulation and utilization of lignocellulosic biomass. *Plant Biotechnol. J.* 8, 244–262. doi:10.1111/j.1467-7652.2009.00481.x

Harrison, M. D., Geijskes, J., Coleman, H. D., Shand, K., Kinkema, M., Palupe, A., et al. (2011). Accumulation of recombinant cellobiohydrolase and endoglucanase in the leaves of mature transgenic sugar cane. *Plant Biotechnol. J.* 9, 884–896. doi:10.1111/j.1467-7652.2011.00597.x

Harvey, M., Huckett, B., and Botha, F. (1994). "Use of polymerase chain reaction (PCR) and random amplification of polymorphic DNAs (RAPDs) for the determination of genetic distances between 21 sugarcane varieties," in *Proceedings of the South African Sugar Technologists Association* (Citeseer), 36–40.

Hattori, T., and Morita, S. (2010). Energy crops for sustainable bioethanol production; which, where and how? *Plant Prod. Sci.* 13, 221–234. doi:10.1626/pps.13.221

Henry, R. J. (2010a). Evaluation of plant biomass resources available for replacement of fossil oil. *Plant Biotechnol. J.* 8, 288–293. doi:10.1111/j.1467-7652.2009.00482.x

Henry, R. J. (2010b). *Plant Resources for Food, Fuel and Conservation*. London: Earthscan.

Henry, R. J., Edwards, M., Waters, D. L., Gopala Krishnan, S., Bundock, P., Sexton, T. R., et al. (2012). Application of large-scale sequencing to marker discovery in plants. *J. Biosci.* 37, 829–841. doi:10.1007/s12038-012-9253-z

Higuchi, T. (1981). "Biosynthesis of lignin," in *Plant Carbohydrates II*, eds W. Tanner and F. Loewus (Berlin: Springer), 194–224.

Himmel, M., Ding, S., Johnson, D., Adney, W., Nimlos, M., Brady, J., et al. (2007). Biomass recalcitrance: engineering plants and enzymes for biofuels production. *Science* 315, 804–807. doi:10.1126/science.1137016

Hirschhorn, J. N., and Daly, M. J. (2005). Genome-wide association studies for common diseases and complex traits. *Nat. Rev. Genet.* 6, 95–108. doi:10.1038/nrg1521

Hoarau, J.-Y., Grivet, L., Offmann, B., Raboin, L.-M., Diorflar, J.-P., Payet, J., et al. (2002). Genetic dissection of a modern sugarcane cultivar (*Saccharum spp.*). II. Detection of QTLs for yield components. *Theor. Appl. Genet.* 105, 1027–1037. doi:10.1007/s00122-002-1047-5

Hotta, C., Lembke, C., Domingues, D., Ochoa, E., Cruz, G. Q., Melotto-Passarin, D., et al. (2010). The biotechnology roadmap for sugarcane improvement. *Trop. Plant Biol.* 3, 75–87. doi:10.1007/s12042-010-9050-5

Huang, E., Aitken, K., and George, A. (2010). "Association Studies," in *Genetics, Genomics and Breeding of Sugarcane*, eds. R. Henry and C. Kole (Enfield, NH: Science Publishers), 43–68.

Ingelbrecht, I. L., Irvine, J. E., and Mirkov, T. E. (1999). Posttranscriptional gene silencing in transgenic sugarcane. Dissection of homology-dependent virus resistance in a monocot that has a complex polyploid genome. *Plant Physiol.* 119, 1187–1198. doi:10.1104/pp.119.4.1187

Ingvarsson, P. K., and Street, N. R. (2011). Association genetics of complex traits in plants. *New Phytol.* 189, 909–922. doi:10.1111/j.1469-8137.2010.03593.x

Jakob, K., Zhou, F., and Paterson, A. (2009). Genetic improvement of C4 grasses as cellulosic biofuel feedstocks. *In vitro Cell. Dev. Biol. Plant* 45, 291–305. doi:10.1007/s11627-009-9214-x

Jannoo, N., Grivet, L., Chantret, N., Garsmeur, O., Glaszmann, J. C., Arruda, P., et al. (2007). Orthologous comparison in a gene-rich region among grasses reveals stability in the sugarcane polyploid genome. *Plant J.* 50, 574–585. doi:10.1111/j.1365-313X.2007.03082.x

Johansson, T. B., and Burnham, L. (1993). *Renewable Energy: Sources for Fuels and Electricity*. Washington, DC: Island Press.

Joyce, P., McQualter, R., Handley, J., Dale, J., Harding, R., and Smith, G. (1998). "Transgenic sugarcane resistant to sugarcane mosaic virus," in *Proceedings-Australian Society of Sugar Cane Technologists* (Salisbury: Watson Ferguson and Company), 204–210.

Jung, H. G. (1989). Forage lignins and their effects on fiber digestibility. *Agronomy Journal* 81, 33–38. doi:10.2134/agronj1989.00021962008100010006x

Jung, J. H., Fouad, W. M., Vermerris, W., Gallo, M., and Altpeter, F. (2012). RNAi suppression of lignin biosynthesis in sugarcane reduces recalcitrance for biofuel production from lignocellulosic biomass. *Plant Biotechnol. J.* 10, 1067–1076. doi:10.1111/j.1467-7652.2012.00734.x

Kim, C., Lee, T. H., Compton, R. O., Robertson, J. S., Pierce, G. J., and Paterson, A. H. (2013). A genome-wide BAC end-sequence survey of sugarcane elucidates genome composition, and identifies BACs covering much of the euchromatin. *Plant Mol. Biol.* 81, 139–147. doi:10.1007/s11103-012-9987-x

Lam, E., Shine, J., Da Silva, J., Lawton, M., Bonos, S., Calvino, M., et al. (2009). Improving sugarcane for biofuel: engineering for an even better feedstock. *Glob. Change Biol. Bioenergy* 1, 251–255. doi:10.1111/j.1757-1707.2009.01016.x

Le Cunff, L., Garsmeur, O., Raboin, L. M., Pauquet, J., Telismart, H., Selvi, A., et al. (2008). Diploid/polyploid syntenic shuttle mapping and haplotype-specific chromosome walking toward a rust resistance gene (Bru1) in highly polyploid sugarcane (2n ~ 12x ~ 115). *Genetics* 180, 649–660. doi:10.1534/genetics.108.091355

Li, C., Sun, L., Simmons, B., and Singh, S. (2013). Comparing the recalcitrance of *Eucalyptus*, *Pine*, and *Switchgrass* using ionic liquid and dilute acid pretreatments. *Bioenerg. Res.* 6, 14–23. doi:10.1007/s12155-012-9220-4

Li, X., Ximenes, E., Kim, Y., Slininger, M., Meilan, R., Ladisch, M., et al. (2010). Lignin monomer composition affects *Arabidopsis* cell-wall degradability after liquid hot water pretreatment. *Biotechnol. Biofuels* 3, 27. doi:10.1186/1754-6834-3-27

Lima, M. L., Garcia, A. A., Oliveira, K. M., Matsuoka, S., Arizono, H., De Souza, C. L. Jr., et al. (2002). Analysis of genetic similarity detected by AFLP and coefficient of parentage among genotypes of sugar cane (*Saccharum spp.*). *Theor. Appl. Genet.* 104, 30–38. doi:10.1007/s001220200003

Loureiro, M., Barbosa, M. P., Lopes, F. F., and Silvério, F. (2011). "Sugarcane breeding and selection for more efficient biomass conversion in cellulosic ethanol," in *Routes to Cellulosic Ethanol*, eds M. S. Buckeridge and G. H. Goldman (New York, NY: Springer), 199–239.

Lupoi, J., Singh, S., Simmons, B., and Henry, R. (2013). Assessment of lignocellulosic biomass using analytical spectroscopy: an evolution to high-throughput techniques. *Bioenerg. Res.* 7, 1–23. doi:10.1007/s12155-013-9352-1

Lupoi, J. S., Singh, S., Parthasarathi, R., Simmons, B. A., and Henry, R. J. (2015). Recent innovations in analytical methods for the qualitative and quantitative assessment of lignin. *Renew. Sustain. Energ. Rev.* 49, 871–906. doi:10.1016/j.rser.2015.04.091

Ma, H., Albert, H. H., Paull, R., and Moore, P. H. (2000). Metabolic engineering of invertase activities in different subcellular compartments affects sucrose accumulation in sugarcane cells. *Funct. Plant Biol.* 27, 1021–1030. doi:10.1071/PP00029

Ma, H. M., Schulze, S., Lee, S., Yang, M., Mirkov, E., Irvine, J., et al. (2004). An EST survey of the sugarcane transcriptome. *Theor. Appl. Genet.* 108, 851–863. doi:10.1007/s00122-003-1510-y

Macrelli, S., Galbe, M., and Wallberg, O. (2014). Effects of production and market factors on ethanol profitability for an integrated first and second generation ethanol plant using the whole sugarcane as feedstock. *Biotechnol. Biofuels* 7, 26. doi:10.1186/1754-6834-7-26

Macrelli, S., Mogensen, J., and Zacchi, G. (2012). Techno-economic evaluation of 2nd generation bioethanol production from sugar cane bagasse and leaves integrated with the sugar-based ethanol process. *Biotechnol. Biofuels* 5, 22. doi:10.1186/1754-6834-5-22

Manickavasagam, M., Ganapathi, A., Anbazhagan, V. R., Sudhakar, B., Selvaraj, N., Vasudevan, A., et al. (2004). *Agrobacterium*-mediated genetic transformation and development of herbicide-resistant sugarcane (*Saccharum species* hybrids) using axillary buds. *Plant Cell Rep.* 23, 134–143. doi:10.1007/s00299-004-0794-y

Manners, J., McIntyre, L., Casu, R., Cordeiro, G., Jackson, M., Aitken, K., et al. (2004). "Can genomics revolutionize genetics and breeding in sugarcane," in *Proceedings of the 4th International Crop Science Congress* (Brisbane, QLD).

Matsuoka, S., Ferro, J., and Arruda, P. (2009). The Brazilian experience of sugarcane ethanol industry. *In vitro Cell. Dev. Biol. Plant* 45, 372–381. doi:10.1007/s11627-009-9220-z

McIntyre, C. L., Goode, M. L., Cordeiro, G., Bundock, P., Eliott, F., Henry, R. J., et al. (2015). Characterisation of alleles of the sucrose phosphate synthase gene family in sugarcane and their association with sugar-related traits. *Mol. Breed.* 35, 1–14. doi:10.1007/s11032-015-0286-5

McIntyre, C. L., Jackson, M., Cordeiro, G. M., Amouyal, O., Hermann, S., Aitken, K. S., et al. (2006). The identification and characterisation of alleles of sucrose phosphate synthase gene family III in sugarcane. *Mol. Breed.* 18, 39–50. doi:10.1007/s11032-006-9012-7

McIntyre, C. L., Whan, V. A., Croft, B., Magarey, R. and Smith, G. R. (2005) Identification and validation of molecular markers associated with pachymetra root rot and brown rust resistance in sugarcane using Map- and association-based approaches. *Molecular Breeding* 16, 151–161. doi:10.1007/s11032-005-7492-5

Mertes, F., Elsharawy, A., Sauer, S., Van Helvoort, J. M. L. M., Van Der Zaag, P. J., Franke, A., et al. (2011). Targeted enrichment of genomic DNA regions for next-generation sequencing. *Brief. Funct. Genomics* 10, 374–386. doi:10.1093/bfgp/elr033

Ming, R., Liu, S.-C., Lin, Y.-R., Da Silva, J., Wilson, W., Braga, D., et al. (1998). Detailed alignment of *Saccharum* and *Sorghum* chromosomes: comparative organization of closely related diploid and polyploid genomes. *Genetics* 150, 1663–1682.

Ming, R., Liu, S. C., Moore, P. H., Irvine, J. E., and Paterson, A. H. (2001). QTL analysis in a complex autopolyploid: genetic control of sugar content in sugarcane. *Genome Res.* 11, 2075–2084. doi:10.1101/gr.198801

Ming, R., Moore, P. H., Wu, K.-K., D'Hont, A., Glaszmann, J. C., Tew, T. L., et al. (2010). "Sugarcane improvement through breeding and biotechnology," in *Plant Breeding Reviews*, ed. J. Janick (Oxford, UK: John Wiley & Sons, Inc), 15–118. doi:10.1002/9780470650349.ch2

Moore, P. H. (2009). "Sugarcane biology, yield, and potential for improvement," in *Workshop BIOEN on Sugarcane Improvement* (San Pablo).

Mosier, N., Wyman, C., Dale, B., Elander, R., Lee, Y., Holtzapple, M., et al. (2005). Features of promising technologies for pretreatment of lignocellulosic biomass. *Bioresour. Technol.* 96, 673–686. doi:10.1016/j.biortech.2004.06.025

Mutwil, M., Debolt, S., and Persson, S. (2008). Cellulose synthesis: a complex complex. *Curr. Opin. Plant Biol.* 11, 252–257. doi:10.1016/j.pbi.2008.03.007

Nagaraj, S. H., Gasser, R. B., and Ranganathan, S. (2007). A hitchhiker's guide to expressed sequence tag (EST) analysis. *Brief. Bioinformatics* 8, 6–21. doi:10.1093/bib/bbl015

Neale, D. B., and Savolainen, O. (2004). Association genetics of complex traits in conifers. *Trends Plant Sci.* 9, 325–330. doi:10.1016/j.tplants.2004.05.006

Pandey, A., Soccol, C. R., Nigam, P., and Soccol, V. T. (2000). Biotechnological potential of agro-industrial residues. I: sugarcane bagasse. *Bioresour. Technol.* 74, 69–80. doi:10.1016/s0960-8524(99)00142-x

Panje, R. R. and Babu, C. N. (1960). Studies in saccharum spontaneum distribution and geographical association of chromosome numbers. *Cytologia* 25, 152–72, doi:10.1508/cytologia.25.152

Park, J. W., Benatti, T. R., Marconi, T., Yu, Q., Solis-Gracia, N., Mora, V., et al. (2015). Cold responsive gene expression profiling of sugarcane and *Saccharum spontaneum* with functional analysis of a cold inducible *Saccharum* homolog of NOD26-like intrinsic protein to salt and water stress. *PLoS ONE* 10:e0125810. doi:10.1371/journal.pone.0125810

Paterson, A., Bowers, J., Bruggmann, R., Dubchak, I., Grimwood, J., Gundlach, H., et al. (2009). The *Sorghum bicolor* genome and the diversification of grasses. *Nature* 457, 551–556. doi:10.1038/nature07723

Paterson, A., Souza, G., Sluys, M. V., Ming, R., D'Hont, A., Henry, R., et al. (2010). "Structural genomics and genome sequencing," in *Genetics, Genomics and Breeding of Sugarcane*, eds. R. Henry and C. Kole (Enfield, NH: Science Publishers), 149–165.

Pauly, M., Gille, S., Liu, L., Mansoori, N., De Souza, A., Schultink, A., et al. (2013). Hemicellulose biosynthesis. *Planta* 238, 627–642. doi:10.1007/s00425-013-1921-1

Pereira, S. C., Maehara, L., Machado, C. M. M., and Farinas, C. S. (2015). 2G ethanol from the whole sugarcane lignocellulosic biomass. *Biotechnol. Biofuels* 8, 44. doi:10.1186/s13068-015-0224-0

Pinto, L. R., Oliveira, K. M., Ulian, E. C., Garcia, A. A., and De Souza, A. P. (2004). Survey in the sugarcane expressed sequence tag database (SUCEST) for simple sequence repeats. *Genome* 47, 795–804. doi:10.1139/g04-055

Piperidis, G., D'Hont, A., and Hogarth, D. (2001). "Chromosome composition analysis of various *Saccharum* interspecific hybrids by genomic in situ hybridisation (GISH)," in *International Society of Sugar Cane Technologists. Proceedings of the XXIV Congress*, Vol. 2 ed. D. M. Hogarth (Brisbane: Australian Society of Sugar Cane Technologists), 565–566.

Piperidis, G., Piperidis, N., and D'Hont, A. (2010). Molecular cytogenetic investigation of chromosome composition and transmission in sugarcane. *Mol. Genet. Genomics* 284, 65–73. doi:10.1007/s00438-010-0546-3

Rabemanolontsoa, H., and Saka, S. (2013). Comparative study on chemical composition of various biomass species. *RSC Adv.* 3, 3946–3956. doi:10.1039/c3ra22958k

Raboin, L. M., Pauquet, J., Butterfield, M., D'Hont, A., and Glaszmann, J. C. (2008). Analysis of genome-wide linkage disequilibrium in the highly polyploid sugarcane. *Theor. Appl. Genet.* 116, 701–714. doi:10.1007/s00122-007-0703-1

Ramos, R. L. B., Tovar, F. J., Junqueira, R. M., Lino, F. B., and Sachetto-Martins, G. (2001). Sugarcane expressed sequences tags (ESTs) encoding enzymes involved in lignin biosynthesis pathways. *Genet. Mol. Biol.* 24, 235–241. doi:10.1590/S1415-47572001000100031

Renewable Fuels Association. (2015). *World Fuel Ethanol Production*. Available at: http://ethanolrfa.org/pages/World-Fuel-Ethanol-Production. Accessed 25 May 2015

Roach, B. (1989). "Origin and improvement of the genetic base of sugarcane," in *Proceedings Australian Society of Sugar Cane Technologists*, eds. B. T. Egan (Salisbury: Ferguson and Company), 34–47.

Saathoff, A., Sarath, G., Chow, E., Dien, B., and Tobias, C. (2011). Downregulation of cinnamyl-alcohol dehydrogenase in switchgrass by RNA silencing results in enhanced glucose release after cellulase treatment. *PLoS ONE* 6:e16416. doi:10.1371/journal.pone.0016416

Santos Brito, M., Nobile, P., Bottcher, A., Dos Santos, A., Creste, S., De Landell, M., et al. (2015). Expression profile of sugarcane transcription factor genes involved in lignin biosynthesis. *Trop. Plant Biol.* 8, 19–30. doi:10.1007/s12042-015-9147-y

Sato, K., Motoi, Y., Yamaji, N., and Yoshida, H. (2011). 454 sequencing of pooled BAC clones on chromosome 3H of barley. *BMC Genomics* 12:246. doi:10.1186/1471-2164-12-246

Saxena, I. M., and Brown, R. M. Jr. (2000). Cellulose synthases and related enzymes. *Curr. Opin. Plant Biol.* 3, 523–531. doi:10.1016/S1369-5266(00)00125-4

Schubert, C. (2006). Can biofuels finally take center stage? *Nat. Biotechnol.* 24, 777–784. doi:10.1038/nbt0706-777

Seabra, J. E. A., Tao, L., Chum, H. L., and Macedo, I. C. (2010). A techno-economic evaluation of the effects of centralized cellulosic ethanol and co-products refinery options with sugarcane mill clustering. *Biomass Bioenergy* 34, 1065–1078. doi:10.1016/j.biombioe.2010.01.042

Shi, J., Gladden, J. M., Sathitsuksanoh, N., Kambam, P., Sandoval, L., Mitra, D., et al. (2013). One-pot ionic liquid pretreatment and saccharification of switchgrass. *Green Chem.* 15, 2579–2589. doi:10.1039/C3GC40545A

Simpson, T. (2009). Biofuels: the past, present, and a new vision for the future. *Bioscience* 59, 926–927. doi:10.1525/bio.2009.59.11.2

Sims, R. E. H., Hastings, A., Schlamadinger, B., Taylor, G., and Smith, P. (2006). Energy crops: current status and future prospects. *Glob. Chang. Biol.* 12, 2054–2076. doi:10.1111/j.1365-2486.2006.01163.x

Somerville, C., Youngs, H., Taylor, C., Davis, S. C., and Long, S. P. (2010). Feedstocks for lignocellulosic biofuels. *Science* 329, 790–792. doi:10.1126/science.1189268

Souza, A., Leite, D. C., Pattathil, S., Hahn, M., and Buckeridge, M. (2013). Composition and structure of sugarcane cell wall polysaccharides: implications for second-generation bioethanol production. *Bioenerg. Res.* 6, 564–579. doi:10.1007/s12155-012-9268-1

Souza, G. M., Berges, H., Bocs, S., Casu, R., D'Hont, A., Ferreira, J. E., et al. (2011). The sugarcane genome challenge: strategies for sequencing a highly complex genome. *Trop. Plant Biol.* 4, 145–156. doi:10.1007/s12042-011-9079-0

Sreedhar, A., and Collins, A. K. (2010). "Molecular genetic linkage mapping in saccharum," in *Genetics, Genomics and Breeding of Sugarcane*, eds. R. Henry and C. Kole (Enfield, NH: Science Publishers), 69–96. doi:10.1201/EBK1578086849-6

Sreenivasan, T. V., Ahloowalia, B. S., and Heinz, D. J. (1987). "Chapter 5 - cytogenetics," in *Developments in Crop Science*, ed. J. H. Don (Amsterdam: Elsevier), 211–253.

Steuernagel, B., Taudien, S., Gundlach, H., Seidel, M., Ariyadasa, R., Schulte, D., et al. (2009). De novo 454 sequencing of barcoded BAC pools for comprehensive gene survey and genome analysis in the complex genome of barley. *BMC Genomics* 10:547. doi:10.1186/1471-2164-10-547

Sticklen, M. (2006). Plant genetic engineering to improve biomass characteristics for biofuels. *Curr. Opin. Biotechnol.* 17, 315–319. doi:10.1016/j.copbio.2006.05.003

Sugesi. (2015). *The Sugarcane Genome Sequencing Initiative*. Available at: http://cnrgv.toulouse.inra.fr/fr/Projets/Canne-a-sucre/The-Sugarcane-Genome-Sequencing-Initiative-SUGESI-Strategies-for-Sequencing-a-Highly-Complex-Genome. Accessed on 25 May 2015

Sun, N., Liu, H., Sathitsuksanoh, N., Stavila, V., Sawant, M., Bonito, A., et al. (2013). Production and extraction of sugars from switchgrass hydrolyzed in ionic liquids. *Biotechnol. Biofuels* 6, 39. doi:10.1186/1754-6834-6-39

Tabasum, S., Khan, F. A., Nawaz, S., Iqbal, M. Z., and Saeed, A. (2010). DNA profiling of sugarcane genotypes using randomly amplified polymorphic DNA. *Genet. Mol. Res.* 9, 471–483. doi:10.4238/vol9-1gmr709

Tai, P., and Miller, J. (2001). A core collection for *Saccharum spontaneum* L. from the world collection of sugarcane. *Crop Sci.* 41, 879–885. doi:10.2135/cropsci2001.413879x

Taylor, L. E. II, Dai, Z., Decker, S. R., Brunecky, R., Adney, W. S., Ding, S. Y., et al. (2008). Heterologous expression of glycosyl hydrolases in planta: a new departure for biofuels. *Trends Biotechnol.* 26, 413–424. doi:10.1016/j.tibtech.2008.05.002

Tew, T., and Cobill, R. (2008). "Genetic improvement of sugarcane (*Saccharum* spp.) as an energy crop," in *Genetic Improvement of Bioenergy Crops*, ed. W. Vermerris (New York, NY: Springer), 273–294.

Todd, J., Wang, J., Glaz, B., Sood, S., Ayala-Silva, T., Nayak, S., et al. (2014). Phenotypic characterization of the Miami world collection of sugarcane (*Saccharum* spp.) and related grasses for selecting a representative core. *Genet. Resour. Crop Evol.* 61, 1581–1596. doi:10.1007/s10722-014-0132-3

Tomkins, J. P., Yu, Y., Miller-Smith, H., Frisch, D. A., Woo, S. S., and Wing, R. A. (1999). A bacterial artificial chromosome library for sugarcane. *Theor. Appl. Genet.* 99, 419–424. doi:10.1007/s001220051252

Van Der Weijde, T., Alvim Kamei, C. L., Torres, A. F., Vermerris, W., Dolstra, O., Visser, R. G., et al. (2013). The potential of C4 grasses for cellulosic biofuel production. *Front. Plant Sci.* 4:107. doi:10.3389/fpls.2013.00107

Van Dijk, E. L., Auger, H., Jaszczyszyn, Y., and Thermes, C. (2014). Ten years of next-generation sequencing technology. *Trends Genet.* 30, 418–426. doi:10.1016/j.tig.2014.07.001

Vargas, L., Santa Brígida, A. B., Mota Filho, J. P., De Carvalho, T. G., Rojas, C. A., Vaneechoutte, D., et al. (2014). Drought tolerance conferred to sugarcane by association with *Gluconacetobacter diazotrophicus*: a transcriptomic view of hormone pathways. *PLoS ONE* 9:e114744. doi:10.1371/journal.pone.0114744

Verheye, W. (2010). "Growth and production of sugarcane," in *Soils, Plant Growth and Production Volume II in Encyclopedia of Life Support Systems (EOLSS), Developed under the Auspices of the UNESCO*, Vol. 2. Paris: Eolss Publishers. Available at: www.eolss.net/sample-chapters/c10/e1-05a-22-00.pdf [Accessed 15 July 15].

Vettore, A. L., Da Silva, F. R., Kemper, E. L., Souza, G. M., Da Silva, A. M., Ferro, M. I., et al. (2003). Analysis and functional annotation of an expressed sequence tag collection for tropical crop sugarcane. *Genome Res.* 13, 2725–2735. doi:10.1101/gr.1532103

Vettore, A. L., Silva, F. R. D., Kemper, E. L., and Arruda, P. (2001). The libraries that made SUCEST. *Genet. Mol. Biol.* 24, 1–7. doi:10.1590/S1415-47572001000100002

Viikari, L., Vehmaanpera, J., and Koivula, A. (2012). Lignocellulosic ethanol: from science to industry. *Biomass Bioeng.* 46, 13–24. doi:10.1016/j.biombioe.2012.05.008

Vogel, J. (2008). Unique aspects of the grass cell wall. *Curr. Opin. Plant Biol.* 11, 301–307. doi:10.1016/j.pbi.2008.03.002

Waclawovsky, A. J., Sato, P. M., Lembke, C. G., Moore, P. H., and Souza, G. M. (2010). Sugarcane for bioenergy production: an assessment of yield and regulation of sucrose content. *Plant Biotechnol. J.* 8, 263–276. doi:10.1111/j.1467-7652.2009.00491.x

Wang, J., Roe, B., Macmil, S., Yu, Q., Murray, J. E., Tang, H., et al. (2010). Microcollinearity between autopolyploid sugarcane and diploid *Sorghum* genomes. *BMC Genomics* 11:261. doi:10.1186/1471-2164-11-261

Wang, X., and Paterson, A. (2013). "Comparative genomic analysis of C4 photosynthesis pathway evolution in grasses," in *Genomics of the Saccharinae*, ed. A. H. Paterson (New York, NY: Springer), 447–477.

Wei, X., Jackson, P. A., Hermann, S., Kilian, A., Heller-Uszynska, K., and Deomano, E. (2010). Simultaneously accounting for population structure, genotype by environment interaction, and spatial variation in marker-trait associations in sugarcane. *Genome* 53, 973–981. doi:10.1139/g10-050

Wei, X., Jackson, P. A., McIntyre, C. L., Aitken, K. S., and Croft, B. (2006). Associations between DNA markers and resistance to diseases in sugarcane and

effects of population substructure. *Theor. Appl. Genet.* 114, 155–164. doi:10.1007/s00122-006-0418-8

Whetten, R., and Ron, S. (1995). Lignin biosynthesis. *Plant Cell* 7, 1001–1013. doi:10.2307/3870053

Whittaker, A., and Botha, F. C. (1997). Carbon partitioning during sucrose accumulation in sugarcane internodal tissue. *Plant Physiol.* 115, 1651–1659.

Xu, Y., Lu, Y., Xie, C., Gao, S., Wan, J., and Prasanna, B. (2012). Whole-genome strategies for marker-assisted plant breeding. *Mol. Breed.* 29, 833–854. doi:10.1007/s11032-012-9699-6

Yan, L., Xu, C., Kang, Y., Gu, T., Wang, D., Zhao, S., et al. (2013). The heterologous expression in *Arabidopsis thaliana* of *Sorghum* transcription factor SbbHLH1 downregulates lignin synthesis. *J. Exp. Bot.* 64, 3021–3032. doi:10.1093/jxb/ert150

Yoshida, K., Sakamoto, S., Kawai, T., Kobayashi, Y., Sato, K., Ichinose, Y., et al. (2013). Engineering the *Oryza sativa* cell wall with rice NAC transcription factors regulating secondary wall formation. *Front. Plant Sci.* 4:383. doi:10.3389/fpls.2013.00383

Yuan, J. S., Tiller, K. H., Al-Ahmad, H., Stewart, N. R., and Stewart, C. N. Jr. (2008). Plants to power: bioenergy to fuel the future. *Trends Plant Sci.* 13, 421–429. doi:10.1016/j.tplants.2008.06.001

Zhang, J. S., Nagai, C., Yu, Q. Y., Pan, Y. B., Ayala-Silva, T., Schnell, R. J., et al. (2012). Genome size variation in three *Saccharum* species. *Euphytica* 185, 511–519. doi:10.1007/s10681-012-0664-6

Zhang, L., Xu, J., and Birch, R. G. (1999). Engineered detoxification confers resistance against a pathogenic bacterium. *Nat. Biotechnol.* 17, 1021–1024. doi:10.1038/13721

Zhang, S.-Z., Yang, B.-P., Feng, C.-L., Chen, R.-K., Luo, J.-P., Cai, W.-W., et al. (2006). Expression of the *Grifola frondosa* trehalose synthase gene and improvement of drought-tolerance in sugarcane (*Saccharum officinarum* L.). *J. Integr. Plant Biol.* 48, 453–459. doi:10.1111/j.1744-7909.2006.00246.x

Zhu, C., Gore, M., Buckler, E. S., and Yu, J. (2008). Status and prospects of association mapping in plants. *Plant Genome* 1, 5–20. doi:10.3835/plantgenome2008.02.0089

Conflict of Interest Statement: The authors declare that the research was conducted in the absence of any commercial or financial relationships that could be construed as a potential conflict of interest.

4

Development of a high throughput platform for screening glycoside hydrolases based on Oxime-NIMS

Kai Deng[1,2*], Joel M. Guenther[1,2], Jian Gao[3], Benjamin P. Bowen [3], Huu Tran[1,2], Vimalier Reyes-Ortiz[1,3], Xiaoliang Cheng[1,3], Noppadon Sathitsuksanoh[1,3†], Richard Heins[1,2], Taichi E. Takasuka[4], Lai F. Bergeman[4], Henrik Geertz-Hansen[1], Samuel Deutsch[3,5], Dominique Loqué[1,3], Kenneth L. Sale[1,2], Blake A. Simmons[1,2], Paul D. Adams[1,3,6], Anup K. Singh[1,2], Brian G. Fox[4,7] and Trent R. Northen[1,3*]

[1] US Department of Energy Joint BioEnergy Institute, Emeryville, CA, USA, [2] Sandia National Laboratories, Livermore, CA, USA, [3] Lawrence Berkeley National Laboratory, Berkeley, CA, USA, [4] US Department of Energy Great Lakes Bioenergy Research Center, University of Wisconsin, Madison, WI, USA, [5] Joint Genome Institute, Walnut Creek, CA, USA, [6] Department of Bioengineering, University of California Berkeley, Berkeley, CA, USA, [7] Department of Biochemistry, University of Wisconsin, Madison, WI, USA

Edited by:
Bo Hu,
University of Minnesota, USA

Reviewed by:
Chiranjeevi Thulluri,
Jawaharlal Nehru Technological
University Hyderabad, India
Jiwei Zhang,
University of Minnesota, USA

*Correspondence:
Kai Deng
kdeng@sandia.gov;
Trent R. Northen
trnorthen@lbl.gov

†Present address:
Noppadon Sathitsuksanoh,
University of Louisville, Louisville, KY,
USA

Cost-effective hydrolysis of biomass into sugars for biofuel production requires high-performance low-cost glycoside hydrolase (GH) cocktails that are active under demanding process conditions. Improving the performance of GH cocktails depends on knowledge of many critical parameters, including individual enzyme stabilities, optimal reaction conditions, kinetics, and specificity of reaction. With this information, rate- and/ or yield-limiting reactions can be potentially improved through substitution, synergistic complementation, or protein engineering. Given the wide range of substrates and methods used for GH characterization, it is difficult to compare results across a myriad of approaches to identify high performance and synergistic combinations of enzymes. Here, we describe a platform for systematic screening of GH activities using automatic biomass handling, bioconjugate chemistry, robotic liquid handling, and nanostructure-initiator mass spectrometry (NIMS). Twelve well-characterized substrates spanning the types of glycosidic linkages found in plant cell walls are included in the experimental workflow. To test the application of this platform and substrate panel, we studied the reactivity of three engineered cellulases and their synergy of combination across a range of reaction conditions and enzyme concentrations. We anticipate that large-scale screening using the standardized platform and substrates will generate critical datasets to enable direct comparison of enzyme activities for cocktail design.

Keywords: cellulase, NIMS, oxime bioconjugation, high throughput screening, enzyme assays

Abbreviations: AFEX-SG, ammonia fiber expansion pretreated switchgrass; CBM, carbohydrate binding module; CelE, broad specificity GH family 5 (GH5) domain from *C. thermocellum* Cthe_0797; DA-SG, diluted acid pretreated switchgrass; GH, glycoside hydrolase; IL-SG, ionic liquid pretreated switchgrass; NIMS, nanostructure-initiator mass spectrometry UT-SG, untreated switchgrass.

INTRODUCTION

Lignocellulosic biomass (Carroll and Somerville, 2009) is a renewable source of energy, capable of providing the nation with clean, renewable transportation fuels. To convert biomass into biofuels, one key step is enzymatic saccharification, which is known to be inefficient and expensive (Klein-Marcuschamer et al., 2012). Thus, low-cost, robust, high-performance enzymes or enzyme cocktails are needed to reduce the overall cost of biofuel production. There are many sources of inedible plant biomass that can serve as feedstocks for biofuel production, including agricultural wastes [corn stover, switchgrass (SG), wood trimming], municipal solid wastes, and emerging bioenergy crops. However, these various feedstocks have different glycan composition, bond linkages, and individual sugar contents, which complicates the development of cost-effective saccharification approaches. Moreover, the content and structure of glycans and lignin from the same biomass may respond in different ways to the prerequisite pretreatments (Li et al., 2010). These variations contribute to the observation that there is no universal enzyme or enzyme cocktail for all substrates and biofuels processes.

Given these process constraints, it is important to characterize saccharification enzymes against a wide range of biomass compositions, pretreatments, and processing conditions to generate data that can help enable customized optimal enzyme cocktails for saccharification (Banerjee et al., 2010; Walton et al., 2011). Given the large number of potential combinations of substrate-pretreatment reaction conditions, it is desirable to have a reliable high throughput enzyme assay methods and standardized panels of substrates and conditions.

Currently, several high throughput glycosyl hydrolyze (GH) assays are available to characterize enzyme activity. The majority of these are based on detection of colorimetric or fluorescent products. For example, the 2,4-dinitrosalicyclic acid (DNS) reducing sugar assay (Decker et al., 2009) can provide rapid analysis of large enzyme libraries. However, it is a non-specific method that can only provide total reducing sugar content. Fluorescence-based enzyme assays using surrogate substrates, e.g., 4-methylumbelliferyl-β-glucopyranoside (van Tilburgh et al., 1982) are also available for the evaluation of β glycosidases, are also available. Surrogate substrate methods require preparation of each substrate for each enzyme type, which may be laborious, and the reactivity may be biased versus the native substrate. Moreover, care is required to avoid interference from background absorbance/fluorescence.

To address some of these limitations, we developed a mass spectrometry-based enzyme assay platform called Nimzyme (Northen et al., 2008; Reindl et al., 2011; Greving et al., 2012). The first-generation Nimzyme platform was based on soluble model substrates, which were synthesized chemically (Deng et al., 2012). This approach provided the specificity, sensitivity, and high throughput needed to screen a GH1 library of 175 enzymes. Several high-performance enzymes were identified with desired bioprocessing conditions (70°C, 20% ionic liquid) (Heins et al., 2014). Another manuscript in this volume reports use of the first-generation Nimzyme platform for numerical analysis of reactions with cellotetraose-nanostructure-initiator mass spectrometry

(NIMS) (Deng et al., 2014). Results of this work demonstrate diagnostic behaviors of several classes of GH enzymes.

However, the use surrogate substrates does not allow interrogation of the myriad of bonding types present in plant biomass or the three-dimensional arrangements of these bonds present on the plant biomass, and so does not represent a fully realistic approach for the study of enzyme reactions.

To overcome this limitation, we developed an oxime-Nimzyme probe (Deng et al., 2014) to directly study enzyme hydrolysis of plant biomass. In this approach, soluble oligosaccharide products are captured in a stable oxime linkage and then delivered to the NIMS chip for subsequent analysis, while inclusion of ^{13}C-labeled monosaccharide standards (glucose and xylose) allows quantitation of the derivatized glycans. Besides the diagnostic detection of solubilized products, this next-generation Nimzyme approach also allows quantitative studies of the time-dependence of product formation, and dissection of individual apparent rates for reactions of individual enzymes with plant biomass.

To further advance application of the Nimzyme approach, here we report the development of a process platform and a panel of 12 diverse glycan substrates. These substrates were selected to represent the diversity of plant glycosidic bond linkages and the sugar compositions that are relevant to biofuel production. We use this panel of substrates to characterize three previously described enzymes from *Clostridium thermocellum* that have been engineered to perform outside of their natural cellulosomal location. These are fusions of the catalytic domains of CelA (gene locus Cthe_0269, GH8 endoglucanase family), CelR (gene locus Cthe_0578, GH9 cellotetrahydrolase), and CelE (gene locus Cthe_0797, GH5 endoglucanase family) to the CBM3a domain from scaffoldin protein CipA. The CBM3a domain used in this work comes from the CipA scaffoldin protein from *C. thermocellum*. CBM3a is a well-studied carbohydrate-binding module helps to promote binding of the enzyme onto the polysaccharide, thus plays a key role in promoting the efficient hydrolysis of cellulose (Yaniv et al., 2012). We envision that the standardization and automation of GH assays enabled by this approach will be valuable in providing large datasets of GH performance needed to select enzymes for improved biomass deconstruction.

MATERIALS AND METHODS

Materials

1,4-β-D-cellotetraose ~95% was purchased from Megazyme (Ireland) Cat. No. O-CTE100, Lot No. 130604; 1,4-β-D-xylotetraose ~95% was purchased from Megazyme (Ireland), Cat. No. O-XTE, Lot No. 120204; 1,4-β-D-mannotetraose ~95% was purchased from Megazyme (Ireland), Cat. No. O-MTE, Lot No. 111004; Arabinoxylan was purchased from Megazyme (Ireland), Cat. No. P-WAXYI, Lot No. 120801a; Carob galactomannan was purchased from Megazyme (Ireland) Cat. No. P-GALML, Lot No. 10501b; beechwood xylan was purchased from Sigma-Aldrich (St. Louis, MO, USA), Cat. No. X4252-100G, Lot No. BCBL2915V.

Switchgrass has emerged as a potential bioenergy crop because it is perennial, resource-efficient, and requires low-inputs for maintenance (Keshwani and Cheng, 2009). Untreated switchgrass (UT-SG), also named as Putnam SG, was obtained from Daniel Putnam at UC Davis; diluted acid pretreated switchgrass (DA-SG) was obtained by mixing Putnam SG with 1% sulfuric acid at 190°C for 0.5 min at NREL. Ammonium fiber expansion pretreated switchgrass (AFEX-SG) was prepared at Michigan State University. Ionic liquid pretreated switchgrass (IL-SG) was prepared at 140°C for 3 h with 15% solid loading in [EMIM][OAc] (1-ethyl-3-methylimidazolium acetate) at JBEI. All four SG substrates (IL-SG, AFEX-SG, DA-SG, and UT-SG) were milled by Thomas Wiley Mill (Model 3383 L1) for 20 min and sieved (passage through a 20-mesh sieve and retention by an 80-mesh sieve). The particle size was found in the range of 200–450 μm. Avicel PH-101 cellulose was requested from FMC Biopolymer (Philadelphia, PA, USA), Lot No. P112824596. Phosphoric acid swollen cellulose (PASC) was prepared from Avicel PH-101 cellulose (FMC Biopolymer) with 85% phosphoric acid as reported (Zhang et al., 2006).

Synthesis

The synthesis of O-alkyloxyamine fluorous tag has been reported previously (Deng et al., 2014).

Enzymes

The catalytic domains of the enzymes studied were obtained from the following gene loci: CelA (Cthe_0269); CelR (Cthe_0578); and CelE (Cthe_0797). Each of these catalytic domains was fused to the CBM3a domain from the scaffoldin CipA (Cthe_3077). Additional information on these genes can be found at Uniprot (Apweiler et al., 2011). All genes were prepared by PCR using C. thermocellum ATCC 27405 genomic DNA as template and cloned into the Escherichia coli expression vector pEC_CBM3a to create enzyme_CBM3a fusion proteins, e.g., CelAcc_CBM3a. The vector pEC_CBM3a is a hybrid of pEU_HSBC_CBM3a and pVP65K (Takasuka et al., 2014; Aceti et al., 2015) that yields fusion proteins having an N-terminal enzyme catalytic domain fused by an ~40 aa linker sequence to the CBM3a domain from Cthe_3077. Methods for PCR amplification, capture, and sequence verification of protein coding sequences, transformation into E. coli 10G competent cells (Lucigen, Middleton, WI, USA) for DNA manipulations and E. coli B834 for protein expression were as previously reported (Takasuka et al., 2014). Additional details of the properties and methods for use of pEU is described elsewhere (Aceti et al., 2015).

Enzyme Plate Construction

Three enzymes (CelAcc-CBM3a, CelRcc-CBM3a, and CelEcc-CBM3a) were chosen to construct an enzyme plate with varied enzyme concentrations (microgram per microliter) shown in **Table 1**. Each enzyme was present in triplicate as shown for CelAcc-CBM3a (columns A, B, C), CelRcc-CBM3a (columns D, E, F), and CelEcc-CBM3a (columns G, H, I). Some enzyme combinations were also included to investigate the potential synergy among these three enzymes as shown in columns J, K, L.

TABLE 1 | Enzyme plate construction.

Unit	CelAcc-CBM3a A	CelAcc-CBM3a B	CelAcc-CBM3a C	CelRcc-CBM3a D	CelRcc-CBM3a E	CelRcc-CBM3a F	CelEcc-CBM3a G	CelEcc-CBM3a H	CelEcc-CBM3a I	CelAcc-CBM3a + CelRcc-CBM3a J	CelRcc-CBM3a + CelEccCBM3a K	CelAcc-CBM3a + CelEcc-CBM3a L
μg/μL 1	0.01	0.01	0.01	0.01	0.01	0.01	0.01	0.01	0.01	0.005 + 0.005	0.005 + 0.005	0.005 + 0.005
μg/μL 2	0.05	0.05	0.05	0.05	0.05	0.05	0.05	0.05	0.05	0.025 + 0.025	0.025 + 0.025	0.025 + 0.025
μg/μL 3	0.1	0.1	0.1	0.1	0.1	0.1	0.1	0.1	0.1	0.05 + 0.05	0.05 + 0.05	0.05 + 0.05
μg/μL 4	0.25	0.25	0.25	0.25	0.25	0.25	0.25	0.25	0.25	0.125 + 0.125	0.125 + 0.125	0.125 + 0.125
μg/μL 5	0.5	0.5	0.5	0.5	0.5	0.5	0.5	0.5	0.5	0.25 + 0.25	0.25 + 0.25	0.25 + 0.25
μg/μL 6	1.25	1.25	1.25	1.25	1.25	1.25	1.25	1.25	1.25	0.625 + 0.625	0.625 + 0.625	0.625 + 0.625
μg/μL 7	2.5	2.5	2.5	2.5	2.5	2.5	2.5	2.5	2.5	1.25 + 1.25	1.25 + 1.25	1.25 + 1.25
μg/μL 8	5	5	5	5	5	5	5	5	5	2.5 + 2.5	2.5 + 2.5	2.5 + 2.5

Procedures for Handling Insoluble Substrates with Labman Solids-Handling Robot

To minimize static electricity, all plastic labware was prophylactically treated using a Tabletop Ionizing Transport System Model IT-7000 (Electrostatics Incorporated, 90610-03610, Harleysville, PA, USA). Commercially sourced substrates – Avicel PH-101, arabinoxylan, carob galactomannan, and beechwood xylan – were received as fine powders and aliquoted without further processing. The Labman (North Yorkshire, UK) solids-handling robot at JBEI was used to aliquot insoluble substrates (**Table 2**, numbers 4 through 12) from 2-mL Sarstedt vials (72.694.007) into 340-µL 96-well thermal cycler plates (plate: VWR 82006-636, holder: Axygen R-96-PCR-FY) with a target mass of 2 ± 0.25 mg per well. Feeder vibration power was set to 40% for all substrates and feeding durations were optimized for each to minimize overfeeding. Average feeding durations ranged from 3 to 5 h per 96-well plate depending on the substrate. After aliquoting, substrate plates were sealed using peelable heat seal (Agilent 24210-001) using an Agilent PlateLoc heat sealer set to 175°C for 2.5 s and stored at 4°C.

Procedures for Handling Insoluble Substrates with Biomek FX Robot

For all nine insoluble solid substrates, all the liquid handling for enzymatic hydrolysis reactions and bioconjugation chemistry were performed using a Biomek FX robot equipped with an AP96 multichannel pod (Beckman Coulter). For the enzymatic hydrolysis step, the Biomek FX transferred 180 µL of 50 mM phosphate buffer, pH 6.0, into 96-well PCR plates containing ~2 mg solid substrate that was previously aliquoted using the Labman robot (plate A). Then 20 µL of enzyme solution was transferred from the enzyme plate into plate A. After sealing the 96-well plate with PlateLoc peelable seal (2.5 s, 175°C), the plate was incubated at 60°C for 18 h in a shaker with shaker speed set to 200 rpm (HT INFORS). The recipe for a typical NIMS bioconjugation solution was a mixture of the following: (1) 1 mL of probe solution (150 mM in 1:1 (v/v) H$_2$O:MeOH); (2) 6 mL of 100 mM glycine buffer, pH 1.3; 3) 0.5 mL of 5 mM

^{13}C glucose aqueous solution; (4) 0.5 mL of 5 mM ^{13}C xylose aqueous solution; (5) 2 mL of acetonitrile; (6) 1 mL of methanol. Plate B is a 96-well PCR plate prepared by the Biomek robot containing with 22.2 µL of NIMS tagging solution in each well. After plate A (enzymatic reaction) was cooled to room temperature, a 4 µL aliquot from plate A was transferred into plate B by the Biomek robot. Plate B was sealed with PlateLoc peelable heat seal (2.5 s, 175°C) and left at room temperature for 16 h. Samples (12 µL) from plate B was transferred into the assay plate (Greiner bio-one, 384 well µ clear-plate, coc black, Lo base, 10 pcs/bag, Lot No. E11060DN; Cat. No. 788876, Made in Germany).

Procedures for Handling Soluble Substrates with Biomek FX Robot

For the three soluble substrates, 1,4-β-D-cellotetraose (G4), 1,4-β-D-xylotetraose (X4), and 1,4-β-D-mannotetraose (M4), 10 mM aqueous solutions were prepared as stock solutions. The Biomek robot was used to transfer 40 µL of 50 mM phosphate, pH 6.0, into each well of a 96-well PCR plate (plate A). Then 5 µL of soluble substrates [G4 (5 mM), X4 (5 mM), or M4 (5 mM)] was transferred to plate A. After that, Biomek transferred 5 µL of enzymes from the enzyme plate to plate A. After sealing the 96-well plate with PlateLoc peelable heat seal (2.5 s, 175°C), the whole plate was incubated at 60°C for 18 h with 200 rpm in a shaker (HT INFORS). After this step, the Biomek robot was used to perform all subsequent liquid-handling steps as indicated above for insoluble substrates.

Acoustic Printing of Sample Arrays

Samples from the 384-well Greiner plate (1 µL) were acoustically transferred by ATS Acoustic Liquid Dispenser (EDC Biosystems) onto a 2 × 2 inch NIMS chip. Individual reaction spots on the NIMS chip were ionized by a laser and products were detected by a time-of-flight mass spectrometer (TOF/TOF 5800 MALDI systems, AB Sciex).

Nanostructure-Initiator Mass Spectrometry Imaging

The 4800 imaging acquisition software was used in these experiments. Chips were loaded using a modified standard MALDI plate. Instrument was set with laser intensity at 2,550 and 15 shots per sub-spectrum. The detector voltage multiplier was set as 0.77. For imaging, the step size was set up at 50 µm for both x and y direction.

Mass Spectrometry Imaging Data Processing (Open MSI)

The imaging file was uploaded to openmsi.nersc.gov by Globus Connect Personal. The converted file was analyzed by a draggable points notebook written in Python. Signal intensities were identified for the ions of the tagging products. Enzyme activities were determined by measuring the concentration of glycan products using either [U]-^{13}C glucose or [U]-^{13}C xylose as an internal standard.

TABLE 2 | Standardized substrates list.

Soluble substrates	Insoluble solid substrates
(1) Cellotetraose	(4) Acid swollen cellulose (PASC)
	(5) Avicel PH-101 cellulose
	(6) Arabinoxylan
(2) Xylotetraose	(7) Carob galactomannan
	(8) Beechwood xylan
	(9) IL-switchgrass
(3) Mannotetraose	(10) AFEX-switchgrass
	(11) DA-switchgrass
	(12) UT-switchgrass

(1) Soluble substrates 1–3 can be dissolved in water and bypass the solid dispersion step by Labman. (2) Insoluble solid substrates 4–12 are being dispersed into a 96-well PCR plate by Labman, then Biomek was used for liquid handling.

RESULTS AND DISCUSSION

Substrate Panel

Table 2 lists some of the properties of the substrates included in the substrate panel. More detailed rationale for selection of individual substrates is described in the following experimental sections. The following general principles were used to assemble the substrate panel: (1) substrates should be readily available; (2) some substrates should have known structures (e.g., cellotetraose, xylotetraose, mannotetraose) so that enzyme specificity can be rapidly determined; (3) some substrates should be plant biomass substrates such as SG; and (4) examples of different pretreatments of the same biomass should be included. Based on evolving needs, other substrates can be added to the panel using similar principles.

After selection of these 12 substrates, we sourced large quantities of each substrate so as to permit detailed analytical characterizations and extended experimentation. Each of the substrates was acquired and then divided such that our two institutions (GLBRC and JBEI) each had large aliquots of each substrate.

Enzymes Selected for Platform Testing

To test this assay platform, we used catalytic domains from three *C. thermocellum* enzymes. In earlier studies (Deng et al., 2014; Takasuka et al., 2014), we showed that CelAcc, CelRcc, and CelEcc were able to release a variety of oligosaccharide products from IL- and AFEX-pretreated SG. As these enzymes are normally included in the cellulosome, we removed the dockerin domains and instead fused them to CBM3a (Yaniv et al., 2012). This engineered addition of CBM3a targets the catalytic domain to a polysaccharide surface. Our studies also showed that CelEcc was a multifunctional GH5 catalytic domain that was reacted with cellulose, xylan, and mannan (classified as CMX). This broad specificity of reaction provides opportunities to understand contributions of an identical active site to biomass hydrolysis. By contrast, CelRcc_CBM3a reacted only with cellulose (classified as C), while CelAcc_CBM3a reacted with cellulose and only weakly with xylan (classified as CX), and so provided more specific enzyme reactions.

Automated Enzyme Assay Platform

Efforts have been made to establish a simple automated workflow that enables high-throughput and that minimizes assay error. In a preparatory step (**Figure 1**), insoluble substrates (**Table 2**, 4 through 12) were aliquoted into 96-well thermal cycler plates for activity assays using a Labman solids-handling robot and stored at 4°C until use. Immediately before initiating the activity assays, enzymes were manually aliquoted into 96-well plates.

The individual steps of the workflow are as follows. In step 1, reactions were prepared using a Biomek FX liquid-handling robot to add enzyme to substrate. In step 2, reactions were incubated at 60°C for 18 h with shaking. In step 3, an aliquot from the reaction plate was combined with the oxime-NIMS tagging solution using the liquid-handling robot, after which the NIMS tag reacted with reducing sugar to form a stable oxime linkage during overnight incubation. In step 4, the solution containing NIMS-tagged glycans was transferred from tagging reactions onto a NIMS chip mass spectrometry surface by sequentially

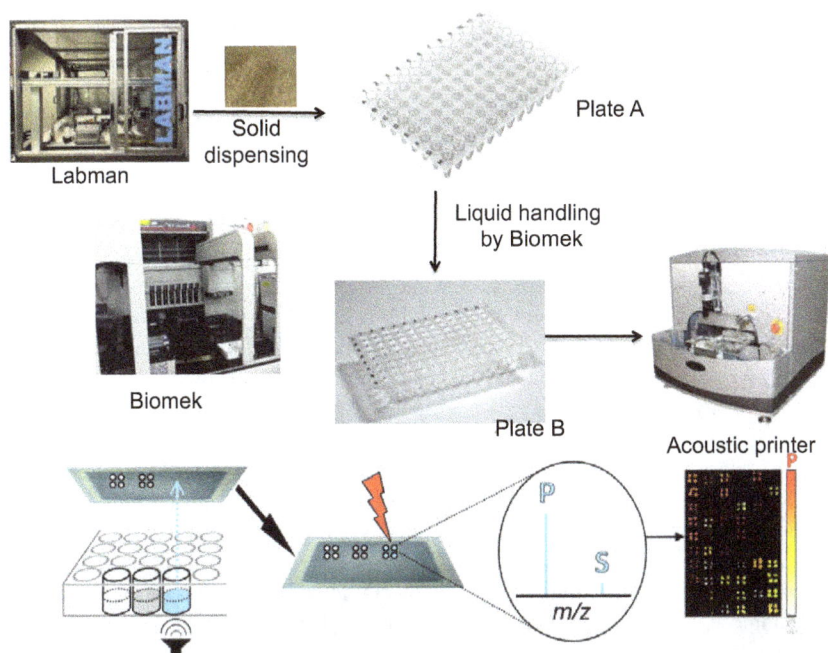

FIGURE 1 | Workflow of oxime-NIMS automation developed to study GHs. (1) Solid biomass was dispersed by Labman. (2) Liquid handling was performed by Biomek Automation Workstation, including setup of enzyme and oxime bioconjugation reactions. (3) Sample array on the NIMS chip was generated by an acoustic printer (ATS Acoustic Liquid Dispenser). (4) Mass spectrometry imaging (MSI) provided the readout of experimental assay results.

using (1) the liquid-handling robot to reformat aliquots from the reactions in 96-well plates into 384-well plates followed by (2) using an ATS Acoustic Liquid Dispenser to print 1-nL droplets onto the chip via non-contact dispensing. In step 5, an AB SCIEX TOF/TOF 5800 mass spectrometer was used to image the NIMS chip to collect data on the tagged glycans. In step 6, the data were processed using OpenMSI (a free web-based visualization, analysis and management package for mass spectrometry imaging (MSI) data) to quantify the tagged glycans in order to compute enzyme activities.

The different characteristics of the substrates introduced challenges for automated sample handling. Thus, while soluble samples can be dispensed using standard liquid handling (**Figure 1**), insoluble samples were more challenging and required application of solid dispensing using a Labman system. As shown in **Table 2**, only 3 out of the 12 substrates were soluble and could processed using simple liquid handling, whereas the nine other substrates were insoluble and required Labman to carry out relatively slow protocol to generate batches of substrate-filled 96-wells PCR plates that were then stored at −20°C freezer until required.

Reactions with Soluble Substrates

Cellotetraose (G4), xylotetraose (X4), and mannotetraose (M4) are purified oligosaccharides that can be used to study cellulase, xylanase, and mannase activities, respectively. These soluble substrates also permit simple liquid-handling approaches. However, it is important to recognize that some GH enzymes require longer oligosaccharide chains to show effective catalysis.

The reactions of G4, X4, and M4 were screened using a 96-well enzyme plate containing individual wells of CelAcc_CBM3a, CelRcc_CBM3a, and CelEcc_CBM3a, and mixtures of these three (**Table 1**). All three endoglucanases produced cellobiose as the major product from G4 under various concentrations (Tables S1–S3 in Supplementary Material). No products were observed at the lowest enzyme concentration tested (1 ng/µL) for all three enzymes. When the enzyme concentration was increased (5 ng/µL), CelEcc_CBM3a gave complete hydrolysis of the cellotetraose present into smaller oligosaccharides, while CelRcc_CBM3a hydrolyzed about 80% of G4 and CelAcc_CBM3a hydrolyzed about 13% of G4. For CelAcc_CBM3a, increasing the enzyme concentration above 0.01 µg/µL was needed to obtain complete hydrolysis. Thus, the concentration screening can give a preliminary assessment of the affinity of an enzyme for the oligosaccharide. In the hydrolysis of cellotetraose, the following effective concentrations of enzyme were determined: CelAcc_CBM3a (10 ng/µL), CelRcc_CBM3a (10 ng/µL), and CelEcc_CBM3a (5 ng/µL).

Xylotetraose (X4) is an oligosaccharide with β 1,4-linked xylose unit, and cellulases CelAcc_CBM3a and CelRcc_CBM3a were not reactive with xylotetraose. However, the multifunctional enzyme CelEcc_CBM3a was active with xylotetraose. Increasing the enzyme concentration increased oligomer and monomer products (cellotriose, cellobiose, and xylose, Table S4 in Supplementary Material). For CelEcc_CBM3a with xylotetraose, the minimum amount of enzyme needed for complete hydrolysis of X4 was 0.25 µg/µL. This is a 50-fold increase in the amount of enzyme needed to hydrolyze xylotetraose versus cellotetraose, suggesting higher affinity for the hexose oligosaccharide.

None of the enzymes studied reacted with mannotetraose (M4). Since CelAcc_CBM3a and CelRcc_CBM3a are cellulases, this was not surprising. However, since CelEcc_CBM3a has been shown to react with mannan and glucomannan, the lack of reaction with M4 was not expected. This may arise from the possibility that mannotetraose is not long enough to bind effectively in the catalytic channel of CelEcc_CBM3a.

Reactions with Avicel and PASC

Phosphoric acid swollen cellulose (Zhang et al., 2006) has become a very popular substrate to study cellulase activities because it has a relatively easily hydrolyzed amorphous habit (Sharrock, 1988; Wood, 1988; Wood and Bhat, 1988). Solid-state cross-polarization magic angle spinning ^{13}C NMR was used to determine the crystallinity (Park et al., 2010) of both Avicel and PASC used in this work. The NMR results demonstrated that Avicel had crystallinity of 53% while PASC had crystallinity of less than 5% (Figure S1 in Supplementary Material). Compared with the microcrystalline cellulose (Avicel), amorphous PASC has the advantages of practical solubility and increased accessibility to the cellulases. Therefore, it is more easily hydrolyzed by cellulases. As expected, all three cellulase enzymes produced multiple times more soluble hexose products from PASC than Avicel (4 times for CelEcc_CBM3a, 2.5 times for CelRcc_CBM3a, and 3 times more for CelAcc_CBM3a) (Tables S5 and S6 in Supplementary Material). For the hydrolysis of Avicel, CelRcc_CBM3a performed better than CelEcc_CBM3a and CelAcc_CBM3a. Interestingly, the enzyme combination of CelEcc_CBM3a and CelRcc_CBM3a worked better than the combination of either CelAcc_CBM3a and CelRcc_CBM3a or CelEcc_CBM3a and CelAcc_CBM3a (**Figure 2**). Since the amount of glycan products produced by the combination of CelEcc_CBM3a and CelRcc_CBM3a was greater than the scaled contributions of the individual enzymes, this combination is demonstrated to have a synergistic effect in reaction.

Reactions with Beechwood Xylan and Arabinoxylan

Beechwood xylan has a high xylose content (~84%) with a majority of β-1,4 linkages. Arabinoxylan consists of a mixture of arabinose and xylose in an approximate 40:60 ratio with β-1,4- and β-1,6-linkages. Our screening results showed that CelRcc_CBM3a was unable to hydrolyze either beechwood xylan or arabinoxylan, even at high enzyme loading. By contrast, CelAcc_CBM3a showed weak activity with beechwood xylan, and at the highest enzyme loading tested (50 mg/g xylan), about 20% of the xylan was hydrolyzed (Table S8 in Supplementary Material). Furthermore, CelEcc_CBM3a reacted well with beechwood xylan, and at the highest enzyme concentration tested (50 mg/g xylan), about 40% of the xylan was hydrolyzed into soluble pentose products. Both CelAcc_CBM3a and CelEcc_CBM3a had only weak activities with arabinoxylan, even at the highest enzyme loading tested (50 mg/g xylan).

Reactions with Galactomannan

Galactomannans are polysaccharides consisting of a mannose backbone with galactose side groups. Mannans are important constituent of hemicellulose in some plant biomass. (Malherbe

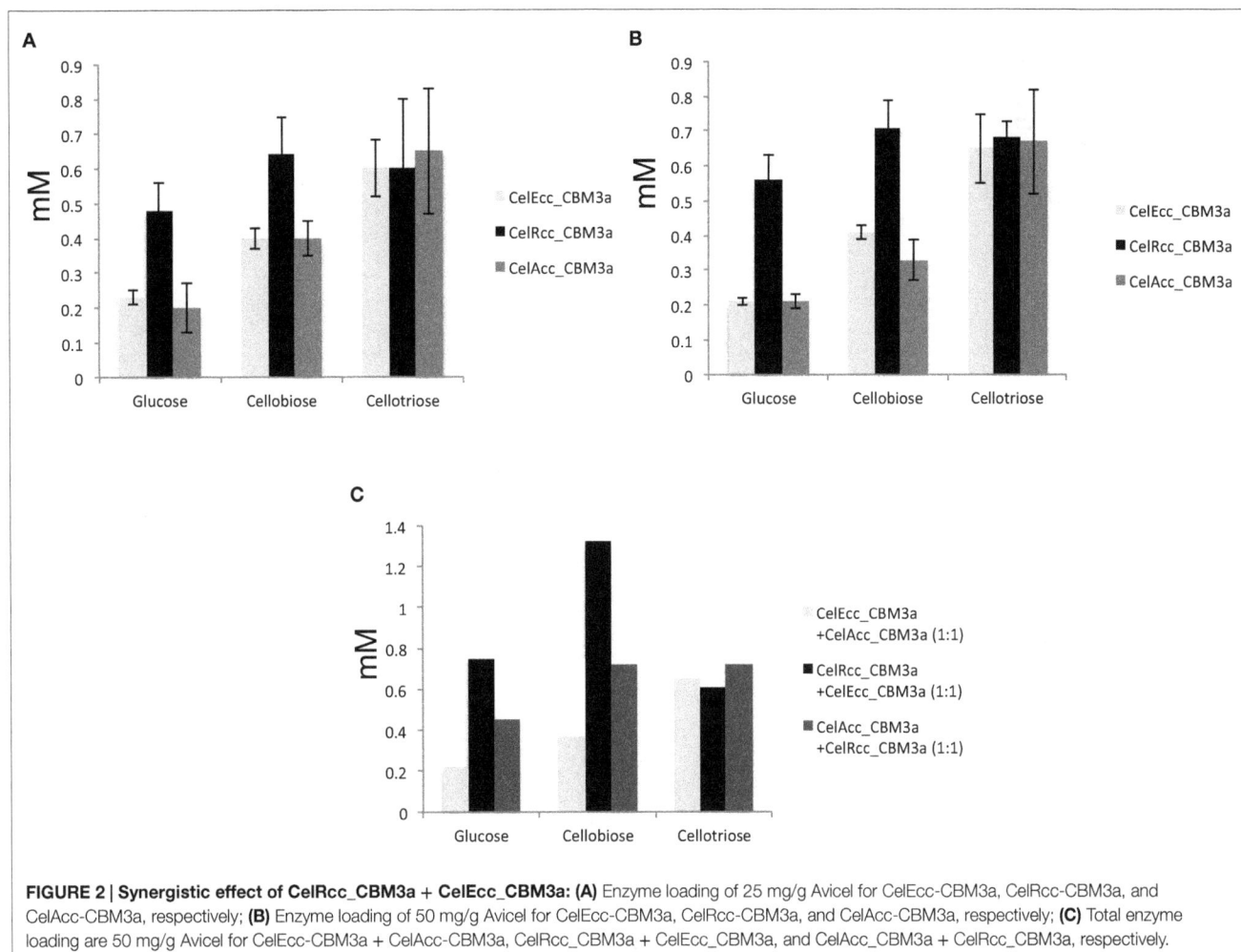

FIGURE 2 | Synergistic effect of CelRcc_CBM3a + CelEcc_CBM3a: (A) Enzyme loading of 25 mg/g Avicel for CelEcc-CBM3a, CelRcc-CBM3a, and CelAcc-CBM3a, respectively; **(B)** Enzyme loading of 50 mg/g Avicel for CelEcc-CBM3a, CelRcc-CBM3a, and CelAcc-CBM3a, respectively; **(C)** Total enzyme loading are 50 mg/g Avicel for CelEcc-CBM3a + CelAcc-CBM3a, CelRcc_CBM3a + CelEcc_CBM3a, and CelAcc-CBM3a + CelRcc_CBM3a, respectively.

et al., 2014) For example, softwoods contain 15–20% (w/w) mannans (Rodríguez-Gacio et al., 2012) and legume seeds can contain more than 30% of mannans (Buckeridge, 2010). For hydrolysis of these types of biomass into simple monosaccharides, it is important to find enzymes that can efficiently degrade mannans. Consequently, galactomannan was included in the substrate panel to test for mannanase activities. The screening results show that cellulases (CelAcc_CBM3a and CelRcc_CBM3a) could not deconstruct galactomannan at all. However, CelEcc_CBM3a (Table S9 in Supplementary Material) was able to hydrolyze galactomannan to hexose products (Fox et al., 2014).

It is interesting to compare the results of CelEcc_CBM3a reaction with galactomannan and its lack of reaction with mannotetraose. Besides the length of the oligosaccharide suggested above, it is also plausible that CelEcc_CBM3a may prefer reaction with the galactose-substituted mannans.

Reactions with Switchgrass

The structural features of SG, including surface area, crystallinity, the contents of cellulose, hemicellulose, and lignin vary considerably depending on pretreatment (Xu and Huang, 2014).

For example, cellulose is changed from cellulose I (untreated) to cellulose II after ionic liquid pretreatment (Cui et al., 2014). Other factors, like degree of delignification, hemicellulose solubilization (especially in diluted acid pretreatment), changes in porosity, and others affect enzyme accessibility so that different glycan products are produced for different pretreated SG. The compositional analysis of these four pretreated samples demonstrates these significant differences caused by pretreatment (Table S10 in Supplementary Material). These compositional differences imply the need for customized enzyme cocktails.

Four types of pretreated SG were included in the substrate panel to permit comparative studies of the consequences of pretreatment on enzymatic saccharification. These are UT-SG as control, IL-SG (Li et al., 2010), AFEX-SG (Bals et al., 2010), and DA-SG (Pu et al., 2013). This selection covers the three predominant pretreatment methods under investigation by the US-DOE funded Bioenergy Research Centers (Singh et al., 2015).

Screening results show that neither the three individual enzymes (CelAcc_CBM3a, CelRcc_CBM3a, and CelEcc_CBM3a) nor their combinations could deconstruct UT-SG. This is consistent with the substantial value of pretreatment before enzymatic saccharification.

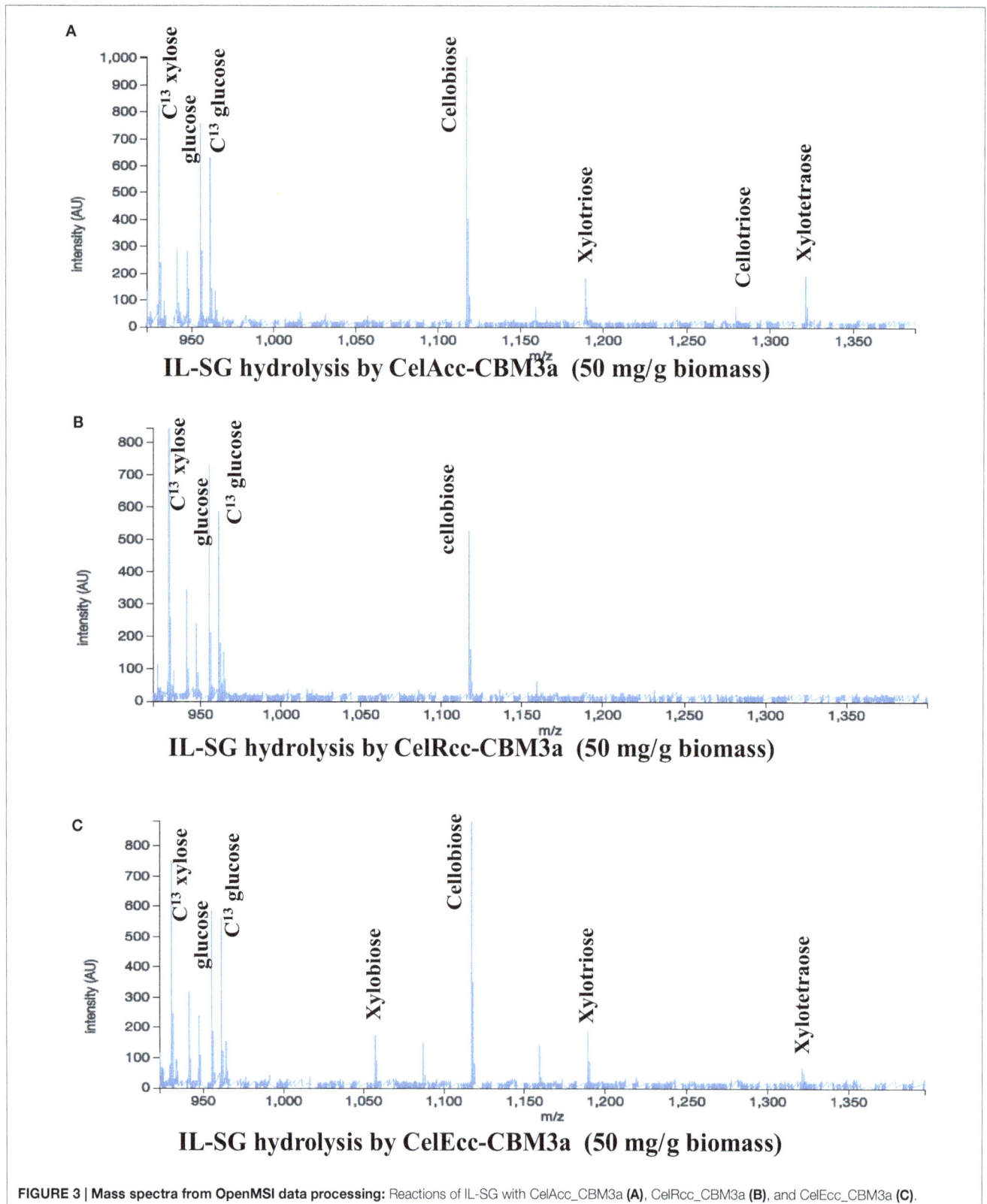

FIGURE 3 | Mass spectra from OpenMSI data processing: Reactions of IL-SG with CelAcc_CBM3a **(A)**, CelRcc_CBM3a **(B)**, and CelEcc_CBM3a **(C)**.

For DA-SG, all assays across the breadth of enzyme concentrations gave similar, low yields for both hexose and pentose products. The relatively low reactivity of these individual enzymes with crystalline cellulose is consistent with the low yield of hexose product. Presumably, the low amount of xylan remaining in DA-SG (4.58%) is not easily accessible to further enzyme hydrolysis. For example, although CelEcc_CBM3a reacts well with pure xylan and the hemicellulose fractions in AFEX-SG and IL-SG (see below), it did not react with the remaining xylan in DA-SG. Whether this is a result of depletion of reactive substructures or other aspects of the diluted acid pretreatment is not clear.

For reactions with AFEX-SG, both CelRcc_CBM3a and CelAcc_CBM3a produced little soluble glycan and no soluble pentose. By contrast, CelEcc_CBM3a performed much better, and yielded both hexose and pentose products. CelEcc_CBM3a was especially reactive with the hemicellulose portion of AFEX-SG, yielding about 2.5× more pentose products than in its reaction with IL-SG (Table S11 in Supplementary Material).

For IL-SG, all three enzymes (**Figure 3**) produce significant amount of hexose products (Table S12 in Supplementary Material), which can be attributed to the ability of the IL pretreatment to reduce the crystallinity of cellulose. CelAcc_CBM3a produced the most hexose products among these three enzymes, and under the highest enzyme loading of 50 mg/g biomass a conversion of 24% of the glycan was observed. A comparison of the reaction of CelEcc_CBM3a with either IL-SG or AFEX-SG under the same experimental conditions (enzyme loading of 50 mg/g biomass) showed that CelEcc_CBM3a produced 8× more of hexose products with IL-SG than AFEX-SG. This result from the automated platform matches the conclusion that cellulose in IL-SG is much easier to be accessed and digested by CelEcc_CBM3a obtained in earlier oxime-NIMS studies (Deng et al., 2014).

CONCLUSION

In this work, we described a panel of 12 substrates and automated platform for characterization of GH enzymes using the oxime-NIMS approach. Standardization is an important step toward more in-depth comparison of GH enzymes activities, both from natural environments and from engineered systems. These studies are consistent with our earlier assignment that CelEcc_CBM3a

is a multifunctional enzyme that has cellulase, mannanase, and hemicellulase activities. By contrast, CelAcc_CBM3a has cellulase and only weak hemicellulase while CelRcc-CBM3a only has cellulase activity. This platform automates the handling of both solid biomass and soluble substrates, the introduction of enzymes as individuals or combinations, and the recovery of products for high sensitivity and high-resolution mass spectral analysis. Simplex optimization of the ratios of natural or engineered enzyme combinations produced by systems biology approaches such as gene synthesis and robotic cell-free translation, as well as optimization of the pretreatment conditions can be readily undertaken using this platform.

AUTHOR CONTRIBUTIONS

KD, BF, and TN designed experiments. KD, JMG, JG, HT, VR-O, XC, NS, RH, TT, LB, HG-H, and SD carried out experimental work, KD, BB, BF, and TN analyzed results, and KD, JMG, BF, and TN prepared the manuscript. DL, KS, BS, and PA supervised the study. All authors read and approved the final manuscript.

ACKNOWLEDGMENTS

The DOE Joint BioEnergy Institute and DOE Great Lakes Bioenergy Research Center are supported by the US Department of Energy, Office of Science, Office of Biological and Environmental Research, through contract DE-AC02-05CH11231 and through contract DE-FC02-07ER64494, respectively. Sandia is a multiprogram laboratory operated by Sandia Corporation, a Lockheed Martin Company, for the United States Department of Energy's Nuclear Security Administration under contract DE-AC04-94AL85000.

REFERENCES

Aceti, D. J., Bingman, C. A., Wrobel, R. L., Frederick, R. O., Makino, S., Nichols, K. W., et al. (2015). Expression platforms for producing eukaryotic proteins: a comparison of E. coli cell-based and wheat germ cell-free synthesis, affinity and solubility tags, and cloning strategies. *J. Struct. Funct. Genomics* 16, 67–80. doi:10.1007/s10969-015-9198-1

Apweiler, R., Martin, M. J., O'Donovan, C., Magrane, M., Alam-Faruque, Y., Antunes, R., et al. (2011). Ongoing and future developments at the Universal Protein Resource. *Nucleic Acids Res.* 39, D214–D219. doi:10.1093/nar/gkq1020

Bals, B., Rogers, C., Jin, M. J., Balan, V., and Dale, B. (2010). Evaluation of ammonia fibre expansion (AFEX) pretreatment for enzymatic hydrolysis of switchgrass harvested in different seasons and locations. *Biotechnol. Biofuels* 3, 1. doi:10.1186/1754-6834-3-1

Banerjee, G., Car, S., Scott-Craig, J. S., Borrusch, M. S., and Walton, J. D. (2010). Rapid optimization of enzyme mixtures for deconstruction of diverse pretreatment/biomass feedstock combinations. *Biotechnol. Biofuels* 12, 22. doi:10.1186/1754-6834-3-22

Buckeridge, M. S. (2010). Seed cell wall storage polysaccharides: models to understand cell wall biosynthesis and degradation. *Plant Physiol.* 154, 1017–1023. doi:10.1104/pp.110.158642

Carroll, A., and Somerville, C. (2009). Cellulosic biofuels. *Annu. Rev. Plant Biol.* 60, 165–182. doi:10.1146/annurev.arplant.043008.092125

Cui, T., Li, J., Yan, Z., Yu, M., and Li, S. (2014). The correlation between the enzymatic saccharification and the multidimensional structure of cellulose changed by different pretreatments. *Biotechnol. Biofuels* 7, 134. doi:10.1186/s13068-014-0134-6

Decker, S. R., Brunecky, R., Tucker, M. P., Himmel, M. E., and Selig, M. J. (2009). High-throughput screening techniques for biomass conversion. *Bioenerg. Res.* 2, 179–192. doi:10.1007/s12155-009-9051-0

Deng, K., George, K. W., Reindl, W., Keasling, J. D., Adams, P. D., Lee, T. S., et al. (2012). Encoding substrates with mass tags to resolve stereospecific reactions using Nimzyme. *Rapid Commun. Mass Spectrom.* 26, 611–615. doi:10.1002/rcm.6134

Deng, K., Takasuka, T. E., Heins, R., Cheng, X., Bergeman, L. F., Shi, J., et al. (2014). Rapid kinetic characterization of glycosyl hydrolases based on oxime derivatization and nanostructure-initiator mass spectrometry (NIMS). *ACS Chem. Biol.* 9, 1470–1479. doi:10.1021/cb5000289

Fox, B. G., Takasuka, T., and Bianchetti, C. M. (2014). *Multifunctional Cellulase and Hemicellulase*. US Patent Application US 20140079683 A1 (Washington, DC: U.S. Patent and Trademark Office).

Greving, M., Cheng, X. L., Reindl, W., Bowen, B., Deng, K., Louie, K., et al. (2012). Acoustic deposition with NIMS as a high-throughput enzyme activity assay. *Anal. Bioanal. Chem.* 403, 707–711. doi:10.1007/s00216-012-5908-8

Heins, R., Cheng, X., Nath, S., Deng, K., Bowen, B. P., Chivian, D. C., et al. (2014). Hylogenomically guided identification of industrially relevant GH1 β-glucosidases through DNA synthesis and nanostructure-initiator mass spectrometry. *ACS Chem. Biol.* 9, 2082–2091. doi:10.1021/cb500244v

Keshwani, D. R., and Cheng, J. J. (2009). Switchgrass for bioethanol and other value-added applications: a review. *Bioresour. Technol.* 100, 1515–1523. doi:10.1016/j.biortech.2008.09.035

Klein-Marcuschamer, D., Oleskowicz-Popiel, P., Simmons, B. A., and Blanch, H. W. (2012). The challenge of enzyme cost in the production of lignocellulosic biofuels. *Biotechnol. Bioeng.* 109, 1083–1087. doi:10.1002/bit.24370

Li, C. L., Knierim, B., Manisseri, C., Arora, R., Scheller, H. V., Auer, M., et al. (2010). Comparison of dilute acid and ionic liquid pretreatment of switchgrass: biomass recalcitrance, delignification and enzymatic saccharification. *Bioresour. Technol.* 101, 4900–4906. doi:10.1016/j.biortech.2009.10.066

Malherbe, A. R., Rose, S. H., Vijoen-Bloom, M., and van Zyl, W. H. (2014). Expression and evaluation of enzymes required for the hydrolysis of galacto-mannan. *J. Ind. Microbiol. Biotechnol.* 41, 1201–1209.

Northen, T. R., Lee, J. C., Hoang, L., Raymond, J., Hwang, D. R., Yannone, S. M., et al. (2008). A nanostructure-initiator mass spectrometry-based enzyme activity assay. *Proc. Natl. Acad. Sci. U.S.A.* 105, 3678–3683. doi:10.1073/pnas.0712332105

Park, S., Baker, J. O., Himmel, M. E., Parilla, P. A., and Johnson, D. K. (2010). Cellulose crystallinity index: measurement techniques and their impact on interpreting cellulose performance. *Biotechnol. Biofuels* 3, 10. doi:10.1186/1754-6834-3-10

Pu, Y., Hu, F., Huang, F., Davison, B. H., and Ragauskas, A. J. (2013). Assessing the molecular structure basis for biomass recalcitrance during dilute acid and hydrothermal pretreatments. *Biotechnol. Biofuels* 6, 15.

Reindl, W., Deng, K., Gladden, J. M., Cheng, G., Wong, A., Singer, S. W., et al. (2011). Colloid-based multiplexed screening for plant biomass-degrading glycoside hydrolase activities in microbial communities. *Energy Environ. Sci.* 4, 2884–2893. doi:10.1039/c1ee01112j

Rodríguez-Gacio, M. C., Iglesias-Fernández, R., Carbonero, P., and Matilla, A. J. (2012). Softening-up mannan-rich cell walls. *J. Exp. Bot.* 63, 3976–3988. doi:10.1093/jxb/ers096

Sharrock, K. R. (1988). Cellulase assay methods: a review. *J. Biochem. Biophys. Methods* 17, 81–105. doi:10.1016/0165-022X(88)90040-1

Singh, S., Cheng, G., Sathitsuksanoh, N., Wu, D., Varanasi, P., George, A., et al. (2015). Comparison of different biomass pretreatment techniques and their impact on chemistry and structure. *Front. Energy Res.* 2:62. doi:10.3389/fenrg.2014.00062

Takasuka, T. E., Walker, J. A., Bergeman, L. F., Vander Meulen, K. A., Makino, S.-I., Elsen, N. L., et al. (2014). Cell-free translation of biofuels enzymes. *Methods Mol. Biol.* 1118, 71–95. doi:10.1007/978-1-62703-782-2_5

van Tilbeurgh, H., Claeyssens, M., and de Bruyne, C. K. (1982). The use of 4-methylumbelliferyl and other chromophoric glycosides in the study of cellulolytic enzymes. *FEBS Lett.* 149, 152–156. doi:10.1016/0014-5793(82)81092-2

Walton, J., Banerjee, G., and Car, S. (2011). GENPLAT: an automated platform for biomass enzyme discovery and cocktail optimization. *J. Vis. Exp.* 56, 3314. doi:10.3791/3314

Wood, T. M. (1988). Preparation of crystalline, amorphous and dyed cellulase substrate. *Meth. Enzymol.* 166, 19–45. doi:10.1016/0076-6879(88)60103-0

Wood, T. M., and Bhat, K. M. (1988). Methods for measuring cellulase activities. *Meth. Enzymol.* 160, 87–117. doi:10.1016/0076-6879(88)60109-1

Xu, Z., and Huang, F. (2014). Pretreatment methods for bioethanol production *Appl. Biochem. Biotechnol.* 174, 43–62. doi:10.1007/s12010-014-1015-y

Yaniv, O., Frolow, F., Levy-Assraf, M., Lamed, R., and Bayer, E. A. (2012). Interactions between family 3 carbohydrate binding modules (CBMs) and cellulosomal linker peptides. *Meth. Enzymol.* 510, 247–259. doi:10.1016/B978-0-12-415931-0.00013-6

Zhang, Y.-H., Cui, J., Lynd, L. R., and Kuang, L. R. (2006). A transition from cellulose swelling to cellulose dissolution by o-phosphoric acid: evidence from enzymatic hydrolysis and supramolecular structure. *Biomacromolecules* 7, 644–648. doi:10.1021/bm050799c

Conflict of Interest Statement: Kai Deng and Trent R. Northen are co-inventors on a patent application that covers the oxime-NIMS assay. Taichi E. Takasuka and Brian G. Fox are co-inventors on a patent application that covers use of multi-functional enzymes. The remaining authors have no conflict of interest to declare.

5

Pyrolysis of algal biomass obtained from high-rate algae ponds applied to wastewater treatment

Fernanda Vargas e Silva* and Luiz Olinto Monteggia

Institute of Hydraulic Research, Federal University of Rio Grande do Sul, Porto Alegre, Brazil

This work presents the results of the pyrolysis of algal biomass obtained from high-rate algae ponds treating sewage. The two high-rate algae ponds (HRAP) were built and operated at the São João Navegantes Wastewater Treatment Plant. The HRAP A was fed with raw sewage while the HRAP B was fed with effluent from an upflow anaerobic sludge blanket (UASB) reactor. The HRAP B provided higher productivity, presenting total solids concentration of 487.3 mg/l and chlorophyll a of 7735 mg/l. The algal productivity in the average depth was measured at 41.8 $g \cdot m^{-2}$ day^{-1} in pond A and at 47.1 $g \cdot m^{-2}$ day^{-1} in pond B. Algae obtained from the HRAP B were separated by the process of coagulation/flocculation and sedimentation. In the presence of alum, a separation efficiency in the range of 97% solid removal was obtained. After centrifugation the biomass was dried and comminuted. The biofuel production experiments were conducted via pyrolysis in a tubular quartz glass reactor which was inserted in a furnace for external heating. The tests were carried out in an inert nitrogen atmosphere at a flow rate of 60 ml/min. The system was operated at 400, 500, and 600°C in order to determine the influence of temperature on the obtained fractional yields. The studies showed that the pyrolysis product yield was influenced by temperature, with a maximum liquid phase (bio-oil and water) production rate of 44% at 500°C, 45% for char and around 11% for gas.

Edited by:
*Umakanta Jena,
Desert Research Institute, USA*

Reviewed by:
*Sandeep Kumar,
Old Dominion University, USA
Kaushlendra Singh,
West Virginia University, USA*

*Correspondence:
*Fernanda Vargas e Silva,
Instituto de Pesquisas Hidráulicas,
Bento Gonçalves 9500, Porto Alegre,
Rio Grande do Sul, Brazil
fervs@globo.com*

Keywords: high-rate algae ponds, pyrolysis, biofuels, wastewater treatment, bioremediaiton

Introduction

Biomass is considered worldwide as an important source of renewable energy, including electricity, automobile fuel, and as a source of heat for industrial equipment.

Cultures commonly used for energy production are sugarcane, corn, beans, beets, and many others. There are two main factors that define when a culture is appropriate for this process: good dry matter yield per unit of land (dry ton/ha), low area requirement for cultivation, and low costs of energy production from biomass (Dermibas et al., 2009).

However, some research has condemned the use of biofuels, associating its production with possible high food prices. Algae, among the aquatic biomass feedstocks, are considered one of the most promising sources of biofuels due to their unique characteristics. They can accumulate lipids that can be converted into biofuels, present fast proliferation, have the ability to sequester CO_2 from the atmosphere for growth and do not require agricultural land or freshwater for growth or higher water consumption, and also the whole plant matter can be used in converting biofuels processes

(Dismukes et al., 2008; Brennan and Owende, 2010; Jena and Das, 2011; Pate et al., 2011; Yanik et al., 2013; Zhou et al., 2014; Hognon et al., 2015).

Wastewater treatment associated with algae cultivation can offer an alternative way for sustainable renewable biofuels, since the large amount of freshwater needed for algae cultivation can be saved, becoming an environmentally friendly process (Zhou et al., 2014).

In a sewage treatment system, high-rate ponds are characterized by having high algal biomass generation which is an undesirable byproduct for the environment. Its presence in water bodies decreases water quality.

High-rate algae ponds are raceway-type ponds, in which water, algae, and nutrients are continually mixed. A paddle wheel generates a mean horizontal water velocity of approximately 0.15–0.3 m/s. This movement is necessary to avoid sedimentation and stratification. The maximum biomass production (mostly algae) is achieved through better use of lighting per volume. This is ensured by the low depth of the ponds and the constant movement of biomass through mechanical mixing (Nascimento, 2001; Chisti, 2007).

Usually the algal biomass productivity is determined by the measurement of solids found in the ponds.

The algal biomass production costs are mainly covered by the costs of treatment when using wastewater high-rate ponds resulting in lower environmental impacts in terms of water, energy, and fertilizer needs.

The biomass in high-rate algae ponds assimilates the nutrients needed for its growth and becomes responsible for the removal of nutrients from wastewater. This has the advantage of controlling pollution of water resources which contributes to the sustainable use of this technology on an industrial scale (Park et al., 2011; Passos et al., 2013).

The biomass separation process requires an increase in algal suspension concentration typically from 0.02 to 0.06% total suspended solids (TSS) to approximately 2 to 7% solids, which may be higher depending on the target process objective (Uduman et al., 2010).

The algal cells have reduced size, sometimes <30 μm and their density is similar to water with a low sedimentation rate, so to be successful in separation, it is necessary to aggregate the cells. Generally, the process comprises of two steps: the first involving destabilization of algal cells using coagulation followed by sedimentation or flotation. In the second step of the process, it is necessary to increase the biomass content, which is often done by filtration, centrifugation, or thermal processes (Molina Grima et al., 2003; Granados et al., 2012; Cai et al., 2013; Udom et al., 2013).

The algae cell has a negative surface charge, which prevents aggregation. This charge may be reduced or neutralized by the addition of flocculants or multivalent cations, such as cationic polymers that change the zeta potential, which is a measure of particle stability, reducing the repulsive forces. So the action of the attractive Van der Waals forces allows algae agglutination (Wessler et al., 2003; Granados et al., 2012). Salts used for this purpose should be non-toxic, low cost, and have high effectiveness at low concentrations (Molina Grima et al., 2003).

Another advantage of using coagulation/flocculation process is nutrient removal. The presence of nutrients in wastewater,

particularly nitrogen and phosphorus, is a serious environmental problem and is receiving increasing attention. Nitrogen in the form of ammonia can be volatilized and cause air pollution. Phosphorus can permeate into the soil and cause damage to the underground water (Chen et al., 2012). When there are excessive levels of nutrients in the wastewater, they cause eutrophication of water sources, possibly damaging the ecosystem (Cai et al., 2013).

The algal biomass, after thickening, may reach 5–15% solid content, and, being perishable, it must be processed as soon as possible. Essential processes such as thickening and drying usually involve high operational costs. Thus, these steps are considered determining factors regarding the economical feasibility analysis of the overall process (Brennan and Owende, 2010; Uduman et al., 2010). The methods commonly used for thickening biomass are centrifugation and filtration followed by different drying techniques, such as natural, oven, spray, and fluidized bed drying.

There are three basic components in algae biomass: proteins, carbohydrates, and lipids. These oils can then be extracted and converted in to biofuels (Um and Kim, 2009).

The pyrolysis process appears to be an excellent alternative for energy conversion, it presents the advantage of using different sources of organic matter, not being limited by the lipid content, as with biodiesel production processes. The pyrolysis process is based on decomposition of organic compounds present in the total biomass under a controlled environment in the absence of oxygen and atmospheric pressure, resulting in different phases: liquid (bio-oil), gas, and solid (char). It is an endothermic reaction that occurs at a temperature of 300–700°C depending on the characteristics of the material to be pyrolyzed (Martini, 2009; Hognon et al., 2015).

Biomass pyrolysis is considered a renewable process, because biomass is turned in several gases when pyrolyzed. Carbon dioxide, one of the gases formed, is absorbed by the algae for its growth, making the process self-sustainable with no serious contribution to greenhouse effect. The relative yield of each phase generated in the process depends on operating parameters (temperature, heating rate, residence time, and flow rate of inert gas), properties of the biomass (the particle size as well as its moisture), and type of pyrolysis used (slow, fast, or flash pyrolysis) (Balat et al., 2009; Martini, 2009; Akhtar and Amin, 2012; Yanik et al., 2013; Hognon et al., 2015).

In order to obtain high yields of aqueous products, fast pyrolysis is normally used, which is characterized by higher heating rates (1000°C/min) and lower residence times of volatiles (10–20 s). In order to favor solid char formation, slow pyrolysis process with lower heating rates (5–80°C/min) and longer residence times (5–30 min) must be used (Van de Velden et al., 2010; Jena and Das, 2011; Yanik et al., 2013).

The bio-oil generated by biomass pyrolysis is generally cleaner than that from fossil fuels, due to its lower nitrogen and sulfur content. The biomass vaporizes, passes through a process of cracking and condensation, producing a dark brown liquid, consisting of a complex mixture of many different hydrocarbons. This process is most successful in fluidized bed reactors due to high heating rates, rapid devolatilization and easy control (Doshi et al., 2005; Martini, 2009).

Materials and Methods

Biomass Production

Two high-rate algae ponds were constructed in the IPH/UFRGS experimental wastewater treatment unit, at São João Navegantes Wastewater Treatment Plant, This plant is responsible for handling the sewage of the north area of Porto Alegre/RS.

The ponds were operated in closed circuit with the following dimensions: overall height: 0.9 m, length of the straight sections: 30 m, width: 5 m (at the upper edge of the slope) and surface area 320 m², as can be seen in **Figure 1**.

The high-rate algae ponds were operated under two feeding conditions: pond A was fed with raw sewage after pretreatment (screening and grit removal) and pond B was fed with effluent from an upflow anaerobic sludge blanket (UASB) reactor. In order to maximize the process of biomass production, the operating parameters of the ponds were useful depth (Hu): 0.3 m, longitudinal flow speed: 0.3 m/s, and hydraulic detention time (HDT): 3 days.

The pond samples were collected in 20 l plastic containers, directly from the body of the ponds, to provide enough biomass for the pyrolysis experiments.

In order to determine algae biomass productivity, total solids, turbidity, and chlorophyll a were measured weekly.

All experiments to determine these parameters were carried out according to Standard Methods for the Examination of Water and Wastewater [American Public Health Association (APHA) and Awwa (2005)].

Algae Separation and Thickening

Experiments of coagulation/flocculation were performed using the effluent from pond B, which showed better performance in terms of algal biomass production.

The equipment used was VELP Jar Test model F.6/S, composed of 6 jars of 2000 ml each, with agitation and controlled independently.

To evaluate the separation process and the removal of nutrients two coagulants were used, Aluminum Sulfate and Ferric Chloride and two flocculants, Sulfloc 1001 and Tanfloc SL. Their concentration ranges are shown in **Table 1**.

After the separation and removal of all the supernatant from the jars, the algae sludge was submitted to centrifugation for 20 min at 2500 rpm to obtain a sample of about 15–20% of dried solids. After

centrifugation, the biomass was dried at 105°C. Finally, the dried algae were ground in a mortar and stored separately according to the reagent used in the separation process.

Nutrient Removal

Experiments were performed to determine the concentration of nitrogen and phosphorus in effluent ponds before and after coagulation/flocculation. Thus, it was possible to determine the effect of algae upon the separation in the removal of nutrients.

Biomass Pyrolysis

The experiments obtaining biofuel via biomass pyrolysis were performed in a tubular quartz reactor, with the dimensions described in **Figure 2**.

The experiments were run in batches, to allow solid char removal. The process flow used in this work, presented in **Figure 3**, was based on Zhang et al. (2011).

In the process, the inert atmosphere was generated by nitrogen gas (1) and the heating process was provided by an external furnace (2). The condensation was performed in (3), where two condensers in series were immersed in an ice bath. The exit of non-condensable gases was in (4).

The pyrolysis reactor was fed manually with 7 g of dried and ground biomass obtained from the previous step of the process. The biomass was inserted in the reactor using an aluminum foil capsule. After it has been charged, the reactor was closed and the inert atmosphere was provided by a 0.06 l/min nitrogen gas flow.

The pyrolysis runs were started by placing the reactor in a programable tubular furnace, with a heating rate of 20°C/min. All the runs were made in two steps: heating the sample and an isothermal reaction step, maintaining the desired temperature (400, 500, and 600°C) for 60 min.

The vapors generated passed through two condensers in series, immersed in ice baths maintained at a temperature of about 0°C. At the end of each experiment, the aqueous phase generated by condensation was collected, combined, weighed, and stored. The non-condensable gases were measured by difference. Following the 60 min reaction time, the system was turned off and cooled to room temperature.

After reaching room temperature, the reactor was opened and the solid fraction (char) was collected, weighed, and stored. The

FIGURE 1 | High-rate algae ponds.

TABLE 1 | Concentration range used.

Product	FeCl₃ 10%	Al₂(SO₄)₃ 10%	Sulfloc 20%	Tanfloc 10%
Concentration range (mg/l)	200–300	100–150	250–300	50–100

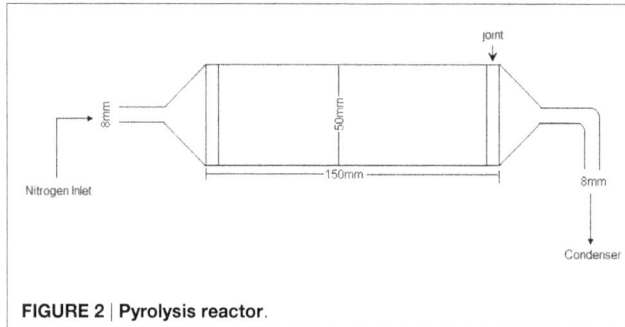

FIGURE 2 | Pyrolysis reactor.

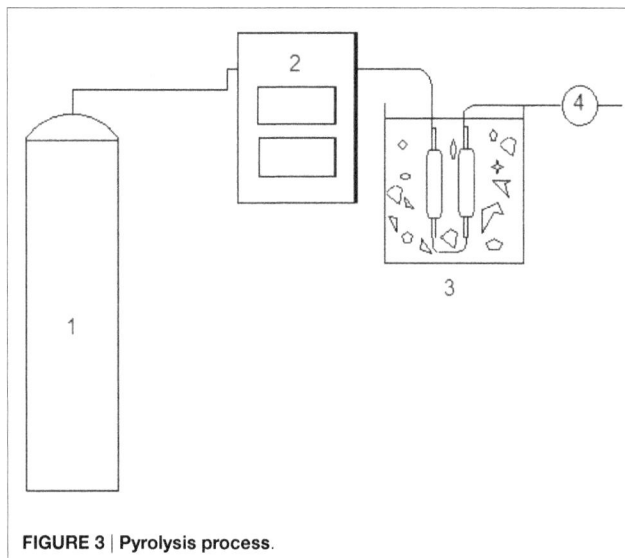

FIGURE 3 | Pyrolysis process.

reactor final mass was also determined, in order to measure the losses by wall adhesion.

The pyrolysis evaluation was performed through yields measurement. Each fraction was determined from the ratio of the weight of respective fraction to initial weight of biomass, expressed as percentage yield, according to Eq. (1).

$$\text{Yield (\%)} = \frac{\text{Fraction mass obtained after pyrolysis}}{\text{Initial algae biomass}} \times 100\% \quad (1)$$

Results and Discussions

Biomass Production

The results of solids, turbidity, chlorophyll a and productivity are shown in **Table 2**, comparing the performance of both biomass production ponds.

In the experiments, we considered the concentration of solids present in effluents and turbidity caused only by the presence of algae. From **Table 2**, it can be noted that pond B in all evaluation parameters showed higher values than those obtained from pond A. Such behavior is explained by the fact that effluent from UASB

TABLE 2 | High-rate ponds performance.

Analysis	Pond A	SD	Pond B	SD
Total solids (mg/l)	433.2	59.2	487.3	56.1
Turbidity (NTU)	41.9	8.9	63.3	13.4
Chlorophyll a (mg/l)	2338	NA	7735	NA
Productivity (g·m⁻² day⁻¹)	41.8	NA	47.1	NA

NA, not applicable.

TABLE 3 | Algae biomass productivity.

Authors	System	Biomass productivity (g·m⁻² day⁻¹)
Nascimento (2001)	HRAP	21.8
Riaño et al. (2012)	Photobioreactor	1.54
Sturm and Lamer (2011)	Open ponds	12
Terigar and Theegala (2014)	Open tanks	43.4

TABLE 4 | Crops productivity [adapted from Trzeciak et al. (2008)].

Crops	Harvest (month/ year)	Biomass productivity (g·m⁻² day⁻¹)
Cotton	3	0.38
Peanut	3	0.55
Canola	3	0.60
Sunflowers	3	0.55
Dendê (*Elaeis guineensis*)	12	6.84
Mamona (*Ricinus communis L.*)	3	0.41

reactor provided low solid concentration, which facilitated higher solar irradiation in the body of the pond, an essential factor for biomass growth. Thus, the effluent selected for tests of separation, thickening, and the tests for obtaining biofuels was collected from the pond B. **Table 3** shows a comparison among biomass productivity obtained in this work and others presented in the literature.

Table 4 shows the comparison between the productivity of crops commonly used in the biofuels production.

As we can see from both tables, high-rate algae ponds can be a competitive source of biomass, with higher productivities and without need of arable land and fresh water. This system high-rate algae pond (HRAP) presents no seasonality and the biomass can be harvested all year, without competition with food crops.

Algae Separation, Thickening, and Nutrient Removal

The results of algae separation are shown in **Table 5**, based on separation efficiency related to the chemical dosage used. This table also shows the evaluation of nutrient removal for each product.

Thus, according to the results shown in **Table 5**, the biomass separated with aluminum sulfate, which was selected as the most convenient chemical due to lower dosage requirement, showed better separation and nutrient removal. The biomass was dried and crushed to be used in the pyrolysis experiments. The efficiency of N and P removal were similar when using Sulfloc 20%, but the dosage required was higher than with sulfate.

Biomass Pyrolysis

The influence of temperature (400, 500, and 600°C) on pyrolysis results are shown in **Figure 4**. According to the results, the

TABLE 5 | Removal obtained and dosage used.

Product	Ferric chloride 10%	Al sulfate 10%	Sulfloc 20%	Tanfloc 10%
Maximum separation (%)	88.4	97.9	94.5	97.5
Concentration (mg/l)	300	150	290	100
P Removal (%)	100	100	100	37.9
N Removal (%)	–	5.5	5.5	–

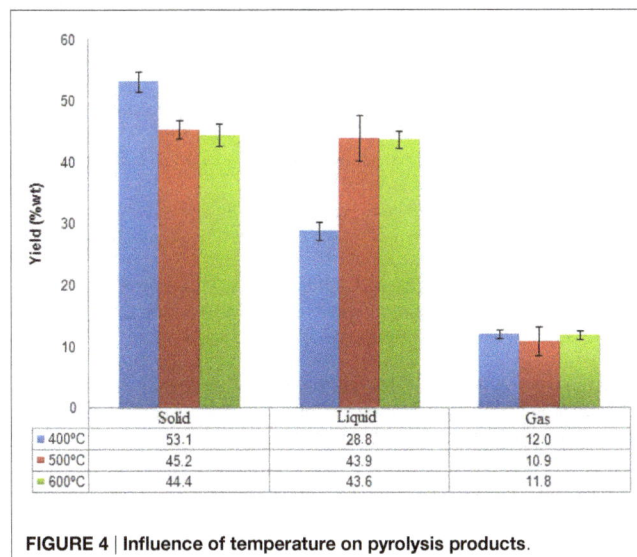

	Solid	Liquid	Gas
400°C	53.1	28.8	12.0
500°C	45.2	43.9	10.9
600°C	44.4	43.6	11.8

FIGURE 4 | Influence of temperature on pyrolysis products.

temperature of 400°C favors solid phase formation, with an average yield of 53.1%. The aqueous and gaseous phases obtained average yields of 28.8 and 12%, respectively. At 500°C, the yields for solid and aqueous phases were similar, however aqueous phase formation was slightly higher, composed of bio-oil and water. The average yield was 43.9 for solid phase 45.2 for the aqueous phase and 10.9% in the gas phase. At 600°C we can see similar yields between solid and liquid formation, 44.4 and 43.6%, respectively. The average for gas formation at this temperature was 11.8%.

The liquid phase, comprising of bio-oil and water, has a reddish brown color, with a strong and distinctive smoky smell, which confirms the information in the literature about products obtained in the pyrolysis (Jena and Das, 2011; Yanik et al., 2013).

As described in the literature, temperature plays an important role on the yield of the fractions obtained in the pyrolysis process. Studies show that temperatures between 450 and 550°C maximize the yield of bio-oil and, and at very high temperatures, secondary reactions of the volatiles may occur, thus decreasing the yield of the liquid phase, which can be seen in **Figure 4**; at 600°C, a small decrease in the aqueous phase yield was observed (Yanik et al., 2013).

For related data, we use an ANCOVA analysis (Analysis of Covariance) with a fixed factor (oven temperature) and a covariate (initial mass of algae) to identify differences in the char mass production. Five replicates were performed for each factor and the software used was SPSS version 18.

The data do not present heteroscedasticity, using the Levene test (p-value of 0.235), the tested factor was significant at a p-value of 0.01. So we went to the *post hoc* analysis, which showed a significant difference between the means of groups, the 400°C group is different from other groups and the 500°C and 600°C are not statistically different from each other.

Conclusion

In this work, the association of wastewater treatment and biofuel production through pyrolysis of algal biomass obtained in high-rate algae ponds was studied. The algal productivity, at the average depth was measured as 41.8 g·m^{-2} day^{-1} for pond A and as 47.1 g·m^{-2} day^{-1} for pond B. The algae were pyrolyzed in a tubular furnace system with external heating at different temperatures. Studies have shown that the pyrolysis process is efficient and the fractions yields are greatly influenced by temperature. Operating under mild conditions, it was possible to obtain maximum yields of 45% at 500°C for aqueous phase (bio-oil and water), 44% for char, and about 11% for gas. As we can see, through this process, it is possible to offer a promising alternative for environmental pollution control with potential economic return.

Acknowledgments

The authors acknowledge CNPq and CAPES for the financial support to this project.

References

Akhtar, J., and Amin, N. S. (2012). A review on operating parameters for optimum liquid oil yield in biomass pyrolysis. *Renew. Sustain. Energ. Rev.* 16, 5101–5109. doi:10.1016/j.rser.2012.05.033

American Public Health Association (APHA) and Awwa, W. E. F. (2005). *Standard Methods for the Examination of Water and Waste Water*, 21st Edn. American Public Health Association, 4–108; 4–147.

Balat, M., Balat, M., Kirtay, E., and Balat, H. (2009). Main routes for thermo-conversion of biomass into fuels and chemicals. Part1: pyrolysis systems. *Energy Convers. Manag.* 50, 3147–3157. doi:10.1016/j.enconman.2009.08.014

Brennan, L., and Owende, P. (2010). Biofuels from microalgae – a review of technologies for production, processing and extractions of biofuels and co-products. *Renew. Sustain. Energ. Rev.* 14, 557–577. doi:10.1016/j.rser.2009.10.009

Cai, T., Park, S. Y., and Li, Y. (2013). Nutrient recovery from wastewater streams by microalgae: status and prospects. *Renew. Sustain. Energ. Rev.* 19, 360–369. doi:10.1016/j.rser.2012.11.030

Chen, R., Li, R., Deitz, L., Liu, Y., Stevenson, R. J., and Liao, W. (2012). Freshwater cultivation with animal waste for nutrient removal and biomass production. *Biomass Bioenergy* 39, 128–138. doi:10.1016/j.biombioe.2011.12.045

Dermibas, M. F., Balat, M., and Balat, H. (2009). Potential contribution of biomass to the sustainable energy development. *Energy Convers. Manag.* 50, 1746–1760. doi:10.1016/j.enconman.2009.03.013

Dismukes, G. C., Carrieri, D., Bennette, N., Ananyev, G. M., and Posewitz, M. C. (2008). Aquatic phototrophs: efficient alternatives to land-based crops for biofuels. *Curr. Opin. Biotechnol.* 19, 235–240. doi:10.1016/j.copbio.2008.05.007

Doshi, V. A., Vuthaluru, H. B., and Bastow, T. (2005). Investigations into the control of odor and viscosity of biomass oil derived from pyrolysis of sewage sludge. *Fuel Process. Technol.* 86, 885–897. doi:10.1016/j.fuproc.2004.10.001

Granados, M. R., Acién, F. G., Gómez, C., Fernandez-Sevilla, J. M., and Molina Grima, E. (2012). Evaluation of flocculants for the recovery of freshwater microalgae. *Bioresour. Technol.* 118, 102–110. doi:10.1016/j.biortech.2012.05.018

Hognon, C., Delrue, F., Texier, J., Gateau, M., Thiery, S., Miller, S., et al. (2015). Comparison of pyrolysis and hydrothermal liquefaction of *Chlamydomonas reinhardti*. Growth studies on the recovered hydrothermal aqueous phase. *Biomass Bioenergy* 73, 23–31. doi:10.1016/j.biombioe.2014.11.025

Jena, U., and Das, K. C. (2011). Comparative evaluation of thermochemical liquefaction and pyrolysis for bio-oil production from microalgae. *Energy Fuels* 25, 5472–5482. doi:10.1021/ef201373m

Martini, P. R. R. (2009). *Conversão Pirolítica de Bagaço Residual da Indústria de Suco de Laranja e Caracterização Química dos Produtos*. Dissertação de Mestrado, PPGQ; Universidade Federal de Santa Maria, Santa Maria.

Molina Grima, E., Belarbi, E. H., Acién Fernández, F. G., Robles Medina, A., and Chisti, Y. (2003). Recovery of microalgal biomass and metabolites: process options and economics. *Biotechnol. Adv.* 20, 491–515. doi:10.1016/S0734-9750(02)00050-2

Nascimento, J. R. S. (2001). *Lagoas de Alta Taxa de Produção de Algas Para Pós-Tratamento de Efluentes de Reatores Anaeróbios*. Dissertação de Mestrado, Instituto de Pesquisas Hidráulicas, UFRGS, Porto Alegre.

Park, J. B. K., Craggs, R. J., and Shilton, A. N. (2011). Wastewater treatment high rate algal ponds for biofuel production. *Bioresour. Technol.*, 102, p. 35–42. doi:10.1016/j.biortech.2010.06.158

Passos, F., Solé, M., García, J., and Ferrer, I. (2013). Biogas production from microalgae grown in wastewater: effect on microwave pretreatment. *Appl. Energy* 108, 168–175. doi:10.1016/j.watres.2013.10.013

Pate, R., Klise, G., and Wu, B. (2011). Resource demand implications for US algae biofuels production scale-up. *Appl. Energy* 88, 3377–3388. doi:10.1016/j.apenergy.2011.04.023

Riaño, B., Hérnandez, D., and Garcia-González, M. C. (2012). Microalgal-based systems for wastewater treatment: effect of applied organic and nutrient loading rate on biomass composition. *Ecol. Eng.* 49, 112–117. doi:10.1016/j.ecoleng.2012.08.021

Sturm, B. S. M., and Lamer, S. L. (2011). An energy evaluation of coupling nutrient removal from wastewater with algal biomass production. *Appl. Energy* 88, 3499–3506. doi:10.1016/j.apenergy.2010.12.056

Terigar, B. C., and Theegala, C. S. (2014). Investigating the interdependence between cell density, biomass productivity, and lipid productivity to maximize biofuel feedstock production from outdoor microalgal cultures. *Renew. Energy* 64, 238–243. doi:10.1016/j.renene.2013.11.010

Trzeciak, M. B., das Neves, M. B., da Silva Vinholes, P., and Amaral Villela, F. (2008). Utilização de sementes de species oleaginosas para produção de biodiesel. *Inf. Abrates* 18, 30–38.

Udom, I., Zaribaf, B. H., Halfhide, T., Gillie, B., Dalrymple, O., Zhang, Q., et al. (2013). Harvesting microalgae grown on wastewater. *Bioresour. Technol.* 139, 101–106. doi:10.1016/j.biortech.2013.04.002

Uduman, N., Ying, Q., Danquah, M. K., Forde, G. M., and Hoadley, A. (2010). Dewatering of microalgal cultures: a major bottleneck to algae-based fuels. *J. Renew. Sustain. Energy* 2. doi:10.1063/1.3294480

Um, B. H., and Kim, Y. S. (2009). Review: a chance for Korea to advance algal-biodiesel technology. *J. Ind. Eng. Chem.* 15, 1–7. doi:10.1016/j.jiec.2008.08.002

Van de Velden, M., Baeyens, J., Brems, A., Janssens, B., and Dewil, R. (2010). Fundamentals, kinetics and endothermicity of the biomass pyrolysis reaction. *Renew. Energy* 35, 232–242. doi:10.1016/j.renene.2009.04.019

Wessler, R. A., Amorim, S., and Cavalli, V. (2003). *Estudo da Viabilidade Técnica e Econômica da Utilização de um Polímero Orgânico Natural Catiônico em Substituição ao Sulfato de Alumínio Convencionalmente Utilizado em Estações de Tratamento de Água (ETA'S). Artigo Técnico*. Santo André: 33 Assembleia Nacional dos Serviços Municipais de Saneamento; ASSEMAE. Available at: http://www.semasa.sp.gov.br/Documentos/ASSEMAE/Trab_29.pdf

Yanik, J., Stahl, R., Troeger, N., and Sinag, A. (2013). Pyrolysis of algal biomass. *J. Anal. Appl. Pyrolysis* 103, 134–141. doi:10.1016/j.jaap.2012.08.016

Chisti, Y. (2007). Biodiesel from microalgae. Research review paper. *Biotechnol. Adv.* 25, 294–306. doi:10.1016/j.biotechadv.2007.02.001

Zhang, B., Xiong, S., Xiao, B., Yu, D., and Jia, X. (2011). Mechanism of wet sludge pyrolysis in a tubular furnace. *Int. J. Hydrogen Energy* 36, 355–363. doi:10.1016/j.ijhydene.2010.05.100

Zhou, W., Chen, P., Min, M., Ma, X., Wang, J., Griffith, R., et al. (2014). Environment-enhancing algal biofuel production using wastewaters. *Renew. Sustain. Energ. Rev.* 36, 256–269. doi:10.1016/j.rser.2014.04.073

Conflict of Interest Statement: The authors declare that the research was conducted in the absence of any commercial or financial relationships that could be construed as a potential conflict of interest.

6

Estimation and comparison of bio-oil components from different pyrolysis conditions

Gaojin Lyu[1,2], Shubin Wu[2]* and Hongdan Zhang[2]

[1] Key Lab of Pulp and Paper Science and Technology of Ministry of Education, Qilu University of Technology, Jinan, China,
[2] State Key Lab of Pulp and Paper Engineering, South China University of Technology, Guangzhou, China

In the case of development and utilization of bio-oils, a quantitative chemical characterization is necessary to evaluate its actual desired characteristics for downstream production. This paper describes an analytical approach for the determination of families of lightweight chemicals from bio-oils by using GC-MS techniques. And on this basis, new explorations in the field of influence factors, such as feedstocks, pyrolysis temperatures, and low-temperature pretreatment, on the composition and products yields of bio-oil were further investigated. Up to 40% (wt.%) of the bio-oil is successfully quantified by the current method. Chemical functionalities in the bio-oil correlate strongly with the original feedstocks because of their different chemical compositions and structure. Pyrolysis temperature plays a vital role in the yields of value-added compounds, both overall and individually. Higher temperature favored the generation of small aldehydes and acids, accompanied by a reduction of phenols. The optimal temperatures for maximum furans and ketones yields were 520 and 550°C, respectively. The low-temperature pretreatment of biomass has a good enrichment for the lightweight components of the bio-oils. In this case, much higher amounts of compounds, such as furans, ketones, and phenols were produced. Such a determination would contribute greatly to a deeper understanding of the chemical efficiency of the pyrolysis reaction and how the bio-oils could be more properly utilized.

Keywords: pyrolysis, bio-oil, characterization, quantification, GC-MS

Edited by:
Junye Wang,
Athabasca University, Canada

Reviewed by:
Charilaos Xiros,
Bern University of Applied Sciences,
Switzerland
Tianju Chen,
Chinese Academy of Sciences, China

*Correspondence:
Shubin Wu,
State Key Lab of Pulp and Paper
Engineering, South China
University of Technology,
381 Wushan Road, Guangzhou,
Guangdong 510640, China
shubinwu@scut.edu.cn

Introduction

Among the biomass thermochemical conversion processes and technologies, pyrolysis is a particularly promising route to produce liquid fuels and value-added chemicals from solid biomass feedstock and is now widely studied (Bridgwater et al., 1999; Dinesh et al., 2006). The resulting "pyrolysis oil" or "bio-oil" can be obtained with yields up to 70–80 wt.% (dry feed basis) depending on the relative amount of cellulose and lignin in the wood material. As bio-oil is significantly denser than its parent biomass, it can be more economically and efficiently transported to a centralized location to be used as a feedstock for further processing, such as by gasification or Fischer–Tropsch synthesis, etc. to produce transportation fuels (Demirbas, 2009a,b; Lu et al., 2009).

Bio-oils are composed of differently sized molecules derived primarily from the depolymerization and fragmentation reactions of three key biomass building blocks, i.e., cellulose, hemicellulose, and lignin, resulting in its composition and properties of considerable difference from those of petroleum-based fuel oils (Yang et al., 2007). The physical properties of bio-oils are well-described in the literature (Dobele et al., 2007; Abdelnur et al., 2013). Despite its dark brown viscous appearance, elemental analysis of bio-oil reveals that it contains relatively little sulfur and nitrogen, but has a high oxygen content, typically near 40 wt.% and as high as 50 wt.% including water. Bio-oils are highly oxygenated, complex mixtures, viscous, relatively unstable, and susceptible to aging. The lower heating value (LHV) of bio-oils is only 40–45% of that exhibited by hydrocarbon fuels. The high water content and the low LHV are detrimental for ignition. Moreover, organic acids in the bio-oils are highly corrosive to common construction materials. These inherent drawbacks make it hardly available to fuel application directly (Helena and Ralph, 2001; Mckendry, 2002).

Another point of view (Jean et al., 2001) is that industrial production of bio-oil should focus in the short term on its utilization for the manufacture of higher value chemicals and/or materials other than fuels. Recent detailed analyses have shown that pyrolysis bio-oils contain more than 400 compounds. And these chemical functionalities in the bio-oil correlate strongly with the feed composition and the pyrolysis processing conditions. Even though it remains a challenge to identify every compound in a bio-oil sample, a large number of previous studies have been consistent with the qualitative analysis of the bio-oils. From a chemical point of view, bio-oil is an extremely complex mixture of organic components, including various types of oxygen-containing organic compounds derived from biomass components of cellulose, hemicellulose, and lignin pyrolysis, such as organic acids, esters, alcohols, aldehydes, ketones, furans, phenols, and dehydrated carbohydrates (Serdar, 2004; Demirbas, 2009a,b).

Because of the complexity, some simplified analytical methods were used to characterize bio-oil. Chemical characterization of pyrolysis oils has generally been based on the fractionation of the oils into different classes of chemical functionality by using solvent extraction, adsorption chromatography, molecular distillation, etc. and then the different fractions obtained were further characterized using more than one analytical techniques, such as GC-MS, HPLC, thermogravimetric techniques, gel permeation chromatography (GPC), etc. with focusing on its different chemical information (Kai et al., 1998; Chiaberge et al., 2014; Lindfors et al., 2014). According to Meier (1999), typical portions of the important fractions in bio-oils are: around 20 wt.% water, around 40 wt.% GC-detectable compounds, around 15 wt.% non-volatile HPLC detectable compounds, and around 15 wt.% high molar mass non-detectable compounds. Garcia et al. (2007) described an analytical approach to identify the chemical composition of bio-oils in terms of macro-chemical families. The bio-oils were first fractionated using different polar organic solvents, and then the fractions were analyzed by GPC, GC-MS, and elemental analyzer. A thorough description of bio-oil composition as a mixture of water, monolignols, polar compounds with moderate volatility, sugars, extractive-derived compounds, heavy polar and non-polar compounds, MeOH–toluene insolubles, and volatile organic were obtained. Guo et al. (2010) separated bio-oil into three fractions by molecular distillation, i.e., a light fraction, a middle fraction, and a heavy fraction. The chemical composition of the three fractions and the crude bio-oil was analyzed by GC–MS, and the pyrolysis characteristics were measured by a thermogravimetric analyzer coupled with Fourier transform infrared spectroscopy (TG–FTIR). Different physical and chemical characteristics were noted and illustrated among the three fractions. By using nuclear magnetic

resonance spectroscopy (NMR), including ^{1}H, ^{13}C, and DEPT spectra, Charles et al. (2009) characterized and compared fast-pyrolysis bio-oils from six different feedstocks. NMR spectroscopy provides important information not only of the kinds of chemicals in bio-oils but also of their relative concentrations, especially for the highly substituted aromatic groups that were not detected by other means.

Although a lot of progress has been made in the field of bio-oil fractionation and the subsequent analysis, most of these studies narrowed their interest to qualitative analysis. Quantitative chemical characterization of bio-oils has always been a challenging undertaking, and the literature on this issue is very scarce apart from a few chemical species were examined and reported. Branca et al. (2003) quantified 40 compounds of bio-oils from low-temperature pyrolysis of wood by the internal standard method, using fluoranthene as internal standard. And the yields of a significant number of compounds on dependence of the reaction temperature and heating rate were also provided. The major compositions of the pyrolysis liquids were considered from acetic acid, hydroxy-propanone, hydroxyacetaldehyde, levoglucosan, formic acid, syringol, and 2-furaldehyde. de Wild et al. (2009) analyzed most of lightweight organic species that were representative for typical thermal degradation products from lignocellulosic biomass with GC-MS. The weight of those unknown components was estimated by a semi-quantitative calibration using the GC-data of the internal standards with the nearest retention time on the GC column. The results show that the yields of individual chemicals are generally below 1 wt.% (based on dry feedstock weight), and only certain groups of thermal degradation products like C_2–C_4 oxygenates and phenols are formed in higher yields up to 3 wt.%. By employing a series of improved analytical methods, the International Energy Agency-European Union (IEA-EU) round robin test was carried out in 12 laboratories with focus on comparing the accuracy of methods to provide instructions for handling and analysis of bio-oils (Meier, 1999; Oasmaa and Meier, 2005). The results showed that even though the repeatability of the physical analyses was good; there were very large quantitative differences between different laboratories for chemical characterization. Relevant researchers thus suggested that proper standard solutions with known amounts of compounds have to be used for quantitative analyses, and a lot of experimental work and methods adjustment should be needed to achieve accurate and consistent results between laboratories.

Based on the above literature review of the research and development advances in studies on pyrolysis liquids, it can be concluded that the present state of knowledge has been focused solely on the fractionation, qualitative characterization, and even quantification was limited to a few species for the optimal conditions of fast pyrolysis (corresponding to maximum yields of bio-oils). However, in the current state of art, an accurate and relatively fast method is urgently needed for a quantitative analyses of the most valuable components of bio-oils. By this method, it may help us to evaluate and determine what kind of bio-oil has particular desired characteristics for downstream production of fuels or chemicals.

The experimental investigations in our present work are therefore set to serve two main purposes: (a) to develop an effective quantitative analysis method for lightweight components of bio-oils by using GC-MS techniques. Thus, yields up to 40 wt.% of the bio-oils can be quantified, and this part usually contains the highest commercial value either used as fuel or chemicals and (b) to determine the yields of a significant number of compounds on dependence of the feedstocks, pyrolysis temperatures, and especially staged pyrolysis conditions, in the view of providing evaluation criteria for the selection of the most appropriate feedstocks and pyrolysis conditions to produce the desired composition and yields of the products.

Materials and Methods

Materials

Four kinds of biomass feedstocks, i.e., bagasse, corncob, spruce, and pine, were selected as pyrolysis materials. The bagasse was collected from a sugar cane plant located in Jiangmen, China. The corncob was obtained from a farm in Changzhi, China. The spruce and pine wood used in this study were purchased from a local sawing mill in Guangzhou, China. All four feedstocks were first air dried and grounded, and then were sieved to a mean particle size of 0.180–0.425 mm. Ultimate analysis of the samples was determined by using an elemental analyzer (Vario EL, ELEMENTAR, Germany). For proximate analysis, the ash content and volatile content were determined based on Chinese National Standards GB/T 2677.3-1993 and GB/T 212-2001, while fixed carbon was calculated by difference. The acid insoluble lignin (klason lignin) and acid soluble lignin were determined based on GB/T2677.8-1994 and GB/T10337-2008, respectively. The extractives (benzene/ethanol 2:1, v/v), hemicellulose and cellulose content were analyzed according to GB/T10741-2008 and methods in reference (Shi and He, 2003). The results were shown in **Table 1**. All these samples were dried at 105°C for 8 h in an oven to about 6% moisture content before the pyrolysis experiment.

Preparation of Bio-Oils

Bio-Oils from Direct Pyrolysis

Direct fast pyrolysis of biomass was carried out in a self-designed fluidized-bed reactor in Guangzhou Institute of Energy Conversion. The schematic drawing of the fluidized-bed reactor was previously reported (Chang et al., 2011; Zheng et al., 2012). Briefly, the reactor uses nitrogen as the carrier gas, quartz sand as bed material, a screw feeder for feeding biomass, and a thermocouple for testing the reaction temperature that is heated by electricity. All the reaction system was online controlled by a computer. After reaction, the pyrolysis vapors

were purged by nitrogen through a two-stage gas-solid cyclone separator to remove ashes and solid particles, then the vapors were condensed in a tubular heat exchanger, in which most of the light components were condensed and collected (Liquid 1) while the remaining volatiles were further isolated by a two-stage gas–liquid cyclone separator to obtain the Liquid 2. The mixed gases were further subjected to an absorption device, in order to ensure complete absorption of the liquid components and to purify the pyrolysis gases. Liquid 1 and Liquid 2 were combined and designated as bio-oil for further analysis. All pipes before tubular heat exchanger as well as the gas–solid cyclone separator were insulated by using heating tape to avoid the pre-condensing of the pyrolysis vapors.

After a number of preliminary experiments, the following optimum pyrolysis parameters were used: the feed rate was 10 kg h^{-1}, the nitrogen flow was 11 m^3 h^{-1}, the particle size of quartz sand was 60–80 meshes, and its density was 2.6 g cm^{-3} with the bed height of 12 cm. The direct pyrolysis was performed at 480, 500, 520, 550, and 580°C, respectively, to explore the effect of pyrolysis temperature on the composition of bio-oils.

Bio-Oils from Staged Pyrolysis

Bio-oils, obtained by conventional rapid pyrolysis, are usually multicomponent mixtures of carbohydrate and lignin thermal decomposition products with low pH, high viscosity, and high water content, which limit its large-scale application as fuel. Based on different thermal stabilities of the main biomass constituents, staged pyrolysis, a new pyrolysis-based conversion route, was designed and applied to improve the quality of bio-oils and to generate more value-added chemicals from biomass. In the present work, staged pyrolysis experiments were first conducted with pine in an auger reactor at the temperature range of 260–320°C, and then the solid residues obtained were further pyrolyzed in a fluidized-bed reactor (10 kg h^{-1}) at moderate temperature of 520°C. As the chemical composition and pyrolysis behavior of the resulting torrefied biomass (torrefaction treatment in the range of 260–320°C) were examined in detail in the previous work (Chang et al., 2012), what we now focus on is the impact of low-temperature pretreatment on changes of the composition of bio-oils as well as the enrichment effect on high value-added chemicals.

Low-temperature pyrolysis of pine was conducted in a stainless steel auger (screw) reactor, which mainly consists of a feed bunker (capacity of 6 kg), an electrically heated (3.2 kW) auger pyrolysis reactor, a collection bin, a temperature control unit, and a cooling system. The torrefaction temperatures were performed at 260, 280,

---

TABLE 1 | Ultimate, proximate, and chemical composition analysis of four biomasses.

Sample	Ultimate analysis (w_{daf}%)					Proximate analysis (w_{ad}%)			Component analysis (w_{ad}%)			
	C	H	O^a	N	S	Volatiles	Fixed carbon	Ash	Hemicellulose	Cellulose	Lignin^b	Extractives
Bagasse	47.74	5.60	46.52	0.08	0.06	85.96	11.36	2.68	27.98	45.92	20.72	3.73
Corncob	47.23	5.95	46.34	0.47	0.01	80.98	15.87	3.15	31.60	37.63	20.77	8.10
Spruce	49.11	6.14	44.62	0.08	0.05	83.77	16.02	0.21	13.28	47.15	36.03	1.87
Pine	49.18	5.83	44.95	0.02	0.02	86.10	12.60	1.30	21.66	46.97	27.65	2.70

daf, dry and ash-free basis; ad, air dry basis.
^{a}Calculated by difference.
^{b}Total of acid insoluble lignin and acid soluble lignin.

300, and 320°C, respectively. When the reactor reached the set temperature, the samples were transported down the length of the reactor tube [600 mm (L) × 100 mm (I.D.)] at a fixed speed (2 kg h⁻¹) by means of a screw to maintain the residence time of the sample in the reactor pipe for 10 min. During the process, a nitrogen gas flow of 7 L min⁻¹ was used as a carrier gas to eliminate the presence of oxygen and remove the volatiles out of the reactor. The torrefied samples were then collected in the collection bin and were used as feedstock for the second stage pyrolysis in a fluidized-bed reactor.

The second stage pyrolysis temperature was 520°C, at which the pine got the maximum bio-oil yield. The other operation parameters were consistent with that of the direct pyrolysis. Each experiment was repeated three times under the same conditions to confirm the reproducibility of each test and to obtain accurate results.

Analytical Methods
Qualitative Analysis
Qualitative analysis of bio-oils was conducted using a gas chromatography–mass spectrometer (GC–MS, 7890A-5975C, Agilent Technology, USA). About 200 mg bio-oils were dissolved in 5 mL of acetone. Before injection, diluted samples were first dehydrated with 1 g anhydrous sodium sulfate in a beaker, and then filtered through a 0.22 μm syringe-driven filter. The chromatographic separation was performed using a HP-INNO Wax capillary column (30 m × 0.25 mm × 0.25 μm). Helium (99.9995%) was used as carrier gas at a flow rate of 1 mL min⁻¹. The injector (7683B ALS) and the GC–MS interface were kept at a constant temperature of 250°C. A sample volume of 1 μL was injected in the split mode with a split ratio of 20:1. The GC oven temperature ramps for the work were as follows: begin heating at 50°C that was first held for 2 min, and then raised to 90°C at a rate of 10°C/min, and after raised to 120°C at a rate of 4°C/min, and then raised to 230°C at a rate of 8°C/min, and this final temperature was held for 10 min.

The mass selective detector was operated in electron impact (EI) ionization mode. MS source and MS quadrupole temperatures were kept at 230 and 150°C, respectively. EI ionization energy was 70 eV and an m/z range from 33 to 500 was scanned. Solvent cut time of 2.8 min was used to avoid solvent influence. The identification of the peaks was based on computer matching of the mass spectra with the NIST 08 MS library and/or on the retention times of known species injected in the chromatographic column.

Quantitative Analysis
From the results of the qualitative analysis, the major compositions that displaying a relatively high content (based on the area%) in the bio-oils were identified. The standard compounds of these major compositions (19 compounds) were purchased from Sigma-Aldrich and directly used for qualitative or quantitative analysis.

These major 19 compounds of bio-oil were quantified by the external standard method. For each of the quantified compounds, calibration curves were prepared by injection of five standard solutions (40, 200, 400, 1000, and 2000 μg mL⁻¹). For quantitative analysis of the other bio-oil compositions, several typical compounds, i.e., 1-hydroxy-2-propanone, acetic acid, furfural, and 2-methoxy-phenol were further employed as internal standards

for quantifying ketones, acids, furans, and phenols compounds of bio-oils, respectively. The GC test parameters were the same as the above qualitative analysis. Injections of each standard or bio-oil samples were made at least in triplicate until acceptable peak reproducibility was achieved by overlapping and comparison of TIC. The product yield was then expressed as the mean of three replications ± SD for each sample are shown in Tables in Supplementary Material.

Evaluation by Standard Addition Method
Considering a situation that the bio-oil sample may have a matrix effect, which may also contribute to the analytical signals and affect the results, the standard addition method was used for quantitation and comparison to further evaluate the effectiveness and impact of the current calibration curve approach.

Experimentally, the chromatogram of the diluted pine bio-oil was recorded first. Then, a known and different amounts of each standard of interest (10 standards were tested) were added and all were diluted to the same volume as the blank one. The samples were tested under the same GC–MS conditions and the chromatogram was recorded. Based on the increase in the peak area and the results plotted, the original concentration of 10 compounds in bio-oil can be computed, which were designated as the actual concentration. The calculating concentration of 10 compounds was obtained based on the chromatogram of the original diluted pine bio-oil and the established external standard calibration curves. Comparison of actual concentration and calculating concentration of 10 compounds and the resulted relative error were listed in Table S1 in Supplementary Material, from which thorough investigation of the influence of the methodology on the quantitation results was available.

Results

Qualitative Analysis and Quantitative Approach
Although new approaches for analysis and characterization of the composition and structure of bio-oils are still being explored, especially for the relatively heavy components, such as dimers or trimers, it has reached a consensus that the main components of bio-oil is a variety of small molecules with oxygen-containing functional groups such as aldehydes, ketones, carboxylic acids, furans, phenols, dehydration carbohydrates, etc. (Dobele et al., 2007; Hyeon et al., 2010). **Figure 1** shows the total ion chromatogram of bio-oil from pine pyrolysis at 520°C, and the main compositions (peak area percentage >1%) corresponding labels in **Figure 1** were identified by computer matching and listed in **Table 2**.

From the qualitative analysis results, it could be seen that acetic acid is the principal acidic components of bio-oil, and furfural and furfuralcohol are major furans products. Ketones mainly composed of 1-hydroxy-2-propanone and a large number of cyclopentanones and cyclopentadiones. Levoglucosan, a typical pyrolysis product from cellulose pyrolysis, represents incomplete cracking of carbohydrates. A large number of methylated or methoxylated phenolic compounds that derived from lignin pyrolysis were also detected, such as 2-methoxy-phenol, 2-methoxy-4-methyl-phenol, 2-methoxy-4-vinylphenol, eugenol, vanillin, etc. and each of them occupies a relatively large percentage.

Based on an overall consideration of the qualitative results of bio-oils from different raw materials pyrolysis, 19 compounds that presented abundantly were identified and selected as standards, i.e., 1-hydroxy-2-propanone, acetic acid, furfural, 2-furanmethanol, 5-hydroxymethylfurfural, 3-methyl-1,2-cyclopentanedione, phenol, 3-methyl-phenol, 4-methyl-phenol, 2-methoxy-phenol, 2-methoxy-4-methyl-phenol, 4-ethyl-phenol, eugenol, (E)-isoeugenol, vanillin, 1,2-benzenediol, 3-methyl-1,2-benzenediol, 2,3-dihydro-benzofuran, and levoglucosan. The standard curves results show that at the concentration range of the first four standard solutions, i.e., from 40 to 1000 µg mL⁻¹, calibration lines of each compound was excellent linear correlation ($R^2 \geq 0.992$). While the concentration of the standard was above 2000 µg mL⁻¹, the calibration lines of some compound began to seem like quadratic polynomial. So the concentration range of 40–1000 µg mL⁻¹ was used to prepare the injection samples to ensure the accuracy of the quantification.

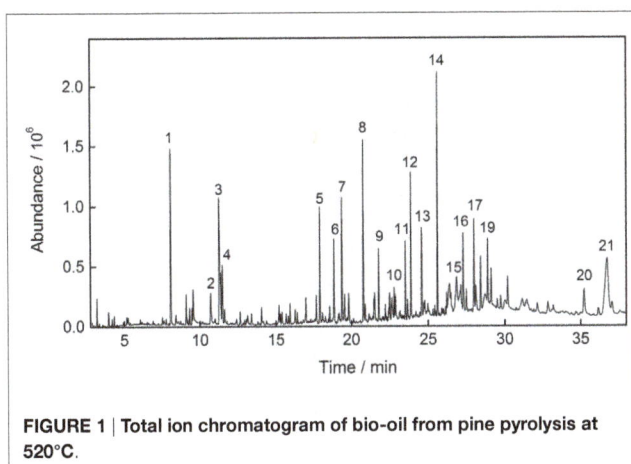

FIGURE 1 | Total ion chromatogram of bio-oil from pine pyrolysis at 520°C.

TABLE 2 | Identification of the main components (peak area >1%).

Peak	RT/min	Library ID	Formula	M_w	Area/%
1	8.053	1-hydroxy-2-propanone	$C_3H_6O_2$	74	4.947
2	10.706	Butanedial	$C_4H_6O_2$	86	1.361
3	11.259	Acetic acid	$C_2H_4O_2$	60	5.679
4	11.474	Furfural	$C_5H_4O_2$	96	2.231
5	17.934	1,2-cyclopentanedione	$C_5H_6O_2$	98	3.320
6	18.868	3-methyl-1,2-cyclopentanedione	$C_6H_8O_2$	112	2.516
7	19.384	2-methoxy-phenol	$C_7H_8O_2$	124	3.338
8	20.780	2-methoxy-4-methyl-phenol	$C_8H_{10}O_2$	138	4.621
9	21.784	4-ethyl-2-methoxy-phenol	$C_9H_{12}O_2$	152	1.862
10	22.487	4-ethyl-phenol	$C_8H_{10}O$	122	1.049
11	23.518	Eugenol	$C_{10}H_{12}O_2$	164	1.959
12	23.878	2-methoxy-4-vinylphenol	$C_9H_{10}O_2$	150	3.793
13	24.582	(Z)-isoeugenol	$C_{10}H_{12}O_2$	164	3.141
14	25.613	(E)-isoeugenol	$C_{10}H_{12}O_2$	164	5.897
15	26.397	4-hydroxy-3-methoxycinnamaldehyde	$C_{10}H_{10}O_3$	178	2.556
16	27.283	5-hydroxymethylfurfural	$C_6H_6O_3$	126	2.573
17	27.992	Vanillin	$C_8H_8O_3$	152	2.944
18	28.427	2-methoxy-4-propyl-phenol	$C_{10}H_{14}O_2$	166	1.777
19	28.883	3-methoxy-4-hydroxyacetophenone	$C_9H_{10}O_3$	166	2.040
20	35.220	4-hydroxy-3-methoxy-benzeneacetic acid	$C_9H_{10}O_4$	182	2.091
21	36.734	Levoglucosan	$C_6H_{10}O_5$	162	8.943

There was a big difference in the coefficient of linear calibration curve of different compounds (see Figure S1 in Supplementary Material), showing the differences of quantitative correction factor from the chromatogram concept, such as that of 1-hydroxy-2-propanone, furfural, and phenol. Therefore, for chemical quantitation of bio-oil compositions, it might be necessary to calibrate the chromatographic systems by preparing standard solutions with as many of known standard compounds as possible. On the other hand, the linear coefficient of the standard compounds with similar structure and chemical properties had little difference, 2-methoxy-phenol and 2-methoxy-4-methyl-phenol, for example. This provides the basis of quantitation of other bio-oil components by using selected compound as internal standards.

Effect of Biomass Feedstock

After a large number of fluidized-bed pyrolysis experiments and the corresponding process adjustment and optimization, the bio-oils from bagasse, corncob, spruce, and pine pyrolysis that with the maximum yield at 500, 520, 520, and 520°C, respectively, were qualitatively and quantitatively analyzed and compared. The content comparison of bio-oils from different feedstock is shown in **Figure 2**. The identification and quantitation results of each single component are listed in Table S2 in Supplementary Material.

It can be seen from the results that the total lightweight organic components quantified by gas chromatography were 10–20 wt.% (yields always expressed as percent of the bio-oil mass). Among these, bio-oils from grasses material pyrolysis contain relatively low organic content, while that of spruce and pine pyrolysis oils have higher organic content. These differences in both major categories and each single compound content of the bio-oils may attribute to the differences in chemical composition and structure of the biomass feedstocks (Akwasi et al., 2008; Lv and Wu, 2012). Herbaceous biomass generally contains more hemicellulose and less lignin than woody biomass, resulting in more non-condensable gaseous products. In addition, the mineral matter of biomass, in combination with the organic composition, plays an important role in pyrolysis product yields and distribution (Raveendran et al., 1995). Higher mineral

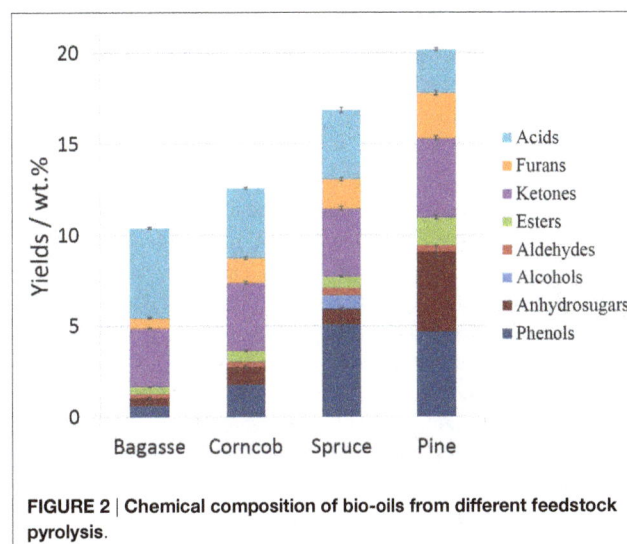

FIGURE 2 | Chemical composition of bio-oils from different feedstock pyrolysis.

salts (ash content) may accelerate secondary depolymerization of pyrolysis volatiles and lead to the formation of low molecular gas species (Pushkaraj et al., 2010). Thus, the over yields of GC-detectable products from corncob and bagasse pyrolysis bio-oil were lower than that of spruce and pine.

On the whole of **Figure 2**, the major categories of four kinds of bio-oils were acids, ketones, phenols, and furans, followed by anhydrosugars, while alcohols, aldehydes, and esters were relatively low. The major organic acid was acetic acid, and it usually possessed the largest share. Bagasse bio-oil had the highest acetic acid content of 4.8 wt.% (Table S2 in Supplementary Material), while acetic acid content of spruce and pine pyrolysis oil was relatively low, around 2.3 wt.%. It has been accepted that the formation of acetic acid was mainly due to the breakage and removal of acetyl groups that originally linked to the xylose unit on C-2 position, and second, attributed to the ring-scission of uronic acid residues after the elimination of the carbonyl and O-methyl groups (Lv et al., 2010). Therefore, the differences in the content and the chemical structure of hemicellulose are the main cause of differences in acetic acid content of bio-oil. It had been noted that the acetyl content of softwood was about 1%, while that of hardwood and grasses was between 3 and 6% (Badal, 2003). The acetyl content of bagasse and corncob was higher than that of spruce and pine, resulting in higher acid content in pyrolysis bio-oil.

All the four feedstocks pyrolysis produced a large proportion of ketones, of which the total content was about 3–4.5 wt.%. And among these, 1-hydroxy-2-propanone has the highest yield and could reach up to 2.2 wt.%. The formation of small molecule ketones was complicated, while the cyclopentanones and cyclopentenones, whose molecular weight were a little larger, were mainly originated from the decomposition of sugar units and then recombination of the opened bonds. All these ketones could be derived from the pyrolysis of either cellulose and/or hemicellulose.

Furans, a common class of substances in bio-oil, can be attributed mainly to holocellulose decomposition. Although both cellulose and hemicellulose pyrolysis can produce furan compounds, our previous work has shown that hemicellulose definitely makes the greatest contribution to the formation of furfural while cellulose is more dedicated to the products of hydroxymethylfurfural and 2-furylmethanol (Lv et al., 2013). From the quantitative results, corncob pyrolysis oil has the highest content of furfural (0.5 wt.%), which is due to the highest hemicellulose content in corncob. The most abundance of 2-furylmethanol and hydroxymethylfurfural appeared in pine pyrolysis oil.

The identified and quantified anhydrosugar compounds were mainly levoglucosan and its dehydration or isomeric forms, which were generated from the direct decomposition of cellulose. The levoglucosan yield was ranging from 0.3 to 4.2 wt.% depending on the type of feedstock, among which corncob pyrolysis oil occupies the lowest and pine pyrolysis oil takes up the highest.

A considerable amount of alkylated, oxyalkylated phenols, which is mainly formed by the fracture of ether linkages and C–C bonds contained in the side chains of the lignin polymer, occupies a large proportion of the bio-oil, especially for wood feedstock. The total phenols yields of spruce and pine bio-oil were 5.1 and 4.7 wt.%, respectively, while corncob and bagasse bio-oils had relatively low content of phenolic compounds, i.e., 1.8 and 0.6 wt.%, respectively.

The type and content of phenolic compounds in bio-oil were relevant to the content of lignin and its structural characteristics of the biomass feedstock. Based on its structural features, the phenolic compounds can be further divided into G-phenols (derived from guaiacyl), S-phenols (derived from syringyl), and H-phenols (derived from p-hydroxyphenyl) (Table S2 in Supplementary Material). The dependence of lignin subunits distribution from different feedstocks on phenols was illustrated in **Figure 3**.

It could be seen from **Figure 3** that the phenolic compounds in spruce and pine pyrolysis bio-oils were mainly G-phenols, each account for more than 80% of the total phenols with major contributions from 2-methoxy-phenol, 2-methoxy 4-methyl-phenol, 2-methoxy-4-vinylphenol, isoeugenol, etc. This may be because of the softwood lignin that almost exclusively contains G-units. On the other hand, grasses lignin also contains more syringyl and p-hydroxyphenyl structural units than softwood lignin. This explains the phenomenon that e bagasse and corncob bio-oil contained a small amount of S-phenols while spruce and pine bio-oils were not.

Effect of Temperature

Although the fast-pyrolysis tests of four kinds of feedstocks have been executed for reactor temperatures in the range 480–580°C, in view of the large amount of data involved, the current work only chose pine as raw material for discussion of the effect of pyrolysis temperature on the composition distribution of bio-oils. The total yields and the trends of the lumped component classes, which have been grouped as AAE (alcohols, aldehydes, and esters), acids, furans, ketones, phenols, and anhydrosugars, were given in **Figure 4**. The details of the classification of 55 compounds and each single compound yield were listed in Table S3 in Supplementary Material.

As showed in **Figure 4**, the total yields of 55 chemical compounds quantified by GC–MS first increased and then decreased gradually with increasing temperature, with the maximum value emerged at 550°C, which correspond to 22.0 wt.% of the bio-oils. Among these, the most abundant species are hydroxypropanone (1.9–2.5 wt.%), acetic acid (2.1–3.3 wt.%), furfural (0.4–0.5 wt.%), 5-hydroxymethyl furfural (0.4–0.7 wt.%), 2-methoxy-4-methyl-phenol (0.2–0.8 wt.%), (E)-isoeugenol (0.2–1.1 wt.%), and

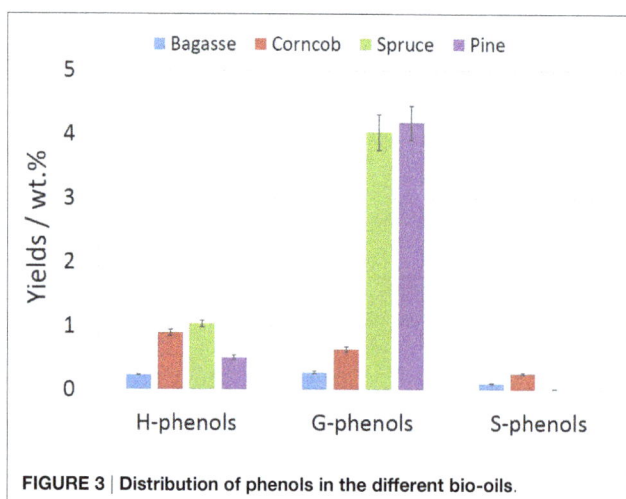

FIGURE 3 | Distribution of phenols in the different bio-oils.

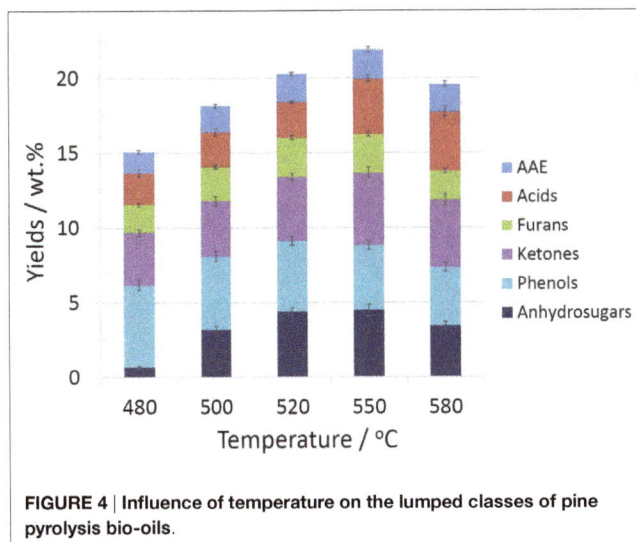

FIGURE 4 | Influence of temperature on the lumped classes of pine pyrolysis bio-oils.

levoglucosan (0.4–4.2 wt.%). By varying the pyrolysis temperature (480–580°C), the content of small molecule aldehydes, acids showed an increasing trend as temperature increased, while the phenols content decreased gradually from 5.5 to 3.9 wt.%. The yields of furans, ketones, and anhydrosugars varying as a function of pyrolysis temperature were similar to that of the total compounds, with the maximum yield of 2.6, 4.9, and 4.5 wt.%, respectively, emerged at either 520 or 550°C.

Figures 5A,B show the major carbohydrates derived compounds as functions of the pyrolysis temperature, including several major acids, aldehydes, ketones, and anhydrosugars. It could be seen that the yields of relatively larger molecular weight compounds, such as levoglucosan, 1,2-cyclopentanedione, 3-methyl-1,2-cyclopentanedione, etc. were first increased and then decreased with the maximum value at about 520–550°C, while that of the lower molecular compounds, such as acetic acid, butanedial, 2,3-butanedione, showed a continuous increasing trend with temperature elevated. This proved that as temperature increased, the relatively larger molecular furanones, cyclopentanones, and cyclopentenones in hemicellulose and cellulose were prone to bond breaking and reforming to generate lower molecular furan ring or cyclopentanone structures (Patrick and Paul, 1996; Kawamoto et al., 2009). It was worth noting that there was a big difference in the content of levoglucosan under different pyrolysis conditions by using different feedstocks. The content of levoglucosan in the bio-oils from bagasse, corncob, and spruce pyrolysis was between 0.3 and 0.9 wt.% in our previous work, while the pine pyrolysis oil contained much higher levoglucosan, which could reach up to 4 wt.%. This may be related to the different content of salts and metal ions in the feedstock. It has also been reported that the reaction paths of levoglucosan were highly dominated by the ash content. High content of salts and metal ions in feedstock were confirmed to favor the secondary decomposition of levoglucosan and further generation of small molecule aldehydes and ketones (Shen and Gu, 2009).

Figure 5C gives the details of the yields of major furans varied with temperature. It can be observed that, pyrolysis temperature had a positive impact on the formation of furfural, i.e., its content

showed an increasing trend with the increase of temperature. While for 5-hydroxymethylfurfural, after a maximum yield of 0.7 wt.% at 520°C was attained, as the temperature increased further, a sharp decrease in the amounts occurred because of its weak thermal stability and secondary vapor-phase degradation. This result confirms the hypothesis proposed by Shen and Gu (2009) that the furfural was produced from the secondary reaction of 5-hydroxymethylfurfural at high temperature, along with the formation of formaldehyde through the dehydroxymethylation reaction of the side chain of the furan ring.

Figures 5D,E show the H-phenols and G-phenols content varied as functions of pyrolysis temperature. Apparently, the temperature was indeed a critical parameter that affected not only the phenols yields but also the structure of phenolic compounds. At low temperatures, lignin began to depolymerize to the products that maintained its monomeric structure as much as possible, for example, (E)-isoeugenol, 4-methylguaiacol, p-vinylguaiacol, etc. all got their maximum yields at 480°C (**Figure 5D**). While the temperature elevated, the side chains of these aromatics were further broken and the methoxy group was removed, thereby generating more H-phenols (**Figure 5E**). These confirmed that some guaiacols and syringols were intermediate species, which gave their decline at higher temperature, and occurred secondary degradation to further form H-phenols and small molecule gases.

Effect of Low-Temperature Pretreatment

As the physicochemical properties of the torrefied biomass and the resulting bio-oils were examined previously (Chang et al., 2012), our main attentions in this paper were then paid to the influence of the low-temperature pretreatment on the chemical composition changes of the staged pyrolysis bio-oils compared to that of direct pyrolysis, especially for the content variation of furfural, levoglucosan, and phenols products, which represented the typical pyrolysis products of hemicellulose, cellulose, and lignin, respectively. Currently, only the staged pyrolysis of pine was discussed because of space limitation. Torrefaction experiments of pine were performed in an auger reactor at 260, 280, 300, and 320°C, separately; the resulting torrefied pine was used as feedstock for flash pyrolysis in fluidized-bed reactor at moderate temperature of 520°C. The obtained bio-oils were analyzed and compared. Determination of each single product was listed in Table S4 in Supplementary Material. The yields of various types of products compared to that of direct pyrolysis were plotted in **Figure 6**.

From **Figure 6** and Table S4 in Supplementary Material, it could be seen that, the low-temperature pretreatment or low-temperature torrefaction of biomass had a good enrichment for the light weight components of the bio-oils compared with the direct pyrolysis. The total yield of the chromatographic quantitative component was improved by 50–90%, i.e., reached to 31–38 wt.% based on the weight of the obtained bio-oils. The main categories of compounds such as phenols, ketones, furans, and anhydrosugars were all greatly improved. This is mainly due to the low-temperature pretreatment that removed most of the free water and partially bound water in the pine raw material, and the hemicellulose degraded to varying degrees depending on the

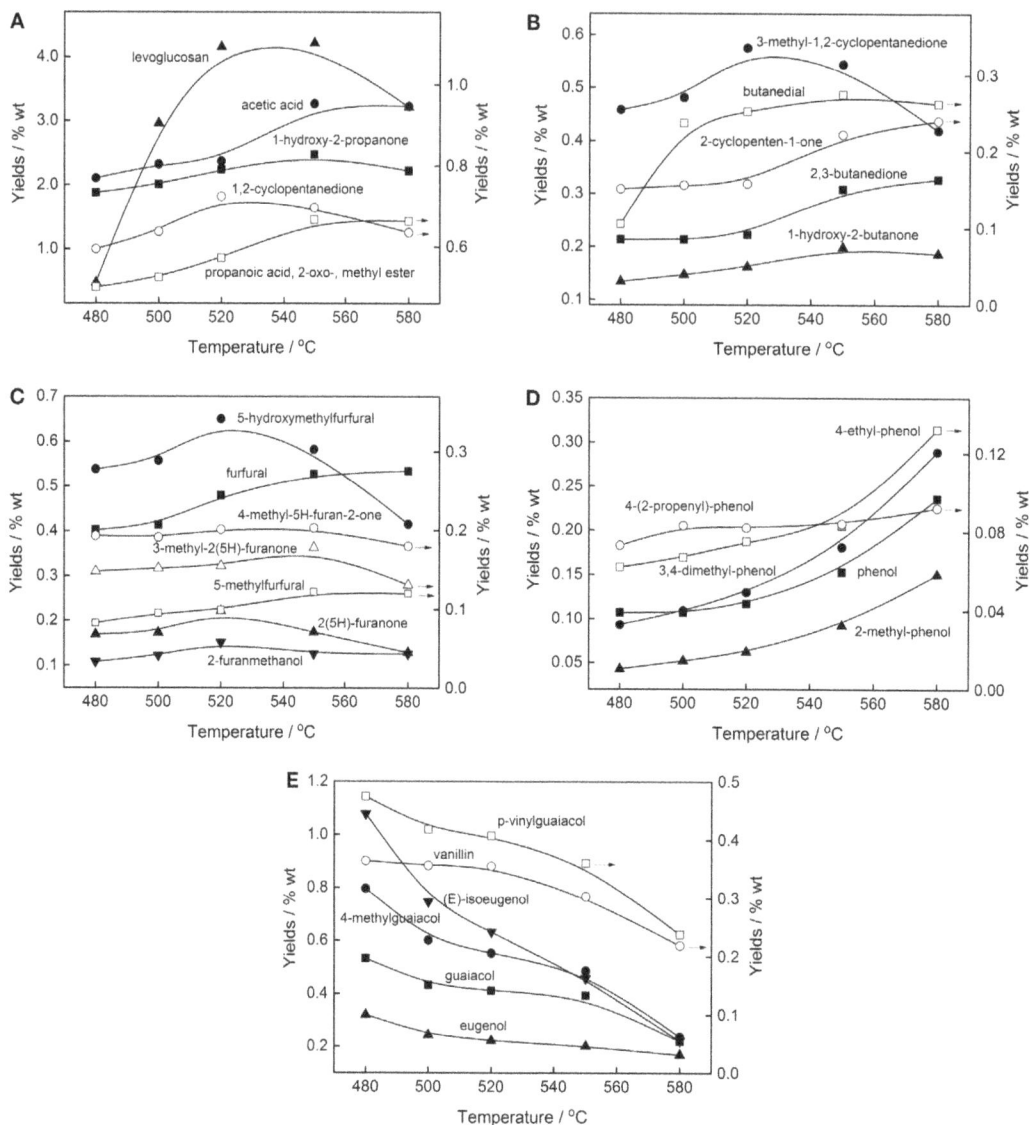

FIGURE 5 | Influence of temperature on the yield of (A) major carbohydrates, (B) minor carbohydrates, (C) furans, (D) H-phenols, and (E) G-phenols of pine pyrolysis bio-oils.

treatment temperature, generating a small amount of methanol, acetic acid, 1-hyroxy-2-propanone, etc. Thus, the physicochemical properties of the raw material were improved after treatment. When the torrefied feedstock was further pyrolyzed, the water content of the obtained bio-oil decreased sharply, resulting in the increase of major organic chemicals content (Bourgois and Guyonnet, 1988). In addition, the decomposition of cellulose and lignin in torrefied biomass that was accelerated and deepened by torrefaction treatment was the most predominant reason for the enrichment of bio-oil components (Wu et al., 2010). On the other hand, the major components of bio-oil were not always increased linearly as the pretreatment increased further. For example, the total yield of low weight chemicals of bio-oil obtained by staged pyrolysis that torrefied first at 320°C was lower than that of pretreated at 300°C. Among them, the aldehydes, esters, ketones,

acids, and furans compounds have similar trend as pretreatment temperature elevated. Higher pretreatment temperature may lead to the excessive depolymerization of the biomass feedstock and the resulted torrefied biomass component condensation or crosslinking, resulting in the decreased yield of bio-oil, and even the reduced content of the main compounds (Dobele et al., 2007).

Figure 7 provides the influence of pretreatment temperature on the yields of major compounds. **Figures 7A,B** show the effect on the yields of major carbohydrates derived products. It can be seen that higher pretreatment temperature has a positive impact on the yield of levoglucosan, and its increment was more obvious with the increase of pretreatment temperature. Compared with direct pyrolysis, LG content in the bio-oil obtained after pretreatment at 320°C increased about 60%, i.e., reached up to 6.7 wt.% of the total bio-oil.

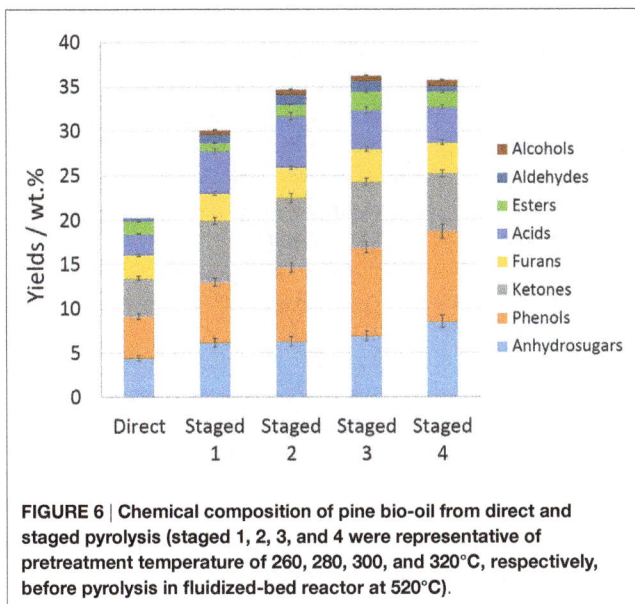

FIGURE 6 | Chemical composition of pine bio-oil from direct and staged pyrolysis (staged 1, 2, 3, and 4 were representative of pretreatment temperature of 260, 280, 300, and 320°C, respectively, before pyrolysis in fluidized-bed reactor at 520°C).

The acids, especially acetic acid, increased after low-temperature treatment; while the pretreatment temperature was higher than 280°C, the acids began to decrease. There are similar trends for the products such as 1-hydroxy-2-propanone, butanedial, 1-hydroxy-2-butanone, etc., i.e., all got maximum yield when the pretreatment temperature was 280°C. This explains why the pH value of the staged pyrolysis bio-oil showed no increase or increased marginally compared to that by direct pyrolysis. The staged pyrolysis makes the water content of the bio-oil decreased because of torrefaction of biomass feedstock, thus all kinds of compounds, such as acetic acid, etc. were enriched to some extent.

Figure 7C shows the yields of furans varied with pretreatment temperature. As expected, furans products in the staged pyrolysis bio-oils such as furfural, 5-hydroxymethylfurfural, 2-furanmethanol, etc. also had a significant increase because of the low-temperature torrefaction. Wherein, the furfural got the maximum content of 0.7 wt.% when the pine was pretreated at 280°C first, while 5-hydroxymethylfurfural and 2-furanmethanol obtained the maximum increase when pretreated at 300°C, which accounted for 1.0 and 0.4 wt.%, respectively.

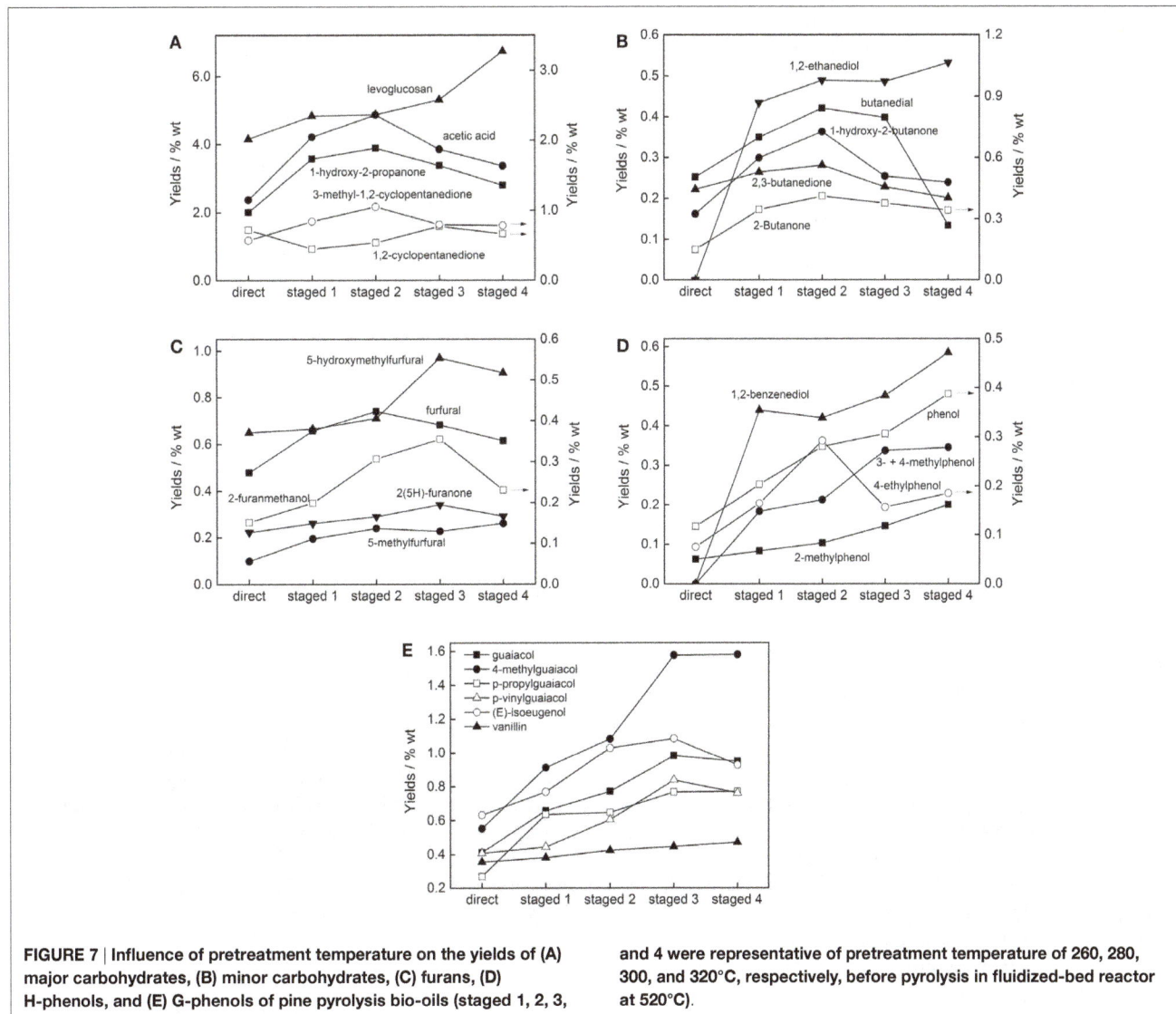

FIGURE 7 | Influence of pretreatment temperature on the yields of (A) major carbohydrates, (B) minor carbohydrates, (C) furans, (D) H-phenols, and (E) G-phenols of pine pyrolysis bio-oils (staged 1, 2, 3, and 4 were representative of pretreatment temperature of 260, 280, 300, and 320°C, respectively, before pyrolysis in fluidized-bed reactor at 520°C).

Figures 7D,E show the influence of staged pyrolysis on the yields of major H-phenols and G-phenols. With a low-temperature torrefaction, the phenolic compounds in the obtained bio-oils were greatly increased, and most of the phenols yields can be increased one to two times, such as guaiacol, 4-methylguaiacol, phenol, 1,2-benzenediol, etc. This was because the chemical composition of the biomass changed during torrefaction. When pretreated at low-temperature, primarily hemicellulose and even small amount of cellulose may be degraded to some extent depending on the temperature used, while the thermal stability of lignin was relatively higher and was rarely depolymerized (de Wild et al., 2009). Thus, the chemical composition of the resulting torrefied biomass was changed and the lignin content was increased, which further resulting in an increase of phenols yields in the staged pyrolysis bio-oils.

Discussion

The prior arts relating to bio-oil component quantification were more inclined to employ the internal standard method. Even this can reduce the workload to a certain extent, but it also may lead to large quantitative differences between different laboratories. Actually, finding a suitable internal standard for accurate chromatographic quantitative bio-oil components is difficult since the composition of bio-oil is so complex. So, proper standard solutions with known amounts of compounds need to be used for accurate quantitative analyses, especially for accurate quantification of the major compositions of the bio-oils. By comparing the coefficient of linear calibration curve of different compounds (Figure S1 in Supplementary Material), we found it necessary to calibrate the chromatographic systems with as many of known standard compounds as possible, while for the compounds with similar structure and chemical properties, their approximate linear coefficient makes it possible to be roughly calculated.

The comparison of our results obtained from the present methodology with that of elaborate standard addition method shows that most of our present experimental results were slightly lower than its actual yields due to the matrix effects of complex bio-oil (see Table S1 in Supplementary Material). In spite of this, the relative error of present quantitation was usually <5% except for vanillin and 1,2-benzenediol. This proves that the quantitation method here could easily provide us more accurate quantitative information about bio-oil components once we establish plenty of external standard curves.

According to the methodological steps, yields up to 40% (wt.% of the bio-oil) can be quantified; more over, the effects of pyrolysis conditions on the yields of different lightweight chemical families were easier and more realistically revealed. The overall yields of GC-detectable products from corncob and bagasse pyrolysis bio-oil were lower than that of spruce and pine, which should be mainly attributed to the more mineral salts, hemicellulose while less lignin content in grasses biomass than in woody biomass. These are more in line with commonly accepted expectations and published reports (Raveendran et al., 1995; Pushkaraj et al., 2010).

The influence of temperature on product distribution profiles as depicted in **Figures 4** and **5** can give us optimal pyrolysis temperatures for maximum yields of each chemical families and individual components. Varies in the effect of temperatures confirmed the occurrence of secondary reactions during biomass pyrolysis. Higher temperature makes the relatively larger molecular furans, cyclopentanones, and phenols further bond breaking and reforming to similar kinds while simpler compounds, such as 5-hydroxymethylfurfural to furfural, G-phenols and S-phenols to H-phenols, accompanied by the generation of small molecule aldehydes, acids, etc. One thing we have to keep in mind is, when further extending the residence time at high temperature, the repolycondensation of pyrolysis products may dominate the secondary reactions. Future direction of characterizing heavy components of bio-oils will help to confirm these postulate theories.

As speculated, the observation also revealed that low-temperature pretreatment of biomass had a good enrichment effect for the light weight components of the bio-oils. The total yield of the chromatographic quantitative component was improved by 50–90%. It is rationalized as the low-temperature pretreatment removed most of the free water and partially bound water in the raw material. However, the deep-seated root cause of these enrichment effects has yet to be explored by mechanism investigation of staged pyrolysis of model compounds.

Conclusion

The knowledge of quantitative chemical characterization of bio-oils is valuable for designing pyrolysis reaction and evaluating its strategic importance as an extended range of applicable fuels and chemicals. To this end, 19 kinds of standard compounds (i.e., major components of bio-oil) were used as external standard for quantification, and standard addition method was further used to evaluate the effectiveness and impact of the current calibration curve approach. Consequently, this methodology was adopted for a new campaign to quantify lightweight components of bio-oils obtained under a wide range of pyrolysis parameters. The results showed more than 50 compounds and yield up to 40% (wt.% of the bio-oil) can be quantified by the present method. Yields of each chemical families and individual component correlate strongly with the original feedstocks and pyrolysis temperatures. In addition, the low-temperature pretreatment of biomass has a good enrichment for the light weight components of the bio-oils, which the total yield was improved by 50–90% depending on pretreatment temperature. The final data set of the obtained results will be used as the basis for tuning pyrolysis conditions and exploring proper utilization of different bio-oils.

Acknowledgments

We acknowledge support provided by the National Key Basic Research Program of China (No. 2013CB228101), the National High Technology Research and Development Program of China (No.2012AA101806), the National Science Foundation of China (31400517), and the Open Foundation of SKLPPE (No. 201437).

References

Abdelnur, P. V., Vaz, B. G., Rocha, J. D., de Almeida, M. B. B., Teixeira, M. A. G., and Pereira, R. C. L. (2013). Characterization of bio-oils from different pyrolysis process steps and biomass using high-resolution mass spectrometry. *Energy Fuels* 27, 6646–6654. doi:10.1021/ef400788v

Akwasi, A. B., Charles, A. M., Neil, G., and Kevin, B. H. (2008). Production of bio-oil from alfalfa stems by fluidized-bed fast pyrolysis. *Ind. Eng. Chem. Res.* 47, 4115–4122. doi:10.1021/ie800096g

Badal, C. S. (2003). Hemicellulose bioconversion. *J. Ind. Microbiol. Biotechnol.* 30, 279–291. doi:10.1007/s10295-003-0049-x

Bourgois, J., and Guyonnet, R. (1988). Characterization and analysis of torrefied wood. *Wood Sci. Technol.* 22, 143–155. doi:10.1007/BF00355850

Branca, C., Giudicianni, P., and Di Blasi, C. (2003). GC/MS characterization of liquids generated from low-temperature pyrolysis of wood. *Ind. Eng. Chem. Res.* 42, 3190–3202. doi:10.1021/ie030066d

Bridgwater, A. V., Meier, D., and Radlein, D. (1999). An overview of fast pyrolysis of biomass. *Org. Geochem.* 30, 1479–1493. doi:10.1016/j.biortech.2012.06.016

Chang, S., Zhao, Z., Zhang, W., Zheng, A., Wu, W., and Li, H. (2011). Comparison of chemical composition and structure of different kinds of bio-oils. *J. Fuel Chem. Technol.* 39, 746–753.

Chang, S., Zhao, Z., Zheng, A., He, F., Huang, Z., and Li, H. (2012). Characterization of products from torrefaction of sprucewood and bagasse in an auger reactor. *Energy Fuels* 26, 7009–7017.

Charles, A. M., Gary, D. S., and Akwasi, A. B. (2009). Characterization of various fast-pyrolysis bio-oils by NMR spectroscopy. *Energy Fuels* 23, 2707–2718. doi:10.1021/ef800774w

Chiaberge, S., Leonardis, I., Fiorani, T., Cesti, P., Reale, S., and Angelis, F. D. (2014). Bio-oil from waste: a comprehensive analytical study by soft-ionization FTICR mass spectrometry. *Energy Fuels* 28, 2019–2026. doi:10.1021/ef402452f

de Wild, P. J., Uil, H. D., Reith, J. H., Kiel, J. H. A., and Heeres, H. J. (2009). Biomass valorisation by staged degasification: a new pyrolysis-based thermochemical conversion option to produce value-added chemicals from lignocellulosic biomass. *J. Anal. Appl. Pyrolysis* 85, 124–133. doi:10.1016/j.jaap.2008.08.008

Demirbas, A. (2009a). Biorefineries: current activities and future developments. *Energy Convers. Manag.* 50, 2782–2801. doi:10.1016/j.enconman.2009.06.035

Demirbas, M. F. (2009b). Biorefineries for biofuel upgrading: a critical review. *Appl. Energy* 86, S151–S161. doi:10.1016/j.apenergy.2009.04.043

Dinesh, M., Charles, U. P. J., and Philip, H. S. (2006). Pyrolysis of wood/biomass for bio-oil: a critical review. *Energy Fuels* 20, 848–889. doi:10.1021/ef0502397

Dobele, G., Urbanovich, I., Volpert, A., Kampars, V., and Samulis, E. (2007). Fast pyrolysis—effect of wood drying on the yield and properties of bio-oil. *Bioresources* 2, 699–706.

Garcia, P. M., Chaala, A., Pakdel, H., Kretschmer, D., and Roy, C. (2007). Characterization of bio-oils in chemical families. *Biomass Bioenergy* 31, 222–242. doi:10.1016/j.biombioe.2006.02.006

Guo, X. J., Wang, S. R., Guo, Z. G., Liu, Q. A., Luo, Z. Y., and Cen, K. F. (2010). Pyrolysis characteristics of bio-oil fractions separated by molecular distillation. *Appl. Energy* 87, 2892–2898. doi:10.1016/j.apenergy.2009.10.004

Helena, L. C., and Ralph, P. O. (2001). Biomass and renewable fuels. *Fuel Process. Technol.* 71, 187–195. doi:10.1016/S0378-3820(01)00146-1

Hyeon, S. H., Hyun, J. P., Young, K. P., Changkook, R., Dong, J. S., Young, W. S., et al. (2010). Bio-oil production from fast pyrolysis of waste furniture sawdust in a fluidized bed. *Bioresour. Technol.* 101, S91–S96. doi:10.1016/j.biortech.2009.06.003

Jean, N. M., Hooshang, P., and Christian, R. (2001). Separation of syringol from birch wood-derived vacuum pyrolysis oil. *Sep. Purif. Technol.* 24, 155–165. doi:10.1016/S1383-5866(00)00225-2

Kai, S., Eeva, K., Leena, F., and Anja, O. (1998). Characterization of biomass-based flash pyrolysis oil. *Biomass Bioenergy* 14, 103–113. doi:10.1016/S0961-9534(97)10024-1

Kawamoto, H., Morisaki, H., and Saka, S. (2009). Secondary decomposition of levoglucosan in pyrolytic production from cellulosic biomass. *J. Anal. Appl. Pyrolysis* 85, 247–251. doi:10.1016/j.jaap.2008.08.009

Lindfors, C., Kuoppala, E., Oasmaa, A., Solantausta, Y., and Arpiainen, V. (2014). Fractionation of bio-oil. *Energy Fuels* 28, 5785–5791. doi:10.1021/ef500754d

Lu, Q., Li, W. Z., and Zhu, X. F. (2009). Overview of fuel properties of biomass fast pyrolysis oils. *Energy Convers. Manag.* 50, 1376–1383. doi:10.1016/j.enconman.2009.01.001

Lv, G., and Wu, S. (2012). Analytical pyrolysis studies of corn stalk and its three main components by TG-MS and Py-GC/MS. *J. Anal. Appl. Pyrolysis* 97, 11–18. doi:10.1016/j.jaap.2012.04.010

Lv, G., Wu, S., and Lou, R. (2010). Characteristics of corn stalk hemicellulose pyrolysis in a tubular reactor. *Bioresources* 5, 2051–2062.

Lv, G., Wu, S., Yang, G., Chen, J., Liu, Y., and Kong, F. (2013). Comparative study of pyrolysis behaviors of corn stalk and its three components. *J. Anal. Appl. Pyrolysis* 104, 185–193. doi:10.1016/j.jaap.2013.08.005

Mckendry, P. (2002). Energy production from biomass (part 1): overview of biomass. *Bioresour. Technol.* 83, 37–46. doi:10.1016/S0960-8524(01)00118-3

Meier, D. (1999). "New methods for chemical and physical characterization and round robin testing," in *Fast Pyrolysis of Biomass: A Handbook*, Vol. 1, ed. D. Meier (Newbury: CPL Press), 92–101.

Oasmaa, A., and Meier, D. (2005). Norms and standards for fast pyrolysis liquids1. round robin test. *J. Anal. Appl. Pyrolysis* 73, 323–334. doi:10.1016/j.jaap.2005.03.003

Patrick, A. H., and Paul, T. W. (1996). Influence of temperature on the products from the flash pyrolysis of biomass. *Fuel* 75, 1051–1059. doi:10.1016/0016-2361(96)00081-6

Pushkaraj, R. P., Justinus, A. S., Robert, C. B., and Brent, H. S. (2010). Influence of inorganic salts on the primary pyrolysis products of cellulose. *Bioresour. Technol.* 101, 4646–4655. doi:10.1016/j.biortech.2010.01.112

Raveendran, K., Anuradda, G., and Kartic, C. K. (1995). Influence of mineral matter on biomass pyrolysis characteristics. *Fuel* 74, 1812–1822. doi:10.1016/0016-2361(95)80013-8

Serdar, Y. (2004). Pyrolysis of biomass to produce fuels and chemical feedstocks. *Energy Convers. Manag.* 45, 651–671. doi:10.1016/j.biotechadv.2012.01.016

Shen, D. K., and Gu, S. (2009). The mechanism for thermal decomposition of cellulose and its main products. *Bioresour. Technol.* 100, 6496–6504. doi:10.1016/j.biortech.2009.06.095

Shi, S., and He, F. (2003). *Analysis and Detection of Pulping and Papermaking*, First Edn. Beijing: Chinese Light Industry Press.

Wu, Y., Zhao, Z., Chang, S., and Li, H. (2010). Low temperature pyrolysis characteristics of corn cob and eucalyptus. *Trans. Chin. Soc. Agric. Eng.* 26, 254–258.

Yang, H., Yan, R., Chen, H., Lee, D. H., and Zheng, C. (2007). Characteristics of hemicellulose, cellulose and lignin pyrolysis. *Fuel* 86, 1781–1788. doi:10.1016/j.fuel.2006.12.013

Zheng, A. Q., Zhao, Z. L., Jiang, H. M., Zhang, W., Chang, S., Wu, W. Q., et al. (2012). Effect of pretreatment temperature of pine on bio-oil characteristics. *J. Fuel Chem. Technol.* 40, 29–36.

Conflict of Interest Statement: The authors declare that the research was conducted in the absence of any commercial or financial relationships that could be construed as a potential conflict of interest.

7

The marine microalga, *Heterosigma akashiwo*, converts industrial waste gases into valuable biomass

Jennifer J. Stewart[1]*, Colleen M. Bianco[2], Katherine R. Miller[3] and Kathryn J. Coyne[1]

[1] College of Earth, Ocean, and Environment, University of Delaware, Lewes, DE, USA, [2] Department of Microbiology, University of Illinois at Urbana-Champaign, Urbana, IL, USA [3] Department of Chemistry, Salisbury University, Salisbury, MD, USA

Edited by:
Umakanta Jena,
Desert Research Institute, USA

Reviewed by:
Alberto Scoma,
Ghent University, Belgium
Probir Das,
Qatar University, Qatar
Liz M. Diaz,
University of Puerto Rico, USA

*Correspondence:
Jennifer J. Stewart,
College of Earth, Ocean, and
Environment, University of Delaware,
700 Pilottown Road, Lewes, DE
19958, USA
jen@udel.edu

Heterosigma akashiwo is an excellent candidate for growth on industrial emissions since this alga has the ability to metabolize gaseous nitric oxide (NO) into cellular nitrogen via a novel chimeric protein (NR2-2/2HbN) and also tolerates wide fluctuations in temperature, salinity, and nutrient conditions. Here, we evaluated biomass productivity and composition, photosynthetic efficiency, and expression of *NR2-2/2HbN* for *Heterosigma* growing on simulated flue gas containing 12% CO_2 and 150 ppm NO. Biomass productivity of *Heterosigma* more than doubled in flue gas conditions compared to controls, reflecting a 13-fold increase in carbohydrate and a 2-fold increase in protein productivity. Lipid productivity was not affected by flue gas and the valuable omega-3 fatty acids, eicosapentaenoic acid and docosahexaenoic acid, constituted up to 16% of total fatty acid methyl esters. Photochemical measurements indicated that photosynthesis in *Heterosigma* is not inhibited by high CO_2 and NO concentrations, and increases in individual fatty acids in response to flue gas were driven by photosynthetic requirements. Growth rates and maximum cell densities of *Heterosigma* grown on simulated flue gas without supplemental nitrogen, along with a significant increase in *NR2-2/2HbN* transcript abundance in response to flue gas, demonstrated that nitrogen derived from NO gas is biologically available to support enhanced CO_2 fixation. Together, these results illustrate the robustness of this alga for commercial-scale biomass production and bioremediation of industrial emissions.

Keywords: bioremediation, biofuel, algae, carbon dioxide, nitric oxide, raphidophyte, biomass

Introduction

The globally distributed algal species *Heterosigma akashiwo* (Y. Hada) Y. Hada ex Y. Hara & M. Chihara (Hara and Chihara, 1987) is a unicellular chromophyte alga within the class Raphidophyceae and is well known for forming dense blooms in coastal and estuarine systems worldwide (Zhang et al., 2006; Martínez et al., 2010). *Heterosigma* has been identified as a promising candidate for the production of high quality biodiesel and is capable of achieving a higher total lipid content than several other microalgal species traditionally used in biodiesel production (Fuentes-Grünewald et al., 2013; 2012; 2009). This robust organism tolerates wide fluctuations in temperature, salinity, and nutrient conditions (Martínez et al., 2010), suggesting that this alga would be a viable option for commercial-scale biomass production. *Heterosigma* is also an excellent candidate for growth on industrial emissions (i.e., "flue gases"), since this alga has the ability to metabolize gaseous NO into cellular nitrogen via a novel chimeric protein, NR2-2/2HbN (Stewart and Coyne, 2011).

Developing innovative CO_2 utilization strategies is essential for overcoming the barriers to economic and sustainable algal biomass production, since CO_2 supplementation is a requirement for commercial biomass production due to carbon limitation at atmospheric CO_2 levels (Benemann, 2013). Utilizing flue gas CO_2 would have the dual advantage of simultaneously decreasing biomass production costs while also mitigating the effects of harmful greenhouse gas emissions. CO_2 accounts for 82% of anthropogenically derived greenhouse gas emissions in the United States (EPA, 2012), and it is widely accepted that remediation of CO_2 emissions is essential for mitigating global climate change. In addition, flue gas also contains cytotoxic nitrogen oxides (NOx, > 90% as nitric oxide), and the reaction products of NOx emissions, ozone (O_3) and nitrous oxide (N_2O), are also potent greenhouse gases (EPA, 2014). Harnessing both CO_2 and NOx emissions from flue gas as nutrient sources for *Heterosigma* growth could theoretically reduce operating costs for biomass production by 50% (Douskova et al., 2009; Nagarajan et al., 2013).

Utilization of industrial CO_2 for algal growth has been investigated for a variety of algal species, including *Spirulina* sp. (Chen et al., 2012), *Chlorella* sp. (Doucha et al., 2005; Douskova et al., 2009; Borkenstein et al., 2011; Chiu et al., 2011), and *Dunaliella* sp. (Harter et al., 2013). The green alga, *Scenedesmus* sp., is able to grow in high CO_2 and NOx environments and is currently being evaluated for growth on industrial emissions (Jin et al., 2008; Santiago et al., 2010; Basu et al., 2013; Jiang et al., 2013; Lara-Gil et al., 2014; Wilson et al., 2014). Continued identification and characterization of algal species that thrive in these harsh conditions is a critical step toward economically viable production of algal biofuels and bioproducts. To address this critical research gap, we evaluated biomass productivity, cellular composition, photosynthetic efficiency, and expression of *NR2-2/2HbN* for *H. akashiwo* growing on simulated flue gas containing 12% CO_2 and 150 ppm NO. Results of this work will support the development of commercial platforms for cultivating algal biomass on flue gas for the biofuels and bioproducts industries.

Materials and Methods

Strains and Experimental Culture Conditions

H. akashiwo CCMP 2393 (NCMA; Boothbay Harbor, ME, USA) was maintained in seawater diluted to a salinity of 20 ppt and amended with f/2 nutrients (-Si) (Guillard, 1975), buffered with 20 mM HEPES (pH = 7.35), and grown at room temperature and an irradiance of ~80 μmol quanta m^{-2} s^{-1} on a 12:12 h light:dark cycle. Light provided by cool white fluorescent bulbs was measured using an LI-250A light meter (LI-COR Biosciences, Lincoln, NE, USA) placed against the external wall of the culture vessel at a point closest to the light source. Cultures were grown in 1 L narrow-mouth polycarbonate bottles (diameter = 99 mm; Thermo Fisher Scientific, Waltham, MA, USA) sealed with screw caps retrofitted with inlet and outlet ports attached to PTFE tubing (3/16″ ID, 1/4″ OD, 1/32″ wall). Cultures were bubbled continuously (2 mL min^{-1}) with compressed air (control) or a simulated flue gas mixture consisting of 12% CO_2 and 150 ppm NO balanced in N_2 through rigid PTFE tubing that extended to the bottom of the vessel. The headspace was vented through a short piece of PTFE tubing stuffed with cotton fitted to the outlet port.

Cultures were maintained in batch growth under experimental conditions for five cycles (35 days) by replacing culture with fresh media every 7 days to achieve an initial density of 180,000 cells/mL at the start of each batch cycle. During the sixth cycle, replicate cultures (500 mL; n = 4) were sampled during mid-log growth for analysis of total carbohydrate, protein and lipid content, lipid profiles, gene expression, seawater chemistry, particulate carbon and nitrogen, and photochemistry as described below.

In a separate experiment, cells acclimated to growth on either air or simulated flue gas for five cycles were used to seed replicate cultures (500 mL; n = 4) and cultivated in modified f/2 media containing either 0 or 220 μM sodium nitrate ($NaNO_3$). Growth was monitored daily for 12 days using an improved neubauer hemocytometer (Thermo Fisher Scientific) to calculate cell density. Specific growth rate (μ) was calculated using the following equation:

$$\mu = [Ln (N_2 \div N_1)] \div (t_2 - t_1) \qquad (1)$$

where N_2 and N_1 are cell densities (cells/mL) at t_2 and t_1, respectively.

Cell Counts, Cell Size, and Cell Weight

Cell counts and cell sizes were determined using a Multisizer 3 Coulter Counter (Beckman Coulter, Indianapolis, IN, USA). Dry weight (DW) was determined from a calibration curve of DW (mg/L) versus cell counts (cells/mL) of serially diluted *H. akashiwo* harvested during exponential growth. Specifically, triplicate samples of each dilution were filtered onto pre-combusted and pre-weighed GF/F glass fiber filters (Whatman/GE Lifesciences, Pittsburg, PA, USA) and washed with 0.5 M ammonium bicarbonate to remove salts. Samples were dried at 90°C to constant weight to obtain DW. Cell concentrations (cells/mL) obtained during the experiment were then converted to DW (mg/L) equivalents. Volumetric productivity (mg L^{-1} day^{-1}) for biomass and biochemical constituents was calculated using the following equation:

$$Productivity = (N_2 - N_1) \div (t_2 - t_1) \qquad (2)$$

where N_2 and N_1 are biomass or biochemical constituent concentrations (mg L^{-1}) at t_2 and t_1, respectively.

Carbon Chemistry

Culture samples were collected in glass vials fitted with conical caps, preserved with 5% $HgCl_2$, and stored at 4°C until analysis. Dissolved inorganic carbon (DIC) was determined by the method of Sharp et al. (2009) using a custom built acid sparging instrument described in Friederich et al. (2002) fitted with a high precision flow control infrared analyzer (LI-COR Biosciences). Partial pressure of CO_2 was calculated from DIC and pH using the CO_2calc application version 1.0.3. Calculations were preformed using the GEOSECS (Li et al., 1969) option for acidity constants, the borate acidity constant of Dickson (1990), and the seawater pH scale.

Expression of *NR2-2/2HbN*

Cultures were filtered on 3-µm polycarbonate membrane filters and immediately submerged in Buffer RLT (Qiagen, Germantown, MD, USA) for gene expression analysis. Total RNA was extracted using the RNEasy Plant Mini Kit (Qiagen) and resuspended in RNase-free water. The purity of total RNA was analyzed spectroscopically (Nanodrop, Thermo Fisher Scientific) and RNA was treated with DNase I (Invitrogen/Life Technologies, Grand Island, NY, USA) as previously described (Coyne and Cary, 2005). Approximately 1 µg of DNase-treated total RNA was reverse transcribed with oligodT primer using the Superscript III First Strand Synthesis System (Invitrogen). Duplicate reactions for each DNase-treated RNA sample without reverse transcriptase were also evaluated by PCR. Transcript abundances for nitrate reductase (NR2-2/2HbN) and glyceraldehyde 3-phosphate dehydrogenase (HaGAP, as a reference gene) were determined by quantitative real time-PCR using the Stratagene MX3005P Sequence Detection System (Agilent Technologies, Santa Clara, CA, USA) as previously described (Stewart and Coyne, 2011).

Total Lipid, Protein, and Carbohydrate Quantification

Culture was centrifuged for 5 min at 4000 RPM using a swinging bucket rotor centrifuge (Thermo Fisher Scientific). Total lipid content was determined using the colorimetric sulfo-phosphovanillin assay for microalgae as described by Cheng et al. (2011) and optimized during this study for *H. akashiwo*. Lipids were extracted from centrifuged algal cells using the method developed by Folch et al. (1957). Briefly, the frozen algal pellet was homogenized with a 2:1 chloroform-methanol mixture then washed with 0.2 volumes of 0.05 M NaCl in deionized water, making a final critical ratio of 2:1:0.8 chloroform-methanol-sodium chloride solution. For the assay, 100 µL of the lower phase containing the pure lipid extract or corn oil standards containing 5–160 µg lipids in chloroform were added directly to a 96-well PCR plate. Methanol was added to each well to obtain a 2:1 chloroform-methanol ratio. The solvent was evaporated by placing the plate in a warm water bath, and then 100 µL of concentrated sulfuric acid was added to each well. The plate was then incubated at 90°C for 20 min and cooled on ice for 2 min. Equal volumes of samples and standards were transferred to a 96-well polypropylene microplate (Costar, Corning Life Sciences, Tewksbury, MA, USA) and background absorbance was measured at 540 nm. Vanillin-phosphoric acid reagent (0.2 mg/mL vanillin in 17% phosphoric acid) was immediately added to obtain a final vanillin concentration of 0.06 mg/mL. After 5 min of color development, the absorbance was measured at 540 nm on an Omega Star Microplate Reader (BMG LABTECH, Ortenburg, Germany) and total lipid content was determined by linear regression using corn oil standards.

Proteins were extracted from the centrifuged algal cells by sonication in 200 mM potassium phosphate buffer. Total protein was measured using the BCA Protein Assay Kit (Pierce, Rockford, IL, USA) according to manufacturer instructions and protein content was determined by linear regression analysis.

Carbohydrate content was determined using the phenol-sulfuric acid colorimetric method described by Dubois et al. (1956). Centrifuged algal cells were re-suspended in deionized water, and phenol and sulfuric acid were added to give a final concentration of 0.66% and 13.0 M, respectively. Samples were incubated in a room temperature water bath for 30 min, and then transferred to a 96-well plate and the absorbance was measured at 482 nm on an Omega Star Microplate Reader (BMG LABTECH). Carbohydrate content was determined by linear regression using a standard curve of known glucose concentrations (range 0–3 mM).

Particulate Carbon and Nitrogen Analysis

Particulate organic carbon and particulate organic nitrogen were quantified using a particulate autoanalyzer (Costech Elemental Analyzer, Costech Analytical Technologies, Valencia, CA, USA) as described by Hutchins et al. (2002). Briefly, 5 mL of culture were filtered onto pre-combusted GF/F Whatman glass-fiber filters, stored at -80°C and dried in an 80°C oven prior to analysis. Phenylalanine and ethylenediaminetetraacetic acid were used as standards.

Photosynthetic Physiology

Cultures were filtered onto GF/A glass fiber filters and chlorophyll *a* (chl *a*) was extracted in 5 mL of 90% acetone for 24 h at $-20°C$. Chl *a* fluorescence was measured on a Turner 10-AU fluorometer. A 1 mL subsample of culture was held under low light conditions for 20 min and dark adapted for 2 min (Hennige et al., 2013), prior to measuring fluorescence with a Fast Repetition Rate Fluorometer (FRRf; Chelsea Technologies Group, West Molesey, UK). The following photosynthetic parameters were quantified (Cosgrove and Borowitzka, 2011): maximum photochemical efficiency of PSII (F_v/F_m), the functional absorption cross section of PSII ( ), energy transfer between PSII units (p), the time constant for reoxidation of Q_A acceptor in the PSII reaction center ( ), minimum fluorescence (F_o), and maximum (F_m) fluorescence.

FAME Analysis

Fatty acid methyl esters (FAMEs) were prepared by acid catalyzed direct transesterification (Ichihara and Fukubayashi, 2010). Briefly, the lyophilized cells were re-suspended in 0.2 mL toluene. Then 1.5 mL methanol and 0.3 mL 8% (w/v) HCl in methanol solution were added to the mixture. This solution was incubated at 45°C overnight. FAMEs were subsequently extracted in 1 mL hexane. Tridecanoic acid (C13:0; final concentration of 30.1 µM) was added as an internal standard. Extracted FAMEs were stored at -20°C until analysis. FAMEs were analyzed by gas chromatography on a Hewlett Packard HP 5890 Series equipped with a flame ionization detector and a Zebron ZB-Wax column (60 m × 0.32 mm × 0.25 µm, Phenomenex, Torrance, CA, USA). Supelco 37 component FAME mix (Sigma Aldrich, St. Louis, MO, USA) was used as a standard for fatty acid identification and quantification. FAMEs were resolved using splitless injection and heating the column as follows: initial oven temperature 190°C, increased by a 15°C/min to 250°C, and held at 250°C for 25 min.

Estimation of Biodiesel Parameters

The following equations were used to estimate saponification number (SN, Eq. 3), iodine number (IN, Eq. 4), and cetane number (CN, Eq. 5) (Lei et al., 2012):

$$SN = \sum (560 \times P_i) \div MW_i \qquad (3)$$

$$IN = \sum (254 \times D \times P_i) \div MW_i \qquad (4)$$

$$CN = 46.3 + 5458 \div SN - 0.225 \times IN \qquad (5)$$

where P_i is the weight percent of each FAME, MW_i is the molecular weight of each FAME, and D is the number of double bonds in each FAME.

Statistical Analysis

Statistical analysis was performed using JMP Pro v11.2 software (SAS, Cary, NC, USA). Prior to comparison of means, data were assessed for normality and equality of variance. Raw data that did not meet assumptions of equal variance (by Levene's test) and/or normality (by the Kolmogorov-Smirnov test) were transformed prior to statistical analysis. Differences were determined to be statistically significant when $P < 0.05$.

For mean comparisons of total carbohydrate, protein and lipid content, lipid profiles, gene expression, seawater chemistry, particulate carbon and nitrogen, and photochemistry between air and simulated flue gas cultures, SDs were calculated from the average of replicates ($n = 4$) and means were compared using the Student's t-test. In cases where transformed data did not meet assumptions of equal variance and normality, the non-parametric Wilcoxon Rank-Sum test was used to compare means.

For mean comparisons of growth rate and maximum cell density between combinations of air, flue gas, $0\,\mu M$ $NaNO_3$, and $220\,\mu M$ $NaNO_3$, interaction effects were tested using a full factorial two-way ANOVA. There were no significant interaction effects for this dataset, so means were compared using a one-way ANOVA with Tukey HSD *post hoc* analysis.

Results

Carbon Chemistry

pH was significantly lower for flue gas (6.916 ± 0.037) versus air (7.336 ± 0.155) cultures ($P < 0.01$), which increased the proportion of dissolved CO_2 in the total DIC pool from 4.0% to 9.6% (**Table 1**). Flue gas treatment resulted in a large increase in both DIC and pCO_2 levels in cultures. Maximum pCO_2 levels in cultures treated with a model flue gas (12% CO_2) were 82-fold higher than cultures treated with atmospheric levels of CO_2.

Expression of *NR2-2/2HbN*

Relative expression of the *NR2-2/2HbN* transcript significantly differed between air grown cultures (1.4 ± 0.8) and flue gas cultures (6.2 ± 1.8), resulting in a 4.4-fold increase in transcript abundance in response to treatment conditions ($P < 0.02$).

Productivity

Total biomass productivity was significantly higher for flue gas cultures [18.2 (± 2.6) mg L^{-1} day^{-1}] versus air cultures [7.0 (± 1.9) mg L^{-1} day^{-1}, $P < 0.001$]. In addition to an increase in growth rate, a significant increase in average cell diameter from

TABLE 1 | Carbon chemistry of *Heterosigma akashiwo* cultures bubbled with air (control) or a simulated flue gas containing 12% CO_2 and 150 ppm NO.

	Air	Flue Gas
DIC (μM)	223 (± 53)	7767 (± 259)
pCO_2 (μatm)	282 (± 112)	23,247 (± 1317)
HCO_3^- (μmol/kg SW)	210 (± 49)	6980 (± 276)
CO_3^{2-} (μmol/kg SW)	3 (± 1)	38 (± 5)
CO_2 (μmol/kg SW)	9 (± 4)	748 (± 42)

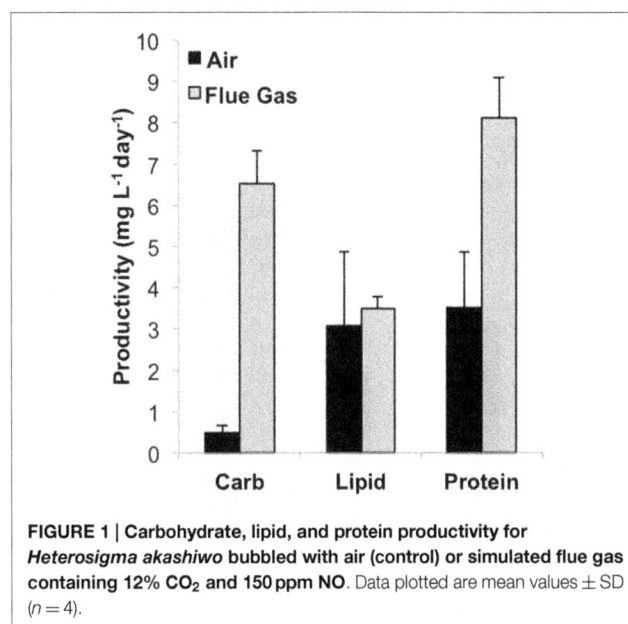

FIGURE 1 | Carbohydrate, lipid, and protein productivity for *Heterosigma akashiwo* bubbled with air (control) or simulated flue gas containing 12% CO_2 and 150 ppm NO. Data plotted are mean values $\pm$ SD ($n = 4$).

$10.5\,\mu m$ in air to $12.0\,\mu m$ in flue gas cultures was also observed ($P < 0.03$). The effect of flue gas on the productivity of biochemical constituents varied (**Figure 1**). While there was a 13-fold increase in carbohydrate ($P < 0.001$) and a 2-fold increase in protein ($P < 0.001$), both total lipid productivity and cellular lipid composition (pg/cell basis, data not shown) were unaffected. Chlorophyll *a* content also increased in flue gas cultures compared to controls (1.09 ± 0.05 and 0.92 ± 0.09 pg/cell, respectively, $P < 0.001$). The simultaneous increase in both carbohydrates and proteins was reflected in the maintenance of C:N ratios between air (7.2 ± 0.7) and flue gas (7.8 ± 0.9) cultures.

Photosynthetic Physiology

Dark-adapted photosynthetic measurements are summarized in **Table 2**. Maximum photochemical efficiency of PSII (F_v/F_m) and the rate of PSII re-oxidation (t) were both significantly lower in flue gas cultures ($P < 0.001$), while the functional absorption cross section (S_{PSII}) did not change in response to flue gas. Both minimum (F_o) and maximum (F_m) raw fluorescence significantly increased in flue gas cultures ($P < 0.001$), which coincided with a significant increase in chl *a* content as previously noted.

FAME Analysis

The following fatty acids were the predominant constituents in all growth conditions: C18:4, C16:0, C20:5n3, and C16:1. In

TABLE 2 | Maximum photochemical efficiency of PSII (F_v/F_m), functional absorption cross section (S_{PSII}), rate of PSII re-oxidation (t), minimum (F_o) and maximum (F_m) fluorescence measured in the dark for _Heterosigma akashiwo_ grown on air (control) or simulated flue gas (treatment).

	Air	Flue Gas
F_v/F_m	0.498 (±0.009)	0.458 (±0.002)
S_{PSII} (Å^2 quantum^{-1})	1.10 (±0.03)	1.10 (±0.02)
t (µs)	708 (±8)	621 (±20)
F_o (RFU)	6144 (±1)	11,758 (±1)
F_m (RFU)	12,250 (±1)	21,764 (±1)

Error represents SD (n = 4).

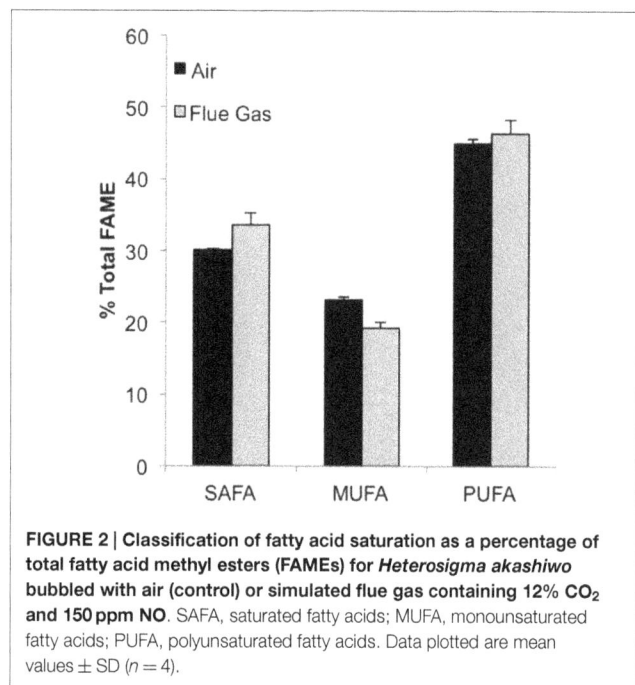

FIGURE 3 | Fatty acid methyl ester (FAME) profile for _Heterosigma akashiwo_ bubbled with air (control) or simulated flue gas containing 12% CO_2 and 150 ppm NO. (A) Individual FAMEs are reported as mean percent of total FAMEs ± SD ($n = 4$). (B) Individual FAMEs are reported on a per cell basis as mean pg/cell ± SD ($n = 4$).

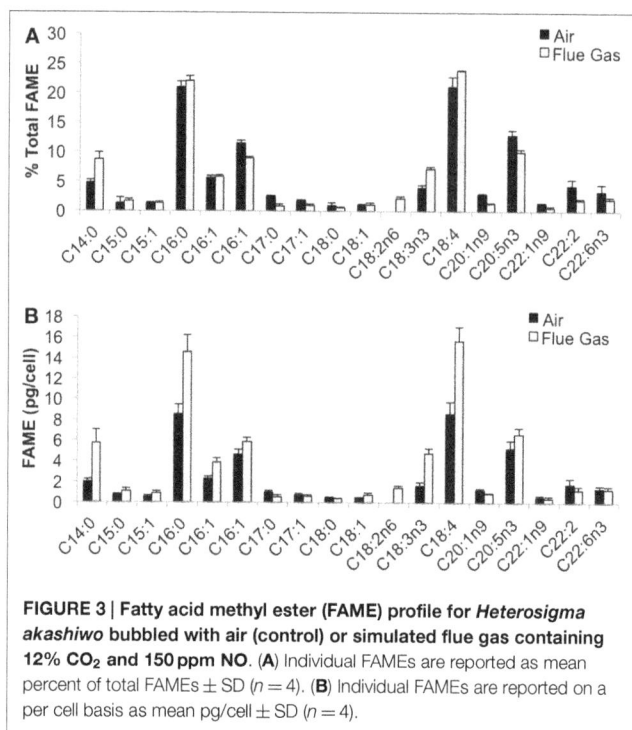

FIGURE 2 | Classification of fatty acid saturation as a percentage of total fatty acid methyl esters (FAMEs) for _Heterosigma akashiwo_ bubbled with air (control) or simulated flue gas containing 12% CO_2 and 150 ppm NO. SAFA, saturated fatty acids; MUFA, monounsaturated fatty acids; PUFA, polyunsaturated fatty acids. Data plotted are mean values ± SD ($n = 4$).

TABLE 3 | Calculated values for estimated biodiesel quality based on FAME profiles for _Heterosigma akashiwo_ grown on air (control) or simulated flue gas (treatment).

	Pre-extraction		Post-extraction (70%)		Post-extraction (100%)	
	Air	Flue gas	Air	Flue gas	Air	Flue gas
SN	197 (±1)	200 (±1)	178 (±2)	185 (±2)	170 (±2)	179 (±2)
IN	177 (±4)	175 (±5)	133 (±4)	141 (±2)	114 (±5)	126 (±1)
CN	34 (±1)	34 (±1)	47 (±1)	44 (±0.2)	53 (±2)	48 (±0.1)

SN, saponification number; IN, iodine number; CN, cetane number were calculated for biomass under three scenarios: without EPA/DHA extraction, with 70% extraction recovery of EPA/DHA, and with total recovery of EPA/DHA. Error represents SD (n = 4).

response to flue gas, the proportion of saturated fatty acids (SAFA) significantly increased ($P < 0.01$), the proportion of mono-unsaturated fatty acids (MUFA) significantly declined ($P < 0.001$), while the proportion of polyunsaturated fatty acids (PUFA) remained constant (**Figure 2**). The proportional composition of individual FAMEs in the total FAME pool showed that the significant increase in total SAFA was driven by a 1.8-fold increase in C14:0 ($P < 0.001$; **Figure 3A**). Conversely, the significant decrease in total MUFA was attributable to a decline in C17:1 and C22:1n9 ($P < 0.02$ in both cases). Total PUFA remained constant due to balanced increases and decreases in several individual PUFAs. Notably, the presence of C18:2n6 was only detected at a low concentration in one air (control) replicate, but represented 2.2 (± 0.3)% of total FAMEs in response to flue gas. In addition, the proportion of C18:4 also increased in response to flue gas ($P < 0.05$) while both C20:5n3 and C22:2 declined ($P < 0.005$ and $P < 0.04$, respectively).

When FAMEs are plotted on a cellular basis, additional patterns emerged (**Figure 3B**). The predominant fatty acids (C18:4, C16:0, C20:5n3, and C16:1) are all significantly increased on a per cell basis in response to flue gas. There was a 1.83-fold increase in

C18:4 ($P < 0.001$), a 1.70-fold increase in C16:0 ($P < 0.002$), a 1.25-fold increase in C20:5n3 ($P < 0.05$), and a total 1.97-fold increase in total isomers of C16:1 ($P < 0.01$).

Estimation of Biodiesel Parameters

Calculated values for estimated biodiesel quality are summarized in **Table 3**. Both air and flue gas cultures have an estimated cetane number (CN) of 34. Eicosapentaenoic acid (EPA) and docosahexaenoic acid (DHA) are long chain, polyunsaturated fatty acids, and their presence resulted in high estimated saponification number (SN) and iodine number (IN) values, which contributed to a decrease in estimated CN. Therefore, CN was also calculated after extraction of EPA/DHA.

Effect of Nitrate Levels on Growth in Air Versus Simulated Flue Gas

Figure 4 summarizes specific growth rate (µ) and maximum cell density (cells/mL) achieved in cultures grown under the following

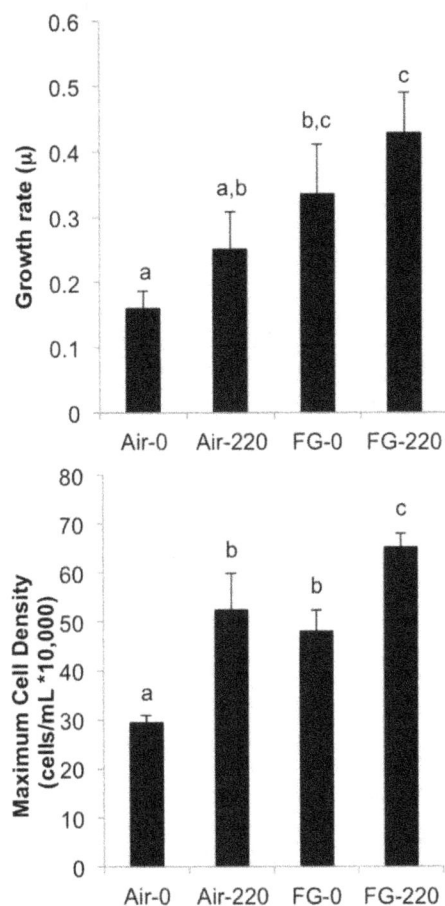

FIGURE 4 | Specific growth rate (μ) and maximum cell density (cells/mL) for cultures grown under the following conditions: air and 0 μM NaNO₃ (Air-0), air and 220 μM NaNO₃ (Air-220), flue gas and 0 μM NaNO₃ (FG-0), and flue gas and 220 μM NaNO₃ (FG-220). Data plotted are mean values ± SD ($n = 4$). Letters denote significant differences at the $P < 0.05$ level.

conditions: air and 0 μM NaNO₃ (Air-0), air and 220 μM NaNO₃ (Air-220), flue gas and 0 μM NaNO₃ (FG-0), and flue gas and 220 μM NaNO₃ (FG-220). There was not a significant interaction effect of gas type and nitrate concentration on growth rate or maximum cell density. FG-0 grew at the same rate as both Air-220 and FG-220 cultures. All cultures grew faster than Air-0, which was observed to be the only treatment to form resting cysts during stationary phase. FG-220 achieved the highest maximum cell density, while maximum cell densities for FG-0 and Air-220 were not significantly different.

Discussion

Growth on a simulated flue gas with relevant NOx levels more than doubled the biomass productivity of *H. akashiwo*. Flue gas treatment drastically increased the amount of bioavailable carbon in the media (**Table 1**), and the accumulation of large amounts of storage carbohydrates suggests that *Heterosigma* is effective at fixing this excess CO₂. Maintenance of the C:N ratio along with

an increase in protein content suggests that CO₂ assimilation in flue gas cultures was not limited by nitrogen availability. Photochemical measurements also indicated that photosynthesis in *Heterosigma* is not inhibited by high CO₂ and NO concentrations (**Table 2**). For example, the rate of PSII re-oxidation (t) was 87 μs faster, suggesting that photosynthetic electron transport was enhanced during growth on flue gas and that cells maintained overall photosynthetic efficiency despite an observed decline in dark-acclimated F_v/F_m. Growth rates and maximum cell densities of *Heterosigma* grown on simulated flue gas without supplemental nitrogen (FG-0) demonstrated that nitrogen derived from NO gas is biologically available to support enhanced CO₂ fixation (**Figure 4**). In addition, the increase in *NR2-2/2HbN* transcript abundance observed here supports the hypothesis that this chimeric enzyme is involved in maintaining cell growth in the presence of NO (Stewart and Coyne, 2011). Collectively, these results support the hypothesis that *Heterosigma* is an ideal candidate for the commercial production of algal biomass using industrial emissions containing high levels of CO₂ and NO.

Heterosigma biomass was subsequently analyzed for its potential to become a biofuel feedstock. In the conventional algae to biofuel pathway, algal lipids are extracted and upgraded to biodiesel (Davis et al., 2012). The residual biomass, which accounts for 50–75% of the total biomass, is then processed into non-fuel products, such as animal feeds and high value chemicals, or is subjected to anaerobic digestion (Davis et al., 2012). In the present study, growth on simulated flue gas did not reduce total lipid yields, so changes in lipid profiles were investigated to assess the potential for converting this lipid fraction to quality biodiesel. The American Society for Testing and Materials standards specify that the CN must be a minimum of 47 for B100 biodiesel and a minimum of 40 for blended B6 to B20 biodiesels (ASTM International, 2014). CNs based on total lipid composition for both air and flue gas cultures were well below this standard unless at least 70% of EPA/DHA was theoretically extracted prior to the transesterification of lipids (**Table 3**). The omega-3 fatty acids, eicosapentaenoic acid (EPA, C20:5n3) and docosahexaenoic acid (DHA, C22:6n3), are potential value-added products from this species with an estimated bulk wholesale value of $12,540/kg for >70% pure oil (J Edwards International, Inc., personal communication). In this study, EPA and DHA constituted approximately 12–16% of total FAMEs. Since the removal of these polyunsaturated fatty acids from the lipid mixture before conversion to biodiesel also enhances the quality of the resulting fuel, purification of these lipids prior to further processing could be economically feasible (Molina Grima et al., 2003; Chauton et al., 2014).

Interestingly, the predominant fatty acids profiled here (C18:4, C16:0, C20:5n3, and C16:1) are the main fatty acyl substituents of the thylakoid sulfolipid, sulfoquinovosyl diacylglycerol (SQDG), which is produced at high levels in *Heterosigma* (Keusgen et al., 1997). Enhanced synthesis of these fatty acids in the presence of high CO₂ and NO (**Figure 3**) likely functioned to support an increase in thylakoid number or surface area as suggested by a significant 17% increase in chlorophyll content (Benning, 1998; Minoda et al., 2002). This subsequently explains the increase in C14:0, which is linked to the fatty acid synthase enzyme prior to the two carbon elongation cycle that produces C16:0, where C16:0 is then released from fatty acid synthase as the precursor to long chain saturated and unsaturated fatty acid synthesis. Here, it appears that changes in individual fatty acid content on a cellular basis were driven by photosynthetic requirements.

In this study, lipid production did not benefit from growth on flue gas, whereas carbohydrate and protein output was significantly enhanced. A recent advancement in algal biomass processing provides an alternative to focusing solely on lipids as the primary feedstock for biofuel production. Whole algae hydrothermal liquefaction can produce fuels using the entire biomass in lieu of separating its biochemical components and has been characterized as a technically and economically feasible processing scheme (Biddy et al., 2013). Under this scheme, the primary goal is to maximize overall biomass productivity, with increases in growth rates and/or increases in the storage of any cellular product contributing to the target outcome. With hydrothermal liquefaction, the large increase in carbohydrates and protein in response to flue gas observed in the present study (**Figure 1**) is advantageous for increasing total biofuel yields. In contrast to green algae, *Heterosigma* stores photosynthetic energy not only in the form of lipids, but also as water-soluble carbohydrates stored within vacuoles (Chiovitti et al., 2006),

and the dramatic increase in carbohydrates seen here indicates that *Heterosigma* is metabolically suited to fix large amounts of anthropogenic CO_2 into valuable biomass. Together, these results illustrate the robustness of this alga and support continued efforts to assess the viability of this species for commercial-scale biomass production and bioremediation of industrial emissions.

Acknowledgments

This work was supported by the United States Department of Agriculture (USDA-NIFA-2011-67012-31175 to JS), the National Oceanic and Atmospheric Administration through Delaware Sea Grant (Award #NA10OAR4170084-12 to JS and KC and Award #NA14OAR4170087 to JS and KC), and the National Science Foundation (Award #1314003 to JS). We would like to acknowledge Catherine Fitzgerald at Salisbury University for her assistance in developing the GC protocol for the resolution of long chain FAMEs. We would also like to thank Mark Warner (University of Delaware) and Josée Nina Bouchard (Algenol, Fort Myers, FL, USA) for technical support.

References

ASTM International. (2014). *Standard Specification for Biodiesel Fuel Blend Stock (B100) for Middle Distillate Fuels: Designation D6751-14*. West Conshohocken, PA: ASTM International.

Basu, S., Roy, A. S., Mohanty, K., and Ghoshal, A. K. (2013). Enhanced CO_2 sequestration by a novel microalga: *Scenedesmus obliquus* SA1 isolated from bio-diversity hotspot region of Assam, India. *Bioresour. Technol.* 143, 369–377. doi:10.1016/j.biortech.2013.06.010

Benemann, J. (2013). Microalgae for biofuels and animal feeds. *Energies* 6, 5869–5886. doi:10.3390/en6115869

Benning, C. (1998). Biosynthesis and function of the sulfolipid sulfoquinovosyl diacylglycerol. *Annu. Rev. Plant Physiol. Plant Mol. Biol.* 49, 53–75. doi:10.1146/annurev.arplant.49.1.53

Biddy, D., Jones, R., Zhu, S., Bilal, K; Pacific Northwest National Laboratory, and National Renewable Energy Laboratory. (2013). *Whole Algae Hydrothermal Liquefaction Technology Pathway: Technical Report* NREL/TP-5100-58051. Washington, DC: U.S. Department of Energy.

Borkenstein, C. G., Knoblechner, J., Frühwirth, H., and Schagerl, M. (2011). Cultivation of *Chlorella emersonii* with flue gas derived from a cement plant. *J. Appl. Phycol.* 23, 131–135. doi:10.1007/s10811-010-9551-5

Chauton, M. S., Reitan, K. I., Norsker, N. H., Tveterås, R., and Kleivdal, H. T. (2014). A techno-economic analysis of industrial production of marine microalgae as a source of EPA and DHA-rich raw material for aquafeed: research challenges and possibilities. *Aquaculture* 436, 95–103. doi:10.1016/j.aquaculture.2014.10.038

Chen, H. W., Yang, T. S., Chen, M. J., Chang, Y. C., Lin, C. Y., Wang, E. I., et al. (2012). Application of power plant flue gas in a photobioreactor to grow *Spirulina* algae, and a bioactivity analysis of the algal water-soluble polysaccharides. *Bioresour. Technol.* 120, 256–263. doi:10.1016/j.biortech.2012.04.106

Cheng, Y., Zheng, Y., and VanderGheynst, J. S. (2011). Rapid quantitative analysis of lipids using a colorimetric method in a microplate format. *Lipids* 46, 95–103. doi:10.1007/s11745-010-3494-0

Chiovitti, A., Ngoh, J. E., and Wetherbee, R. (2006). 1, 3- -d-glucans from *Haramonas dimorpha* (Raphidophyceae). *Botanica Marina* 49, 360–362. doi:10.1515/BOT.2006.045

Chiu, S. Y., Kao, C. Y., Huang, T. T., Lin, C. J., Ong, S. C., Chen, C. D., et al. (2011). Microalgal biomass production and on-site bioremediation of carbon dioxide, nitrogen oxide and sulfur dioxide from flue gas using *Chlorella* sp. cultures. *Bioresour. Technol.* 102, 9135–9142. doi:10.1016/j.biortech.2011.06.091

Cosgrove, J., and Borowitzka, M. A. (2011). "Chlorophyll fluorescence terminology: an introduction," in *Chlorophyll a Fluorescence in Aquatic Sciences: Methods and Applications*, eds D. J. Suggett, O. Prasil, and M. A. Borowitzka (New York, NY: Springer), 1–18.

Coyne, K. J., and Cary, S. C. (2005). Molecular approaches to the investigation of viable dinoflagellate cysts in natural sediments from estuarine environments. *J. Euk. Microbiol.* 52, 90–94. doi:10.1111/j.1550-7408.2005.05202001.x

Davis, R., Fishman, D., Frank, E. D., Wigmosta, M. S., et al. (2012). *Renewable Diesel from Algal Lipids: An Integrated Baseline for Cost, Emissions, and Resource Potential from a Harmonized Model. NREL/TP-5100-55431*. Washington, DC: U.S. Department of Energy.

Dickson, A. G. (1990). Standard potential of the reaction – AGCL(S)+1/2H-2(G)=AG(S)+HCL(AQ) and the standard acidity constant of the ion HSO4- in synthetic sea-water from 273.15-k to 318.15-k. *J. Chem. Thermodyn.* 22, 113–127. doi:10.1016/0021-9614(90)90074-Z

Doucha, J., Straka, F., and Lívanský, K. (2005). Utilization of flue gas for cultivation of microalgae *Chlorella* sp. in an outdoor open thin-layer photobioreactor. *J. Appl. Phycol.* 17, 403–412. doi:10.1007/s10811-005-8701-7

Douskova, I., Doucha, J., Livansky, K., Machat, J., Novak, P., Umysova, D., et al. (2009). Simultaneous flue gas bioremediation and reduction of microalgal biomass production costs. *Appl. Microbiol. Biotechnol.* 82, 179–185. doi:10.1007/s00253-008-1811-9

Dubois, M., Gilles, K., Hamilton, J., Rebers, P., and Smith, F. (1956). Colorimetric method for determination of sugars and related substances. *Anal. Chem.* 28, 350–356. doi:10.1021/ac60111a017

EPA. (2012). *Overview of Greenhouse Gases: Carbon Dioxide Emissions*. Available at: http://www.epa.gov/climatechange/ghgemissions/gases/co2.html

EPA. (2014). *Ground Level Ozone: Basic Information*. Available at: http://www.epa.gov/climatechange/ghgemissions/gases/co2.html

Folch, J., Lees, M., Stanley, G. H. S., Han, J. I., and Park, J. K. (1957). A simple method for the isolation and purification of total lipids from animal tissues. *J. Biol. Chem.* 226, 497–509.

Friederich, G. E., Walz, P. M., Burczynski, M. G., and Chavez, F. P. (2002). Inorganic carbon in the central California upwelling system during the 1997 – 1999 el niño – la niña event. *Prog. Oceanogr.* 54, 185–203. doi:10.1016/S0079-6611(02)00049-6

Fuentes-Grünewald, C., Garcés, E., Alacid, E., Rossi, S., and Camp, J. (2013). Biomass and lipid production of dinoflagellates and raphidophytes in indoor and outdoor photobioreactors. *Mar. Biotechnol.* 15, 37–47. doi:10.1007/s10126-012-9450-7

Fuentes-Grünewald, C., Garcés, E., Alacid, E., Sampedro, N., Rossi, S., and Camp, J. (2012). Improvement of lipid production in the marine strains *Alexandrium minutum* and *Heterosigma akashiwo* by utilizing abiotic parameters. *J. Ind. Microbiol. Biotechnol.* 39, 207–216. doi:10.1007/s10295-011-1016-6

Fuentes-Grünewald, C., Garcés, E., Rossi, S., and Camp, J. (2009). Use of the dinoflagellate *Karlodinium veneficum* as a sustainable source of biodiesel production. *J. Ind. Microbiol. Biotechnol.* 36, 1215–1224. doi:10.1007/s10295-009-0602-3

Guillard, R. R. L. (1975). "Culture of phytoplankton for feeding marine invertebrates," in *Culture of Marine Invertebrate Animals*, eds W. L. Smith and M. H. Chanley (New York, NY: Plenum Press), 26–60.

Hara, Y., and Chihara, C. (1987). Morphology, ultrastructure and taxonomy of the raphidophycean alga, *Heterosigma akashiwo. Bot. Mag.* 100, 151–163. doi:10.1007/BF02488320

Harter, T., Bossier, P., Verreth, J. A. J., Bode, S., van der Ha, D., Deebeer, A. E., et al. (2013). Carbon and nitrogen mass balance during flue gas treatment with *Dunaliella salina* cultures. *J. Appl. Phycol.* 25, 359–368. doi:10.1007/s10811-012-9870-9

Hennige, S. J., Coyne, K. J., Macintyre, H., Liefer, J., and Warner, M. E. (2013). The photobiology of *Heterosigma akashiwo*: photoacclimation, diurnal periodicity, and its ability to rapidly exploit exposure to high light. *J. Phycol.* 49, 349–360. doi:10.1111/jpy.12043

Hutchins, D. A., Hare, C. E., Weaver, R. S., Zhang, Y., Firme, G. F., DiTullio, G. R., et al. (2002). Phytoplankton iron limitation in the Humboldt current and Peru upwelling. *Limnol. Oceanogr.* 47, 997–1011. doi:10.4319/lo.2002.47.4.0997

Ichihara, K., and Fukubayashi, Y. (2010). Preparation of fatty acid methyl esters for gas-liquid chromatography. *J. Lipid Res.* 51, 635–640. doi:10.1194/jlr.D001065

Jiang, Y., Zhang, W., Wang, J., Chen, Y., Shen, S., and Liu, T. (2013). Utilization of simulated flue gas for cultivation of *Scenedesmus dimorphus. Bioresour. Technol.* 128, 359–364. doi:10.1016/j.biortech.2012.10.119

Jin, H. F., Santiago, D. E. O., Park, J., and Lee, K. (2008). Enhancement of nitric oxide solubility using Fe(II)EDTA and its removal by green algae *Scenedesmus* sp. *Biotechnol. Bioprocess Eng.* 13, 48–52. doi:10.1007/s12257-007-0164-z

Keusgen, M., Curtis, J. M., Thibault, P., Walter, J. A., Windust, A., and Ayer, S. W. (1997). Sulfoquinovosyl diacylglycerols from the alga *Heterosigma carterae. Lipids* 32, 1101–1112. doi:10.1007/s11745-997-0142-9

Lara-Gil, J. A., Álvarez, M. M., and Pacheco, A. (2014). Toxicity of flue gas components from cement plants in microalgae CO_2 mitigation systems. *J. Appl. Phycol.* 26, 357–368. doi:10.1007/s10811-013-0136-y

Lei, A., Chen, H., Shen, G., Hu, Z., Chen, L., and Wang, J. (2012). Expression of fatty acid synthesis genes and fatty acid accumulation in *Haematococcus pluvialis* under different stressors. *Biotechnol. Biofuels* 5, 1–11. doi:10.1186/1754-6834-5-18

Li, Y. H., Takahash, T., and Broecker, W. S. (1969). Degree of saturation of CACO3 in oceans. *J. Geophys. Res.* 74, 5507. doi:10.1371/journal.pone.0016069

Martínez, R., Orive, E., Laza-Martínez, A., and Seoane, S. (2010). Growth response of six strains of *Heterosigma akashiwo* to varying temperature, salinity and irradiance conditions. *J. Plankton Res.* 32, 529–538. doi:10.1093/plankt/fbp135

Minoda, A., Sato, N., Nozaki, H., Okada, K., Takahashi, H., Sonoike, K., et al. (2002). Role of sulfoquinovosyl diacylglycerol for the maintenance of photosystem II in *Chlamydomonas reinhardtii. Eur. J. Biochem.* 269, 2353–2358. doi:10.1046/j.1432-1033.2002.02896.x

Molina Grima, E., Belarbi, E. H., Acién Fernández, F. G., Robles Medina, A., and Chisti, Y. (2003). Recovery of microalgal biomass and metabolites: process options and economics. *Biotechnol. Adv.* 20, 491–515. doi:10.1016/S0734-9750(02)00050-2

Nagarajan, S., Chou, S. K., Cao, S., Wu, C., and Zhou, Z. (2013). An updated comprehensive techno-economic analysis of algae biodiesel. *Bioresour. Technol.* 145, 150–156. doi:10.1016/j.biortech.2012.11.108

Santiago, D. E. O., Jin, H. F., and Lee, K. (2010). The influence of ferrous-complexed EDTA as a solubilization agent and its auto-regeneration on the removal of nitric oxide gas through the culture of green alga *Scenedesmus* sp. *Process Biochem.* 45, 1949–1953. doi:10.1016/j.procbio.2010.04.003

Sharp, J. H., Yoshiyama, K., Parker, A. E., Schwartz, M. C., Curless, S. E., Beauregard, A. Y., et al. (2009). A biogeochemical view of estuarine eutrophication: seasonal and spatial trends and correlations in the Delaware estuary. *Estuaries Coast.* 32, 1023–1043. doi:10.1007/s12237-009-9210-8

Stewart, J. J., and Coyne, K. J. (2011). Analysis of raphidophyte assimilatory nitrate reductase reveals unique domain architecture incorporating a 2/2 hemoglobin. *Plant Mol. Biol.* 77, 565–575. doi:10.1007/s11103-011-9831-8

Wilson, M. H., Groppo, J., Placido, A., Graham, S., Morton, S. A. III, Santillan-Jimenez, E., et al. (2014). CO_2 recycling using microalgae for the production of fuels. *Appl. Petrochem. Res.* 4, 41–53. doi:10.1007/s00253-012-4362-z

Zhang, Y., Fu, F.-X., Whereat, E., Coyne, K. J., and Hutchins, D. A. (2006). Bottom-up controls on a mixed-species HAB assemblage: a comparison of sympatric *Chattonella subsalsa* and *Heterosigma akashiwo* (Raphidophyceae) isolates from the Delaware Inland Bays, USA. *Harmful Algae* 5, 310–320. doi:10.1016/j.hal.2005.09.001

Conflict of Interest Statement: Jennifer J. Stewart and Kathryn J. Coyne. (2013). Novel Nitrate Reductase Fusion Proteins and Uses Thereof. U.S. Patent No. 8,409,827. Washington, DC: U. S. Patent and Trademark Office. The other co-authors declare that the research was conducted in the absence of any commercial or financial relationships that could be construed as a potential conflict of interest.

8

Efficient Eucalypt Cell Wall Deconstruction and Conversion for Sustainable Lignocellulosic Biofuels

Adam L. Healey[1]*, David J. Lee[2,3], Agnelo Furtado[1], Blake A. Simmons[1,4,5] and Robert J. Henry[1]

[1] Queensland Alliance for Agriculture and Food Innovation, University of Queensland, St. Lucia, QLD, Australia, [2] Forest Industries Research Centre, University of the Sunshine Coast, Maroochydore, QLD, Australia, [3] Department of Agriculture and Fisheries, Forestry and Biosciences, Agri-Science Queensland, Gympie, QLD, Australia, [4] Joint BioEnergy Institute, Lawrence Berkeley National Laboratory, Emeryville, CA, USA, [5] Biological and Engineering Sciences Center, Sandia National Laboratories, Livermore, CA, USA

Edited by:
Subba Rao Chaganti,
University of Windsor, Canada

Reviewed by:
Chiranjeevi Thulluri,
Jawaharlal Nehru Technological
University Hyderabad, India
Maria Carolina Quecine,
University of São Paulo, Brazil
Brahmaiah Pendyala,
University of Toledo, USA

*Correspondence:
Adam L. Healey
a.healey1@uq.edu.au

In order to meet the world's growing energy demand and reduce the impact of greenhouse gas emissions resulting from fossil fuel combustion, renewable plant-based feedstocks for biofuel production must be considered. The first-generation biofuels, derived from starches of edible feedstocks, such as corn, create competition between food and fuel resources, both for the crop itself and the land on which it is grown. As such, biofuel synthesized from non-edible plant biomass (lignocellulose) generated on marginal agricultural land will help to alleviate this competition. Eucalypts, the broadly defined taxa encompassing over 900 species of *Eucalyptus*, *Corymbia*, and *Angophora* are the most widely planted hardwood tree in the world, harvested mainly for timber, pulp and paper, and biomaterial products. More recently, due to their exceptional growth rate and amenability to grow under a wide range of environmental conditions, eucalypts are a leading option for the development of a sustainable lignocellulosic biofuels. However, efficient conversion of woody biomass into fermentable monomeric sugars is largely dependent on pretreatment of the cell wall, whose formation and complexity lend itself toward natural recalcitrance against its efficient deconstruction. A greater understanding of this complexity within the context of various pretreatments will allow the design of new and effective deconstruction processes for bioenergy production. In this review, we present the various pretreatment options for eucalypts, including research into understanding structure and formation of the eucalypt cell wall.

Keywords: eucalypts, biotechnology, pretreatment, lignocellulosic biofuel, bioenergy

INTRODUCTION

Currently, approximately 40% of the world's transportation fuels (fossil fuels) are derived from non-renewable sources, the combustion of which directly contributes to global climate change (Simmons et al., 2008; González-García et al., 2012). As such, renewable plant-based feedstocks for fuel synthesis, aptly referred to as "biofuels," are under consideration to alleviate these concerns. The first generation of feedstocks used for biofuel synthesis was mainly derived from sugarcane and corn, as their energy storage polysaccharides are readily available and easily hydrolyzed into monosaccharides for microbial fermentation. However, as these feedstocks are important links within the human food chain, generation of biofuel from these crops creates a direct competition for resources. Furthermore, in the USA alone, the maximum biofuel yield from the first-generation biofuel feedstocks is roughly 30% of the renewable fuel target (Perlack et al., 2005), creating a large gap that must be filled with alternatives. Plant cell wall structural polysaccharides, although more complex than starch molecules, represent the most abundant biopolymers in the world, containing large stores of carbon for conversion into liquid fuels, such as ethanol and butanol (Wyman, 1999). As structural polysaccharides represent the non-edible portions of plants, fuel synthesized from cellulose and hemicellulose can help alleviate the competition between energy and agriculture. Crops intended for this purpose are known as the second-generation biofuel feedstocks.

There are numerous advantages to using the second-generation feedstocks as a source of renewable energy. Combustion of fossil fuels adds carbon dioxide to the atmosphere, the main contributor to the greenhouse effect and subsequent climate change. Biofuel crops help to mitigate the effect of CO_2 by sequestering more carbon within their biomass than is released during biofuel combustion, thus creating a net reduction in CO_2 levels (Rubin, 2008; Shepherd et al., 2011; Soccol et al., 2011). High production grassy species, such as those belonging to the *Miscanthus* and *Saccharum* genera, are high-value bioenergy crops due to their exceptional growth rate and desirable biomass composition that is relatively easy to deconstruct for polysaccharides using mild pretreatments (Rubin, 2008). However, high production crops, such as these require nutrient rich soils, normally reserved for intensive agriculture. This indirect competition for land and soil between food or fuel crops can be avoided through the cultivation of feedstocks that grow well on marginal land, of which there is approximately 1.4 billion hectares available globally (Carroll and Somerville, 2009; Somerville et al., 2010). Fuel production from woody (or lignocellulosic) biomass also offers several advantages over grassy biomass. Growing trees for energy production allows biomass to be "stored on the stump" to be harvested when needed (Shepherd et al., 2011), a luxury not afforded by grasses, which must be harvested at particular times during the year and must be processed immediately before fungal degradation begins. Also, woody biomass can be transported to processing facilities more economically, as it more energy dense than grassy biomass which requires greater amounts of fuel to move the biomass than can be generated from its fibers (Kaylen et al., 2000; Somerville et al., 2010).

Eucalypts, a native Australian taxon that includes genera *Eucalyptus, Corymbia,* and *Angophora,* are an attractive prospective biofuel crop, being the most widely planted hardwood trees in the world (Myburg et al., 2007; Grattapaglia and Kirst, 2008). Having adapted to the terrestrial environment of Australia, eucalypts are well suited for plantations in a wide variety of climates, soil types, and rainfall conditions (Ladiges et al., 2003; Myburg et al., 2007; Grattapaglia and Kirst, 2008). They are grown commercially in over 100 countries with well-established silviculture practices already in place, such as clonal propagation, allowing plantations to achieve high rates of productivity, up to 25 dry

tonnes/hectare/year (Stricker et al., 2000; Rockwood et al., 2008), more than double the required productivity rate estimated by the US Department of Energy for a long-term renewable energy crop (Hinchee et al., 2009). Furthermore, many eucalypt species also regenerate shoots after harvesting, which ensures ease of management by potentially eliminating the need for re-planting (Shepherd et al., 2011).

Eucalypts, due to differences in flowering times (protantry) and self-incompatibility, are predominately out-crossing species which maintains high levels of heterozygosity in their genomes and encourages genetic diversity and phenotypic variation (Horsley and Johnson, 2007; Grattapaglia and Kirst, 2008). This variation is exploited by breeders through selection and combination of desirable traits for industrial application, such as controlled-cross hybrids that combine the high cellulose and fiber content of *Eucalyptus globulus* with the growth rate and form of *E. grandis* (Poke et al., 2005; Grattapaglia, 2008). Phenotypic traits that are desirable for efficient biofuel production are closely aligned with those sought by the pulp and paper industry. High-quality wood pulp is primarily composed of cellulosic fibers, which upon enzymatic hydrolysis releases monomeric glucose subunits, which serve as the main substrate for microbial fermentation and conversion to liquid fuel (Hisano et al., 2009; Wegrzyn et al., 2010).

Despite the advantages of lignocellulosic biofuel crops, woody biomass conversion into a source of renewable energy in hindered through its natural complexity and recalcitrance to deconstruction (Ramos and Saddler, 1994; Blanch et al., 2011). Many of the options for deconstruction require harsh and expensive chemicals (such as acids and alkalis) or energy intensive methods, such as grinding and ball milling. An increased understanding of eucalypt biomass will allow the engineering of more cost-effective pretreatments that can increase fuel production efficiency, while lessening the formation and impact of inhibitory compounds produced during conversion. In this review, we present an overview of the major contributing components of eucalypt cell wall recalcitrance, and the current research surrounding eucalypt biomass pretreatment for fuel production.

CHALLENGES TO LIGNOCELLULOSIC BIOFUEL CONVERSION

Efficient conversion of lignocellulosic biomass to biofuel requires pretreatment, saccharification, and fermentation, each presenting unique challenges (**Figure 1**). Pretreatment breaks down and separates each of the major components of biomass (cellulose, hemicellulose, and lignin), either through mechanical or through chemical means to reduce cellulose crystallinity, increase surface area, and remove lignin, the largest barrier to efficient enzymatic saccharification (Furtado et al., 2014). **Table 1** summarizes common lignocellulose pretreatments and highlights pros and cons of each process.

Despite the low cost of producing lignocellulosic biomass, the economic cost of producing biofuel remains high (Lange, 2007). Pretreatment, a required process for increasing saccharification is costly, requiring large amounts of energy or expensive chemicals

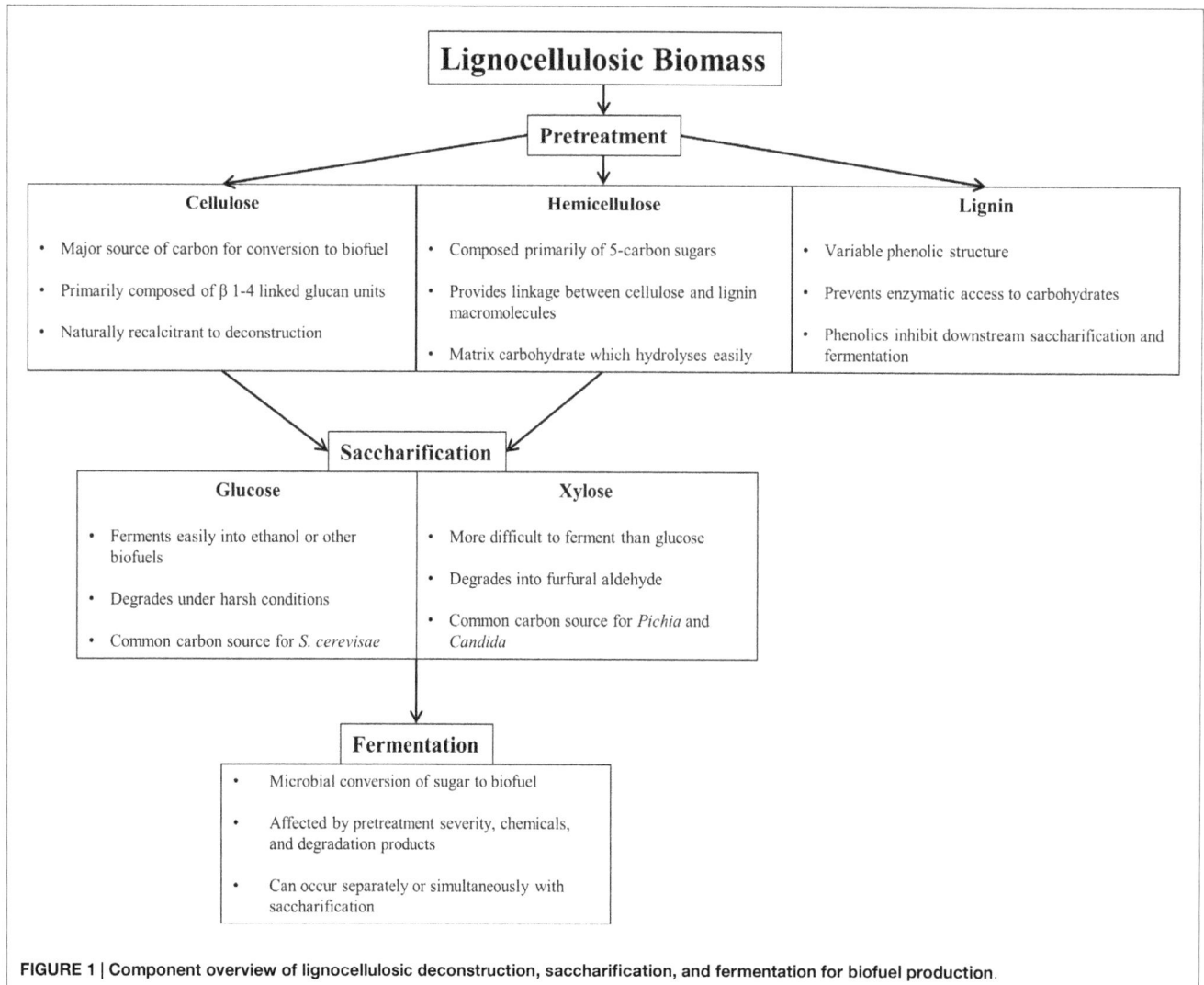

FIGURE 1 | Component overview of lignocellulosic deconstruction, saccharification, and fermentation for biofuel production.

(e.g., sulfuric acid) to promote enzymatic access to polysaccharides. Fermentation also represents a significant cost to biofuel production as the production of enzymes is expensive, and the efficiency at which microorganisms can convert sugars into fuel is dependent on pretreatment (Hamelinck et al., 2005). Therefore, harsh pretreatments that are used for biomass deconstruction in other industrial processes (e.g., pulping) may not be appropriate for biofuel production. There are also significant operational costs associated with biofuel conversion, including capital costs, labor, and waste water processing. As such, the development of simple, cost-effective, and environmentally safe pretreatments is critical for large-scale sustainable production. Given that pretreatment and fermentation represent the highest costs of producing biofuel, feedstock selection is also critical for fuel production as well.

The simplest pretreatment option is grinding and milling of biomass to increase reactive surface area for hydrolysis. However, the energy required to generate small enough particles is often too high to be a cost-effective option (Zheng et al., 2009; Talebnia et al., 2010). A more common pretreatment is acid hydrolysis, where strong acids (e.g., H_2SO_4) solubilize the hemicellulose

polysaccharide matrix, leaving behind cellulose and lignin (Galbe and Zacchi, 2007). Although effective, acid pretreatment generates compounds that inhibit downstream biomass conversion processes through reduction of microbial growth and enzymatic release (Jönsson et al., 2013). For instance, while the majority of lignin present in the cell wall is acid insoluble (Klason lignin), upon pretreatment, a small portion hydrolyzes releasing phenolics, such as vanillin, trans-cinnamic acid, and 4-hydrobenzoic acid (Palmqvist and Hahn-Hägerdal, 2000; Ximenes et al., 2010). Additionally, monomeric subunits of cellulose and hemicellulose degrade in low pH conditions, generating aldehydes [furfural and 5-hydroxymethyl-2-furaldehyde (5-HMF)] and organic acids. Formation of these degradation products inhibits fermentation by reducing available sugars and limiting microbial growth (Zheng et al., 2009; Soccol et al., 2011; Puri et al., 2012). Similarly, organosolv pretreatment combines an organic solvent (e.g., ethanol) with an inorganic acid catalyst (e.g., sulfuric acid) to destroy internal lignin and hemicellulose bonds, resulting in effective recovery of high-quality cellulose and lignin portions of biomass. Although an effective pretreatment for both hardwood and

TABLE 1 | Summary and assessment of common pretreatment options for lignocellulose.

Pretreatment	Summary	Pros	Cons
Grinding and milling	Mechanical disruption of biomass to increase surface area	No chemicals required No degradation products generated	Energy inefficient Lignin structure remains
Concentrated acid	Relatively complete hydrolysis of biomass with hydrochloric or sulfuric acid	Complete biomass hydrolysis Low inhibitory product formation under low temperature conditions	High cost and loss of acid High environmental impact Phenolic release Inhibition of fermentation
Dilute acid	Combination of acid and high temperature to solubilize hemicellulose	Low acid concentrations required (<1%) Short reaction times	Sugar degradation and loss Release of phenolics
Alkaline	Cleaves linkages within lignin and between hemicellulose and lignin	Swells biomass Established pulping practice Works with various feedstocks Low temperature, low pressure reaction	High environmental impact Low recovery Requires neutralization
Organosolv	Aqueous/organic solvent at high temperatures break hemicellulose–lignin bonds	Allows intact lignin recovery Works well across various feedstocks	Organic solvents are expensive and inhibit fermentation High temperatures (250°C) required
Steam explosion	Biomass explosion of biomass by high temperature/pressure coupled with rapid decompression	Solubilization of hemicellulose and reduced cellulose crystallinity Short reaction time	High temperatures generate inhibitory products
Autohydrolysis	Pressurized, high temperature water solubilizes hemicellulose with *in situ* acids	No chemicals needed Low environmental impact	Requires a low lignin feedstock to be efficient High temperature and pressure required
Ionic liquids	Room-temperature organic liquid salts dissolve biomass	Selective precipitation of cellulose Lignin recovery Stable, low volatility chemicals Works well regardless of varying wood properties	High cost of chemicals Inhibition of microbial fermentation

Hendriks and Zeeman (2009), Alvira et al. (2010), and Blanch et al. (2011).

softwood biomass, downstream ethanol production still suffers from the formation of inhibitory products (Sun and Cheng, 2002; Zhu and Pan, 2010).

Alkaline pretreatment, which employs chemicals, such as sodium hydroxide, lime and hydrazine, to disrupt the linkage between hemicellulose and lignin, reduces the formation of inhibitory products but nonetheless remains an expensive option that is dependent on lignin content which determines its efficacy (Blanch et al., 2011). An alternative method, which seeks to work universally well regardless of biomass composition, is ionic liquid (IL) pretreatment. ILs are non-volatile, stable compounds that solubilize lignocellulosic biomass, allowing selective precipitation of components for easy recovery. Once dissolved, cellulose precipitates from solution upon addition of an antisolvent (e.g., water or ethanol) while lignin and other solutes remain intact (Zhu et al., 2006; Singh et al., 2009).

CELLULOSE CRYSTALLINITY

Cellulose, the most abundant biopolymer on earth, is composed of thousands of glucose monomers linked together by β 1–4 glycosidic bonds. Its function within the cell wall is to provide strength and rigidity, while remaining flexible during cell expansion and growth (Mutwil et al., 2008; Mansfield, 2009). Sucrose, generated through photosynthesis, supplies the glucose molecule required for cellulose synthesis, which is phosphorylated by hexokinase, and is incorporated into growing cellulose microfibrils by cellulose synthase (CESA) enzymes (Somerville, 2006; Joshi and Mansfield, 2007; Mohnen et al., 2008). During synthesis, each cellulose microfibril associates with other glucan chain through extensive hydrogen bonding and Van der Waals forces, creating a highly compact polysaccharide. Within the cell wall, cellulose exists in primarily two forms, a highly ordered crystalline structure that lacks surface area and a less ordered, amorphous type (Harris and DeBolt, 2010). The highly compact crystalline structure lends itself toward the natural recalcitrance of woody biomass to deconstruction, as it prevents cellulase enzymes to accessing microfibrils, thus inhibiting efficient saccharification (Mosier et al., 2005; Hall et al., 2010).

Crystalline cellulose formation in eucalypts has been traditionally researched through the formation of tension wood. Tension wood, characterized by the formation a gelatinous layer of crystalline cellulose (G-layer), serves to re-direct a growing stem upwards in response to gravitational stress (Jourez et al., 2001). As tension wood can be artificially induced, Paux et al. (2005) investigated tension wood formation in *E. globulus* by tying the growing stems of 2-year-old trees to the adjacent tree, bending their trunks to a 45° angle. By extracting RNA from the xylem of the bent trees on either side of the bend (tension wood and opposite wood) at various timepoints (0, 6, 24, and 168 h), the authors were able to identify differentially expressed genes during cellulose formation using a xylem complementary DNA (cDNA) array. As evidenced by a much larger bent-stem experiment performed in *Eucalyptus nitens* with 4,900 xylem cDNAs, Qiu et al. (2008) found tension wood, although lacking the characteristic "G-layer," contained high concentrations of cellulose and low amounts of Klason lignin. Additionally, X-ray diffraction of upper and lower bent stems revealed that the cellulose microfibril angle (MFA) on the upper branch was much less than that of the lower branch. MFA, the angle at which cellulose polymers at synthesized within the cell wall affects their tendency to form hydrogen bonds. MFA, which affects wood stiffness (Schimleck et al., 2001), is an indirect biofuel trait as cellulose content negatively correlates with MFA and lignin content (Plomion et al., 2001). Qiu et al. (2008) also found that in tension wood, the highest expression profiles belonged to β-tubulin genes and fasciclin-like arabinogalactan (FLA) proteins. β-Tubulin proteins are responsible for transporting cellulose synthesis machinery to the plasma membrane, which may in-turn affect MFA. FLA genes, known to associate pectic side-chains and other structural polysaccharides also affect MFA, as demonstrated through transformation of *E. nitens* with FLA3, identified from the *E. grandis* genome (Macmillan et al., 2015).

To investigate the effect of tension and opposite wood on saccharification and fermentation, Muñoz et al. (2011) treated *E. globulus* biomass to organosolv (ethanol/water) pretreatment, followed by simultaneous saccharification and fermentation (SSF) (discussed later). The authors found that tension wood (as compared to opposite wood) contained similar glucan content (46–47%), higher xylan amounts (16.0 and 12.0%, respectively), and lower lignin content (22.1 and 26.1%, respectively). Upon pretreatment, remaining residual lignin was lower in tension wood and required less time and cooking (as expressed by H factor, a single variable calculated from the combination of cooking temperature and time) for delignification. Pulp from tension and opposite wood were assayed for glucose conversion by enzymatic hydrolysis, finding that despite similar or higher lignin content, glucan to glucose conversion was more efficient in opposite wood. However, investigation into pulp viscosity showed that tension wood glucans were of higher molecular mass, which may have influenced their rate of conversion. Upon submission of pulps from tension and opposite wood for SSF, the authors found that harsh pretreatment conditions (H factor – 12,500) outperformed milder conditions (H factor – 3,900) to produce 35 and 30 g/L of ethanol, respectively. Considering the maximum theoretical conversion of ethanol from glucose is 51%, these concentrations represent 95 and 85% conversion efficiency, which scales to a yield of 290 L of ethanol/tonne of biomass. Considering the formation of tension wood is undesirable from a timber standpoint and good management practices within plantations dictate that trees of low economic value are removed to increase the growth of high-value trees (McIntosh et al., 2012), ethanol production from eucalypt plantation thinnings is a potential option for bioenergy production, dependent on distance required for biomass transport, growth rate, and stocking rate.

NON-CELLULOSIC POLYSACCHARIDES

Before the formation of the secondary cell wall, the plant primary cell wall is a thin yet flexible structure that resists gravity and internal pressure while allowing growth and expansion (Cosgrove,

2005). Cellulose, being the core of the internal structure, provides the scaffold that non-cellulosic polysaccharides, such as hemicellulose and pectin, surround within a polysaccharide matrix (Carpita and Gibeaut, 1993; Mellerowicz and Sundberg, 2008). Although hemicellulose and pectin are polysaccharides, and thus can hydrolyze into monomeric subunits, these monomers consist mainly of pentose sugars which are more difficult to ferment than glucose. As such, based on their difficulty to ferment and how they reduce access to cellulose, hemicellulose and pectin also contribute to biomass recalcitrance (Himmel et al., 2007; Sticklen, 2008).

Xyloglucan is the most abundant hemicellulose polysaccharide of woody dicot species, with a repeating structure of β 1–4 glucan residues with various side-chains, predominantly unbranched glycosyl residues or α 1–6 xylose. Other side-chain molecules include galactose, fructose, and arabinose (Harris and DeBolt, 2010; Scheller and Ulvskov, 2010). Xyloglucan interacts with cellulose by crosslinking with non-crystalline regions or through hydrogen bonding with the microfibrils themselves (Cosgrove, 2005). For further reinforcement and strength, woody plant cell walls synthesize a secondary cell wall of cellulose, hemicellulose, and lignin. However, unlike the primary cell wall with a repetitive hemicellulose structure, the secondary cell wall polysaccharide matrix is composed of highly variable xylan molecules. This varied structure is highly substituted, with the most common modification in woody dicots being glucuronosyl residues which generates glucuronoxylan (Li et al., 2006; Scheller and Ulvskov, 2010). Given that *E. globulus* is a major source of fiber for the pulp and paper industry, the structure of its non-cellulosic polysaccharides has been extensively researched. Originally, eucalypts were believed to possess glucuronoxylan as found in woody dicot species, but investigations by Shatalov et al. (1999) and Evtuguin et al. (2003) found that *E. globulus* xylan structure was highly substituted by galactosyl and acetyl residues. These residues, although not targets for saccharification, can affect downstream conversion efficiency. Galactose is one of the most difficult sugars to ferment (Lee et al., 2011), while acetyl groups can contribute acetic acid during fermentation conditions which inhibits ethanol production in *Pichia* (Ferrari et al., 1992) and *Saccharomyces* (Taherzadeh and Karimi, 2007).

Acid pretreatment, designed to hydrolyze the hemicellulose matrix surrounding cellulose, requires various acid concentrations, pretreatment times, and temperatures to be effective. To examine these parameters on various eucalypt species, McIntosh et al. (2012) conducted a 3^3 factorial design (acid concentration, temperature, and pretreatment time) to understand sugar solubilization and degradation, enzymatic saccharification in response to pretreatment, and the fermentation of various hydrolyzates. Thinned trees of *Eucalyptus dunnii* and *Corymbia citriodora* subsp. *variegata* at ages 6 and 10 were tested within the factorial design, finding their biomass composition contained approximately 47–48% glucan, 16–17% xylan, 5% minor sugars, and 30% lignin. The authors found that under the mild pretreatment conditions [expressed as a combined severity factor (CSF)], monomeric xylose was the first to solubilize. However, as pretreatment became more severe, recovered xylose yields decreased, likely lost to degradation. Glucose release correlated with CSF increase, with temperature being the main contributing factor, followed by acid concentration and reaction time. In the presence of crude *E. dunnii* hydrolyzate, *Saccharomyces cerevisiae* could be cultured for fermentation, although the time (30 h) at which the organism was able to convert 38 g of glucose into 18 g/L of ethanol (92% efficiency) was double when compared to starch-fed fermentations (Sánchez and Cardona, 2008). This study highlights the cost/benefit analysis of biomass conversion, where more severe treatments will result in greater glucose yields but will generate more degradation products from matrix polysaccharides that inhibit fermentation. The authors also encountered significant differences in saccharification yield between biomass of different ages. After two pretreatment severity conditions (CSF 1.60 and 2.48), 6-year-old eucalypt biomass yielded greater amounts of glucose than their 10-year-old counterparts, despite similar chemical composition. These differences were attributed to changes in cellulose crystallinity, which may be species specific based on similar studies in *Populus* (DeMartini and Wyman, 2011).

Although xylose, the main monosaccharide present within hemicellulose, is more difficult to ferment by fungi due to an overproduction of nicotinamide adenine dinucleotide (NADH) under anaerobic conditions (Bruinenberg et al., 1983), hemicellulose exists as a matrix polysaccharide and is thus far less resistant to pretreatment than cellulose. To demonstrate the ease at which xylose, generated from residual *E. grandis* wood chips during pulp production, could be fermented into fuel, Silva et al. (2011) optimized ethanol production from hemicellulose hydrolyzate, generated from mild acid pretreatment. Dilute sulfuric acid was mixed with the wood chips and was then autoclaved (121°C, 45 min) to allow separation from the hemicellulose hydrolyzate portion from the solids' (cellulose and lignin) portion. The hydrolyzate was then fermented to ethanol by a *Pichia stipitis* strain, known for its ability to ferment xylose, to achieve an ethanol concentration of 15.3 g/L (100 L/tonne of biomass). As a comparison, the solids' portion, which was delignified using an alkaline NaOH pretreatment step (4%, w/v, 121°C, 20 min), was fermented by *S. cerevisiae* by an SSF process yielded a final ethanol concentration of 28.7 g/L.

Although this study demonstrates eucalypt biomass conversion from debarked biomass, bark accounts for approximately 10–12% of tree biomass residue processed from a plantation (Perlack et al., 2005; Zhu and Pan, 2010), which contains considerable levels of glucose (40%) and xylose (10%) (Lima et al., 2013). Given that bark is often not considered or optimized during lignocellulose pretreatment, Lima et al. (2013) tested various options for bark deconstruction from commercial *E. grandis* (EG) and *E. grandis × urophylla* (EGU) trees. The authors tested both one- and two-step acid and alkaline combinations in order to maximize sugar recovery. A combination of acid (1%) and NaOH (4%) pretreatment resulted in a solids fraction containing high concentrations of glucose from EG and EGU (78 and 81% dry weight, respectively); however, only 54.2 and 66.6% of total glucose was actually recovered after treatment. Upon saccharification, 65.4 and 84.5% of glucose was released from the acid + alkaline-treated bark samples. Alternatively, a single NaOH (4%) pretreatment step, while retaining lesser amounts of glucose

within the solids fraction (56 and 62%), resulted in higher total recovered glucose (63.4 and 73.1%) and more efficient enzymatic saccharification (78.5 and 98.6%).

Although alkaline pretreatments are widely used, particularly in the pulp and paper industry, the chemicals required are considered pollutants and require multiple purification steps for removal from hydrolyzate. More recently, ILs, organic salts that are liquid at room temperature act as a solvent to solubilize cellulose, hemicellulose, and lignin without degradation, have been used as an effective pretreatment (Zhu et al., 2006). ILs, although not yet developed for large-scale use, are prized for their stability, recyclability, and low volatility during biomass solubilization (Zhu et al., 2006; Shi et al., 2015). As an emerging, pretreatment option, their exact interaction with biomass during solubilization is not well understood. To examine changes in cell wall structure and composition in woody biomass in response to IL pretreatment, Çetinkol et al. (2010) compared the cell wall of *E. globulus* before and after exposure to IL 1-ethyl-3-methyl imidazolium acetate [C2min][OAc]. Using a variety of imaging and spectroscopy techniques [2-dimensional nuclear magnetic resonance spectroscopy (2D-NMR), Fourier transform infrared spectroscopy, scanning electron microscopy, small angle neutron scattering, and X-ray diffraction], they found IL pretreatment resulted in the deacetylation of xylan, acetylation of lignin, and the selective removal of G lignin monomers thereby increasing the S/G ratio. Subsequent saccharification of the treated biomass showed a significant increase in glucose (5×) yield after 1 h saccharification, which authors attributed to a decrease in cellulose crystallinity. Xylose yield was also increased after IL treatment, which was undetectable after saccharification of untreated biomass.

Depending on their chemistry, ILs interact with biomass differently. Protic ILs (PILs) can be prepared via a one-step process with low-cost acids and bases and preferentially solubilize lignin, while aprotic IL (AIL) preparation is a multistep process and preferentially dissolve carbohydrate macromolecules (Greaves et al., 2006; Zhang et al., 2015). Zhang et al. (2015) developed a concerted IL pretreatment (CIL) for *Eucalyptus* bark, combining pyrrolidinium acetate ([Pyrr][AC]; PIL) with 1-butyl-3-methylimidazolium acetate ([BMIM][AC]; AIL). Compared to untreated bark, each IL pretreatment alone ([Pyrr] or [BMIM]) or separate combinations of each ([Pyrr] and [BMIM]), the CIL pretreatment ([Pyrr]/[BMIM]) resulted in 91% enzymatic hydrolysis of cellulose, as compared to 5, 67, 50, and 77%. The same trend (13, 48, 65, and 79%) was observed during enzymatic hemicellulose hydrolysis as well (untreated biomass, [Pyrr], [BMIM], [Pyrr] and [BMIM], and [Pyrr]/[BMIM]). Reduced lignin content correlated with cellulose conversion, which was further enhanced through the removal of hemicellulose. These strategies of converting underutilized (bark, thinned trees, and hemicellulose hydrolyzate) or undesirable (tension wood) lignocellulose will be a key for the sustainable generation of biofuels through coupling bioenergy production with traditional industrial forestry practices (van Heiningen, 2006).

While acid pretreatment remains a common method of pretreatment due to its effectiveness, strong industrial acids are expensive to generate and difficult to recycle and neutralize

(Menon et al., 2010). An alternative pretreatment method utilizes residues on the xylan backbone to disrupt the structure of lignocellulose. Hot water pretreatment, or autohydrolysis, is a cost-effective pretreatment option that mixes pressurizes hot water with biomass in a reaction vessel, causing acetyl residues on the xylan backbone to generate *in situ* acetic acid. The internal generation of acetic acid reduces the pH of the biomass liquor and accelerates delignification and the solubilization of hemicellulose (Galbe and Zacchi, 2007). To demonstrate the effectiveness of liquid hot water pretreatment for eucalypt biomass, Yu et al. (2010) developed a two-step pretreatment assay (step 1: 180–200°C, 0–60 min and step 2: 180–240°C; 0, 20, 40, and 60 min) to achieve maximize xylose recovery and minimize cellulose degradation. Their results demonstrated that during the first pretreatment step, degradation of xylose to furfural increases linearly with reaction severity, a trend which continues during the second pretreatment step where furfural concentration increases between 180 and 200°C then seemingly decreases through the formation of other aldehyde products. During the second pretreatment step, furfural and 5-HMF production increased steadily over time at constant temperature (200°C), demonstrating that extended pretreatments are detrimental for recovery of monomeric sugars. Temperature had the greatest effect on the formation of inhibitory products, with authors finding that shorter reaction times and lower temperatures (180°C, 20 min; 200°C, 20 min) maximized sugar recovery (96.6%) and enzymatic digestion (81.5%).

Although autohydrolysis pretreatment can effectively solubilize hemicellulose, cellulose will remain in its recalcitrant, crystalline form after pretreatment. To reduce cellulose crystallinity in conjunction with autohydrolysis pretreatment, Inoue et al. (2008) used ball milling to improve saccharification yield from *Eucalyptus* biomass. The authors demonstrated that milling alone for short periods of time (20 min) could dramatically reduce cellulose crystallinity from 59.7 to 7.6%, although only 44.2% of sugars were captured after saccharification. To achieve higher rates of enzymatic saccharification from ball-milled biomass (86.2%), restrictively long milling times were required (120 min). To combat this, the authors combined a hot water pretreatment (160°C, 30 min) and ball milling (20 min) step to yield approximately 70% of total sugars with a low enzyme loading [4 filter paper units (FPU)/g substrate]. By comparison, the same yields were achieved by hot water pretreatment (160°C, 30 min) or ball milling (40 min) separately, each requiring 10× enzyme loading (40 FPU/g). This study demonstrates how combining methods can effectively reduce the severity of the pretreatment required to deconstruct biomass, which will lessen the formation of inhibitory products and the costs associated with enzymatic saccharification.

Traditionally, lignocellulosic biofuel production required separated process vessels where polysaccharide hydrolysis was carried out independently from microbial fermentation. Separate hydrolysis and fermentation (SHF) required additional processing and distilling steps to remove contaminants that prevent biofuel production (Olofsson et al., 2008). To improve biomass conversion efficiency and reduce fuel production costs, SSF processes generate liquid fuel from sugars as they are hydrolyzed from a polysaccharide. The advantages of SSF over SHF include

the use of a single reactor for production to reduce capital costs, lower accumulation of sugars which bolsters saccharification rate and yield, and the presence of ethanol in the reaction vessel helps reduce microbial contamination (Krishna and Chowdary, 2000; Olofsson et al., 2008). To examine the efficiency of organosolv (in this case, ethanol and water) pretreatment with SSF processes, Yáñez-S et al. (2013) pretreated *E. globulus* biomass using an SSF process with various substrate loadings (10 and 15%, w/v), thermostable yeast concentrations (6 and 12 g/L), and enzyme loadings (as expressed as cellulase FPU/β-glucosidase IU [10/20, 20/40, and 30/60]). The authors found that the highest ethanol concentration (42 g/L) was obtained from 15% (w/v) substrate loading, 20 FPU/40 IU enzyme loading, at either yeast concentration. Although higher substrate loading decreased the overall ethanol yield, ethanol concentration within the reaction vessel was increased. Furthermore, mass balance calculation from 15% substrate loading within SSF and SHF processes suggested that greater ethanol amounts could be achieved by SSF (164 and 107 L/tonne, respectively).

The strategy of increasing the solids loading during an SSF is another strategy to further reduce operation costs associated with fuel production. By increasing the weight of solids to 15–20% of the SSF reaction, the energy required to heat and distil the reaction is dramatically reduced (Wang et al., 2011). Of course, this requires optimization of process parameters, such as liquid-to-solid ratio (LSR) and enzyme-to-substrate ratio (Romaní et al., 2011). Optimization of these parameters with *E. globulus* biomass, as well as autohydrolysis pretreatment severity, allowed Romaní et al. (2012) to reach an ethanol concentration of 67.4 g/L, representing 91% conversion of ethanol from cellulose, which scales to 291 L of ethanol per tonne of biomass.

Steam explosion (SE), another cost-effective pretreatment that is similar to autohydrolysis, solubilizes hemicellulose and disrupts the structure of biomass through the breakage of linkages caused by a sudden drop in pressure. SE pretreatment is often combined with alkaline or dilute acid catalysts to increase saccharification through either delignification or increased recovery of xylose (respectively). However, addition of catalysts increase biofuel production costs either through the cost of the chemical itself or through the additional washing and neutralization steps. Thus, optimization of SE pretreatment can provide an environmentally friendly process for biofuel production. Romaní et al. (2013) optimized the temperature (173–216) and pretreatment time ranges (6–34 min) with fixed enzyme loadings (15 FPU/10 IU) to improve ethanol production from *E. globulus* biomass. Using a scanning electron micrograph (SEM) to visualize the biomass after explosion, the authors observed that exposure to a temperature of 210°C for 30 min completely opened up the fibular structure of the biomass. Although, maximum ethanol production of the SE treated material was achieved under less severe conditions (210°C, 10 min) which produced 50.9 g/L from an SSF reactor. This represents again a 91% theoretical conversion of ethanol from cellulose, scaling to 248 L/tonne of biomass.

Microbial fermentation efficiency is another limiting step during lignocellulosic biofuel production. High-fuel production strains of yeast can readily convert glucose to ethanol while withstanding ethanol toxicity but are largely unable to utilize hemicellulose derived pentose sugars (Lange, 2007). Alternate strains, belonging to *Pichia* and *Candida* genera, are capable of xylose fermentation but lack productivity. Metabolic engineering achieved through transformation to generate an organism capable of efficiently utilizing multiple carbon sources will greatly increase lignocellulosic fuel production, particularly one unfettered by high concentrations of ethanol or aldehydes, such as furfural and 5-HMF (Sun and Cheng, 2002; Wen et al., 2009). Despite eucalypt's desirable biofuel characteristics, their preferred climate ranges from cool temperate to tropical rainforest (Grattapaglia and Kirst, 2008; Shepherd et al., 2011). As such, their productivity as an energy crop outside of these climates is limited. To combat this limitation, Castro et al. (2014) investigated *E. benthamii*, a naturally cold resistant species that is commercially grown in Southeast USA, as a potential biofuel feedstock. To maximize biomass conversion, authors used a process known as liquefaction plus simultaneous saccharification and cofermentation (L + SScF), which combines dilute acid SE pretreatment with SSF processes with an inhibitor-resistant *E. coli* strain (SL100) capable of dual glucose/xylose fermentation. In addition, the authors used phosphoric acid instead of sulfuric acid, as it forms fewer inhibitory products during deconstruction and it allows the use of lower grades of stainless steel in reaction vessels, which saves on capital costs. Through optimization of temperature, acid concentration and pretreatment time (combined as a function of CSF), Castro et al. (2014) found that sugar yields were affected primarily by pretreatment time and temperature, with acid concentration having the smallest impact. Within the reaction vessel during fermentation, glucose was completely consumed within 48 h of fermentation, at which point the SL100 strain began fermenting xylose for the remainder of the 96 h fermentation. The cofermentation strategy to utilize all available carbon for conversion was successful, producing 240 g of ethanol/kg of raw biomass (304 L/ tonne). For comparison, average ethanol production from sugarcane bagasse using the same process achieved 270–280 g/ kg (342 SScF 355 L/tonne) (Geddes et al., 2013). Given the low costs of producing woody biomass (Hamelinck et al., 2005), this combination of strategies to employ alternative chemicals, SSF reaction vessels and cofermentation microbial strains that are engineered to withstand the detrimental effects of inhibitors demonstrates the feasibility of using eucalypts as a cost-effective crop for bioenergy production.

While ethanol is the most widely produced biofuel due to its ease of production, butanol is another fermentation product that can be used as a liquid fuel. Butanol is less volatile, hygroscopic, corrosive, and explosive than ethanol, can be transported with current infrastructure, and has similar energy content to gasoline (Antoni et al., 2007; Dürre, 2007; Ezeji et al., 2007; Fortman et al., 2008). Despite its advantages, microbial fermentation to butanol lacks efficiency given butanol's toxicity and often requiring nutrient supplementation which increasing operating costs (Zheng et al., 2015). Zheng et al. (2015) demonstrated the feasibility of acetone–butanol–ethanol (ABE) production from *Clostridium saccharoperbutylacetonicum* from steam exploded *Eucalyptus* biomass without nutrient supplementation. Various glucose concentrations (30–75 g/L) were achieved though varying solid

loadings (6.7–25%) finding that a hydrolyzate loading of 10% (39.5 g/L) generated the highest concentration of ABE (acetone 4.07 g/L, butanol 7.72 g/L, and ethanol 0.467 g/L). However, further optimization of glucose concentration (dilution of 75–45 g/L) produced the highest ABE concentrations (4.27 g/L acetone, 8.16 g/L butanol, and 0.643 g/L ethanol). Solids loading beyond 10% had a detrimental effect on ABE production, likely due to formation of fermentation inhibitors such as 5-HMF and phenolics.

LIGNIN

Lignin, being the second most abundant biopolymer in plant tissue, accounts for roughly 25% of biomass. Its primary role is to provide strength and rigidity to the plant, as well as assisting in vascular water transport and protection from pathogens (Boerjan et al., 2003; Ralph et al., 2004). While providing critical functions for the plant, lignin effectively surrounds structural polysaccharides within the secondary cell well, resulting in inefficient release of fermentable sugars from chemical or enzymatic hydrolysis (Hinchee et al., 2010; Jönsson et al., 2013).

Lignin synthesis begins with the conversion of phenylalanine to trans-cinnamic acid, catalyzed by the enzyme phenylalanine ammonia lyase (PAL). The remaining enzymatic steps have been well-reviewed (Ona et al., 1997; Li et al., 2006; Déjardin et al., 2010; Vanholme et al., 2010), but ultimately this biosynthetic pathway ends with the generation of the main precursors of the lignin molecule: coniferyl, *p*-coumaryl, and sinapyl alcohol (Bonawitz and Chapple, 2010). Upon transportation to the secondary cell wall, each alcohol precursor undergoes an oxidation reaction, mediated by laccase and peroxidase enzymes, which destabilize the monolignol causing it to form a covalent bond with another monolignol. Once bonded, these subunits form ρ-hydroxyphenyl (H), guaiacyl (G), and syringyl (S) lignin (Ralph et al., 2004; Bonawitz and Chapple, 2010; Vanholme et al., 2010). The most common covalent bond to occur, particularly in eucalypt lignin, is the β-θ-4 linkage, which is predominately formed from S lignin monomers. Other linkages are present, such as β–β and β-5 dimers, but β-θ-4 linkages are preferential for pulp and biofuel production as they are less stable than other bonds, branch less frequently, and are more easily broken during alkaline pretreatment (Huntley et al., 2003; Hinchee et al., 2010).

Lignin represents the largest barrier to efficient deconstruction of woody biomass. Studies performed in transgenic lines of alfalfa, poplar and *Arabidopsis* have demonstrated how slight alterations in the quantity and composition of lignin can result in large downstream effects for the saccharification of biomass (Chen and Dixon, 2007; Leplé et al., 2007; Eudes et al., 2012). Given its importance to the survivability of the plant, genetic control of cell lignification is tightly regulated. Using the promoter region of cinnamoyl CoA reductase (*CCR*) from *E. gunnii*, paired with a reporter gene (*GUS*), Lacombe et al. (2000) demonstrated using transgenic tobacco plates that *EgCCR* was highly activated during development and lignification of xylem tissues. Control of the lignin biosynthetic pathway is achieved through AC-rich elements within gene promoters. These AC elements serve as a binding platform for transcription factors (such as LIM and

MYB) that modulate gene expression (Rogers and Campbell, 2004; Zhong and Ye, 2007). The LIM transcription factor, first identified in tobacco, upregulates lignin genes. When silenced in tobacco using antisense *NtLIM1* constructs, transcripts for phenylpropanoid genes *PAL*, 4 coumarate CoA ligase (*4CL*), and cinnamyl alcohol dehydrogenase (*CAD*) were also downregulated, resulting in plants with 27% less lignin than wild type (Kawaoka et al., 2000). Similarly, suppression of the *LIM1* ortholog in *E. camaldulensis* also downregulated the *PAL*, *4CL*, and *CAD* gene pathways, resulting in plants with not only 29% less lignin but also 5% higher structural polysaccharides. The polysaccharide increase could be a result of shifting carbon resources as a result of downregulating the phenylpropanoid pathways (Kawaoka et al., 2006).

The MYB transcription factor, first discovered as a regulator of the lignin pathway in snapdragons, also affects the transcription of the lignin gene pathways. Identified from cDNA libraries of differentiating xylem tissue, the *E. grandis MYB2* gene when overexpressed in tobacco resulted in abnormal secondary cell wall thickening and altered lignin composition. Interestingly, while the expression of phenylpropanoid genes was unaltered, downstream genes responsible for monolignol synthesis [*4CL*, ρ-coumarate 3-hydroxylase (*C3H*), hydroxycinnamoyl:shikimate hydroxycinnamoyl transferase (*HCT*), caffeoyl CoA O-methyltransferase (*CCoAOMT*), ferulate 5-hydroxylase (*F5H*), caffeic acid O-methyltransferase (*COMT*), *CCR*, and *CAD*] were upregulated, increasing the S/G ratio composition of the lignin (Goicoechea et al., 2005). Another MYB transcription factor, identified from *E. grandis*, *EgMYB1*, when overexpressed in poplar and *Arabidopsis* resulted in plants with dwarfed leaves and stems and downregulated lignin and cellulose and hemicellulose transcripts. Given that the upregulation of *EgMYB1* resulted in the alteration of the major components of secondary cell wall structures suggests that MYB1 is a weak activator of lignocellulose genes, and its upregulation outcompetes stronger activators, thereby reducing overall transcription (Rogers and Campbell, 2004; Legay et al., 2010).

To investigate the effects of various wood properties on the enzymatic saccharification of woody biomass, such as lignin content, S/G ratio, cellulose crystallinity, fiber pore size, and enzyme adsorbtion, Santos et al. (2012) characterized the biomass of nine woody plants, including *E. nitens*, *E. globulus*, and *E. urograndis*. Using a Kraft alkaline pretreatment and fixed enzyme loading, the authors found that of all the parameters investigated, lignin content is the most significant contributing factor for saccharification. *E. globulus* biomass conversion resulted in the highest sugar recovery, efficient enzymatic conversion, and least residual lignin (75.2, 97.9, and 6.9%, respectively). However, lignin content alone did not fully explain saccharification yields, as biomass with similar lignin levels released much less glucose than *E. globulus*. Lignin S/G ratios were also found to impact enzymatic hydrolysis, as increased S lignin monomers undergo less frequent branching, producing a more linear polymer which increases enzymatic access to polysaccharides. Although, the effect of S/G ratio on saccharification appears to be dependent on biomass pretreatment, as acid hydrolysis has been shown to have a greater effect on low S/G lignin (Davison et al., 2006) while Papa et al. (2012)

demonstrated using three mutant lines of *E. globulus* with varying S/G ratios (0.94, 1.13, and 2.15) that lignin composition did not affect saccharification after IL pretreatment.

Given that lignin remains the largest barrier to effective deconstruction of woody biomass for fermentation, treatments to increase the efficiency at which it can be removed from biomass will aid biofuel production. To improve enzymatic saccharification of eucalypt biomass, Sykes et al. (2015) generated transgenic *E. grandis* × *urophylla* hybrids with RNA interference (RNAi)-downregulated lignin biosynthetic genes *C3H* and cinnamate 4-hydroxylase (*C4H*). Total lignin content in transgenic lines was reduced by 8–9%, and after hot water pretreatment (designed as a mild, cost-effective method for biomass disruption) and enzymatic saccharification, both *C3H* (94%) and *C4H* (97%) transgenic lines released higher total sugars than control biomass (80% saccharification). However, transgenic lines were dwarfed (*C3H* – 2.0 m and *C4H* – 3.4 m) as compared to controls (6.0 m), a common issue for lignin transgenic plants that could be alleviated through silviculture practices.

Until low lignin transgenic plants are further developed, large-scale biofuel production will depend on harsher pretreatments that inhibit microbial growth and enzymatic action through solubilization of phenolics (Ximenes et al., 2010; Jönsson et al., 2013). An alternative option to aid in delignification of biomass is the addition of laccase to destabilize the lignin network through phenol oxidation. Gutiérrez et al. (2012) and Rico et al. (2014) tested the potential of a laccase enzyme to increase saccharification from *E. globulus* biomass. Tested in the presence of an enzyme mediator, either 1-hydroxybenzotriazole (HBT) or methyl syringate (respectively), both studies reported lignin reduction (~48%) in *E. globulus* substrate and increased glucose and xylose yields after saccharification. Using pyrolysis-gas chromatography/mass spectroscopy to understand the effect of the laccase treatment, authors found an increased S/G composition (4.9 vs. 4.0) within the lignin because of preferential hydrolysis of G lignin subunits, resulting in a less condensed phenolic polymer. Continued investigation of laccase pretreatment with mediators was conducted by Rico et al. (2015) using 2D-NMR to characterize each step of delignification by fungal enzymes with *E. globulus* biomass and cellulolytic lignin. The low redox potential *M. thermophila* laccase enzyme and methyl syringate mediator pretreatment was tested against a high redox potential laccase, isolated from *Pycnoporus cinnabarinus*, with HBT mediator across several stages of pretreatment and alkaline extraction. Though various structural changes occurred throughout each stage of the fungal pretreatments, the most striking effects involved the preferential removal of guaiacyl units, reduced β-0-4 alkyl–aryl ether linkages, and S/G ratio increase. Syringyl lignin subunits underwent C_α oxidation during laccase pretreatment, which were incompletely removed through alkaline extraction. Both fungal enzyme treatments achieved similar delignification results (~50%), although multistage analysis suggests that the rate of oxidation by *P. cinnabarinus* laccase + HBT was greater. The 50% delignification result correlated with a 30% increase in glucose yield after enzymatic saccharification. These results suggest that the largest gains in sugar release from biomass result from total delignification of biomass rather than the alteration of lignin composition.

TABLE 2 | Advances in lignocellulosic biofuel production from eucalypt biomass.

Reference	Strategy	Pretreatment and fermentation conditions	Conclusions	Result
Inoue et al. (2008)	Pretreatments without acids/bases/solvents are cheaper with fewer environmental impacts	Autohydrolysis + milling	Duel pretreatment required 10x less enzyme for saccharification	70% sugar recovery
Yu et al. (2010)	Two-step liquid hot water hydrolysis of biomass	Autohydrolysis	Temperature affects degradation products formation Short reaction times and low temperatures maximize recovery	96.6% sugar recovery; 81.5% saccharification
Çetinkol et al. (2010)	IL pretreatment of biomass	1-Ethyl-3-methyl imidazolium acetate	Deacetylation of xylan Acetylation of lignin Increased S/G ratio	5x glucose yield
Silva et al. (2011)	Hemicellulose deconstruction and fermentation from residual wood chips	Dilute sulfuric acid *P. stipitis* (*S. cerevisiae* fermentation of solids)	Hemicellulose was separated from cellulose and lignin	15.3 g/L ethanol (100 L/tonne biomass) 28.7 g/L ethanol (obtained from solids)
Muñoz et al. (2011)	Fermentation of tension and opposite wood	Organosolv SSF fermentation with *S. cerevisiae*	Tension wood required milder conditions to delignify	35 g/L ethanol (290 L/tonne biomass)
McIntosh et al. (2012)	Optimization of acid concentration, temperature, and pretreatment time	Sulfuric acid *S. cerevisiae*	Hemicellulose solubilizes and degrades first Temperature contributes most to glucose release	18 g/L ethanol
Santos et al. (2012)	Screened various woody feedstocks with varying for wood properties	Alkaline pretreatment	Lignin content, enzyme adsorbtion, and S/G ratio contribute most saccharification	*E. globulus* biomass (low lignin content 7%, 98% saccharification, and 75% sugar recovery)
Papa et al. (2012)	Investigate effects of S/G ratio on IL pretreatment efficiency	1-Ethyl-3-methyl imidazolium acetate	S/G ratio did not affect IL pretreatment efficiency	Glucose yield of 759–897 g/kg cellulose after 24 h saccharification
Yáñez-S et al. (2013)	SSF optimization of substrate loading, yeast concentration, and enzyme loading	Organosolv	Higher substrate loading and midrange enzyme loading maximize yield	42 g/L ethanol (164 L/tonne of biomass)
Romaní et al. (2012)	SSF optimization of substrate and enzyme loading	Autohydrolysis and SSF reaction	91% conversion of cellulose to ethanol	67.4 g/L ethanol (291 L/tonne of biomass)
Romaní et al. (2013)	Optimization of temperature and pretreatment time	Steam explosion and SSF reaction	Maximum ethanol yield is achieved at 210°C for 10 min	50.9 g/L ethanol (248 L/tonne of biomass)
Lima et al. (2013)	Optimization of pretreatment for *Eucalyptus* bark	One/two-step acid/alkaline pretreatment	Single alkaline step recovered most glucose	73.1% glucose recovery and 98.6% saccharification
Castro et al. (2014)	SSF fermentation with inhibitor-resistant cofermentation *E. coli* strain	Steam explosion + phosphoric acid SSF fermentation + cofermentation *E. coli*	Sugar yield is primarily determined by pretreatment time and temperature	240 g ethanol/kg biomass (304 L/tonne biomass)
Rico et al. (2014, 2015)	Fungal laccases with mediator pretreatment	Laccase pretreatment + alkaline extraction	Preferential G unit removal S unit oxidation Increased S/G ratio	~50% lignin reduction and 30% increase in saccharification
Martín-Sampedro et al. (2015)	Screening, isolation, and pretreatment with endophytic fungal laccases	Fungal pretreatment + autohydrolysis	Endophytic fungi outperformed white rot reference *Trametes* strain	3.3 and 2.9x increase in total sugar release after pretreatment
Sykes et al. (2015)	RNAi downregulation of lignin genes *C3H* and *C4H*	Hot water pretreatment	Transgenic lines had less lignin and underwent more efficient saccharification Transgenic plants were dwarfed	*C3H* (94% saccharification) *C4H* (97% saccharification) Control (80% saccharification)
Zhang et al. (2015)	IL pretreatment of eucalyptus bark	Pyrrolidinium acetate and 1-butyl-3-methylimidazolium acetate	IL combinations had a synergistic effect on pretreatment	91% enzymatic hydrolysis of cellulose
Zheng et al. (2015)	ABE production without nutrients	Steam explosion and *Clostridium* fermentation	Solids loading and glucose concentration are critical for microbial inhibition	4.27 g/L acetone, 8.16 g/L butanol, and 0.643 g/L ethanol

IL, ionic liquid; S, syringyl; G, guaiacyl; SSF, simultaneous saccharification fermentation; RNAi, RNA interference; C3H, p-coumarate 3-hydroxylase; C4H, cinnamate 4-hydroxylase; ABE, acetone/butanol/ethanol.

While *M. thermophila* is a commercially available strain, its laccase enzymes may lack specificity when applied to various lignocellulose feedstocks. To investigate novel laccase enzymes from endophytic fungi, occurring in symbiosis with *Eucalyptus* trees, Martín-Sampedro et al. (2015) screened more than 100 strains, selecting five for their ligninolytic enzymes. These strains, tested against a white rot *Trametes* sp. reference, were combined with 10 g of *Eucalyptus* wood chips, before or after mild autohydrolysis pretreatment (selected to minimize the production of fungal inhibitory products). Enzymatic saccharification of each pretreatment released greater sugar yields from combination of treatments (fungal + autohydrolysis) than either pretreatment alone. Endophytic fungi strains *Ulocladium* sp. and *Hormonema* sp. outperformed the *Trametes* sp. reference strain, resulting in 3.3- and 2.9-fold increase of total sugars (compared to a 2.3-fold increase) as compared to autohydrolyzed control biomass (~3 g/L). The authors postulated that the specific activity of the ligninolytic enzymes could be a result of evolutionary processes, and endophytic fungi represent a large reservoir of biodiversity to aid biofuel production.

CONCLUSION

Given the global demand and potential for lignocellulosic biofuels, selection and research into alternative feedstocks is essential. Eucalypts, given their wide range of phenotypic diversity, genetic potential, environmental adaptability, and desirable cell wall chemistry, are excellent candidates for bioenergy production (**Table 2**). While eucalypt biomass is highly prized for other industrial processes, such as pulp and paper and timber production, the most economical way to introduce lignocellulose into the energy supply chain will be in conjunction with other plantation practices where thinned and undesirable trees are removed to promote growth of high-value trees. In addition, the production of fuel from waste wood chips and bark within pulping factories will help convert mills into complete biorefineries. Indeed, as global paper consumption diminishes, alternative uses for eucalypt biomass will require research and development. While pulping plants are efficient at deconstruction, harsh pretreatments are not suitable for downstream microbial conversion of polysaccharides to monosaccharides to fuel. High temperatures and pressures, while effective for deconstruction, generate inhibitory compounds from lignin and carbohydrates that result in sugar losses and inefficient downstream processes. Lignocellulosic fuel will require mild, low-cost pretreatments, coupled with SSF or "one-pot" processes to promote efficient biofuel production.

Genetic and chemical exploitation of eucalypt cell walls has allowed the design of mild and environmentally friendly pretreatments, such as autohydrolysis and SE, relying on *in situ* acid generation to aid deconstruction without expensive and caustic chemicals. Although these pretreatments help to reduce the formation of inhibitory products, aldehydes and phenolics formed from cellulose, hemicellulose, and lignin will likely remain in low concentrations within reaction vessels, necessitating the need for robust fermentive microbial strains. Metabolic engineering to exploit genetic variation has great potential to overcome the largest barriers to fuel conversion. These techniques have already generated dual fermentation stains to utilize all present carbon sources and resist the effects of degradation products within reaction vessels to main productivity. Application of the same principles to feedstocks have downregulated lignin gene pathways, designing plant cell walls that deconstruct with ease under mild conditions. Coupled with screening and isolation of endophytic fungi with specific ligninolytic enzymes, lignin deconstruction and removal from process vessels will maximize enzyme adsorption, sugar recovery, and fermentation.

Ionic liquids are the most promising for biomass pretreatment, given their stability and low volatility, and action at low temperatures. Despite the commercial use of cold resistant *E. benthamii*, eucalypts are not the ideal biofuel feedstock in all climates. ILs work universally well regardless of feedstock composition, solubilizing whole biomass without degradation, and selectively precipitating cellulose upon the addition of an antisolvent. Efficient saccharification of the cellulose precipitate maximizes sugar recovery and maintains intact lignin for alternate chemical processing.

In addition to the biological components of biofuel production, process optimization, such as single reaction SSF and high solids' loading, increases achievable ethanol concentrations and lower capital costs for production. Additional savings will be gained through the combination of pretreatments to reduce energy costs and enzyme loading for efficient saccharification. Increased understanding of eucalypt cell wall formation, particularly lignin formation, will allow the engineering of new and effective pretreatment options to make biofuel production suitable for a wide range of lignocellulosic feedstocks to provide renewable fuels for the future.

ACKNOWLEDGMENTS

This work was part of the DOE Joint BioEnergy Institute (http://www.jbei.org) supported by the U.S. Department of Energy, Office of Science, Office of Biological and Environmental Research, through contract DE-AC02-05CH11231 between Lawrence Berkeley National Laboratory and the U.S. Department of Energy. The United States Government retains and the publisher, by accepting the article for publication, acknowledges that the United States Government retains a non-exclusive, paid-up, irrevocable, world-wide license to publish or reproduce the published form of this manuscript, or allow others to do so, for United States Government purposes.

REFERENCES

Alvira, P., Tomás-Pejó, E., Ballesteros, M., and Negro, M. J. (2010). Pretreatment technologies for an efficient bioethanol production process based on enzymatic hydrolysis: a review. *Bioresour. Technol.* 101, 4851–4861. doi:10.1016/j.biortech.2009.11.093

Antoni, D. W., Zverlov, V. V., and Schwarz, W. H. (2007). Biofuels from microbes. *Appl. Microbiol. Biotechnol.* 77, 23–35.

Blanch, H. W., Simmons, B. A., and Klein-Marcuschamer, D. (2011). Biomass deconstruction to sugars. *Biotechnol. J.* 6, 1086–1102. doi:10.1002/biot.201000180

Boerjan, W., Ralph, J., and Baucher, M. (2003). Lignin biosynthesis. *Annu. Rev. Plant Biol.* 54, 519–546. doi:10.1146/annurev.arplant.54.031902.134938

Bonawitz, N. D., and Chapple, C. (2010). The genetics of lignin biosynthesis: connecting genotype to phenotype. *Annu. Rev. Genet.* 44, 337–363. doi:10.1146/annurev-genet-102209-163508

Bruinenberg, P. M., Bot, P. H. M., Dijken, J. P., and Scheffers, W. A. (1983). The role of redox balances in the anaerobic fermentation of xylose by yeasts. *Eur. J. Appl. Microbiol. Biotechnol.* 18, 287–292. doi:10.1007/BF00500493

Carpita, N. C., and Gibeaut, D. M. (1993). Structural models of primary cell walls in flowering plants: consistency of molecular structure with the physical properties of the walls during growth. *Plant J.* 3, 1–30. doi:10.1111/j.1365-313X.1993.tb00007.x

Carroll, A., and Somerville, C. (2009). Cellulosic biofuels. *Annu. Rev. Plant Biol.* 60, 165–182. doi:10.1146/annurev.arplant.043008.092125

Castro, E., Nieves, I. U., Mullinnix, M. T., Sagues, W. J., Hoffman, R. W., Fernández-Sandoval, M. T., et al. (2014). Optimization of dilute-phosphoric-acid steam pretreatment of *Eucalyptus benthamii* for biofuel production. *Appl. Energy* 125, 76–83. doi:10.1016/j.apenergy.2014.03.047

Çetinkol, ÖP., Dibble, D. C., Cheng, G., Kent, M. S., Knierim, B., Auer, M., et al. (2010). Understanding the impact of ionic liquid pretreatment on *Eucalyptus*. *Biofuels* 1, 33–46. doi:10.4155/bfs.09.5

Chen, F., and Dixon, R. A. (2007). Lignin modification improves fermentable sugar yields for biofuel production. *Nat. Biotechnol.* 25, 759–761. doi:10.1038/nbt1316

Cosgrove, D. J. (2005). Growth of the plant cell wall. *Nat. Rev. Mol. Cell Biol.* 6, 850–861. doi:10.1038/nrm1746

Davison, B. H., Drescher, S. R., Tuskan, G. A., Davis, M. F., and Nghiem, N. P. (2006). Variation of S/G ratio and lignin content in a *Populus* family influences the release of xylose by dilute acid hydrolysis. *Appl. Biochem. Biotechnol.* 130, 427. doi:10.1385/ABAB

Déjardin, A., Laurans, F., Arnaud, D., Breton, C., Pilate, G., and Leplé, J.-C. (2010). Wood formation in angiosperms. *C. R. Biol.* 333, 325–334. doi:10.1016/j.crvi.2010.01.010

DeMartini, J. D., and Wyman, C. E. (2011). Changes in composition and sugar release across the annual rings of *Populus* wood and implications on recalcitrance. *Bioresour. Technol.* 102, 1352–1358. doi:10.1016/j.biortech.2010.08.123

Dürre, P. (2007). Biobutanol: an attractive biofuel. *Biotechnol. J.* 2, 1525–1534. doi:10.1002/biot.200700168

Eudes, A., George, A., Mukerjee, P., Kim, J. S., Pollet, B., Benke, P. I., et al. (2012). Biosynthesis and incorporation of side-chain-truncated lignin monomers to reduce lignin polymerization and enhance saccharification. *Plant Biotechnol. J.* 10, 609–620. doi:10.1111/j.1467-7652.2012.00692.x

Evtuguin, D. V., Tomás, J. L., Silva, A. M. S., and Neto, C. P. (2003). Characterization of an acetylated heteroxylan from *Eucalyptus globulus* Labill. *Carbohydr. Res.* 338, 597–604. doi:10.1016/S0008-6215(02)00529-3

Ezeji, T. C., Qureshi, N., and Blaschek, H. P. (2007). Bioproduction of butanol from biomass: from genes to bioreactors. *Curr. Opin. Biotechnol.* 18, 220–227. doi:10.1016/j.copbio.2007.04.002

Ferrari, M. D., Neirotti, E., Albornoz, C., and Saucedo, E. (1992). Ethanol production from *Eucalyptus* wood hemicellulose hydrolysate by *Pichia stipitis*. *Biotechnol. Bioeng.* 40, 753–759. doi:10.1002/bit.260400702

Fortman, J. L., Chhabra, S., Mukhopadhyay, A., Chou, H., Lee, T. S., Steen, E., et al. (2008). Biofuel alternatives to ethanol: pumping the microbial well. *Trends Biotechnol.* 26, 375–381. doi:10.1016/j.tibtech.2008.03.008

Furtado, A., Lupoi, J. S., Hoang, N. V., Healey, A., Singh, S., Simmons, B. A., et al. (2014). Modifying plants for biofuel and biomaterial production. *Plant Biotechnol. J.* 12, 1246–1258. doi:10.1111/pbi.12300

Galbe, M., and Zacchi, G. (2007). Pretreatment of lignocellulosic materials for efficient bioethanol production. *Adv. Biochem. Eng. Biotechnol.* 108, 41–65. doi:10.1007/10_2007_070

Geddes, C. C., Mullinnix, M. T., Nieves, I. U., Hoffman, R. W., Sagues, W. J., York, S. W., et al. (2013). Seed train development for the fermentation of bagasse from sweet sorghum and sugarcane using a simplified fermentation process. *Bioresour. Technol.* 128, 716–724. doi:10.1016/j.biortech.2012.09.121

Goicoechea, M., Lacombe, E., Legay, S., Mihaljevic, S., Rech, P., Jauneau, A., et al. (2005). EgMYB2, a new transcriptional activator from *Eucalyptus* xylem, regulates secondary cell wall formation and lignin biosynthesis. *Plant J.* 43, 553–567. doi:10.1111/j.1365-313X.2005.02480.x

González-García, S., Moreira, M. T., and Feijoo, G. (2012). Environmental aspects of *Eucalyptus* based ethanol production and use. *Sci. Total Environ.* 438, 1–8. doi:10.1016/j.scitotenv.2012.07.044

Grattapaglia, D. (2008). *Genomics of Eucalyptus, a Global Tree for Energy, Paper, and Wood* (New York: Springer), 259–297.

Grattapaglia, D., and Kirst, M. (2008). *Eucalyptus* applied genomics: from gene sequences to breeding tools. *New Phytol.* 179, 911–929. doi:10.1111/j.1469-8137.2008.02503.x

Greaves, T. L., Weerawardena, A., Fong, C., Krodkiewska, I., and Drummond, C. J. (2006). Protic ionic liquids: solvents with tunable phase behavior and physicochemical properties. *J. Phys. Chem.* 110, 22479–22487. doi:10.1021/jp0634048

Gutiérrez, A., Rencoret, J., Cadena, E. M., Rico, A., Barth, D., Del Río, J. C., et al. (2012). Demonstration of laccase-based removal of lignin from wood and non-wood plant feedstocks. *Bioresour. Technol.* 119, 114–122. doi:10.1016/j.biortech.2012.05.112

Hall, M., Bansal, P., Lee, J. H., Realff, M. J., and Bommarius, A. S. (2010). Cellulose crystallinity – a key predictor of the enzymatic hydrolysis rate. *FEBS J.* 277, 1571–1582. doi:10.1111/j.1742-4658.2010.07585.x

Hamelinck, C. N., Hooijdonk, G. V., and Faaij, A. P. (2005). Ethanol from lignocellulosic biomass: techno-economic performance in short-, middle- and long-term. *Biomass Bioenergy* 28, 384–410. doi:10.1016/j.biombioe.2004.09.002

Harris, D., and DeBolt, S. (2010). Synthesis, regulation and utilization of lignocellulosic biomass. *Plant Biotechnol. J.* 8, 244–262. doi:10.1111/j.1467-7652.2009.00481.x

Hendriks, A. T. W. M., and Zeeman, G. (2009). Pretreatments to enhance the digestibility of lignocellulosic biomass. *Bioresour. Technol.* 100, 10–18. doi:10.1016/j.biortech.2008.05.027

Himmel, M. E., Ding, S.-Y., Johnson, D. K., Adney, W. S., Mark, R., Brady, J. W., et al. (2007). Biomass recalcitrance: engineering plants and enzymes for biofuels production. *Science* 315, 804–807. doi:10.1126/science.1137016

Hinchee, M., Rottmann, W., Mullinax, L., Zhang, C., Chang, S., Cunningham, M., et al. (2009). Short-rotation woody crops for bioenergy and biofuels applications. *In Vitro Cell. Dev. Biol. Plant* 45, 619–629. doi:10.1007/s11627-009-9235-5

Hinchee, M. A. W., Mullinax, L. N., and Rottmann, W. H. (2010). Chapter 7 woody biomass and purpose-grown trees as feedstocks for renewable energy. in *Plant Biotechnology for Sustainable Production of Energy and Co-products*, eds Mascia P. N., Scheffran J., and Widholm J. M. (Berlin: Springer), 155–208.

Hisano, H., Nandakumar, R., and Wang, Z.-Y. (2009). Genetic modification of lignin biosynthesis for improved biofuel production. *In Vitro Cell. Dev. Biol. Plant* 45, 306–313. doi:10.1007/s11627-009-9219-5

Horsley, T. N., and Johnson, S. D. (2007). Is *Eucalyptus* cryptically self-incompatible? *Ann. Bot.* 100, 1373–1378. doi:10.1093/aob/mcm223

Huntley, S. K., Ellis, D., Gilbert, M., Chapple, C., and Mansfield, S. D. (2003). Significant increases in pulping efficiency in C4H-F5H-transformed poplars: improved chemical savings and reduced environmental toxins. *J. Agric. Food Chem.* 51, 6178–6183. doi:10.1021/jf034320o

Inoue, H., Yano, S., Endo, T., Sakaki, T., and Sawayama, S. (2008). Combining hot-compressed water and ball milling pretreatments to improve the efficiency of the enzymatic hydrolysis of *Eucalyptus*. *Biotechnol. Biofuels* 1, 2–2. doi:10.1186/1754-6834-1-2

Jönsson, L. J., Alriksson, B., and Nilvebrant, N.-O. (2013). Bioconversion of lignocellulose: inhibitors and detoxification. *Biotechnol. Biofuels* 6, 16–16. doi:10.1186/1754-6834-6-16

Joshi, C. P., and Mansfield, S. D. (2007). The cellulose paradox – simple molecule, complex biosynthesis. *Curr. Opin. Plant Biol.* 10, 220–226. doi:10.1016/j.pbi.2007.04.013

Jourez, B., Riboux, A., and Leclercq, A. (2001). Comparison of basic density and longitudinal shrinkage in tension wood and opposite wood in young stems of *Populus euramericana* cv. Ghoy when subjected to a gravitational stimulus. *Can. J. For. Res.* 31, 1676–1683. doi:10.1139/cjfr-31-10-1676

Kawaoka, A., Kaothien, P., Yoshida, K., Endo, S., Yamada, K., and Ebinuma, H. (2000). Functional analysis of tobacco LIM protein Ntlim1 involved in lignin biosynthesis. *Plant J.* 22, 289–301. doi:10.1046/j.1365-313x.2000.00737.x

Kawaoka, A., Nanto, K., Ishii, K., and Ebinuma, H. (2006). Reduction of lignin content by suppression of expression of the LIM domain transcription factor in *Eucalyptus camaldulensis*. *Silvae Genet.* 6, 269–277.

Kaylen, M., Van Dyne, D. L., Choi, Y.-S., and Blase, M. (2000). Economic feasibility of producing ethanol from lignocellulosic feedstocks. *Bioresour. Technol.* 72, 19–32. doi:10.1016/S0960-8524(99)00091-7

Krishna, S. H., and Chowdary, G. V. (2000). Optimization of simultaneous saccharification and fermentation for the production of ethanol from lignocellulosic biomass. *J. Agric. Food Chem.* 48, 1971–1976. doi:10.1021/jf991296z

Lacombe, E., Van Doorsselaere, J., Boerjan, W., Boudet, A. M., and Grima-Pettenati, J. (2000). Characterization of cis-elements required for vascular expression of the cinnamoyl CoA reductase gene and for protein-DNA complex formation. *Plant J.* 23, 663–676. doi:10.1046/j.1365-313x.2000.00838.x

Ladiges, P. Y., Udovicic, F., and Nelson, G. (2003). Australian biogeographical connections and the phylogeny of large genera in the plant family *Myrtaceae*. *J. Biogeogr.* 30, 989–998. doi:10.1046/j.1365-2699.2003.00881.x

Lange, J. P. (2007). Lignocellulose conversion: an introduction to chemistry, process and economics. *Biofuels Bioprod. Biorefin.* 1, 39–48. doi:10.1002/bbb

Lee, K. S., Hong, M. E., Jung, S. C., Ha, S. J., Yu, B. J., Koo, H. M., et al. (2011). Improved galactose fermentation of *Saccharomyces cerevisiae* through inverse metabolic engineering. *Biotechnol. Bioeng.* 108, 621–631. doi:10.1002/bit.22988

Legay, S., Sivadon, P., Blervacq, A.-S., Pavy, N., Baghdady, A., Tremblay, L., et al. (2010). EgMYB1, an R2R3 MYB transcription factor from *Eucalyptus* negatively regulates secondary cell wall formation in *Arabidopsis* and poplar. *New Phytol.* 188, 774–786. doi:10.1111/j.1469-8137.2010.03432.x

Leplé, J.-C., Dauwe, R., Morreel, K., Storme, V., Lapierre, C., Naumann, A., et al. (2007). Downregulation of cinnamoyl-coenzyme a reductase in poplar: multiple-level phenotyping reveals effects on cell wall polymer metabolism and structure. *Plant Cell* 19, 3669–3691. doi:10.1105/tpc.107.054148

Li, L., Lu, S., and Chiang, V. (2006). A genomic and molecular view of wood formation. *CRC Crit. Rev. Plant Sci.* 25, 215–233. doi:10.1080/07352680600611519

Lima, M. A., Lavorente, G. B., Da Silva, H. K., Bragatto, J., Rezende, C. A., Bernardinelli, O. D., et al. (2013). Effects of pretreatment on morphology, chemical composition and enzymatic digestibility of *Eucalyptus* bark: a potentially valuable source of fermentable sugars for biofuel production – part 1. *Biotechnol. Biofuels* 6, 75–75. doi:10.1186/1754-6834-6-75

Macmillan, C. P., Taylor, L., Bi, Y., Southerton, S. G., and Evans, R. (2015). The fasciclin-like arabinogalactan protein family of *Eucalyptus grandis* contains members that impact wood biology and biomechanics. *New Phytol.* 206, 1314–1327. doi:10.1111/nph.13320

Mansfield, S. D. (2009). Solutions for dissolution – engineering cell walls for deconstruction. *Curr. Opin. Biotechnol.* 20, 286–294. doi:10.1016/j.copbio.2009.05.001

Martín-Sampedro, R., Fillat, Ú, Ibarra, D., and Eugenio, M. E. (2015). Use of new endophytic fungi as pretreatment to enhance enzymatic saccharification of *Eucalyptus globulus*. *Bioresour. Technol.* 196, 383–390. doi:10.1016/j.biortech.2015.07.088

McIntosh, S., Vancov, T., Palmer, J., and Spain, M. (2012). Ethanol production from *Eucalyptus* plantation thinnings. *Bioresour. Technol.* 110, 264–272. doi:10.1016/j.biortech.2012.01.114

Mellerowicz, E. J., and Sundberg, B. (2008). Wood cell walls: biosynthesis, developmental dynamics and their implications for wood properties. *Curr. Opin. Plant Biol.* 11, 293–300. doi:10.1016/j.pbi.2008.03.003

Menon, V., Prakash, G., and Rao, M. (2010). Value added products from hemicelluloses: biotechnological perspective. *Glob. J. Biochem.* 1, 36–67.

Mohnen, D., Somerville, C., and Bar-Peled, M. (2008). "Biosynthesis of plant cell walls," in *Biomass Recalcitrance*, ed. Himmel M. (Oxford: Blackwell Publishing), 94–187.

Mosier, N., Wyman, C., Dale, B., Elander, R., Lee, Y. Y., Holtzapple, M., et al. (2005). Features of promising technologies for pretreatment of lignocellulosic biomass. *Bioresour. Technol.* 96, 673–686. doi:10.1016/j.biortech.2004.06.025

Muñoz, C., Baeza, J., Freer, J., and Mendonça, R. T. (2011). Bioethanol production from tension and opposite wood of *Eucalyptus globulus* using organosolv pretreatment and simultaneous saccharification and fermentation. *J. Ind. Microbiol. Biotechnol.* 38, 1861–1866. doi:10.1007/s10295-011-0975-y

Mutwil, M., Debolt, S., and Persson, S. (2008). Cellulose synthesis: a complex complex. *Curr. Opin. Plant Biol.* 11, 252–257. doi:10.1016/j.pbi.2008.03.007

Myburg, A. A., Potts, B. M., Marques, C. M., Kirst, M., Gion, J.-M., and Grima-Pettenatti, J. (2007). "Eucalypts," in *Genome Mapping and Molecular Breeding in Plants, Volume 7, Forest Trees*, ed. Kole C. (Berlin: Springer-Verlag), 115–160.

Olofsson, K., Bertilsson, M., and Lidén, G. (2008). A short review on SSF – an interesting process option for ethanol production from lignocellulosic feedstocks. *Biotechnol. Biofuels* 1, 7–7. doi:10.1186/1754-6834-1-7

Ona, T., Sonoda, T., Itoh, K., and Shibata, M. (1997). Relationship of lignin content, lignin monomeric composition and hemicellulose composition in the same trunk sought by their within-tree variations in *Eucalyptus camaldulensis* and *E. globulus*. *Holzforschung* 51, 396–404. doi:10.1515/hfsg.1997.51.5.396

Palmqvist, E., and Hahn-Hägerdal, B. (2000). Fermentation of lignocellulosic hydrolysates. II: inhibitors and mechanisms of inhibition. *Bioresour. Technol.* 74, 25–33. doi:10.1016/S0960-8524(99)00161-3

Papa, G., Varanasi, P., Sun, L., Cheng, G., Stavila, V., Holmes, B., et al. (2012). Exploring the effect of different plant lignin content and composition on ionic liquid pretreatment efficiency and enzymatic saccharification of *Eucalyptus globulus* L. mutants. *Bioresour. Technol.* 117, 352–359. doi:10.1016/j.biortech.2012.04.065

Paux, E., Carocha, V., Marques, C., Mendes De Sousa, A., Borralho, N., Sivadon, P., et al. (2005). Transcript profiling of *Eucalyptus* xylem genes during tension wood formation. *New Phytol.* 167, 89–100. doi:10.1111/j.1469-8137.2005.01396.x

Perlack, R. D., Wright, L. L., Turhollow, A. F., Graham, R. L., Stokes, B. J., and Erbach, D. C. (2005). *Biomass as a Feedstock for a Bioenergy and Bioproducts Industry: The Technical Feasibility of a Billion-Ton Annual Supply*. Oak Ridge National Lab TN.

Plomion, C., Leprovost, G., and Stokes, A. (2001). Wood formation in trees. *Plant Physiol.* 127, 1513–1523. doi:10.1104/pp.010816.1

Poke, F. S., Vaillancourt, R. E., Potts, B. M., and Reid, J. B. (2005). Genomic research in *Eucalyptus*. *Genetica* 125, 79–101. doi:10.1007/s10709-005-5082-4

Puri, M., Abraham, R. E., and Barrow, C. J. (2012). Biofuel production: prospects, challenges and feedstock in Australia. *Renew. Sustain. Energ. Rev.* 16, 6022–6031. doi:10.1016/j.rser.2012.06.025

Qiu, D., Wilson, I. W., Gan, S., Washusen, R., Moran, G. F., and Southerton, S. G. (2008). Gene expression in *Eucalyptus* branch wood with marked variation in cellulose microfibril orientation and lacking G-layers. *New Phytol.* 179, 94–103. doi:10.1111/J.1469-8137.2008.02439.X

Ralph, J., Lundquist, K., Brunow, G., Lu, F., Kim, H., Schatz, P. F., et al. (2004). Lignins: natural polymers from oxidative coupling of 4-hydroxyphenyl- propanoids. *Phytochem. Rev.* 3, 29–60. doi:10.1023/B:PHYT.0000047809.65444.a4

Ramos, L. P., and Saddler, J. N. (1994). "Bioconversion of wood residues mechanisms involved in pretreating and hydrolyzing lignocellulosic materials," in *Enzymatic Conversion of Biomass for Fuels Production*, ed. Himmel M. (Washington, DC: ACS Symposium Series; American Chemical Society), 325–341.

Rico, A., Rencoret, J., Del Río, J. C., Martínez, A. T., and Gutiérrez, A. (2014). Pretreatment with laccase and a phenolic mediator degrades lignin and enhances saccharification of *Eucalyptus* feedstock. *Biotechnol. Biofuels* 7, 6–6. doi:10.1186/1754-6834-7-6

Rico, A., Rencoret, J., del Río, J. C., Martínez, A. T., and Gutiérrez, A. (2015). In-depth 2D NMR study of lignin modification during pretreatment of *Eucalyptus* wood with laccase and mediators. *Bioenerg. Res.* 8, 211–230. doi:10.1007/s12155-014-9505-x

Rockwood, D. L., Rudie, A. W., Ralph, S. A., Zhu, J. Y., and Winandy, J. E. (2008). Energy product options for *Eucalyptus* species grown as short rotation woody crops. *Int. J. Mol. Sci.* 9, 1361–1378. doi:10.3390/ijms9081361

Rogers, L. A., and Campbell, M. M. (2004). The genetic control of lignin deposition during plant growth and development. *New Phytol.* 164, 17–30. doi:10.1111/j

Romaní, A., Garrote, G., Ballesteros, I., and Ballesteros, M. (2013). Second generation bioethanol from steam exploded *Eucalyptus globulus* wood. *Fuel* 111, 66–74. doi:10.1016/j.fuel.2013.04.076

Romaní, A., Garrote, G., López, F., and Parajó, J. C. (2011). *Eucalyptus globulus* wood fractionation by autohydrolysis and organosolv delignification. *Bioresour. Technol.* 102, 5896–5904. doi:10.1016/j.biortech.2011.02.070

Romaní, A., Garrote, G., and Parajó, J. C. (2012). Bioethanol production from autohydrolyzed *Eucalyptus globulus* by simultaneous saccharification and fermentation operating at high solids loading. *Fuel* 94, 305–312. doi:10.1016/j.fuel.2011.12.013

Rubin, E. M. (2008). Genomics of cellulosic biofuels. *Nature* 454, 841–845. doi:10.1038/nature07190

Sánchez, ÓJ., and Cardona, C. A. (2008). Trends in biotechnological production of fuel ethanol from different feedstocks. *Bioresour. Technol.* 99, 5270–5295. doi:10.1016/j.biortech.2007.11.013

Santos, R. B., Lee, J. M., Jameel, H., Chang, H.-M., and Lucia, L. A. (2012). Effects of hardwood structural and chemical characteristics on enzymatic hydrolysis for biofuel production. *Bioresour. Technol.* 110, 232–238. doi:10.1016/j.biortech.2012.01.085

Scheller, H. V., and Ulvskov, P. (2010). Hemicelluloses. *Annu. Rev. Plant Biol.* 61, 263–289. doi:10.1146/annurev-arplant-042809-112315

Schimleck, L. R., Evans, R., and Ilic, J. (2001). Estimation of *Eucalyptus delegatensis* wood properties by near infrared spectroscopy. *Can. J. For. Res.* 31, 1671–1675. doi:10.1139/cjfr-31-10-1671

Shatalov, A. A., Evtuguin, D. V., and Pascoal Neto, C. (1999). (2-O-alpha-D-galactopyranosyl-4-O-methyl-alpha-D-glucurono)-D-xylan from *Eucalyptus globulus* Labill. *Carbohydr. Res.* 320, 93–99. doi:10.1016/S0008-6215(99)00136-6

Shepherd, M., Bartle, J., Lee, D. J., Brawner, J., Bush, D., Turnbull, P., et al. (2011). eucalypts as a biofuel feedstock. *Biofuels* 2, 639–657. doi:10.4155/bfs.11.136

Shi, J., George, K. W., Sun, N., He, W., Li, C., Stavila, V., et al. (2015). Impact of pretreatment technologies on saccharification and isopentenol fermentation of mixed lignocellulosic feedstocks. *Bioenerg. Res.* 8, 1004–1013. doi:10.1007/s12155-015-9588-z

Silva, N. L. C., Betancur, G. J. V., Vasquez, M. P., Gomes, E. D. B., and Pereira, N. (2011). Ethanol production from residual wood chips of cellulose industry: acid pretreatment investigation, hemicellulosic hydrolysate fermentation, and remaining solid fraction fermentation by SSF process. *Appl. Biochem. Biotechnol.* 163, 928–936. doi:10.1007/s12010-010-9096-8

Simmons, B. A., Loque, D., and Blanch, H. W. (2008). Next-generation biomass feedstocks for biofuel production. *Genome Biol.* 9, 242–242. doi:10.1186/gb-2008-9-12-242

Singh, S., Simmons, B. A., and Vogel, K. P. (2009). Visualization of biomass solubilization and cellulose regeneration during ionic liquid pretreatment of switchgrass. *Biotechnol. Bioeng.* 104, 68–75. doi:10.1002/bit.22386

Soccol, C. R., Faraco, V., Karp, S., Vandenberghe, L. P. S., Thomaz-Soccol, V., Woiciechowski, A., et al. (2011). *Lignocellulosic Bioethanol: Current Status and Future Perspectives*. San Diego, CA: Elsevier Inc.

Somerville, C. (2006). Cellulose synthesis in higher plants. *Annu. Rev. Cell Dev. Biol.* 22, 53–78. doi:10.1146/annurev.cellbio.22.022206.160206

Somerville, C., Youngs, H., Taylor, C., Davis, S. C., and Long, S. P. (2010). Feedstocks for lignocellulosic biofuels. *Science* 329, 790–792. doi:10.1126/science.1189268

Sticklen, M. B. (2008). Plant genetic engineering for biofuel production: towards affordable cellulosic ethanol. *Nat. Rev. Genet.* 9, 433–443. doi:10.1038/nrg2336

Stricker, J., Rockwood, D. L., Segrest, S. A., Alker, G. R., Prine, G. M., and Carter, D. R. (2000). Short rotation woody crops for Florida. in *Proc. 3rd Biennial Short Rotation Woody Crops Operations Working Group Conference*. p. 15–23.

Sun, Y., and Cheng, J. (2002). Hydrolysis of lignocellulosic materials for ethanol production: a review. *Bioresour. Technol.* 83, 1–11. doi:10.1016/S0960-8524(01)00212-7

Sykes, R. W., Gjersing, E. L., Foutz, K., Rottmann, W. H., Kuhn, S. A., Foster, C. E., et al. (2015). Down-regulation of p-coumaroyl quinate/shikimate 3'-hydroxylase (C3'H) and cinnamate 4-hydroxylase (C4H) genes in the lignin biosynthetic pathway of *Eucalyptus urophylla* × *E. grandis* leads to improved sugar release. *Biotechnol. Biofuels* 8, 128–128. doi:10.1186/s13068-015-0316-x

Taherzadeh, M. J., and Karimi, K. (2007). Acid-based hydrolysis processes for ethanol from lignocellulosic materials: a review. *Bioresources* 2, 472–499.

Talebnia, F., Karakashev, D., and Angelidaki, I. (2010). Production of bioethanol from wheat straw: an overview on pretreatment, hydrolysis and fermentation. *Bioresour. Technol.* 101, 4744–4753. doi:10.1016/j.biortech.2009.11.080

van Heiningen, A. (2006). Converting a kraft pulp mill into an integrated forest biorefinery. *Pulp Pap. Canada* 107, 141–146.

Vanholme, R., Demedts, B., Morreel, K., Ralph, J., and Boerjan, W. (2010). Lignin biosynthesis and structure. *Plant Physiol.* 153, 895–905. doi:10.1104/pp.110.155119

Wang, W., Kang, L., Wei, H., Arora, R., and Lee, Y. Y. (2011). Study on the decreased sugar yield in enzymatic hydrolysis of cellulosic substrate at high solid loading. *Appl. Biochem. Biotechnol.* 164, 1139–1149. doi:10.1007/s12010-011-9200-8

Wegrzyn, J. L., Eckert, A. J., Choi, M., Lee, J. M., Stanton, B. J., Sykes, R., et al. (2010). Association genetics of traits controlling lignin and cellulose biosynthesis in black cottonwood (*Populus trichocarpa*, Salicaceae) secondary xylem. *New Phytol.* 188, 515–532. doi:10.1111/j.1469-8137.2010.03415.x

Wen, F., Nair, N. U., and Zhao, H. (2009). Protein engineering in designing tailored enzymes and microorganisms for biofuels production. *Curr. Opin. Biotechnol.* 20, 412–419. doi:10.1016/j.copbio.2009.07.001.Protein

Wyman, C. E. (1999). Biomass ethanol: technical progress, opportunities, and commercial challenges. *Annu. Rev. Energy Environ.* 24, 189–226. doi:10.1146/annurev.energy.24.1.189

Ximenes, E., Kim, Y., Mosier, N., Dien, B., and Ladisch, M. (2010). Inhibition of cellulases by phenols. *Enzyme Microb. Technol.* 46, 170–176. doi:10.1016/j.enzmictec.2009.11.001

Yáñez-S, M., Rojas, J., Castro, J., Ragauskas, A., Baeza, J., and Freer, J. (2013). Fuel ethanol production from *Eucalyptus globulus* wood by autocatalized organosolv pretreatment ethanol-water and SSF. *J. Chem. Technol. Biotechnol.* 88, 39–48. doi:10.1002/jctb.3895

Yu, Q., Zhuang, X., Yuan, Z., Wang, Q., Qi, W., Wang, W., et al. (2010). Two-step liquid hot water pretreatment of *Eucalyptus grandis* to enhance sugar recovery and enzymatic digestibility of cellulose. *Bioresour. Technol.* 101, 4895–4899. doi:10.1016/j.biortech.2009.11.051

Zhang, C., Xu, W., Yan, P., Liu, X., and Zhang, Z. C. (2015). Overcome the recalcitrance of *Eucalyptus* bark to enzymatic hydrolysis by concerted ionic liquid pretreatment. *Process Biochem.* doi:10.1016/j.procbio.2015.09.009

Zheng, J., Tashiro, Y., Wang, Q., Sakai, K., and Sonomoto, K. (2015). Feasibility of acetone–butanol–ethanol fermentation from *Eucalyptus* hydrolysate without nutrients supplementation. *Appl. Energy* 140, 113–119. doi:10.1016/j.apenergy.2014.11.037

Zheng, Y., Pan, Z., and Zhang, R. (2009). Overview of biomass pretreatment for cellulosic ethanol production. *Int. J. Agric. Biol. Eng.* 2, 51–68. doi:10.3965/j.issn.1934-6344.2009.03.051-068

Zhong, R., and Ye, Z.-H. (2007). Regulation of cell wall biosynthesis. *Curr. Opin. Plant Biol.* 10, 564–572. doi:10.1016/j.pbi.2007.09.001

Zhu, J. Y., and Pan, X. J. (2010). Woody biomass pretreatment for cellulosic ethanol production: technology and energy consumption evaluation. *Bioresour. Technol.* 101, 4992–5002. doi:10.1016/j.biortech.2009.11.007

Zhu, S., Wu, Y., Chen, Q., Yu, Z., Wang, C., Jin, S., et al. (2006). Dissolution of cellulose with ionic liquids and its application: a mini-review. *Green Chem.* 8, 325–325. doi:10.1039/b601395c

Conflict of Interest Statement: The authors declare that the research was conducted in the absence of any commercial or financial relationships that could be construed as a potential conflict of interest.

9

Dynamic modeling of the microalgae cultivation phase for energy production in open raceway ponds and flat panel photobioreactors

Matteo Marsullo[1], Alberto Mian[2], Adriano Viana Ensinas[2,3], Giovanni Manente[1], Andrea Lazzaretto[1]* and François Marechal[2]

[1] Department of Industrial Engineering, University of Padova, Padova, Italy, [2] Industrial Process and Energy System Engineering Group (IPESE), École Polytechnique Fédérale de Lausanne, Lausanne, Switzerland, [3] Universidade Federal do ABC, Santo Andre, Brazil

Edited by:
Jalel Labidi,
University of the Basque Country,
Spain

Reviewed by:
Gergely Forgacs,
University of Bath, UK
Tianju Chen,
Chinese Academy of Sciences, China

*Correspondence:
Andrea Lazzaretto,
Department of Industrial Engineering,
University of Padova, Via Venezia 1,
Padova 35131, Italy
andrea.lazzaretto@unipd.it

A dynamic model of microalgae cultivation phase is presented in this work. Two cultivation technologies are taken into account: the open raceway pond and the flat panel photobioreactor. For each technology, the model is able to evaluate the microalgae areal and volumetric productivity and the energy production and consumption. Differently from the most common existing models in literature, which deal with a specific part of the overall cultivation process, the model presented here includes all physical and chemical quantities that mostly affect microalgae growth: the equation of the specific growth rate for the microalgae is influenced by CO_2 and nutrients concentration in the water, light intensity, temperature of the water in the reactor, and by the microalgae species being considered. All these input parameters can be tuned to obtain reliable predictions. A comparison with experimental data taken from the literature shows that the predictions are consistent and slightly overestimating the productivity in the case of closed photobioreactor. The results obtained by the simulation runs are consistent with those found in literature, being the areal productivity for the open raceway pond between 50 and 70 t/(ha × year) in Southern Spain (Sevilla) and Brazil (Petrolina) and between 250 and 350 t/(ha × year) for the flat panel photobioreactor in the same locations.

Keywords: microalgae cultivation phase, open raceway pond, flat panel photobioreactor, repeated batch cultivation, carbon dioxide

Introduction

Microalgae are single cell organisms, which can be found in colonies or individual cells. Their most interesting characteristic is the ability of realizing a photosynthetic reaction in a single cell. They are extremely resistant and may grow in many different environments, from fresh water to marine and hyper-saline water (Le et al., 2010; Mata et al., 2010). Microalgae can be seen as an interesting alternative to more typical biomass as a source for biofuels production. Studies on microalgae have been carried out since the 1980s, but in recent years their importance has grown fast: as Chisti (2007) claims, the reason is that microalgae appear to be the only source of biodiesel that has the potential to completely displace fossil diesel. Moreover, compared to other biofuels sources, such as traditional crops and wood, microalgae have several advantages: they grow extremely fast, reaching high areal and volumetric productivities; they do not require arable land; they ask for less freshwater than normal crops, being able to use wastewater; they can directly capture CO_2 released by industries and they overtake the food vs. fuel debate; and their ability to grow in a wide range of conditions, resisting to severe temperatures, pH, and salinity (NREL, 0000), makes them even more attractive.

In 1978, the National Renewable Energy Laboratory of the United States of America started a 20-year program (Aquatic Species Program) to develop renewable transportation fuels from algae. The researches explored both the genetic engineering for manipulating the metabolism of microalgae and the engineering of microalgae production systems (NREL, 0000). In the last decades of the twentieth century, many other academic studies have been carried out all over the world, focusing on different aspects: the biological studies try to identify the best strain for a specific purpose, the engineering work tries to define open or closed systems to optimize the growth process and the system analysis works aim at defining the possible impact of the microalgae as a source of biomass for different applications, adopting a holistic approach.

Microalgae can be cultivated with the main purpose of producing lipids that can substitute biodiesel (Rodolfi et al., 2009) or produce higher value products that can be used as base chemicals for the production of biobased chemical products (Hempel et al., 2012). Cultivation of microalgae can be realized in open ponds systems where hydrodynamics (Hadiyanto et al., 2013) and CO_2 supply (Yang, 2011) are key drivers of the growth efficiency. Pollution and contamination related to mass transfer between open systems and environment together with the use of mechanical methods to obtain a proper mixing in the reactor make it impossible for a large number of microalgae to grow: only the most resistant strains might be used in open systems (Borowitzka, 1999). An alternative is the use of closed systems named photobioreactors that have the potential of maximizing the growth rate by controlling the growth conditions (pH, nutrients) and the gas supply at the expense of a much higher (factor ten) energy consumption. Examples are flat panel reactors or tubular reactors (Cuaresma et al., 2011). Studies have been conducted regarding the geometric characterization of flat panel photobioreactors, being the vertical positioning and east–west orientation the most suitable for microalgae proliferation (Sierra et al., 2008). Laboratory scale prototypes have been tested and compared to evaluate the relation between light availability, composition and growth rate of microalgae finding that high light availability leads to a change in the microalgae composition more than in a growth rate increase (Münkel et al., 2013). Tubular reactors need a much higher energy consumption (factor ten) than flat panel photobioreactors to maintain a proper mixing and circulation of the solution in the reactor and to ensure the oxygen removal, since there might be a decrease of the growth rate if oxygen saturation is reached (Molina et al., 2001). Mathematical models for light distribution have shown that photo-inhibition is a key problem for outdoor closed photobioreactors, both for flat panel technology and tubular reactors, being 100–200 W/m^2 the saturation light intensity for most of microalgae species (Fernández et al., 1997).

The land use and the solar efficiency of the microalgae system require adopting a holistic approach for the microalgae production. On the one hand, it is important to assess the overall conversion chain to define the substitution potential of the microalgae with respect of the extracted products. It is, therefore, important to adopt a holistic vision considering the complete life-cycle assessment for both the biobased products and the fossil one that are substituted (Azadi et al., 2014). Recent studies analyze the microalgae cultivation and transformation technologies evaluating their environmental impact in terms of GHG emissions and non-renewable energy consumption (Azadi et al., 2014) through LCA methodology, resulting that microalgae drying phase and water (Stephenson et al., 2010) and nutrient (Collet et al., 2011) reuse are the crucial aspects from an environmental point of view. From techno-economic analysis, irradiation conditions, mixing, medium, and CO_2 costs, together with dewatering technology, are the most important cost factors (Norsker et al., 2011). The opportunity to use microalgae strains capable of sustaining high growth rates and high lipid content give to costs the potential for significant improvement in the future (Davis et al., 2011).

The goal of this paper is to develop a comprehensive modeling framework to assess and compare different microalgae cultivation technologies in a consistent and holistic manner. A dynamic model of microalgae cultivation is developed. It aims at representing the influence of the most significant parameters on the system productivity and consumptions. The model is able to evaluate the variation of the most significant parameters, which influence microalgae growth, and therefore to estimate the microalgae biomass productivity and the energy productivity and consumption. The cultivation technologies taken into account are the open raceway pond and the flat panel photobioreactor: they are considered the most suitable technologies for extensive microalgae cultivation for biofuel production (Jorquera et al., 2010). Microalgae growth rate depends on several factors, such as light accessibility and nutrient availability, temperature, pH and salinity of the water, and CO_2 and O_2 concentration in the water. As all these terms are not constant in time, the model has to consider the dynamic behavior of the system and integrate the possible control strategies to be adopted to optimize the system performance.

The paper aims at including the effect of all the physical and chemical parameters that mostly affect microalgae growth, giving a consistent and robust modeling framework in which the contributions of recent publications are improved to interact together. Repeated batch operating strategy by Radmann et al. (2007) has been applied to microalgae growth model and mass balances by Yang (2011) together with thermal balances by Slegers et al. (2013) and pH control strategy by Sills (2013); reactor geometry has been taken into account for both the flat panel photobioreactor and the open pond to define the effect of solar radiation to microalgae growth, using the Beer–Lambert equations, and geometric evaluations by Duffie and Beckman (2013), included in Slegers et al. (2011).

Materials and Methods

Reactor Geometry

Two photo bioreactor technologies are considered in this work, which are likely to be the most promising technologies for extensive microalgae production.

Open raceway pond is the most used artificial system for microalgae cultivation. The geometry used in this work is a closed oval-shaped channel: the most common values for pond depth are between 0.2 and 0.5 m [0.3 m being the most used value (Chisti, 2007)]. Circulation around the oval ring and mixing are guaranteed with the use of a mechanical rotating device, usually a paddlewheel (Brennan and Owende, 2010). The paddlewheel is in continuous operation to prevent sedimentation, giving the water a speed between 0.15 and 0.25 m/s, the most commonly used velocity range for open pond microalgae cultivations (Doucha and Lívanský, 2006; Hadiyanto et al., 2013). To satisfy the microalgae CO_2 requirement, the open pond is equipped with submerged aerators. The main characteristics of the raceway pond used in the model have been taken from the literature (Borowitzka, 1999; James and Boriah, 2010; Norsker et al., 2011; Yang, 2011; Sompech et al., 2012) and presented in **Table 1** and **Figure 1**. Open cultivation technologies have several problems, such as lower volumetric productivity compared to closed systems [0.1 compared to 1.5 kg m^{-3} day^{-1} (Chisti, 2007)], evaporation of the water through the open surface, contamination which causes a limitation in species sustainability and the need for large land area (Singh and Dhar, 2011).

Closed photobioreactors are designed to overcome the limitations of open systems. They have higher efficiency and biomass productivity, shorter harvesting times (2–3 days compared to 7–10 days for open pond), high surface-to-volume ratios, reduced contamination risks, and can be used to cultivate wider range of algal species than open systems (Bahadar and Bilal Khan, 2013). However, closed systems are more expensive to be constructed since they need high quality materials, and difficult to operate and scale up (Muñoz and Guieysse, 2006).

Flat panel photobioreactor is a flat, transparent vessel, made of glass, Plexiglas or plastic or other transparent material. The mixing of the water is carried out directly with air bubbling, which is introduced via a perforated tube at the bottom of the reactor. Flat panel PBRs are never thicker than 5–6 cm as the light entering the panel would not penetrate deeper in the culture. Only panels with both a height and width of <1 m have been studied (Janssen et al., 2003). The flat panel reactors are positioned vertically, closely spaced, to reach a higher photosynthetic efficiency through self-shading of the panels (Rawat et al., 2013). As for the open raceway pond, the main geometric characteristics have been taken from the literature (Janssen et al., 2003; Sierra et al., 2008; Pruvost et al., 2011; Ruiz et al., 2013; Kochem et al., 2014; Sugai-Guérios et al., 2014) and presented in **Table 2** and **Figure 2**.

Operating Strategy

The reactors can be operated either in continuous or in batch mode. The continuous operation requires a continuous extraction of the microalgae, while the batch operation extracts the produced microalgae after a certain time of production. An inoculum is, therefore, maintained in the reactor to serve as a basis for the new batch. The nutrient feeding strategy changes depending on the operating strategy: in continuous operation, two different reactors in series would be needed for lipid production, while in batch operation, a recipe for the batch feeding with nutrients has to be applied. As explained by Radmann et al. (2007), in the repeated batch cultivation after a certain period in the reactor, a specific culture volume is removed and replaced with an equal amount of fresh medium. Consequently, a part of cultivation medium is kept in the reactor as a starting inoculum. Repeated batch cultivation presents several operational advantages, the most important of which are the maintenance of a constant inoculum and high growth rates.

The operating strategy is also influenced by two considerations regarding the upstream and downstream processes. The downstream processing includes a settler tank where the extracted

TABLE 1 | Open pond characteristics.

Pond surface	1 ha
Length to width ratio	10
Pond depth	30 cm
Speed of the water	0.20 m/s
CO_2 absorption system efficiency	0.9

TABLE 2 | Flat panel photobioreactor characteristics.

Height of the panels	1.5 m
Thickness of the panels	5 cm
Distance between the panels	1 m
Reactor surface	1 ha
Operating temperature	Constant

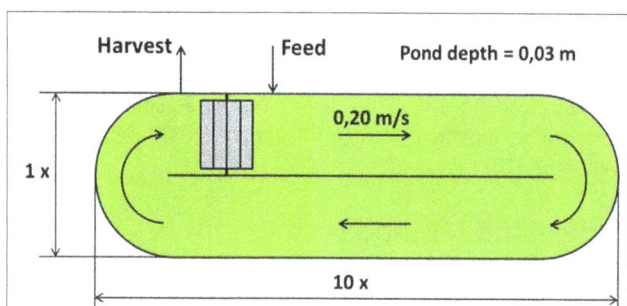

FIGURE 1 | Raceway open pond.

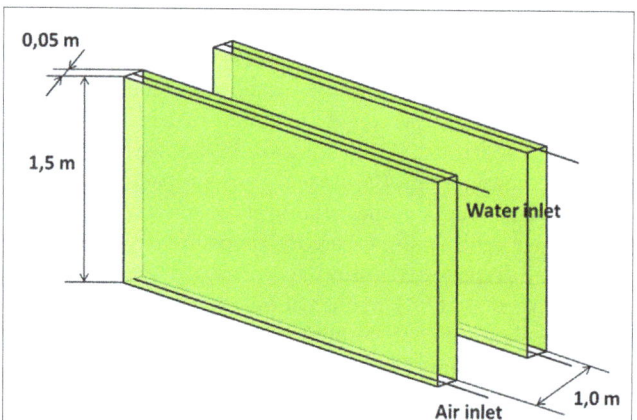

FIGURE 2 | Flat panel photobioreactor.

microalgae undergo a first concentration by sedimentation. The separated water is recycled while the products enter the downstream processing step. In this situation, the time required by the microalgae to grow can be decoupled from the time of the downstream technology. Second, in case of shortage of nutrients, the microalgae growth rate decreases as well as the production of new biomass, while the quantity of lipids in the biomass increases: this growth condition is particularly interesting if the downstream process aims at producing biodiesel. The models presented in this work do not consider the input condition of scarcity of nutrients, which instead are always supplied in excess.

Model
Cultivation Model Structure

The mass and energy produced by the cultivation systems in the form of biomass are calculated by the mathematical model through time-dependent mass and energy balances (**Figure 3**).

Before calculating the results of these balances, it is necessary to introduce some auxiliary equations. The auxiliary equations include the microalgae growth model, the pH control through CO_2 injection, and the equations to evaluate the radiation affecting the microalgae cells.

A set of differential equations is developed: it is solved by an explicit integration scheme with fixed time decomposition. To simplify the resolution of this system, the finite differences approximation was adopted. All the differential equations were treated as finite differences, and the backward difference formula was implemented

$$\frac{dx}{dt} = f(t) \rightarrow \frac{x(t) - x(t-1)}{\Delta t} = f(t) \qquad (1)$$

This expression was preferred to the forward difference formula, being less vulnerable to instability.

If the backward difference formula is applied to matrices and vectors, the following expression is obtained:

$$\bar{x}(t) = F(t) * \Delta t + \bar{x}(t-1) \qquad (2)$$

The model is a differential algebraic equations system that is converted into a set of algebraic equations by defining the proper solving sequence: some of the differential equations depend on one another, making impossible to solve them in a sequential logic: it is, therefore, necessary to solve them altogether as a system. Therefore, all the equations were manipulated to separate the known terms from the unknown variables and to obtain an easily solvable system of algebraic equations.

A dynamic approach to the resolution of energy and mass balances has been preferred to a steady state analysis because the laws describing the microalgae growth are time dependant; moreover, a dynamic model gives the opportunity of a deeper and more accurate evaluation of microalgae production and energy consumption over the time period.

Inputs to the Model

The following inputs are required by the model.

- *Location data*: latitude and longitude of the reactor are needed together with the hour difference between the location of the bioreactor and Greenwich.
- *Weather data* are also required by the model both to calculate the irradiation factor, which influences the microalgae growth and to evaluate the thermal balance. The open raceway pond model requires the global irradiation over a horizontal

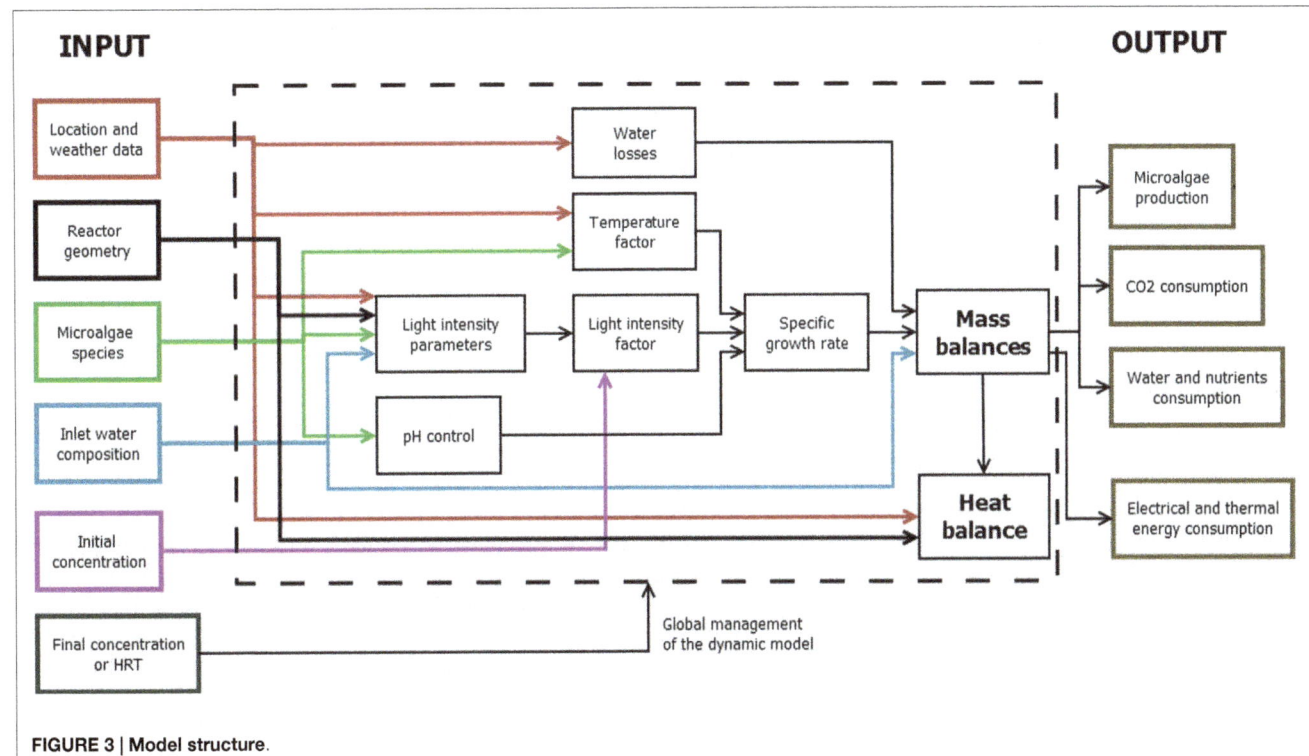

FIGURE 3 | Model structure.

surface, whereas the flat panel model requires the direct and the diffuse radiation over a horizontal surface as two distinct data. Moreover, both the two models require other weather data, such as the atmospheric temperature, wind speed, and relative humidity. Hourly data are used in all models.

- *Microalgae species characterization*: many parameters of the model depend on this input, such as the maximum growth rate, the optimal growth temperature, and the saturation light irradiation. All the characteristics assume different values depending on the microalgae species. **Table 3** presents the model parameters considered for two different microalgae strains. The model assumes a constant pH that is kept at the optimal value for microalgae species using CO_2 injection: to evaluate the correct quantities of CO_2 to be injected, the pH optimal value is represented as an optimal value of alkalinity of the reactor. LHV is the lower heating value for wet microalgae biomass. I_S is the saturation light intensity, μ_{max} is the maximum microalgae growth rate, T_{let} is the lethal temperature, T_{opt} is the optimal temperature for microalgae growth (maximum growth rate), and β is a curve modulating constant for the temperature factor. The decay rate is a coefficient that is used to evaluate the microalgae mass losses. K_C and K_{NA} are the half-saturation constants, i.e., the compound concentration when the specific growth rate [day^{-1}] is $\mu_A = \hat{\mu}_A / 2$. From Yang (2011), K_C is set to $0.001 \, mol_{CO_2} / m^3$ while $K_{NA} = 0.001 \, mol_{NA}/m^3$. When nitrogen and carbon dioxide concentrations are next to the saturation values, these terms assume a value close to 1 and they do not affect the microalgae growth.

- *The composition of the inlet water*, which is used to fill the reactor at the beginning of each cultivation: since each cultivation cycle works as a batch reaction, nutrients must be supplied in sufficient quantities to feed the microalgae for the entire duration of the batch process. The model gives as input a surplus of nutrients in the water to guarantee they are not the limiting factor to microalgae growth. Then, as output, the model gives the exact quantity of nutrients consumed by the microalgae.

- *The operating strategy*, which is used to predict microalgae production. Two different strategies can be adopted: if the objective consists in searching the hydraulic retention time (HRT) required to reach a certain microalgae concentration in the reactor, the initial and the final microalgae

concentrations are to be supplied as input information to the model. If the HRT is fixed along the whole time period, then the input data will be the initial concentration and the HRT, leaving the final concentration in the reactor free to change.

Outputs of the Model

The mass production of microalgae is the most significant output of the cultivation model. In addition to this parameter, it is possible to define two other global performance parameters: the volumetric productivity {expressed in [g/(l × day)] or in [kg/(m³ × day)]} and the areal productivity {expressed in [t/(ha × y)] or in [kg/m² × day)]}. These parameters refer to the volume of the reactor and to the area of the soil covered by the reactor. Moreover, the model is able to give as output the exact quantities of all the compounds consumed or produced in the time period considered by the simulation (usually 1 year). For example, the model gives the CO_2 that is needed to be bubbled in the reactor to maintain a constant pH level and to feed the microalgae. In this way, the model gives the possibility to analyze the carbon footprint of the technology or at least to understand the quantity of CO_2, which can be fixed by the biomass produced, being a part of the injected CO_2 lost to the atmosphere: this quantity can be relevant in the case of open raceway pond.

Fundamental outputs of the model are the total electrical and thermal energies required by the reactor: to evaluate these quantities, the energy balance is not sufficient, as it is necessary to include the electrical energy spent at the boundary of the system, to refill the reactor and to harvest the biomass from the reactor itself; the electrical energy also includes the quantities which are needed for mixing and for the air bubbling in the bioreactor. Finally, another output of the model is the quantity of water required by the reactor, without considering the possible water recirculation coming from the downstream process.

Model Equations
Growth model
Following Yang (2011), the growth rate is calculated by Eq. (3):

$$r_{gA} = \mu_A X_A \tag{3}$$

where X_A is the mass concentration of microalgae [g/m³] and μ_A is the specific growth rate [day^{-1}] that is calculated as:

$$\mu_A = \hat{\mu}_A \left(\frac{CO_{2D}}{K_C + CO_{2D}} \right) \left(\frac{N_T}{K_{NA} + N_T} \right) f_I f_T \tag{4}$$

where the specific growth rate $\hat{\mu}_A$ depends on the microalgae species; CO_{2D}, N_T are the quantities of dissolved nutrients and CO_2 [mol/m³] at time t. Nitrogen is the only nutrient that has been taken into account in the specific growth rate expression; other nutrients, such as phosphorus, are not explicitly considered, as suggested by Yang (2011), under the reasonable assumption that the metabolism of the microalgae is not limited or inhibited by these compounds. Although the growth rate may be affected by the presence of micro nutrients, they have not been introduced in the model to avoid the definition of a more complex reaction scheme with the corresponding experimental data to represent those influences. The specific growth rate can reach higher

TABLE 3 | Microalgae species characteristics.

Typology	P. tricornutum	T. pseudonana
pH_opt	8.3	8.3
alk_opt [meq/l]	0.032	0.032
LHV [kJ/kg]	21,527	21,527
I_S [W/m²]	37.118	21.834
μ_{max} [1/day]	1.392	3.288
T_{let} [°C]	30	31
T_{opt} [°C]	21	24
β	1.57	1.83
Decay rate [1/day]	0.048	0.048
K_{NA} [mol$_{NA}$/m³]	0.001	0.001
K_C [mol$_C$/m³]	0.001	0.001

values thanks to the use of some micronutrients, such as iron and silicon (Chisti, 2007; Çelekli and Yavuzatmaca, 2009; Le et al., 2010; Ak, 2012; Bahadar and Bilal Khan, 2013). Nitrogen and CO_2 are considered limiting nutrients for microalgae growth and so microalgae growth rate dependence is expressed as saturation functions. K_C and K_{NA} are the half-saturation constants, as explained in the previous paragraphs. f_l is the light intensity factor representing the influence of the light on the microalgae production. Following Yang (Yang, 2011; Fernández et al., 2013), it is calculated by the following equation:

$$f_l = \frac{I_a}{I_s} \exp\left(1 - \frac{I_a}{I_s}\right) \tag{5}$$

where I_a is the average light intensity in the volume of the bioreactor at a given time t, while I_s is the saturation light intensity, which depends on the microalgae species being considered; saturation light intensity is usually in the range between 30 and 100 W/m^2 (Kumar et al., 2011). The average light intensity I_a depends on the weather, the turbidity caused by the microalgae and other substances in the water and on the reactor geometry. The model includes the hypothesis of perfect mixing: the whole quantity of microalgae in the water is affected by the same quantity of light radiation. f_T is the temperature factor. This term is equal to one and does not influence microalgae growth when the reactor temperature corresponds to the optimal growth temperature for the given microalgae species: this is what happens in the closed reactor, where a heat exchanger is used to keep a constant optimal temperature. Below and above the optimal growth temperature, the growth is negatively influenced by the water temperature. Above the optimal temperature, the value of f_T decreases fast and reaches 0 for a certain temperature that again depends on the algae species: this value is called lethal temperature (T_{let}). The expression of the temperature factor is shown in the following equation, taken from Slegers et al. (2013):

$$f_T = \left(\frac{T_{let} - T_w}{T_{let} - T_{opt}}\right)^{\beta_{algae}} \exp\left(-\beta_{algae}\left(\frac{T_{let} - T_w}{T_{let} - T_{opt}} - 1\right)\right) \tag{6}$$

β_{algae} depends on the algae species and it is a curve modulating constant.

Figure 4 shows the shape of the temperature factor f_T: it is possible to note that the equation used for the temperature factor suggested by Slegers et al. (2013) overestimates the growth for low temperature values, since the growth is not possible for low temperatures.

pH control strategy

Each microalgae species has an optimal pH value: the model includes a control system to keep the optimal value of pH in the reactor through CO_2 injection. As shown in the following equations from Sills (2013), the pH can be controlled by keeping constant two parameters: dissolved CO_2 concentration and alkalinity. Since alkalinity does not change with time in the model, to keep pH constant, it is necessary to calculate the quantity of dissolved CO_2 which has to be maintained constant during time.

The total concentration (C_T) of carbonate species in solution is defined as

$$C_T = \left[H_2CO_3^*\right] + \left[HCO_3^-\right] + \left[CO_3^{2-}\right] \tag{7}$$

Dissolved CO_2 concentration is $\left[CO_{2(aq)}\right] \cong \left[H_2CO_3^*\right]$. The molar concentration of each carbonate species (as a fraction of C_T) depends on pH, according to the following equations:

$$\left[H_2CO_3^*\right] = \frac{C_T}{1 + \frac{k_1}{[H^+]} + \frac{k_1 k_2}{[H^+]^2}} = \alpha_0 C_T \tag{8}$$

$$\left[HCO_3^-\right] = \frac{C_T}{1 + \frac{[H^+]}{k_1} + \frac{k_2}{[H^+]}} = \alpha_1 C_T \tag{9}$$

From Eqs (8) and (9), α_0 and α_1 are calculated.

$$\left[CO_3^{2-}\right] = \frac{C_T}{1 + \frac{[H^+]}{k_2} + \frac{[H^+]^2}{k_1 k_2}} = \alpha_2 C_T \tag{10}$$

Finally:

$$C_T = \frac{alk - OH^- + H^+}{\alpha_1 + 2\alpha_2} \tag{11}$$

Since pH is the concentration of dissolved $[H^+]$ in the water and $pH = 14 - pOH$, all terms are known and the total concentration of carbonate species can be obtained by

$$C_T = \frac{alk - OH^- + H^+}{\alpha_1 + 2\alpha_2} \tag{12}$$

This term will be a part of the time-dependent CO_2 mass balance, which is used to calculate the CO_2 injection at each time t.

Mean light intensity for open pond

The light intensity factor in the specific growth rate expression contains I_a, that is the average light intensity in the bioreactor at a

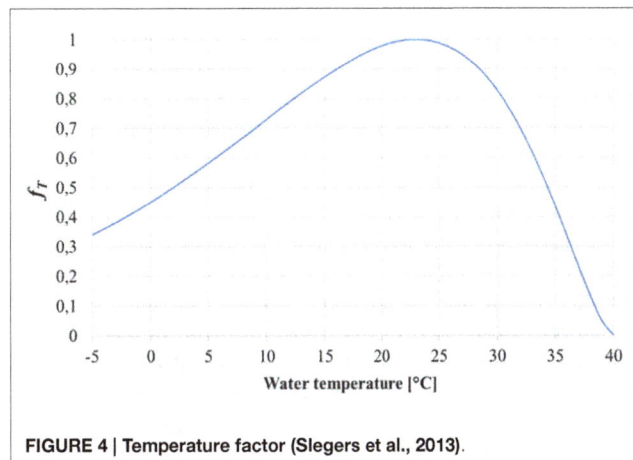

FIGURE 4 | Temperature factor (Slegers et al., 2013).

given time. Starting from global horizontal radiation, the model evaluates I_a using the Beer–Lambert's law, which assumes an exponential decay of the light intensity from the external surface of the cultivation system:

$$I_a(s) = I_0 \exp(-\sigma X_A s) \quad (13)$$

As it is explained by Béchet et al. (2013), $I_a(s)$ is the local light intensity, s is the distance from the external surface of the system to the position under consideration, I_0 is the incident light intensity, σ is the extinction coefficient, and X_A the cell concentration. The Beer–Lambert's law can be applied only if two conditions are verified by the culture medium: first, it must be isotropic (i.e., the optical properties of the broth are independent from light direction); this condition is often verified in well-mixed systems. Second, algae cells should not scatter light: this second condition is not always met, but the model considers both the two requirements always verified. The equations to calculate I_a strictly depend on the geometry of the bioreactor: for the open pond, as suggested by Yang (2011), an integration through the pond depth of the Beer–Lambert's law has been used:

$$I_a = \frac{1}{Z} \int_0^Z I_0 \exp(-K_e z)\, dz \quad (14)$$

where I_0 is obtained directly from the global horizontal radiation I_{GHR}: $I_0 = PAR \times I_{GHR}$. PAR is the photosynthetic active radiation (PAR), i.e., the amount of solar radiation which is used by the microalgae for the photosynthesis and which corresponds to 45% of the total incoming light. K_e is related to algal concentration in the pond, and it is called extinction coefficient:

$$K_e = K_{e1} + K_{e2} X_A \quad (15)$$

Light intensity for flat panel

In case of flat panel, the complexity of the geometry requires the use of different equations from those applied for the open pond. The equations used to calculate the light intensity for flat panels are taken from Slegers et al. (2011). For this typology of reactors, the input data are the direct and diffuse radiation on the horizontal surface; moreover, the equations must take into account also the reflection of the radiation on the surface of the panel. As explained by Slegers et al. (2011), to evaluate the effect of direct light irradiation over a tilted surface (panels positioned vertically), it is necessary to introduce a geometrical parameter for direct radiation:

$$G_{direct}(t) = \frac{\cos\theta}{\cos\theta_z} \quad (16)$$

in which θ is the solar incidence angle, and θ_z is the solar zenith angle. The values assumed by these angles during each day of the year depend on the location, and have been calculated using the equations taken from Duffie and Beckman (2013).

For large scale cultivations, parallel positioned flat panels are used. Parallel placement causes shading, and consequently part of the panels no longer receive direct sky light. The shadow height on vertical reactor panels is given by

$$h_{shadow} = h - \frac{d \tan(90 - \theta_z)}{\cos(\gamma)} \quad (17)$$

which is a function of the reactor height h [m], distance between the reactor panels d [m], solar elevation, which is equal to $90 - \theta_z$ and angle between solar rays and the azimuth of the surface. If $h_{shadow} > 0$, the flat panel is divided into two parts. The upper part receives direct and diffuse radiation, the lower part only diffuse light. The separation between the upper and the lower part varies with the solar position. Parallel placement of the reactors also influences the penetration of diffuse sky light into the space between panels; the light intensity decreases from the top to the bottom. A similar situation can be seen with the penetration of light in urban street canyons (Robinson and Stone, 2004). For these reasons, the geometrical factor for diffuse light depends on the position over the surface of the panel.

At height $y < h_{shadow}$:

$$G_{diffuse} = \frac{1 + \cos(\beta + u)}{2} \quad (18)$$

where $u = atan(y/d)$, β [deg] is the slope of the reactor, i.e., the angle that the surface makes with the surface of the earth.

At height $y > h_{shadow}$:

$$G_{diffuse} = \frac{1 + \cos(\beta)}{2} \quad (19)$$

All panels are treated similarly in the calculations. Moreover, ground reflection is low for parallel placed panels, and is therefore not taken into account.

The total amount of light falling on each reactor surface at a given height y, of a given time t is:

$$I_0(y,t) = G_{direct}(t) I_{direct}(t) + G_{diffuse}(y) I_{diffuse}(t) \quad (20)$$

At this point, it is fundamental to consider the reflected fraction of the irradiation, which does not enter into the reactor and so does not contributes to the microalgae growth. The amount of reflected light on each interface is related to the differences in refractive indices and the incidence angle (Sukhatme and Sukhatme, 1996).

The incidence angle for diffuse radiation which is considered to evaluate the light reflection is assumed to be 60°, as it is suggested by Duffie and Beckman (2013).

Light reflection by the flat panel walls follows Fresnel equations:

$$R_s = \left[\frac{\eta_i \cos(\theta_i) - \eta_t \sqrt{1 - \left(\frac{\eta_i}{\eta_t} \sin(\theta_i)\right)^2}}{\eta_i \cos(\theta_i) + \eta_t \sqrt{1 - \left(\frac{\eta_i}{\eta_t} \sin(\theta_i)\right)^2}} \right]^2 \quad (21)$$

$$R_p = \left[\frac{\eta_i \sqrt{1 - \left(\frac{\eta_i}{\eta_i}\sin(\theta_i)\right)^2} - \eta_t \cos(\theta_i)}{\eta_i \sqrt{1 - \left(\frac{\eta_i}{\eta_i}\sin(\theta_i)\right)^2} + \eta_t \cos(\theta_i)} \right]^2 \qquad (22)$$

where θ_i [°] is the incidence angle, η_i [–] is the refractive index of the material before the interface, and η_t [–] is the refractive index for the material after the interface. Normal sunlight is non-polarized, therefore the overall reflection coefficient equals the average of the reflection coefficients for s-polarized and p-polarized light:

$$R' = \frac{R_s + R_p}{2} \qquad (23)$$

The light reflected within the reactor wall is completely transmitted to the air, introducing the hypothesis of a non-absorbing material for the walls of the reactor. The light transmitted to the culture, which has to be calculated separately for direct and diffuse radiation, is

$$I_i(t) = I_0(t)(1 - R_1' R_2')T_m \qquad (24)$$

T_m [–] is a factor which takes into account a possible low transparency of the material.

The calculation is performed for both the two sides of the reactor and, for parallel positioned panels, for each height. R_1' and R_2' are the reflection coefficients for the air–reactor wall interface and the reactor wall–culture volume interface, respectively. Two light intensity gradients exist in the culture volume. First, as a function of height due to shading and the penetration of diffuse light between parallel positioned panels. Second, in the liquid between the two reactor walls. The second gradient runs from the reactor wall to the center of the reactor and is caused by the absorption of light by the medium and the algae (Slegers et al., 2011). Only the PAR of the spectrum is absorbed by the algae. This accounts for about 45% of the total light. The Lambert–Beer's law is used for the overall light gradient in the culture volume, as it was done for the open pond:

$$I(y,z,t) = I_{front}e^{-(K_{e1}+K_{e2}X_A)z} + I_{back}e^{-(K_{e1}+K_{e2}X_A)(s-z)} \qquad (25)$$

This equation gives the light intensity at location z [m] inside the reactor thickness [m], at a given height y [m] in the reactor at time t.

At this point, to simplify the model, the values of light intensity $I(y,z,t)$ are integrated to find a mean value of irradiation for the whole culture inside the whole reactor at a given time t: with these integrations, a single value of radiation is obtained and used in the growth model, for the whole panel, at a given time t. The hypothesis of a perfect mixing inside the culture at each time t is necessary to integrate the equation in both height and depth.

Mass balances

Mass balances of nitrogen and oxygen inside the pond can be modeled in the same way, suggested by Yang (2011):

$$\frac{dM}{dt} = \mu_A X_A Y_{AM} - k_{Lg}\alpha\left(M - M^*\right) \qquad (26)$$

where M is the concentration of the respective component in the water in the bioreactor, Y_{AM} is the mass of the respective component consumed or generated by the microalgae per unit mass of microalgae produced. The last term of the right-end side of Eq. (26) represents the mass transfer between the atmosphere and the pond, where $k_{Lg}\alpha$ is the mass transfer coefficient for a given element, M^* is the saturation concentration of the associated dissolved element.

For total inorganic carbon, the mass balance assumes a different formulation since the CO_2 is injected continuously in the pond during the growth phase to keep a constant concentration of dissolved CO_2 in the reactor, balancing the losses of CO_2 to the atmosphere and the consumption of CO_2 by the microalgae.

$$\frac{dCO_2}{dt} = \mu_A X_A Y_{ACO_2} + f_{CO_2} - k_{Lg}\alpha\left(CO_2 - CO_2^*\right) \qquad (27)$$

where f_{CO_2} represents the flux of CO_2 introduced by the supply of gas flow into the system. Since the quantity of dissolved CO_2 is kept constant ($dCO_2/dt = 0$), the mass balance can be written again as

$$f_{CO_2} = -\mu_A X_A Y_{ACO_2} + k_{Lg}\alpha\left(CO_2 - CO_2^*\right) \qquad (28)$$

Finally, the mass balance for microalgae species can be written as

$$\frac{dX_A}{dt} = \mu_A X_A - k_{dA}X_A \qquad (29)$$

As it can be seen, all these equations are time dependent, and all strictly depend on one another: this means that they form altogether a system of differential equations. The strategy to solve the equations with the finite difference methodology is used to solve this system in Matlab, and therefore to solve the system of differential equations as a system of algebraic equations.

Thermal balance for open pond

The model includes the thermal balance as it is suggested by Slegers et al. (2013):

$$V_R cp_w \rho_w \frac{dT_w}{dt} = Q_{irr} - Q_{algae} - Q_{rad} - Q_{evap} - Q_{conv} - Q_{cond} \qquad (30)$$

where V_R is the volume of the pond, cp_w the heat capacity of the growth medium, ρ_w the density of the growth medium, T_w the temperature in the pond, Q_{irr} the heat flow rate to the pond by the sunlight, Q_{algae} the light energy flow rate to algae during growth, Q_{rad} the heat flow rate by emission of long-wave radiation in the infrared region, Q_{evap} [W] the heat flow rate caused either by evaporation or condensation, Q_{conv} the heat flow rate by convection, and Q_{cond} the heat flow rate between pond and ground via conduction. The water in the pond is heated by sunlight that enters the culture volume. Solar energy that is not used by algae for growth is considered as thermal energy that is dissipated through the different thermal exchange mechanisms (Q_{rad}, Q_{evap},

Q_{conv}, and Q_{cond}) or as a contribution to temperature growth in the pond. The total heat flow rate by the sunlight is given by

$$Q_{irr} = A_w I_{surface}(t) \tag{31}$$

where A_w [m²] is the water surface area of the pond and $I_{surface}$ [W/m²] is the total light arriving on the pond. Part of this light is absorbed by microalgae for growth:

$$Q_{algae} = h_{comb} \mu_A X_A V_R \tag{32}$$

which is a function of the lower heating value on a dry basis of algae biomass h_{comb} [J/kg], specific growth rate μ_A [s⁻¹] and the biomass concentration X_A [kg/m³].

The water in the pond emits thermal energy by long-wave radiation. The overall long-wave radiation flow between water in the pond and sky is calculated using Duffie and Beckman (2013):

$$Q_{rad} = A_w \varepsilon_w \sigma_{SB} \left((T_w + 273.15)^4 - T_{sky}^4 \right) \tag{33}$$

where ε_w [−] is the emissivity of the water in the infrared region, σ_{SB} [W/(m²K⁴)] the Stefan–Boltzmann constant, and T_{sky} [K] the equivalent sky temperature for clear sky days, which is expressed by Duffie and Beckman (2013) as

$$\begin{aligned} T_{sky} = (T_a + 273.15)(0.711 + 0.0056T_{dew} \\ + 0.000073T_{dew}^2 + 0.013\cos(15t_{solar}))^{0.25} \end{aligned} \tag{34}$$

where T_a [°C] is the air temperature, T_{dew} [°C] the dew point temperature, and t_{solar} [−] the number of hours after solar midnight.

Evaporation has a large effect on the water temperature, especially in locations with low humidity and high wind velocities. The evaporation rate depends on the shape of the water area, wind velocity, and consequently also movement of the water. The evaporation flow is driven by the difference of water vapor pressures between ambient air the saturated water body. The evaporation energy flow is given by

$$Q_{evap} = A_w h_{evap} (p'_s - p'_a) \tag{35}$$

The evaporation flow depends on the heat transfer coefficient for evaporation h_{evap} [W/(m² Pa)], the saturated water pressure p'_s [Pa] at water temperature T_w, and the water pressure of air p'_a [Pa] at air temperature T_a. The evaporation rates have been calculated using the heat transfer coefficient h_{evap} found in Duffie and Beckman (2013):

$$h_{evap} = 0.036 + 0.025v \tag{36}$$

where v [m/s] is the wind speed. The Antoine equation is applied to calculate the saturated water pressure p'_s [Pa] at water temperature T_w and the water pressure of the air p'_a [Pa] at air temperature T_a:

$$p' = RH\, 10^{\left(8.07131 + \log_{10}\left(\frac{101325}{760}\right) \right) - \frac{1730.63}{233.46 + T_a}} \tag{37}$$

where RH [−] is the relative humidity and T_a [°C] the temperature.

Convection and evaporation are related processes, as it is shown in Eq. (37). The flow for passive and forced convection at the water surface mainly depends on the difference between water and air temperature. The convection flow is given by

$$Q_{conv} = C_{Bowen} \frac{p_a (T_w - T_a)}{p_{ref}(p'_s - p'_a)} Q_{evap} \tag{38}$$

where C_{Bowen} is the Bowen constant [Pa/°C], p_a is the ambient pressure [Pa] and p_{ref} the reference pressure [Pa], and p'_s and p'_a are derived using equation

$$p' = RH\, 10^{\left(8.07131 + \log_{10}\left(\frac{101325}{760}\right) \right) - \frac{1730.63}{233.46 + T_a}} \tag{39}$$

Conductive heat transfer takes place between the open pond and the soil. The soil is assumed to be an infinite source for heat transfer. This heat transfer calculation is derived from Fourier's law:

$$Q_{cond} = h_{soil} A_{soil} (T_w - T_{soil}) \tag{40}$$

where h_{soil} [W/(m²°C)] is the heat transfer coefficient of the surrounding soil layer, A_{soil} [m²] is the area of the pond that is embedded in the soil, and T_{soil} [°C] is the temperature of the soil surrounding the pond.

Thermal balance for flat panel

Due to the significant difference in the geometry of the reactor and the temperature strategy being considered, the thermal balance for the flat panel assumes a different form from that implemented for the open raceway pond:

$$V_R c p_w \rho_w \frac{dT_w}{dt} = Q_{irr} - Q_{algae} - Q_{exch} - Q_{conv+cond} \tag{41}$$

In a flat panel, a constant value of the water temperature T_w [°C] is desired, and obtained using a heat exchanger placed at the bottom of the reactor to remove or supply heat.

Thus, the thermal balance can be written as follows:

$$\frac{dT_w}{dt} = 0 \rightarrow Q_{irr} - Q_{algae} - Q_{exch} - Q_{conv+cond} = 0 \tag{42}$$

Q_{irr} [W] is given by the following expression:

$$Q_{irr} = 2A_{panel} I_{mean} \tag{43}$$

where A_{panel} is the surface of the reactor, which is multiplied by two, as it is necessary to consider both the front and the back surfaces; I_{mean} [W/m²] is the radiation previously calculated taking into account both the direct and diffuse radiation and the reactor geometry that is the radiation which interacts with the water in the reactor and with the microalgae. As calculated for the open pond, a part of the incoming heat is used by the microalgae for growing:

$$Q_{algae} = h_{comb} \mu_A X_A V_R \tag{44}$$

which is a function of the combustion energy of algae biomass h_{comb} [J/kg], specific growth rate μ_A [s⁻¹], biomass concentration X_a [kg/m³], and reactor volume V_R [m³].

The model takes into account the natural convection over the panel surface caused by the wind and the glass conductivity:

$$Q_{conv+cond} = 2A_{panel}U_{tot}\left(T_w - T_{atm}\right) \qquad (45)$$

where

$$U_{tot} = U_{cond} + U_{conv} \qquad (46)$$

where U_{cond} is the glass conductivity [W/(m²K)] and U_{conv} is the conductivity of the natural convection that is obtained by Duffie and Beckman (2013). Using the equations above, the quantity of heat that has to be removed or supplied by the heat exchanger at each time t can be calculated. Heat exchange through radiation can be omitted from this balance, since the water temperature is always kept between 20 and 30°C, depending on microalgae growth optimal temperature.

Electrical energy consumption for harvesting, refilling, mixing and bubbling

For both the open pond and flat panel, the harvesting of the water from the reactor and its refilling are carried out in 8 h, during one night: 3.5 h are required for harvesting and 3.5 h for refilling. These operations are performed using a pump. The electrical consumption has been calculated as follows:

$$E_{harv/refil} = P_{pump}t_{harv/refil} \qquad (47)$$

where the power of the pump is

$$P_{pump} = \frac{\rho g Q h}{\eta_{pump}} \qquad (48)$$

where ρ is the water density [kg/m³], g is the acceleration of gravity [m/s²], η_{pump} [−] is the efficiency of the pump, set at 0.85, Q [m³/s] is the volumetric flow rate which has to be pumped, and h [m] is the height difference between the two basins before and after the pump: thanks to the fact that the model includes also the design of a settler positioned after the bioreactor, it is possible to know the exact h for the harvesting, which has been increased to consider the losses in the pipes, whereas for the refilling data have been taken from the literature, considering $h = 1$ m for the open pond and 3 m for the flat panel: the difference between these two values is again associated with the energy losses in pipes.

The power required for mixing in the open pond is

$$P_{mix} = \frac{\rho g Q h}{\eta_{paddlewheel}} \qquad (49)$$

where Q [m³/s] is obtained from the water speed that is desired in the reactor (0.20 m/s) and from the cross-section of the open pond (which depends on the geometry), h is the given height difference before and after the paddle wheel, taken from literature (0.05 m). $\eta_{paddlewheel}$ is lower than the efficiency of a normal pump and is assumed to be equal to 0.25.

Both for the open pond and flat panel, a bubbling system has to be taken into account: for the flat panel, this system should be able both to supply the CO_2 necessary for the photosynthesis of the microalgae

and to guarantee an adequate mixing inside the reactor. For this reason, the amount of air bubbled in the flat panel is higher than the air supplied to the open pond. These quantities are controlled by the CO_2 molar fraction inside the injected air which is 0.04 in the case of open pond and 0.02 for the flat panel. The design and the energy consumption of the bubbling system have been calculated through Belsim Vali modeling software, using a compressor. The result is that it is necessary to supply 4 kJ for each kg of air injected in the reactor.

Results

Open Pond
Tables 4 and 5 show the input and output data in the open pond modeling:

The open pond productivity that is obtained by the model is consistent with the literature: Slegers et al. (2013) estimated an annual biomass production for a 1 ha surface in the Netherlands and Algeria equal to 41.5 and 63.7 t, respectively. Fernández et al. (2013) report a maximum microalgae areal productivity equal to 30 g/(m² day), higher than those resulting from the model calculations, but still of the same order of magnitude. Jiménez et al. (2003) obtained a volumetric productivity equal to 0.05 kg/(m³ day) for a location in Southern Spain (Malaga), by cultivating the microalgae until a maximum concentration of 470 g/m³: the cultivation conditions and the results from the dynamic model are consistent with the results of this work.

In this work, the Net Energy Ratio (NER)

$$\text{NER} = \text{Net Energy Ratio}$$
$$= \frac{\text{total energy requirement for operation}}{\text{total energy production (biomass)}} \qquad (50)$$

expresses the fraction of the energy produced in the cultivation system that is used by the system itself to generate the biomass. If this value is close to 1, the cultivation technology of microalgae is too energy intensive, requiring a big share of the energy produced. From Table 5, it appears that for both the locations analyzed by the model the NER is quite far from 1, showing that the energy demand for the operation of the cultivation system in Petrolina is 10.8% of the energy contained in the biomass produced, while for Sevilla is 6%. These values are quite promising for a potential production of microalgae in these locations, since the NER is far enough from 1: even if the energy content of open pond construction and materials

TABLE 4 | Input data for open pond simulations.

HRT not fixed, X target fixed		HRT fixed, X target not fixed	
Location	Sevilla (SPA) Petrolina (BRA)	Location	Sevilla (SPA) Petrolina (BRA)
Typology	P. tricurnutum T. pseudonana	Typology	P. tricurnutum T. pseudonana
Xa_init [g/m³]	100	Xa_init [g/m³]	100
Xa_target [g/m³]	490	HRT [day]	7
Tw_in [°C]	15	Tw_in [°C]	15
CO₂ rate [%]	0.04	CO₂ rate [%]	0.04
Z_pond [m]	0.3	Z_pond [m]	0.3
LW [−]	10	LW [−]	10

TABLE 5 | Output from open pond simulations.

	HRT not fixed X target fixed		HRT fixed X target not fixed	
	Petrolina	Sevilla	Petrolina	Sevilla
Number of harvesting in 1 year	45	50	–	–
HRT [day]	–	–	7	7
Mass microalgae [t/(ha y)]	57.29	64.68	54.04	67.88
CO_2 captured [t/(ha y)]	140	156	133	163
CO_2 injected [t/(ha y)]	243	260	235	268
CO_2 lost to atmosphere [t/(ha y)]	78.62	77.35	78.14	77.27
CO_2 ratio losses	0.324	0.298	0.33	0.29
CO_2 ratio algae	0.57	0.6	0.56	0.61
N absorbed [t/(ha y)]	5.8	6.55	5.57	6.8
Water injected [t/(ha y)]	109,210	121,640	120,220	119,600
Water evaporation [t/(ha y)]	5.8	5.01	5.83	5.24
Energy microalgae [kWh/(ha y)]	342,610	386,790	323,180	405,930
Electrical energy [kWh/(ha y)]	18,497	19,024	18,522	19,111
Thermal energy [kWh/(ha y)]	–	–	–	–
Volumetric productivity [kg/(m^3 × day)]	0.0523	0.0591	0.0494	0.062
Areal productivity [kg/(m^2 × day)]	0.0157	0.0177	0.0148	0.0186
NER	0.108	0.0603	0.1146	0.0942

are added to the total energy requirement, it seems reasonable to state that the plant would still be convenient from an energetic point of view. Another important output of the model from an environmental point of view is the quantity of CO_2 fixed by the microalgae during the growth process: the downstream process that transforms microalgae biomass into biofuel releases CO_2 to the atmosphere, with a negative environmental impact. Since the cultivation phase captures more CO_2 than the quantity released, using the data of CO_2 fixed in the algae it is possible to calculate a global CO_2 balance, for the entire biofuel production chain: this environmental analysis may lead to a comparison with other biofuel chain production and with traditional fossil fuels, to evaluate which product has a positive or negative environmental impact.

With the first operating strategy (meaning that the inputs are the initial and final concentration), one of the output would be the number of days that are needed to reach the final target concentration. The number of days is not the same along the year, being variable with weather conditions. During wintertime, the harvesting period in Sevilla is extremely long due to low irradiation and low temperatures, which make the microalgae growth in the pond slow and difficult. A possible strategy to overcome this limitation might be to interrupt the production during wintertime: in this case, the productivity during the whole year would decrease, since a part of the biomass would not be produced, but the NER would positively decrease, because of the absence of energy consumption during winter: the production lost is less than the energy saving obtained, and so the overall effect would be positive. For Petrolina, the HRT is more homogeneous along the year, but longer: the reason might be that weather conditions in Petrolina cause a strong increase of pond temperature that may cause a consequent reduction in microalgae productivity when pond temperature is higher than the optimal temperature for microalgae growth.

Results of dynamic simulation runs are reported in **Figures 5–9**, showing the effect of PAR and pond temperature

over the specific growth rate and microalgae growth rate. Input parameters for simulation runs are shown in **Table 6**.

Even if the simulation has been run for the entire year, as it is shown in **Figures 8** and **9**, **Figures 5–7** report a small time range that includes two consequent batch cultivations, for an easier understanding of the light and temperature effect over the microalgae growth. **Figure 5** includes the pond temperature values (A), calculated in each hour of the entire simulation period, and the consequent temperature factor (B). In **Figure 5**, the model does not evaluate the pond temperature during the hours when the pond is emptied. **Figure 6** reports the PAR and the consequent light intensity factor (B). **Figure 7** shows the specific growth rate (A) and the microalgae concentration (B) in the open pond during two consequent batch cultivations. Microalgae growth rate (C) is obtained, as explained in Eq. (3), by the product of the specific growth rate and the microalgae concentration. As shown in **Figure 7B**, each batch cultivation starts with the same initial concentration, while the final concentration depends on each batch: the model waits for the end of the solar radiation to start to empty the pond that takes almost the whole night.

In **Figures 8** and **9**, the time of the entire simulation run has been considered presenting input values (temperature in **Figure 8A** and PAR in **Figure 8B**) and output [specific growth rate in **Figure 9A**, microalgae concentration in the pond in **Figure 9B** and microalgae growth rate (productivity) in **Figure 9C**].

The time to reach the minimum microalgae target concentration to start the harvest in winter time in Sevilla is really long and it might be useful to interrupt the production for some months.

Flat Panel

Table 7 shows the input data for the flat panel model for both the two operating strategies.

The azimuth of the panel is an input, which may vary from 0°, when the panel faces south, to −90°, when the panel faces

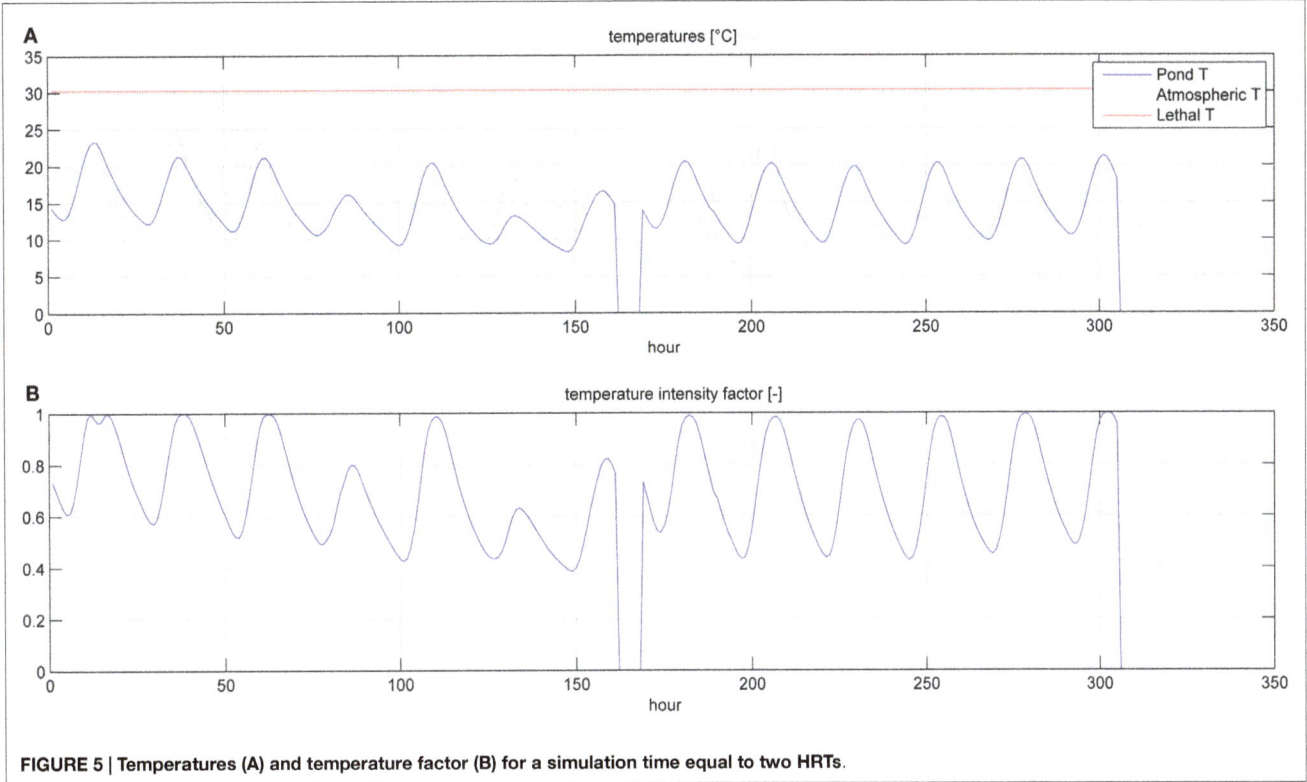

FIGURE 5 | Temperatures (A) and temperature factor (B) for a simulation time equal to two HRTs.

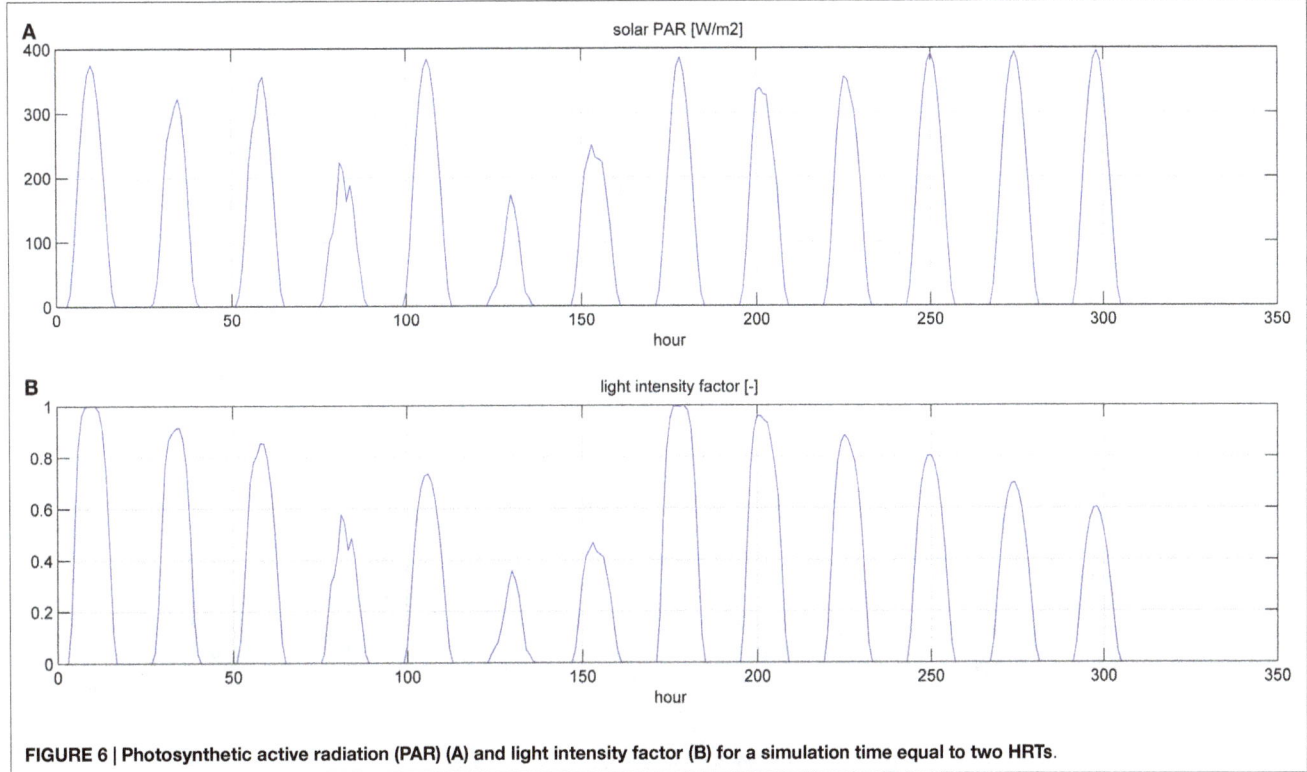

FIGURE 6 | Photosynthetic active radiation (PAR) (A) and light intensity factor (B) for a simulation time equal to two HRTs.

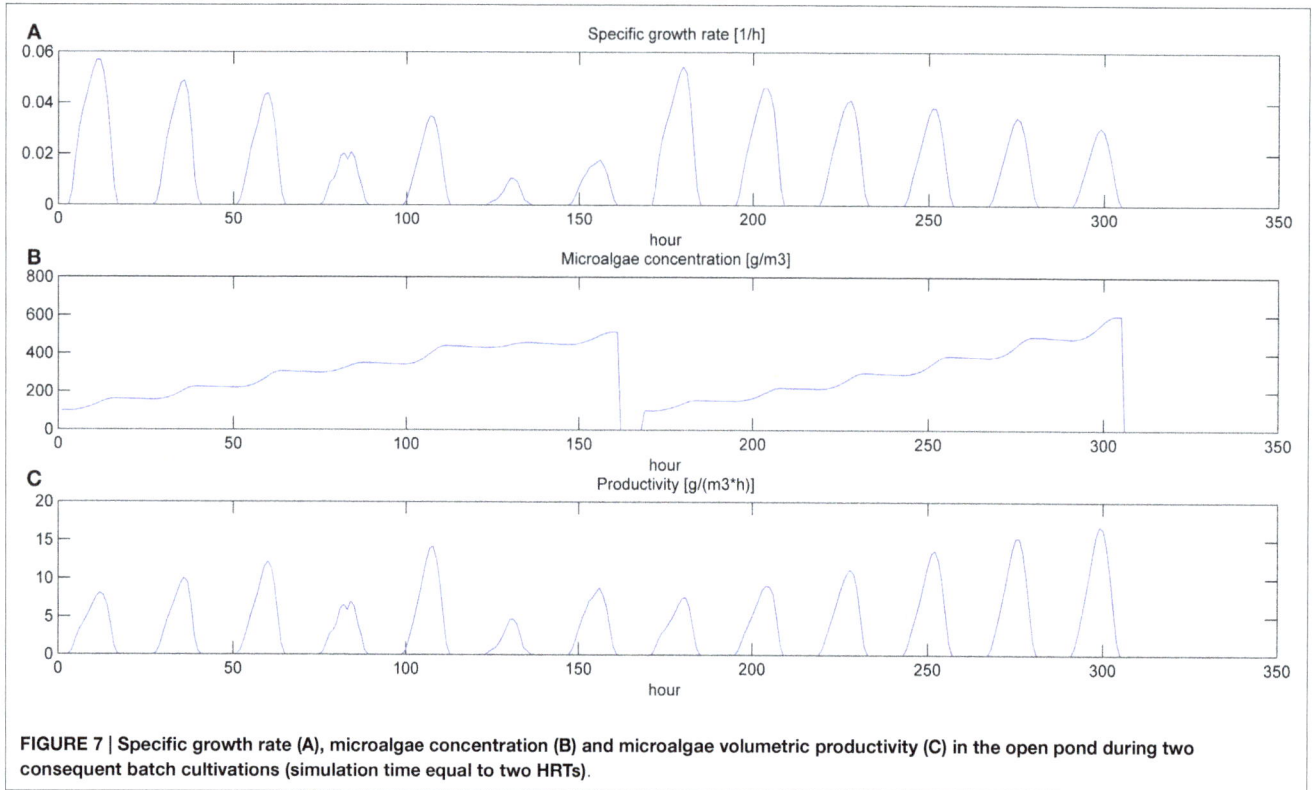

FIGURE 7 | Specific growth rate (A), microalgae concentration (B) and microalgae volumetric productivity (C) in the open pond during two consequent batch cultivations (simulation time equal to two HRTs).

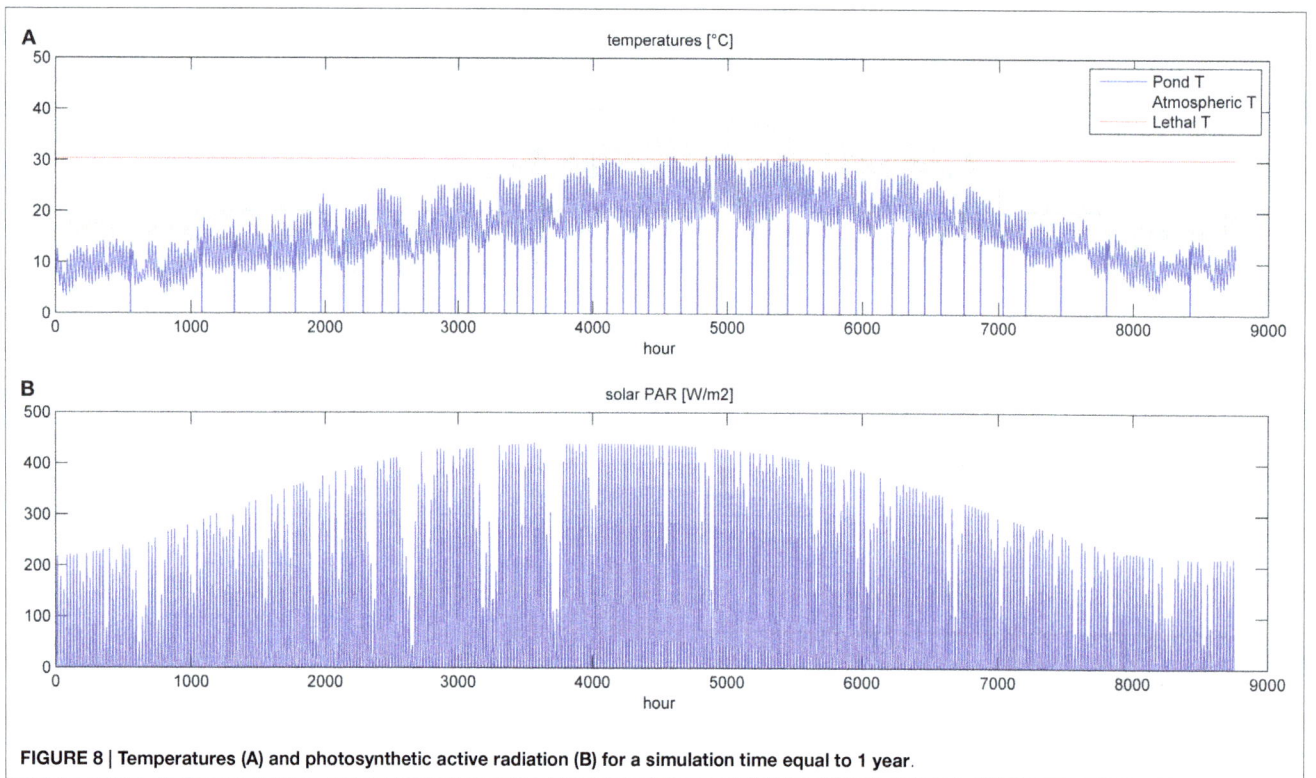

FIGURE 8 | Temperatures (A) and photosynthetic active radiation (B) for a simulation time equal to 1 year.

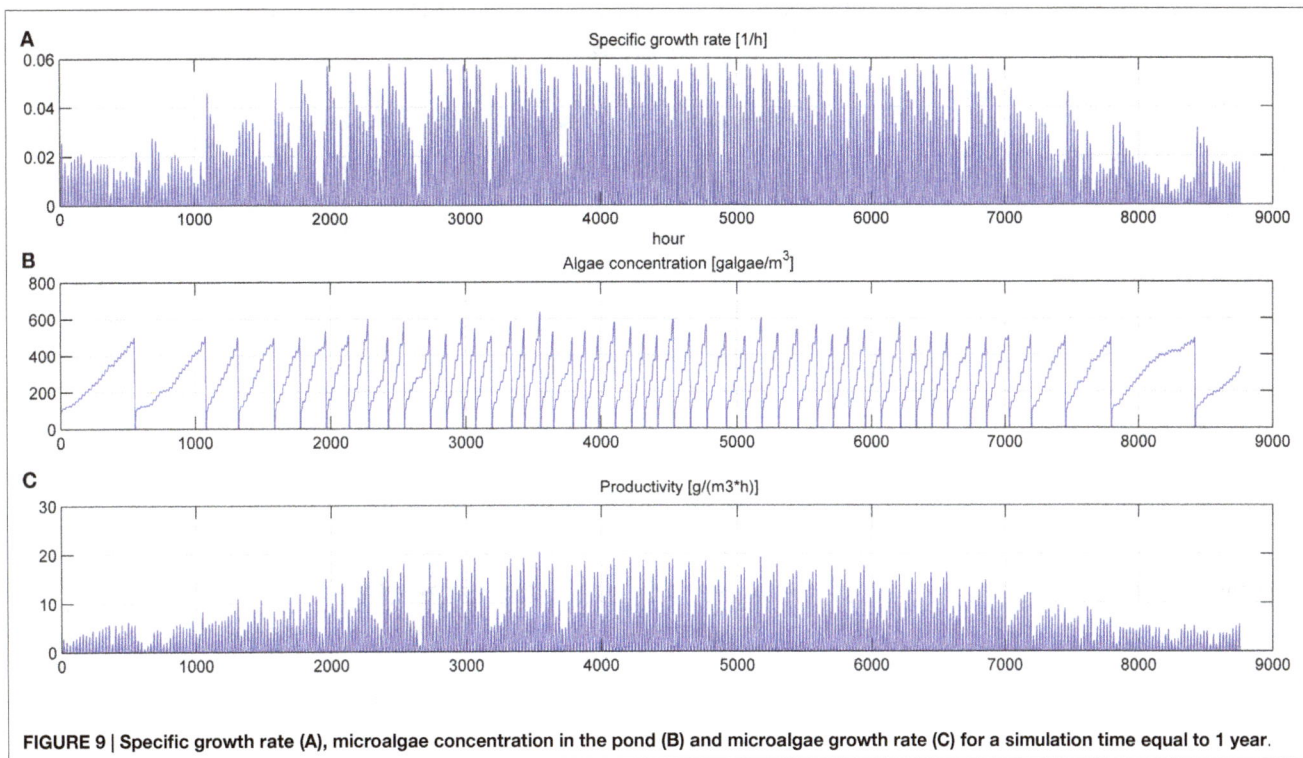

FIGURE 9 | Specific growth rate (A), microalgae concentration in the pond (B) and microalgae growth rate (C) for a simulation time equal to 1 year.

TABLE 6 | Input data for open pond simulation runs.

HRT not fixed, X target fixed	
Location	Sevilla (SPA)
Typology	P. tricurnutum
Xa_init [g/m³]	100
Xa_target [g/m³]	490
Tw_in [°C]	15
CO₂ rate [%]	0.04
Z_pond [m]	0.3
LW [−]	10

TABLE 7 | Input data for flat panel simulations.

HRT not fixed, X target fixed		HRT fixed, X target not fixed	
Location	Sevilla Petrolina	Location	Sevilla Petrolina
Typology	P. tricurnutum T. pseudonana	Typology	P. tricurnutum T. pseudonana
Azimuth [°]	0 −90	Azimuth [°]	0 −90
Slope [°]	90	Slope [°]	90
Xa_init [g/m³]	3000	Xa_init [g/m³]	3000
Xa_target [g/m³]	6000	HRT [day]	4 or 5
CO₂ rate [%]	0.02	CO₂ rate [%]	0.02
h [m]	1.5	h [m]	1.5
s [m]	0.05	s [m]	0.05
d [m]	0.5	d [m]	0.5

east. The model is not able to consider different slopes of the reactor, which may only be positioned vertically: this position is the most favorable for light distribution and dilution, leading to the highest values of productivity (Sierra et al., 2008). CO_2 molar concentration in the injected gases is lower than that of the open raceway pond, since the injected gases in the flat panel photobioreactor do not only have to supply the CO_2 needed by microalgae for photosynthesis, but also to generate the necessary mixing in the reactor: adequate mixing is essential to obtain good levels of productivity in every microalgae cultivation system. Due to the complex geometry of the flat panel photobioreactor, more geometrical data are needed as input if compared to the open raceway pond system. As explained by Fernández et al. (2013), heights <1.5 m and widths <0.10 m are preferred; following this indication and data from Slegers et al. (2011), the distance between the vertical panels has been set equal to 0.5 m, the height of each panel to 1.5 m and the thickness to 5 cm. Both height and distance between panels

have been varied within a range of possible values in a further parametrical analysis. The initial and final concentrations have been suggested by Münkel et al. (2013); the values are higher than in the case of open raceway pond: the more sophisticate closed photobioreactor allows higher concentrations to be reached without compromising the productivity of the cultivation system. This is possible thanks to an optimal light distribution over the whole reactor for the entire operation time that is guaranteed also by an adequate mixing of the medium. The flat panel model, as well as the open raceway pond model, is able to produce the dynamic trend of all the time dependent physical quantities, which are included in the analysis. The global results for the flat panel reactor are shown in **Tables 8** and **9**: **Table 8** reports the results for the first operating strategy, i.e., the case

TABLE 8 | Results from flat panel simulations using the first operating strategy, meaning that input and output microalgae concentrations are known.

	HRT not fixed South		HRT not fixed East	
	Petrolina	Sevilla	Petrolina	Sevilla
N harvesting	65	77	88	81
Mass microalgae [t]	289	345	393	366
CO_2 captured [t]	860	976	1083	1025
CO_2 injected [t]	956	1085	1204	1139
CO_2 ratio algae	0.9	0.9	0.9	0.9
N absorbed [t]	35.92	40.77	45.2	42.77
Water injected [t]	46,023	54,592	62,353	57,742
Energy microalgae [kWh]	1733,700	2063,700	2350,700	2191,100
Electrical energy [kWh]	35,634	40,476	44,906	42,473
Thermal energy [kWh]	1310,300	2124,000	1267,400	2093,900
Volumetric productivity [kg/(m³ × day)]	0.5851	0.6965	0.7934	0.7395
Areal productivity [kg/(m² × day)]	0.0794	0.0946	0.1077	0.1004
NER	1.55	2.09	1.1165	1.95

TABLE 9 | Results from flat panel simulation using the second strategy, meaning that input microalgae concentration and HRT are known.

	HRT fixed South		HRT fixed East	
	Petrolina	Sevilla	Petrolina	Sevilla
HRT [day]	5	5	4	4
Mass microalgae [t]	284	367	390	380
CO_2 captured [t]	844	1041	1081	1052
CO_2 injected [t]	938	1157	1201	1168
CO_2 ratio algae	0.9	0.9	0.9	0.9
N absorbed [t]	35,023	43,044	45.11	43.88
Water injected [t]	46,691	52,620	62,422	57,201
Energy microalgae [kWh]	1701,300	2197,000	2336,600	2273,000
Electrical energy [kWh]	34,973	43,043	44,819	43,542
Thermal energy [kWh]	1292,400	2264,900	1262,700	2113,900
Volumetric productivity [kg/(m³ × day)]	0.5742	0.7415	0.7886	0.7672
Areal productivity [kg/(m² × day)]	0.078	0.1007	0.1071	0.1041
NER	1.56	2.101	1.1192	1.8983
Xa final mean [kg/m³]	5.87	6.707	6.016	6.077

in which both input and output microalgae concentrations are known, whereas **Table 9** corresponds to the second operating strategy, in which the input parameters are the input concentration and the HRT.

For the two operating strategies, results are reported for both the locations (Sevilla and Petrolina) and for the two most significant orientations, that is when the flat panel is facing south (and north) and when it is facing east (and west). The east–west orientation appears to be preferable in both locations because it leads to a higher microalgae production: as explained by Sierra et al. (2008), if the orientation of the two faces of the reactor is east–west, the intercepted radiation is maximum during the first and last solar hours, because of the orientations toward sunrise and sunset. Thus, light availability during the daylight solar cycle is also more homogenous for this configuration. Areal and volumetric productivities are consistent with those found in literature: Chisti (2007) reports a volumetric productivity equal to 1.535 kg m^{-3} day^{-1}, which is higher than the values obtained

from the model; if compared to some other works, the results of the dynamic model, in particular the areal productivity, seem to be optimistic: for example, the microalgae production coming from the model [~400 t/(ha × year)] is two times higher than the production obtained by Slegers et al. (2011) from a flat panel reactor located in Algeria which produces up to 200 t/(ha × year). Moreover, the volumetric productivity for a flat panel reported by Jorquera et al. (2010) is equal to 0.27 kg m^{-3} day^{-1}, whereas the model supplies values equal to 0.8 kg m^{-3} day^{-1}. From a recent work by Münkel et al. (2013), volumetric productivities equal to 1.25 kg m^{-3} day^{-1} have been reached in experimental analyzes. The difference from some values found in literature could be a consequence of a series of related factors: the microalgae species chosen may strongly influence the productivity of the reactor. Moreover, the model created in this work contains a temperature control, which fixes the temperature inside the reactor at the optimal level for microalgae growth: this means that the specific growth rate is not affected by the temperature factor (which is always equal to one), and consequently it is nearer to the maximum growth rate than in the real operating conditions. Furthermore, the locations chosen for the analysis present optimal values of irradiation, next to the saturation irradiation, where the light intensity factor affecting the growth rate is next to 1. From **Tables 8** and **9** the high dependence on the orientation of the panel clearly appears in the productivity of Petrolina: if the two faces of the reactor are oriented toward east and west, the productivity is higher than in case of south and north orientation. A possible explanation might be related to radiation reflection by the panel: when the sun is high in the sky, the radiation hits the flat panel with an incidence angle close to 90°; if the incidence angle is too high, radiation could not enter the reactor, due to glass reflection. For this reason, if the panel is oriented toward south, the largest part of the radiation during summer is lost and does not contribute to microalgae growth: if the orientation if east–west, the radiation is collected during morning and afternoon with an incidence angle next to 0°. The same situation does not take place in Sevilla, because the sun does not reach high elevations during the whole year; the east–west orientation is preferable also in Sevilla, as suggested by Sierra et al. (2008).

In general, higher volumetric and areal productivity values have been reached both in Sevilla and in Petrolina, and the difference between the two locations is less remarkable than it was for the open pond; the flat panel photobioreactor growth model includes a temperature control, which maintains the temperature of the water at a constant level. In spite of the higher productivities, Sevilla appear to be less suitable for a flat panel photobioreactor than Petrolina, as shown by NER. The thermal energy requirement is extremely high in Sevilla, where winter time brings low atmospheric temperatures: the photobioreactor should operate only during summertime.

Microalgae mass production in flat panel reactor is much higher than in open raceway pond, CO_2 absorption is more efficient, but the thermal energy requirement implies a NER value >1. There are different possible strategies to solve this problem: keeping the water in the reactor within a range of suboptimal temperatures where the productivity of microalgae is still high and the thermal energy requirement is lower; leaving the temperature in the reactor without any control during the night: this might be an interesting solution also to limit the microalgae losses due to dark respiration, which are higher for optimal temperature. If water temperature is lower than the optimal value for growth, the metabolic energy required by the microalgae for their maintenance during the night is lower.

References

Ak, İ (2012). Effect of an organic fertilizer on growth of blue-green alga *Spirulina platensis*. *Aquacult. Int.* 20, 413–422. doi:10.1007/s10499-011-9473-5

Azadi, P., Brownbridge, G., Mosbach, S., Smallbone, A., Bhave, A., Inderwildi, O., et al. (2014). The carbon footprint and non-renewable energy demand of algae-derived biodiesel. *Appl. Energy* 113, 1632–1644. doi:10.1016/j.apenergy.2013.09.027

Bahadar, A., and Bilal Khan, M. (2013). Progress in energy from microalgae: a review. *Renew. Sustain. Energ. Rev.* 27, 128–148. doi:10.1016/j.rser.2013.06.029

Béchet, Q., Shilton, A., and Guieysse, B. (2013). Modeling the effects of light and temperature on algae growth: state of the art and critical assessment

Discussion

Unlike the most common existing models in the literature, which deal with a specific part of the overall cultivation process, the models presented in this paper include all physical and chemical quantities that mostly affect microalgae growth. These features allow the model to correctly predict the overall behavior of microalgae cultivation plants for energy production. All input parameters can be tuned and varied to obtain reliable predictions.

This paper aims to present a model for microalgae cultivation where geometric and physical parameters, presented in recent works by Slegers et al. (2011, 2013), Duffie and Beckman (2013), Béchet et al. (2013) and more, receive a robust analysis together with the chemical and biological aspects, as presented by Yang (2011), Chisti (2007), Le et al. (2010), and more.

A comparison with experimental data taken from the literature shows that the predictions are consistent, only slightly overestimating the productivity in case of closed photobioreactor. The reason for the overestimation might be found in the temperature control strategy of the reactor, which is kept constant in the simulation runs. This strategy is currently applied in laboratory scale plants but it is expensive both from an economic and energetic point of you for a large scale cultivation plant.

The model is a part of a wider activity on microalgae cultivation plants for energy production which consider the additional directions of work:

- Further analysis could be conducted at the boundaries of the system taken into consideration by the model, evaluating and modeling the downstream microalgae transformation process and the upstream technologies that generate the input streams entering the cultivation phase.
- The cultivation phase model might be modified to include other less important, but still valuable parameters that influence microalgae growth, such as some other nutrients like phosphorus.
- The microalgae growth rate equation could be compared with other models in the literature and with results coming from real pilot plants to make predictions more reliable.
- The flat panel model should be tested with different temperature control strategies, which might lead to a lower productivity but also to a lower thermal energy consumption, making the technology more interesting from an economic point of view.

for productivity prediction during outdoor cultivation. *Biotechnol. Adv.* 31, 1648–1663. doi:10.1016/j.biotechadv.2013.08.014

Borowitzka, M. A. (1999). Commercial production of microalgae: ponds, tanks, tubes and fermenters. *J. Biotechnol.* 70, 313–321. doi:10.1016/S0168-1656(99)00083-8

Brennan, L., and Owende, P. (2010). Biofuels from microalgae – a review of technologies for production, processing, and extractions of biofuels and co-products. *Renew. Sustain. Energ. Rev.* 14, 557–577. doi:10.1016/j.rser.2009.10.009

Çelekli, A., and Yavuzatmaca, M. (2009). Predictive modeling of biomass production by *Spirulina platensis* as function of nitrate and NaCl concentrations. *Bioresour. Technol.* 100, 1847–1851. doi:10.1016/j.biortech.2008.09.042

Chisti, Y. (2007). Biodiesel from microalgae. *Biotechnol. Adv.* 25, 294–306. doi:10.1016/j.biotechadv.2007.02.001

Collet, P., Hélias, A., Lardon, L., Ras, M., Goy, R.-A., and Steyer, J.-P. (2011). Life-cycle assessment of microalgae culture coupled to biogas production. *Bioresour. Technol.* 102, 207–214. doi:10.1016/j.biortech.2010.06.154

Cuaresma, M., Janssen, M., Vílchez, C., and Wijffels, R. H. (2011). Horizontal or vertical photobioreactors? How to improve microalgae photosynthetic efficiency. *Bioresour. Technol.* 102, 5129–5137. doi:10.1016/j.biortech.2011.01.078

Davis, R., Aden, A., and Pienkos, P. T. (2011). Techno-economic analysis of autotrophic microalgae for fuel production. *Appl. Energy* 88, 3524–3531. doi:10.1016/j.apenergy.2011.04.018

Doucha, J., and Lívanský, K. (2006). Productivity, CO2/O2 exchange and hydraulics in outdoor open high density microalgal (*Chlorella sp.*) photobioreactors operated in a Middle and Southern European climate. *J. Appl. Phycol.* 18, 811–826. doi:10.1007/s10811-006-9100-4

Duffie, J. A., and Beckman, W. A. (2013). *Solar Engineering of Thermal Processes*, 4th edn. New York: A Wiley-Interscience Publication, John Wiley and Sons, Inc.

Fernández, F. G. A., Camacho, F. G., Pérez, J. A. S., Sevilla, J. M. F., and Grima, E. M. (1997). A model for light distribution and average solar irradiance inside outdoor tubular photobioreactors for the microalgal mass culture. *Biotechnol. Bioeng.* 55, 701–714. doi:10.1002/(SICI)1097-0290(19970905)55:5

Fernández, F. G. A., Sevilla, J. M. F., and Grima, E. M. (2013). Photobioreactors for the production of microalgae. *Rev. Environ. Sci. Biotechnol.* 12, 131–151. doi:10.1007/s11157-012-9307-6

Hadiyanto, H., Elmore, S., Van Gerven, T., and Stankiewicz, A. (2013). Hydrodynamic evaluations in high rate algae pond (HRAP) design. *Chem. Eng. J.* 217, 231–239. doi:10.1016/j.cej.2012.12.015

Hempel, N., Petrick, I., and Behrendt, F. (2012). Biomass productivity and productivity of fatty acids and amino acids of microalgae strains as key characteristics of suitability for biodiesel production. *J. Appl. Phycol.* 24, 1407–1418. doi:10.1007/s10811-012-9795-3

James, S. C., and Boriah, V. (2010). Modeling algae growth in an open-channel raceway. *J. Comput. Biol.* 17, 895–906. doi:10.1089/cmb.2009.0078

Janssen, M., Tramper, J., Mur, L. R., and Wijffels, R. H. (2003). Enclosed outdoor photobioreactors: light regime, photosynthetic efficiency, scale-up, and future prospects. *Biotechnol. Bioeng.* 81, 193–210. doi:10.1002/bit.10468

Jiménez, C., Cossio, B. R., Labella, D., and Xavier Niell, F. (2003). The feasibility of industrial production of Spirulina (*Arthrospira*) in Southern Spain. *Aquaculture* 217, 179–190. doi:10.1016/S0044-8486(02)00118-7

Jorquera, O., Kiperstok, A., Sales, E. A., Embiruçu, M., and Ghirardi, M. L. (2010). Comparative energy life-cycle analyses of microalgal biomass production in open ponds and photobioreactors. *Bioresour. Technol.* 101, 1406–1413. doi:10.1016/j.biortech.2009.09.038

Kochem, L. H., Da Fré, N. C., Redaelli, C., Rech, R., and Marcílio, N. R. (2014). Characterization of a novel flat-panel airlift photobioreactor with an internal heat exchanger. *Chem. Eng. Technol.* 37, 59–64. doi:10.1002/ceat.201300420

Kumar, K., Dasgupta, C. N., Nayak, B., Lindblad, P., and Das, D. (2011). Development of suitable photobioreactors for CO2 sequestration addressing global warming using green algae and cyanobacteria. *Bioresour. Technol.* 102, 4945–4953. doi:10.1016/j.biortech.2011.01.054

Le, P. J., Williams, B., and Laurens, L. M. L. (2010). Microalgae as biodiesel & biomass feedstocks: review & analysis of the biochemistry, energetics & economics. *Energy Environ. Sci.* 3, 554–590. doi:10.1039/b924978h

Mata, T. M., Martins, A. A., and Caetano, N. S. (2010). Microalgae for biodiesel production and other applications: a review. *Renew. Sustain. Energ. Rev.* 14, 217–232. doi:10.1016/j.rser.2009.07.020

Molina, E., Fernández, J., Acién, F. G., and Chisti, Y. (2001). Tubular photobioreactor design for algal cultures. *J. Biotechnol.* 92, 113–131. doi:10.1016/S0168-1656(01)00353-4

Münkel, R., Schmid-Staiger, U., Werner, A., and Hirth, T. (2013). Optimization of outdoor cultivation in flat panel airlift reactors for lipid production by *Chlorella vulgaris*. *Biotechnol. Bioeng.* 110, 2882–2893. doi:10.1002/bit.24948

Muñoz, R., and Guieysse, B. (2006). Algal-bacterial processes for the treatment of hazardous contaminants: a review. *Water Res.* 40, 2799–2815. doi:10.1016/j.watres.2006.06.011

Norsker, N.-H., Barbosa, M. J., Vermuë, M. H., and Wijffels, R. H. (2011). Microalgal production – a close look at the economics. *Biotechnol. Adv.* 29, 24–27. doi:10.1016/j.biotechadv.2010.08.005

NREL. (0000). *Biomass Research – Publications*. Available at: http://www.nrel.gov/biomass/publications.html?print

Pruvost, J., Cornet, J. F., Goetz, V., and Legrand, J. (2011). Modeling dynamic functioning of rectangular photobioreactors in solar conditions. *AIChE J.* 57, 1947–1960. doi:10.1002/aic.12389

Radmann, E. M., Reinehr, C. O., and Costa, J. A. V. (2007). Optimization of the repeated batch cultivation of microalga *Spirulina platensis* in open raceway ponds. *Aquaculture* 265, 118–126. doi:10.1016/j.aquaculture.2007.02.001

Rawat, I., Ranjith Kumar, R., Mutanda, T., and Bux, F. (2013). Biodiesel from microalgae: a critical evaluation from laboratory to large scale production. *Appl. Energy* 103, 444–467. doi:10.1016/j.apenergy.2012.10.004

Robinson, D., and Stone, A. (2004). Solar radiation modelling in the urban context. *Solar Energy* 77, 295–309. doi:10.1016/j.solener.2004.05.010

Rodolfi, L., Chini Zittelli, G., Bassi, N., Padovani, G., Biondi, N., Bonini, G., et al. (2009). Microalgae for oil: strain selection, induction of lipid synthesis and outdoor mass cultivation in a low-cost photobioreactor. *Biotechnol. Bioeng.* 102, 100–112. doi:10.1002/bit.22033

Ruiz, J., Álvarez-Díaz, P. D., Arbib, Z., Garrido-Pérez, C., Barragán, J., and Perales, J. A. (2013). Performance of a flat panel reactor in the continuous culture of microalgae in urban wastewater: prediction from a batch experiment. *Bioresour. Technol.* 127, 456–463. doi:10.1016/j.biortech.2012.09.103

Sierra, E., Acién, F. G., Fernández, J. M., García, J. L., González, C., and Molina, E. (2008). Characterization of a flat plate photobioreactor for the production of microalgae. *Chem. Eng. J.* 138, 136–147. doi:10.1016/j.cej.2007.06.004

Sills, D. (2013). *Modeling CO2 Requirements for Cultivation of Microalgae in Open Raceway Pond*.

Singh, N. K., and Dhar, D. W. (2011). Microalgae as second generation biofuel. A review. *Agron. Sustain. Dev.* 31, 605–629. doi:10.1007/s13593-011-0018-0

Slegers, P. M., Lösing, M. B., Wijffels, R. H., van Straten, G., and van Boxtel, A. J. B. (2013). Scenario evaluation of open pond microalgae production. *Algal. Res.* 2, 358–368. doi:10.1016/j.biortech.2012.11.123

Slegers, P. M., Wijffels, R. H., van Straten, G., and van Boxtel, A. J. B. (2011). Design scenarios for flat panel photobioreactors. *Appl. Energy* 88, 3342–3353. doi:10.1016/j.apenergy.2010.12.037

Sompech, K., Chisti, Y., and Srinophakun, T. (2012). Design of raceway ponds for producing microalgae. *Biofuels* 3, 387–397. doi:10.4155/bfs.12.39

Stephenson, A. L., Kazamia, E., Dennis, J. S., Howe, C. J., Scott, S. A., and Smith, A. G. (2010). Life-cycle assessment of potential algal biodiesel production in the United Kingdom: a comparison of raceways and air-lift tubular bioreactors. *Energy Fuels* 24, 4062–4077. doi:10.1021/ef1003123

Sugai-Guérios, M. H., Mariano, A. B., Vargas, J. V. C., de Lima Luz, L. F., and Mitchell, D. A. (2014). Mathematical model of the CO2 solubilisation reaction rates developed for the study of photobioreactors. *Can. J. Chem. Eng.* 92, 787–795. doi:10.1002/cjce.21937

Sukhatme, K., and Sukhatme, S. P. (1996). *Solar Energy: Principles of Thermal Collection and Storage*. Noida, UP: Tata McGraw-Hill Education.

Yang, A. (2011). Modeling and evaluation of CO2 supply and utilization in algal ponds. *Ind. Eng. Chem. Res.* 50, 11181–11192. doi:10.1021/ie200723w

Conflict of Interest Statement: The authors declare that the research was conducted in the absence of any commercial or financial relationships that could be construed as a potential conflict of interest.

Appendix

Nomenclature

A_{soil}: area of the pond that is embedded in the soil [m²]

A_w: water surface area of the pond [m²]

C_{Bowen}: Bowen constant [Pa/°C]

CO_{2D}: dissolved CO_2 molar quantity in the bioreactor [mol/m³]

cp_w: heat capacity of the growth medium [J/(kg°C)]

C_T: total concentration of carbonate species in the bioreactor

d: distance between the panels [m]

$E_{harv/refil}$: electrical energy consumption for microalgae harvesting and reactor refilling [J]

f_{CO_2}: flux of CO_2 introduced by the supply of gas flow into the system [g/(m³ s)]

f_I: light intensity factor [−]

f_T: temperature factor [−]

g: gravitational acceleration [m/s²]

$G_{diffuse}$: geometrical factor for diffuse radiation [−]

G_{direct}: geometric factor for direct radiation [−]

h: height of the panels [m]

h_{comb}: lower heating value on a dry basis of algae biomass [J/kg]

h_{shadow}: height of the shadow on the panel behind [m]

h_{soil}: heat transfer coefficient of the surrounding soil layer [W/(m²°C)]

I_0: incident light intensity [W/m²]

I_a: average light intensity in the volume of the bioreactor at a given time t [W/m²]

I_{back}: radiation light intensity for the back surface [W/m²]

$I_{diffuse}$: diffuse radiation over a horizontal surface [W/m²]

I_{direct}: direct radiation over a horizontal surface [W/m²]

I_{front}: radiation light intensity for the front surface [W/m²]

I_{GHR}: global horizontal radiation [W/m²]

I_s: saturation light intensity for microalgae species [W/m²]

$I_{surface}$: total light arriving on the pond [J/(kg°C)]

K_C, K_{NA}: half-saturation constants [mol/m³]

K_e: light extinction coefficient [m⁻¹]

K_{e1}, K_{e2}: light extinction constants

$k_{Lg}\alpha$: mass transfer coefficient for a given element

k_1, k_2: carbonate species equilibrium constants

M: concentration of the respective component [g/m³]

M^*: saturation concentration of the associated dissolved element [g/m³]

N_T: dissolved nitrogen molar quantity in the bioreactor [mol/m³]

PAR: photosynthetic active radiation [%]

p_a: ambient pressure [Pa]

P_{pump}: power of the installed pump for harvesting and refilling of the reactor [W]

p_{ref}: reference pressure [Pa]

p'_a: water pressure of the air [Pa]

p'_s: saturated water pressure [Pa]

Q: volumetric flow rate [m³/s]

Q_{algae}: light energy flow to algae during growth [W]

Q_{cond}: heat flow between the pond and the ground via conduction [W]

Q_{conv}: heat flow by convection [W]

Q_{evap}: heat flow caused by either evaporation or condensation [W]

Q_{irr}: heat flow to the pond by sunlight [W]

Q_{rad}: heat flow by emission of long-wave radiation in the infrared region [W]

r_{gA}: microalgae growth rate [g/(m³ s)]

RH: relative humidity [−]

R_p: reflection coefficient for p-polarized light [−]

R_s: reflection coefficient for s-polarized light [−]

R': overall reflection coefficient for each interface [−]

R'_1: overall reflection coefficient for the air – reactor interface [−]

R'_2: overall reflection coefficient for the reactor – culture interface [−]

s: position in bioreactor depth [m]

T_a: air temperature [°C]

T_{dew}: dew point temperature [°C]

$t_{harv/refil}$: time for microalgae harvesting and reactor refilling [s]

T_{let}: lethal temperature for microalgae species [K]

T_m: transparency of wall material [−]

T_{opt}: optimal temperature for microalgae species [K]

T_{sky}: equivalent sky temperature for clear sky days [K]

T_{soil}: temperature of the soil surrounding the pond [°C]

t_{solar}: number of hours after solar midnight [−]

T_w: water temperature in the bioreactor [K]

T_w: pond temperature [°C]

V_R: reactor volume [m³]

X_A: microalgae concentration in the bioreactor [g/m³]

y: position in reactor height [m]

Y_{AM}: mass of the respective component consumed or generated by the microalgae per unit mass of microalgae produced [g_M/g_{microalgae}]

z: pond depth [m] or photobioreactor thickness [m]

β: slope of the reactor [°]

β_{algae}: curve modulating constant for the temperature factor: it depends on microalgae species

γ: azimuth angle [°]

ϵ_w: emissivity of the water in the infrared region [−]

η_i: refractive index of the material before the interface [−]

η_{pump}: efficiency of the pump [−]

η_t: refractive index of the material after the interface [−]

θ: solar incidence angle [°]

θ_i: incidence angle [°]

θ_z: solar zenith angle [°]

μ_A: microalgae specific growth rate [s⁻¹]

$\hat{\mu}_A$: maximum microalgae specific growth rate [s⁻¹]

ρ: water density [kg/m³]

ρ_w: density of the growth medium [kg/m³]

σ: light extinction coefficient [m]

σ_{SB}: Stefan–Boltzmann constant [W/(m²K⁴)]

10

Optimization of protein extraction from *Spirulina platensis* to generate a potential co-product and a biofuel feedstock with reduced nitrogen content

Naga Sirisha Parimi[1], Manjinder Singh[1], James R. Kastner[1], Keshav C. Das[1]*, Lennart S. Forsberg[2] and Parastoo Azadi[2]

[1] *College of Engineering, The University of Georgia, Athens, GA, USA,* [2] *Complex Carbohydrate Research Center, The University of Georgia, Athens, GA, USA*

Edited by:
S. Kent Hoekman,
Desert Research Institute, USA

Reviewed by:
John Chandler Cushman,
University of Nevada, USA
Arumugam Muthu,
Council of Scientific and Industrial
Research, India
Mi Li,
Oak Ridge National Laboratory, USA

*Correspondence:
Keshav C. Das,
Driftmier Engineering Center,
University of Georgia, Room 509,
Athens, GA 30602, USA
kdas@engr.uga.edu

The current work reports protein extraction from *Spirulina platensis* cyanobacterial biomass in order to simultaneously generate a potential co-product and a biofuel feedstock with reduced nitrogen content. *S. platensis* cells were subjected to cell disruption by high-pressure homogenization and subsequent protein isolation by solubilization at alkaline pH followed by precipitation at acidic pH. Response surface methodology was used to optimize the process parameters – pH, extraction (solubilization/precipitation) time and biomass concentration for obtaining maximum protein yield. The optimized process conditions were found to be pH 11.38, solubilization time of 35 min and biomass concentration of 3.6% (w/w) solids for the solubilization step, and pH 4.01 and precipitation time of 60 min for the precipitation step. At the optimized conditions, a high protein yield of 60.7% (w/w) was obtained. The protein isolate (co-product) had a higher protein content [80.6% (w/w)], lower ash [1.9% (w/w)] and mineral content and was enriched in essential amino acids, the nutritious -linolenic acid and other high-value unsaturated fatty acids compared to the original biomass. The residual biomass obtained after protein extraction had lower nitrogen content and higher total non-protein content than the original biomass. The loss of about 50% of the total lipids from this fraction did not impact its composition significantly owing to the low lipid content of *S. platensis* (8.03%).

Keywords: *Spirulina platensis*, protein isolate, high-pressure homogenization, response surface methodology, residual biomass, biofuel feedstock

Introduction

The concept of biorefinery which proposes the integration of biofuel production processes with the extraction of co-product(s) such as proteins, pigments, and other high-value compounds is the path forward to improve the sustainability and economic feasibility of microalgal processing technologies. The high protein (and nitrogen) content of algal feedstock is a major limitation to whole biomass to biofuel conversion processes such as hydrothermal liquefaction (HTL) and anaerobic digestion (AD). High-protein feedstocks result in high nitrogen content in the fuel produced from HTL and ammonia toxicity in AD (Chen et al., 2008; López Barreiro et al., 2013). Thus, nitrogen removal through protein extraction could potentially improve the feedstock composition for biofuel applications, while generating a useful co-product. Microalgal proteins are comparable to conventional protein sources such as soymeal and eggs, and hence find potential applications in human nutrition and animal feed (Spolaore et al., 2006; Becker, 2007).

Pre-treatments such as mechanical cell lysis, enzymatic, thermal, and chemical treatments result in improved component extraction by complete or partial degradation of the microalgal cell wall, thus, improving the accessibility of the intra-cellular components. High-pressure homogenization and ultrasonication were reported to enhance microalgal protein solubilization, the former being the most effective method (Gerde et al., 2013; Safi et al., 2014). Autoclaving was reported as an effective pretreatment to improve lipid extraction from microalgae (Prabakaran and Ravindran, 2011).

Protein solubility is pH dependent. Highly acidic and alkaline conditions enhance the solubility of algal proteins by inducing net charges on the amino acid residues (Damodaran, 1996). Proteins are least soluble at their isoelectric pH and precipitate out. Thus, solubilization under alkaline conditions followed by precipitation at isoelectric pH is a useful strategy for obtaining crude protein isolates. Several authors reported protein extraction from green algae and cyanobacteria using this method (Choi and Markakis, 1981; Chronakis et al., 2000; Gerde et al., 2013; Safi et al., 2014; Ursu et al., 2014). Other parameters that could impact protein solubility include extraction (solubilization or precipitation) time, solvent/biomass ratio (biomass concentration), and temperature (Abas Wani et al., 2006). High temperature causes protein denaturing and also increases the energy input for the overall process (Goetz and Koehler, 2005). Hence, heat treatment is undesirable in protein isolation processes.

Process optimization and statistical analysis is necessary to maximize protein extraction and determine the independent and interaction effects of various process parameters on the extraction yields. Response surface methodology (RSM) is a popular statistical method for optimization of process parameters while conducting the least number of experiments (Firatligil-Durmus and Evranuz, 2010). Protein extraction process optimization using RSM for non-algal sources and Chlorella pyrenoidosa (green algae) was reported previously (Quanhong and Caili, 2005; Zhang et al., 2007; Ma et al., 2010; Wang and Zhang, 2012).

The current study dealt with process optimization for maximizing protein extraction from the cyanobacterium (blue-green alga) Spirulina platensis, and the generation of a residual biomass with lower nitrogen content than the original biomass for potential applications as a biofuel feedstock in whole biomass conversion processes such as HTL and AD. Cyanobacteria differ significantly from green algae in cell wall structure and biochemical composition. Unlike the latter which have a recalcitrant cell wall comprising of cellulose and hemicellulose (Payne and Rippingale, 2000), cyanobacteria such as Spirulina and Nostoc sp. have a peptidoglycan-based cell wall (Palinska and Krumbein, 2000). Moreover, they have a higher protein and lower lipid content (Becker, 2007). These differences necessitate the optimization of process parameters for the specific phylum. S. platensis was

chosen in the current study for two reasons. First, it is an edible cyanobacterium and hence its protein isolate is expected to have a high nutritive value. Second, it has a very high protein content (Cohen, 1997) and hence the impact of protein isolation on the biochemical composition of the residual biomass would be very striking in this species compared to those with a lower protein content. Although some reports on extraction of proteins from S. platensis may be found in the literature, major knowledge gaps on process optimization, component fractionation, and product characterization remain (Devi et al., 1981; Chronakis et al., 2000; Safi et al., 2013b). The current work aimed at filling these gaps in order to understand the fate of various cell components as a result of the fractionation process and identify the bottlenecks in the process. Some of the parameters described in the literature to characterize protein isolates such as protein content, amino acid composition, mineral composition, and molecular weight range of the proteins were reported for the protein isolate obtained in this study (Chronakis et al., 2000; Gerde et al., 2013; Safi et al., 2013a). Such knowledge is very useful in assessing the sustainability, scalability, and economic feasibility of the process.

Materials and Methods

Microalgae

Spirulina platensis was obtained from Earthrise Nutritionals LLC (Calipatria, CA, USA) in dry powder form and was stored in sealed, air tight plastic packages at room temperature prior to use. The dry powder was mixed with deionized (DI) water to form biomass slurry at the desired concentration (solids content).

Protein Isolation Process

Spirulina platensis biomass slurry prepared at the desired concentration was subjected to a protein isolation process (Figure S1 in Supplementary Material) which involved pretreatment of the biomass and subsequent extraction of proteins by solubilization at alkaline pH using 1M NaOH followed by precipitation from the supernatant (obtained from the previous step) at acidic pH using either 1M HCl or 1M HCOOH. The solid–liquid separation after the solubilization and the precipitation steps was achieved by centrifugation at 8670 g for 35 min. The pellet and the supernatant from the solubilization step are henceforth referred to as alkali pellet and alkali supernatant, respectively, and those from the precipitation step are referred to as acid pellet and acid supernatant, respectively. The acid pellet was the protein isolate. The combined fraction of the alkali pellet and acid supernatant was the residual biomass.

Selection of Pretreatment

A 6% slurry of S. platensis biomass was subjected to three different pretreatments namely autoclaving, ultrasonication, and high-pressure homogenization. Autoclaving was carried out at 121°C with 103.4 kPa (15 psi) for 30 min. Ultrasonication was carried out using a probe sonicator (Biologics, Inc., VA, USA) at 20% maximum power for 60 min. High-pressure homogenization involved two passes through a high-pressure homogenizer (Constant systems LTD., UK) at 103.4 MPa (15 kpsi). The samples were placed on ice bath during ultrasonication and high-pressure homogenization, and a chiller was attached to the latter unit to minimize sample heating. The control experiment did not involve any pretreatment. Each of the pretreated and control samples was subjected to protein solubilization at pH 11 for 60 min followed by solid–liquid separation. The treatments were compared based on protein recovery in the supernatant fraction. The cells were observed visually under an optical microscope (400 times magnification).

Optimization of Experimental Conditions
Solubility curve determination
A 6% *S. platensis* biomass slurry was subjected to cell disruption by high-pressure homogenization and separated into aliquots. The pH of each aliquot was adjusted to various values in the range of 2–13 (with a step size of 1 U) using either 1M NaOH or 1M HCl and stirred for 30 min before subjecting to solid–liquid separation. A graph of pH versus protein recovery in the supernatant was plotted to obtain the solubility curve.

Statistical optimization
The design of optimization experiments and the statistical analysis was carried out using SAS-based JMP Pro (version 10) statistical software. A Box–Behnken design based on RSM was employed to optimize the process conditions affecting protein solubilization and precipitation. The optimization range for pH for both the steps was chosen based on the solubility curve data. The range for solubilization and precipitation times was 10–60 min. The 60 min maximum was chosen based on the literature which reported that increasing the solubilization time beyond 60 min did not result in a significant increase in the extracted proteins from pH 11 sonicated, non-defatted algae biomass (Gerde et al., 2013). The chosen range for biomass concentration was 2–10% solids, a typical solids range of harvested algal biomass.

Based on the design, set of 15 and 10 experiments were carried out for the solubilization and precipitation steps, respectively (Tables S1 and S2 in Supplementary Material). A second degree polynomial with the following general equation was fit to the data obtained from the solubilization experiments:

$$Y = A_0 + A_1X_1 + A_2X_2 + A_3X_3 + A_{11}X_1{}^2 + A_{22}X_2{}^2 + A_{33}X_3{}^2$$
$$+ A_{12}X_1X_2 + A_{13}X_1X_3 + A_{23}X_2X_3 \qquad (1)$$

where Y was the protein recovery in the alkali supernatant, X_i (i = 1, 2, 3) was the coded dimensionless value of an independent input variable x_i (i = 1, 2, 3) in the range of −1 to 1. The independent input variables were x_1 (pH), x_2 (solubilization time), and x_3 (biomass concentration). A_0 was the constant term, A_i (i = 1, 2, 3), A_{ii} (i = 1, 2, 3), and A_{ij} (i = 1, 2, 3; j = 2, 3; i ≠ j) are the linear, quadratic, and interaction regression coefficients. The variables were coded according to the following equation:

$$X_i = (x_i - x_0)/ \ x_i, i = 1, 2, 3 \qquad (2)$$

where x_0 was the real value of the center point of each input variable and x_i was the step change.

Protein precipitation from the alkali supernatant was carried out using 1M HCOOH obtained at the RSM optimized conditions.

A second degree polynomial with the following general equation was fit to the data obtained from the precipitation experiments:

$$Y = B_0 + B_1X_1 + B_2X_2 + B_{11}X_1{}^2 + B_{22}X_2{}^2 + B_{12}X_1X_2 \qquad (3)$$

where Y was the protein recovery in the acid pellet, X_i (i = 1, 2) was the coded dimensionless value of an independent input variable x_i (i = 1, 2) in the range of −1 to 1. The independent input variables were x_1 (pH) and x_2 (precipitation time). B_0 was the constant term, B_i (i = 1, 2), B_{ii} (i = 1, 2), and B_{ij} (i = 1; j = 2) were the linear, quadratic, and interaction regression coefficients. The input variables were coded in a manner similar to the solubilization step variables.

The coefficient of determination (R^2) and the scattered plots between the experimental and predicted protein recoveries were obtained. The significance of the regression coefficients of the polynomial equations was determined using the Student's *t*-test and *p* value. Optimum process conditions were obtained from the response surface analysis and were experimentally validated.

Analytical Methods
Total Nitrogen, Protein, and Amino Acid Analysis
A HACH high-range total nitrogen assay method (HACH Corporation, Loveland, CO, USA) was used to measure the total nitrogen concentration (mg L^{-1}) in each sample. The nitrogen concentration obtained was multiplied by a factor of 6.25 to obtain the protein concentration (Piorreck et al., 1984; Chronakis et al., 2000; Safi et al., 2013a). A modified Lowry protein assay was used to determine the hydro-soluble protein content (Lowry et al., 1951). Bovine serum albumin (BSA) was used to prepare the standard curve for Lowry protein quantification. Nitrogen content (% N on dry basis) was obtained from the C, H, N, S elemental analysis carried out using a LECO brand analyzer (Model CHNS-932) according to the methods described in ASTM D 5291 and D 3176 (Jena et al., 2011a). Protein content (based on elemental analysis) was determined by multiplying the nitrogen content by the conversion factor of 6.25. Amino acid analysis and quantification was carried out by the University of Missouri Agricultural Experiment Station (Columbia, MO, USA). The proteins in the feed and product fractions were visualized under denatured conditions by SDS-PAGE using a Bio-Rad Miniprotean System™ with Any kD™ gels (Bio-Rad Laboratories, Hercules, CA, USA) (Gerde et al., 2013).

Total Solids and Non-Protein Components Analysis
Total solids content was determined by drying the samples at 105°C for 4 h in a conventional oven (Sluiter et al., 2008a). Lipids were extracted by Folch extraction method using chloroform/methanol mixture (2:1 ratio) (Folch et al., 1957), followed by centrifugation at 2600 g for 10 min. The chloroform-soluble fractions were analyzed for fatty acids by preparing fatty acid methyl esters (FAMES) by methanolysis (1M methanolic HCl, 80°C, 16 h) and subjecting to GC-MS analysis using a non-polar DB-1 capillary column equipped with mass selective detector following procedures as described (York et al., 1986). All extracts were first analyzed without any internal standard, allowing the use of behenic acid (C:22:0, 10 μg) as an appropriate internal standard.

Hydroxy fatty acids were subjected to trimethylsilylation following methanolysis to facilitate GC separation; the response factors of common normal chain saturated and unsaturated fatty acid standards, and 2-hydroxy myristic acid standard were normalized relative to that of behenic acid. Ash content was determined after drying the samples in a conventional oven for 4 h and then incinerating them in a furnace at 575°C for 3 h using a slightly modified version of the NREL procedure (Sluiter et al., 2008b). The rest of biomass which comprises predominantly of carbohydrates and small amounts of other cellular components may simply be considered as the carbohydrate fraction for convenience. Thus, the carbohydrate content was determined by the difference (Valdez et al., 2014).

PG Analysis

The product fractions were delipidated by the Folch lipid extraction method described in Section "Total Solids and Non-Protein Components Analysis" and then subjected to PG component analysis. In order to identify and quantify PG amino acids, a portion of the delipidated samples was hydrolyzed in 6M HCl for 16 h at 105°C followed by methanolysis for 4 h at 80°C to yield methyl esters of amino acids, and finally derivatized with heptafluorobutyric anhydride (HFBA), which yields the *N*-heptafluorobutyrate (and *O*- heptafluorobutyrate for Serine and Threonine) derivatives of the PG-derived amino acids (Pons et al., 2003). The method was modified slightly wherein trans-esterification with isoamylalcohol was not performed and 2-amino adipic acid (25 μg) was used as internal standard. The resulting methyl esterified, HFBA derivatives were analyzed by GC-MS analysis using the DB-1 capillary column programed to 240°C. For PG carbohydrate analysis, a separate aliquot was hydrolyzed in 1M HCl for 2 h at 105°C followed by methanolysis for 6 h at 80°C followed by N-acetylation (acetic anhydride/pyridine in methanol, 1:1:10 v/v, 45 min, 50°C) and trimethylsilylation using "Tri-Sil" reagent (20 min, 80°C) (York et al., 1986). Carbohydrates were measured relative to the internal standard myo-inositol (20 μg). The resulting HFBA-amino acids and TMS-methyl glycosides of monosaccharide sugars were analyzed separately by GC-MS analysis using a 30 m DB-1 capillary column with electron impact mass fragmentation and detection, using temperature programs optimized for separately analyzing the amino acid and carbohydrate derivatives.

Results and Discussion

Protein Isolation Optimization
Comparison of Different Pretreatments

The results indicated that both high-pressure homogenization and ultrasonication resulted in a higher protein recovery in the supernatant compared to control (Figure S2 in Supplementary Material). High-pressure homogenization was the better of the two pretreatments with a protein recovery of 83.5% as opposed to 69.9% in case of ultrasonication. Microscopic observation of the disrupted cells showed greater cell disruption with the former compared to the latter (Figure S3 in Supplementary Material). Similar trend was reported for various algae and cyanobacteria (Safi et al., 2014; Ursu et al., 2014). Cell counting revealed that

high-pressure homogenization resulted in a near-complete cell lysis with disruption efficiency >99%, thus releasing most of the intra-cellular proteins. Autoclave treatment was the worst among all pretreatments with a protein recovery of only 29%, which was slightly lower than the 32.1% in the control. No visible cell disruption was observed under the microscope for the autoclaved *S. platensis* cells, explaining the lack of improvement in protein recovery. Thus, high-pressure homogenizer-based cell disruption was chosen as a pretreatment for all further protein isolation experiments.

Protein Solubility Curve

The solubility curve (**Figure 1**) showed that protein solubility (recovery in the supernatant) decreased with increasing pH in the acidic range of 2–4 and increased steadily in the range of 4–7. Least solubility was observed in the proximity of pH 4. High solubility (>75% recovery) was observed in the alkaline range of 7–12. However, under extremely high alkaline conditions (beyond pH 12) the solubility decreased notably. This could be a result of significant protein denaturation and clustering, rendering the proteins insoluble (Haque et al., 2005). The variation in protein recovery was only about 10% in the entire pH range of 6–12, although the trend was irregular. Highest recovery was obtained at pH 11 and closely followed by pH 8. These results differed from those reported for green algae. For *Chlorella vulgaris* the solubilization after cell lysis was 19% higher at pH 12 compared to pH 7 (Ursu et al., 2014). For *Nannochloropsis* species, protein solubilization was reported to increase with increasing pH all the way until 13 (Gerde et al., 2013). Thus, pH 11 and 8 were further explored under different experimental conditions to determine the better of the two for protein solubilization. A 3% *S. platensis* biomass slurry subjected to cell disruption by high-pressure homogenization and protein solubilization resulted in 87.9% protein recovery at pH 11 as opposed to 77.8% at pH 8. Similarly, *S. platensis* biomass at nearly the same solids content but disrupted using ultrasonication resulted in 58.2% protein recovery at pH 11 while only 38.7% at pH 8. Thus, pH 11 was better than 8 for protein solubilization.

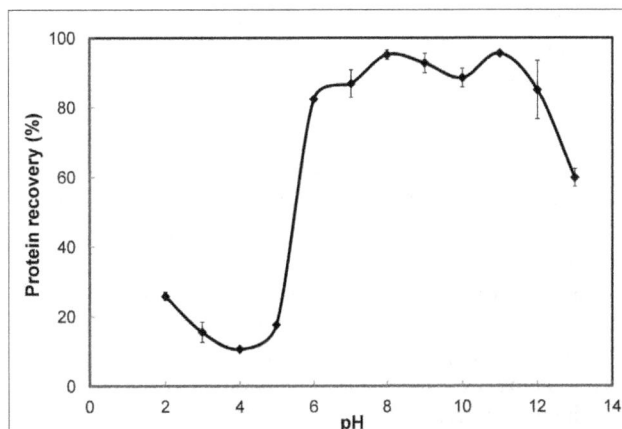

FIGURE 1 | Solubility curve for *S. platensis* biomass. Error bars represent standard deviation of mean.

Optimization of Protein Isolation Using RSM

pH ranges of 10.5–12 and 3–5 that were in the proximity (within 1 U) of the points of highest and least solubility (reported in Section "Protein Solubility Curve") were chosen for the design of protein solubilization and precipitation optimization experiments, respectively. **Figure 2A** shows the scattered plot between experimentally determined and RSM predicted protein recoveries in the alkali supernatant at different levels of the input variables. The experimental recovery varied from 64.87 to 95.6% (data presented in Table S1 in Supplementary Material). The regression coefficients of the second degree polynomial used to fit the protein recovery data, the standard error in their estimation, and the statistical analysis are presented in **Table 1**. The regression equation obtained from the analysis was as follows:

$$Y = 93.03 - 1.54X_1 + 1.32X_2 - 10.36X_3 - 1.72X_1{}^2 - 2.33X_2{}^2$$
$$- 9.16X_3{}^2 - 0.55X_1X_2 - 3.57X_1X_3 + 1.95X_2X_3 \quad (4)$$

The predicted recoveries were highly significant ($p = 0.0027$) and the coefficient of determination (R^2) for this model was 0.97, indicating a good fit. The results from the t-test showed that biomass concentration was a highly significant factor ($p = 0.0001$) in impacting protein recovery. The other two factors, pH, and solubilization time were not significant in the chosen range. However, the interaction of pH and biomass concentration was slightly significant ($p = 0.0522 < 0.1$). Among the quadratic effects, only the quadratic biomass concentration term was highly significant ($p = 0.0015$). The rest of the interaction and quadratic terms were not significant. The optimal values for pH, solubilization time, and biomass concentration determined by RSM were 11.38, 35.32 min, and 3.61% (w/w) solids, respectively, and the predicted value of the response (protein recovery) at these conditions was 96%.

Formic acid is a weak organic acid compared to hydrochloric acid which is a strong inorganic acid. In a comparative study, protein recovery in the acid pellet (protein isolate) was 71.7% when precipitation was carried out using HCl and 71.5% using HCOOH at the same experimental conditions. Thus, the substitution of HCOOH for HCl did not show any significant impact on protein precipitation. The former is more preferable than the latter when the residual biomass is intended to be used for biofuel production processes because chloride ions can corrode reactor vessels in thermochemical processes such as HTL (Kritzer, 2004), and the NaCl formed as a result of NaOH and HCl added during the protein isolation process can be toxic to the microbes in biochemical processes such as AD (Chen et al., 2008). Thus, HCOOH was used for protein precipitation in all further experiments.

Figure 2B shows the scattered plot between experimental and RSM predicted protein recoveries in the acid pellet at different levels of the input variables. The experimental recovery varied from 67 to 74.5% (data presented in Table S2 in Supplementary Material). The regression coefficients of the second degree polynomial used to fit the protein recovery data, the standard error in their estimation, and their statistical analysis are presented in **Table 2**. The regression equation obtained from the analysis is as follows:

$$Y = 73.43 - 0.72X_1 + 1.28X_2 - 5.11X_1{}^2 + 1.49X_2{}^2 - 0.48X_1X_2 \quad (5)$$

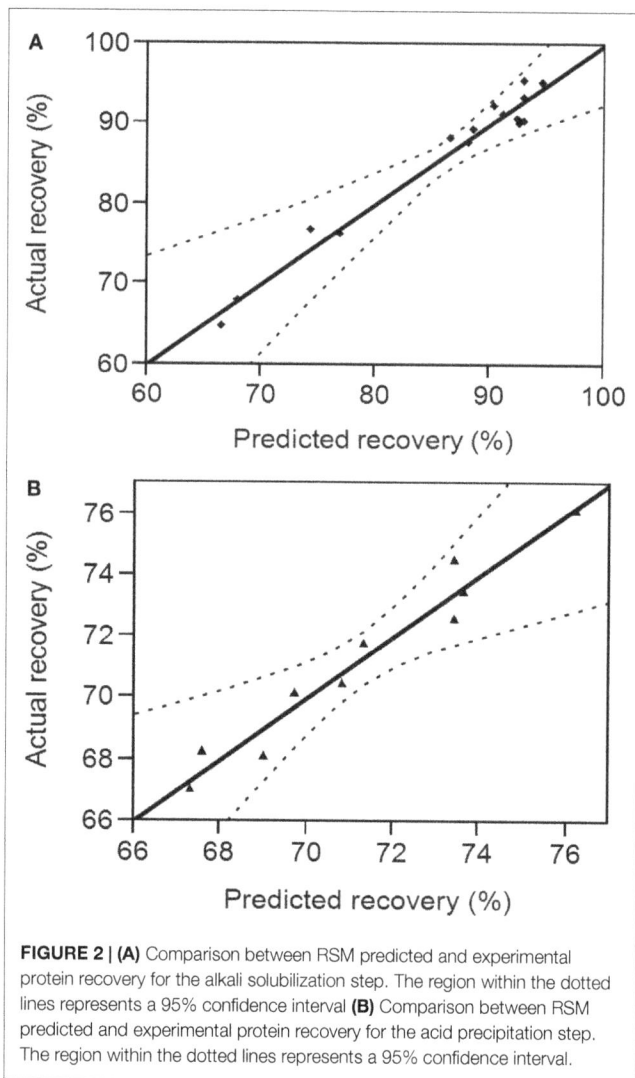

FIGURE 2 | (A) Comparison between RSM predicted and experimental protein recovery for the alkali solubilization step. The region within the dotted lines represents a 95% confidence interval **(B)** Comparison between RSM predicted and experimental protein recovery for the acid precipitation step. The region within the dotted lines represents a 95% confidence interval.

TABLE 1 | Estimate of the regression coefficients for the alkali solubilization optimization model and their statistical significance determined by Student's t-test.

Source	Estimate	SE	t ratio	$p > |t|$
Intercept	93.03	1.6256	57.23	<0.0001*
X_1	−1.54	0.9955	−1.54	0.1834
X_2	1.32	0.9955	1.32	0.2434
X_3	−10.36	0.9955	−10.4	0.0001*
X_1X_2	−0.55	1.4078	−0.39	0.7134
X_1X_3	−3.57	1.4078	−2.54	0.0522
X_2X_3	1.95	1.4078	1.38	0.2253
X_1^2	−1.72	1.4653	−1.17	0.2941
X_2^2	−2.33	1.4653	−1.59	0.1721
X_3^2	−9.16	1.4653	−6.25	0.0015*

*Significant ($p < 0.05$).

The predicted recoveries were significant ($p = 0.01$) and the coefficient of determination (R^2) for this model was 0.95, indicating a reasonably good fit. The results from the t-test showed that precipitation time was a significant factor ($p = 0.03$) in impacting the protein recovery. The quadratic regression term for pH was

TABLE 2 | Estimate of the regression coefficients for the acid precipitation optimization model and their statistical significance determined by Student's t-test.

Source	Estimate	SE	t ratio	$p > |t|$
Intercept	73.43	0.5908	124.28	<0.0001*
x_1	0.72	0.4036	1.78	0.1505
x_2	1.28	0.4036	3.18	0.0036*
$x_1 x_2$	−0.48	0.4943	−0.96	0.391
x_1^2	−5.11	0.6472	−7.89	0.0014*
x_2^2	1.49	0.6472	2.31	0.0823

*Significant (p < 0.05).

highly significant ($p = 0.0014$) but not the linear term, implying a quadratic dependence of protein recovery on pH in the chosen range. The quadratic term for precipitation time was slightly significant ($p = 0.0823 < 0.1$). However, the interaction of pH and time was not significant implying that both of these factors are independent of each other in the chosen range. The model predicted the solution to be a saddle point. However, based on single parameter profiles, the optimum conditions for maximum protein precipitation were determined as pH 4.01 and precipitation time of 60 min. The predicted value of the response (protein recovery) at these values was 76.2%.

The RSM predicted maximum for overall protein yield after the alkali solubilization and acid precipitation steps was calculated as 73.15%. The experimentally determined protein recovery in the alkali supernatant and acid pellet at the RSM optimized process conditions for the solubilization and precipitation steps were 86 and 70.6%, respectively. Although the experimental recoveries for both the steps were lower than the theoretically predicted values, the variation (10.4 and 7.3%, respectively) was within acceptable limits, considering the scale of operation (the amount of biomass used in each of the optimization experiments was 10 times lower than that used in the protein isolation process at the optimized conditions), handling, and instrumental errors. The overall experimental protein yield at the optimum conditions was 60.7%.

Component Fractionation among the Product Fractions

Figure 3A shows the fractionation of various components between the protein isolate and the residual biomass obtained at the RSM optimized process conditions. The overall yield of total nitrogen and hence the yield of total protein in the protein isolate was 60.7%. This value was higher than the yields reported in the literature for proteins extracted using alkali–acid method from green algae (Gerde et al., 2013; Ursu et al., 2014) but lower than the 80% yield reported for S. platensis protein isolates (Devi et al., 1981). The higher yield reported in the latter case was a result of the use of hexane defatted biomass as the starting material and the repeated (three times) aqueous extraction and dialysis steps. Lowry protein assay estimated that 56.9% of soluble proteins were recovered in the protein isolate affecting a lower recovery in the residual biomass. The total solids fractionated almost equally between the two product fractions and so did the total lipids. However, carbohydrate recovery was higher in the residual biomass compared to the protein isolate.

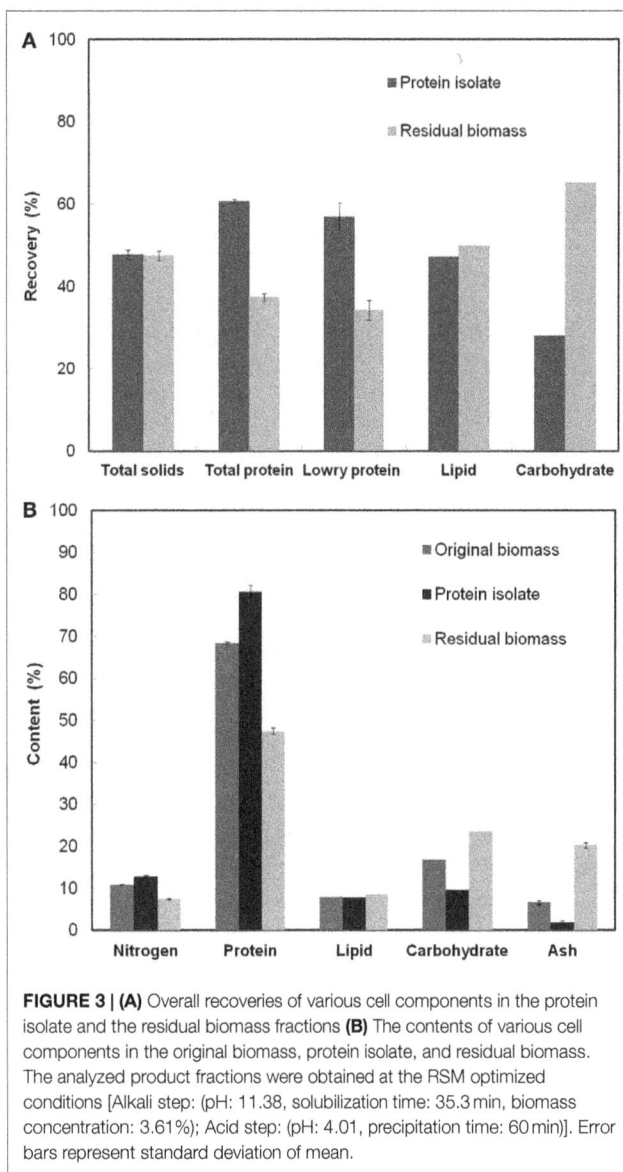

FIGURE 3 | (A) Overall recoveries of various cell components in the protein isolate and the residual biomass fractions **(B)** The contents of various cell components in the original biomass, protein isolate, and residual biomass. The analyzed product fractions were obtained at the RSM optimized conditions [Alkali step: (pH: 11.38, solubilization time: 35.3 min, biomass concentration: 3.61%); Acid step: (pH: 4.01, precipitation time: 60 min)]. Error bars represent standard deviation of mean.

The calculated purity or the protein content (% w/w) in the protein isolate was 80.6%, which was 12.2% higher than S. platensis biomass. This value of protein content was higher than that reported in the literature for the protein isolate obtained from S. platensis using a slightly different procedure (Chronakis et al., 2000). Recovery of non-protein components in the protein isolate due to co-precipitation of insoluble carbohydrates, cell wall PG fragments (composed of amino sugars), and lipids limited the purity of this fraction. The PG fragments from the cell wall of S. platensis did not possibly degrade into their respective sugar and peptide components under the relatively mild pH (=4) condition used in the protein precipitation process resulting in their co-extraction with proteins (Vollmer, 2008). Further, the residual biomass fraction had an undesirably high nitrogen and protein content (7.6 and 47.5%, respectively) indicating incomplete protein extraction, the loss of non-protein components due to co-extraction with proteins and the presence of PG fragments. A PG composition analysis based on the diagnostic markers, diaminopimelic acid (DAP) and N-acetyl muramic acid (NAMA), revealed the presence of PG fragments in both the protein isolate and the residual biomass fractions. Although the latter had a slightly higher proportion of all PG components compared to the protein isolate, their overall contents were very low compared to other cellular components. The contribution of amino sugars toward the total nitrogen and carbohydrate content in both the fractions was also extremely low (0.16 and 1.39% of the total estimated nitrogen in the two fractions, respectively). Thus, a further reduction in the nitrogen content of the residual biomass may be achieved only by repeated protein extractions involving additional processing steps and/or other unit operations. However, such procedures would demand higher processing costs and other resources, and may negatively impact the scalability of the process. Hence, this idea was not investigated in this work.

Initial Biomass, Protein Isolate, and Residual Biomass Characterization

Figure 3B shows the nitrogen and protein (based on elemental analysis), lipid, carbohydrate and ash contents in the original biomass, the protein isolate, and the residual biomass obtained at the RSM optimized conditions. The original *S. platensis* biomass was comprised of 10.95% nitrogen, 68.4% total protein, and 6.7% ash by weight. Analysis of the protein isolate and the residual biomass revealed higher nitrogen and protein contents and lower lipid, carbohydrate, and ash contents in the protein isolate compared to the residual biomass, which was in accordance with the desired outcome. The former was enriched in proteins while the latter was enriched in non-protein components. Although only 50% of the total lipids were recovered in the residual biomass, this did not have a huge impact on its composition due to the low lipid content of the original *S. platensis* biomass (8.03%).

The PG carbohydrate analysis method described in Section "PG Analysis" also quantified non-PG originated sugars present in the biomass in addition to the PG amino sugars. The relative composition (% w/w) of the detected sugars in the protein isolate and the residual biomass fractions are shown in **Figure 4A**. A major proportion of the sugars were glucose, which accounted for 77.50 and 63.84% of the total sugars (by weight) in each of these fractions, respectively. This was expected, given that glucose is the most abundant sugar present in *S. platensis* (Shekharam et al., 1987). Galactose accounted for 8.24 and 12.72% in the protein isolate and the residual biomass, respectively. The PG amino sugars NAMA and GlcNAc accounted for 6.11 and 11.39% of the total sugars, respectively in the residual biomass. In the protein isolate the proportions of these amino sugars were 1.52% NAMA and 5% GlcNAc. Small amounts of mannose, 3-methyl hexose and fucose were also detected in both of these fractions.

Figure 4B shows the relative composition (% w/w) of the fatty acids detected by FAMES analysis in the protein isolate and the residual biomass. C16:0 (Palmitic acid) was the dominant fatty acid in both the fractions, as was the case for original *S. platensis* biomass (Cohen, 1997). However, this fatty acid represented 81.83% of the total fatty acids in the residual biomass but only 46.76% of the protein isolate. The latter contained significant amounts of mono- and poly-unsaturated fatty acids (C16–18) while the residual biomass had very small amounts. These and

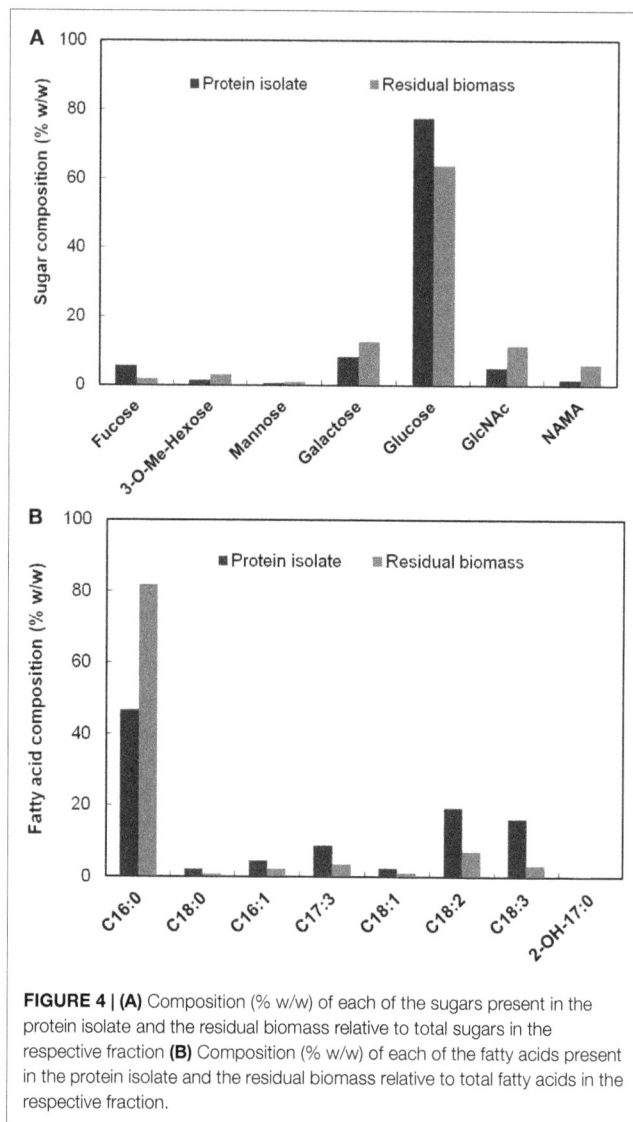

FIGURE 4 | (A) Composition (% w/w) of each of the sugars present in the protein isolate and the residual biomass relative to total sugars in the respective fraction **(B)** Composition (% w/w) of each of the fatty acids present in the protein isolate and the residual biomass relative to total fatty acids in the respective fraction.

other fatty acids typically originate from membrane phospholipids where they are acylated to moieties carrying choline (phosphatidyl choline) and other polar head groups (Hoiczyk and Hansel, 2000). An -hydroxy fatty acid (2-OH-C17:0) was detected in low levels in the protein isolate, but not in the residual biomass. These results clearly indicated that the protein isolate was enriched in poly-unsaturated fatty acids while the residual biomass was enriched in saturated fatty acids. The former had a higher proportion of the essential fatty acid, -linolenic acid (C18:3) compared to the original *S. platensis* biomass. This and other unsaturated fatty acids can be separated from the protein isolate using methods such as supercritical CO_2 extraction and urea complex formation (Cohen et al., 1993; Mendes et al., 2005) to yield high-value co-products. The lower proportion of unsaturated fatty acids in the residual biomass is favorable for biofuel production processes because they could result in lower oxidative stability (rancidification) of the generated biofuel (Gunstone, 1967).

SDS-PAGE analysis revealed that several lighter (low protein concentration) bands observed in the molecular weight range of

TABLE 3 | Amino acid composition (expressed as g/100 g total amino acids) of the original S. platensis biomass and the protein isolate.

Amino acid	Composition	
	Original biomass	Protein isolate
Taurine	0.03	0.03
Hydroxyproline	0.00	0.02
Aspartic acid	10.12	9.86
Threonine[a]	4.92	4.85
Serine	4.32	4.41
Glutamic acid	15.58	13.28
Proline	3.66	3.79
Lanthionine	0.00	0.00
Glycine	5.06	5.24
Alanine	7.48	7.31
Cysteine	1.02	0.99
Valine[a]	6.46	6.91
Methionine[a]	2.38	2.40
Isoleucine[a]	5.85	6.34
Leucine[a]	8.91	9.80
Tyrosine	4.40	5.07
Phenylalanine[a]	4.71	5.16
Hydroxylysine	0.16	0.15
Ornithine	0.09	0.09
Lysine[a]	4.84	4.49
Histidine[a]	1.57	1.69
Arginine	7.30	6.72
Tryptophan[a]	1.14	1.42

[a] Essential amino acids.

25–100 kDa in original S. platensis biomass were not found in the disrupted biomass implying protein degradation as a result of cell disruption by high-pressure homogenization (Figure S4 in Supplementary Material). The bands around 100 and 55 kDa were the most prominent ones among both the protein isolate and residual biomass fractions, although they were lighter in the latter indicating lower concentration of these proteins in this fraction. Thus, a higher proportion of the high molecular weight proteins fractionated into the protein isolate. Some of the bands observed in the disrupted biomass between 15 and 20 kDa were not observed in the protein isolate and the residual biomass fractions suggesting that these low molecular weight proteins degraded into peptide components during the protein isolation process. Further, the small dark band at the bottom of the gel in the original biomass was observed only in the residual biomass and not in the protein isolate, indicating that the low molecular weight peptides and free amino acids typically present in algae remained in the residual biomass.

The protein isolate obtained at a low ash content of 1.9% was freeze dried and further analyzed for amino acid and mineral contents. The results presented in **Table 3** show that the variation

in the composition of a majority of the amino acids between the protein isolate and the edible original S. platensis biomass was low (below 10%). The contents of six out of the eight essential amino acids were slightly higher in the protein isolate. S. platensis biomass has widely been accepted as a rich protein source for humans and animals (Becker, 2004) and hence the protein isolate could potentially be used in these applications. The predominant minerals present in the protein isolate were aluminum, calcium, iron, potassium, magnesium, sodium, phosphorus, sulfur, and silicon (Table S4 in Supplementary Material). Except for sodium, the composition of all the elements was lower than original S. platensis biomass (Jena et al., 2011b) and hence is within agreeable limits for nutritional purposes. The excess sodium originated from the NaOH added during the solubilization step.

Conclusion

In this study, protein isolation from S. platensis cyanobacterium was carried out using the alkali–acid method after cell disruption using high-pressure homogenization. The process conditions were optimized using RSM. At the optimized conditions, the proteins were extracted at a high yield of 60.7% and content of 80.6%. Further improvement of protein extraction was limited by co-fractionation of the non-protein components into the protein isolate and incomplete protein precipitation. The extracted protein isolate was enriched in proteins, essential amino acids, and unsaturated fatty acids, and had a lower ash and mineral content compared to the original biomass. Such a composition is suitable for human food or animal feed applications. The residual biomass had a lower protein and nitrogen content than the original biomass and was enriched in carbohydrates and saturated lipids, a composition better suited for biofuel applications such as HTL and AD.

Acknowledgments

This research was funded in part by the Department of Defense ARO grant W911NF-11-1-0218 (KD, PI) and the Department of Energy grant DE-FG02-93ER20097 (A. Darvill, PI) for the Center for Plant and Microbial Complex Carbohydrates (Complex Carbohydrate Research Center), at the University of Georgia. The authors thankfully acknowledge Dr. David Blum, Paul Volny, and Ron Garrison for their assistance with laboratory analyses and equipment.

References

Abas Wani, A., Sogi, D., Grover, L., and Saxena, D. (2006). Effect of temperature, alkali concentration, mixing time and meal/solvent ratio on the extraction of watermelon seed proteins – a response surface approach. Biosyst. Eng. 94, 67–73. doi:10.1016/j.biosystemseng.2006.02.004

Becker, E. W. (2007). Micro-algae as a source of protein. Biotechnol. Adv. 25, 207–210. doi:10.1016/j.biotechadv.2006.11.002

Becker, E. W. (2004). "Microalgae in human and animal nutrition," in Handbook of Microalgal Culture: Biotechnology and Applied Phycology, ed. A. Richmond (Oxford: Blackwell Publishing), 312.

Chen, Y., Cheng, J. J., and Creamer, K. S. (2008). Inhibition of anaerobic digestion process: a review. Bioresour. Technol. 99, 4044–4064. doi:10.1016/j.biortech.2007.01.057

Choi, Y. R., and Markakis, P. (1981). Blue-green algae as a source of protein. Food Chem. 7, 239–247. doi:10.1016/0308-8146(81)90029-7

Chronakis, I. S., Galatanu, A. N., Nylander, T., and Lindman, B. (2000). The behaviour of protein preparations from blue-green algae (*Spirulina platensis* strain Pacifica) at the air/water interface. *Colloids Surf. A Physicochem. Eng. Asp.* 173, 181–192. doi:10.1016/s0927-7757(00)00548-3

Cohen, Z. (1997). "The chemicals of spirulina," in *Spirulina Platensis Arthrospira: Physiology, Cell-Biology and Biotechnology*, ed. A. Vonshak (London: CRC Press), 175–204.

Cohen, Z., Reungjitchachawali, M., Siangdung, W., and Tanticharoen, M. (1993). Production and partial purification of -linolenic acid and some pigments from *Spirulina platensis*. *J. Appl. Phycol.* 5, 109–115. doi:10.1007/bf02182428

Damodaran, S. (1996). "Amino acids, peptides and proteins," in *Food Chemistry*, ed. O. R. Fennema (New York, NY: Marcel Dekker, Inc.), 321–430.

Devi, M. A., Subbulakshmi, G., Devi, K. M., and Venkataraman, L. V. (1981). Studies on the proteins of mass-cultivated, blue-green alga (*Spirulina platensis*). *J. Agric. Food Chem.* 29, 522–525. doi:10.1021/jf00105a022

Firatligil-Durmus, E., and Evranuz, O. (2010). Response surface methodology for protein extraction optimization of red pepper seed (*Capsicum frutescens*). *LWT Food Sci. Technol.* 43, 226–231. doi:10.1016/j.lwt.2009.08.017

Folch, J., Lees, M., and Sloane-Stanley, G. (1957). A simple method for the isolation and purification of total lipids from animal tissues. *J. Biol. Chem.* 226, 497–509.

Gerde, J. A., Wang, T., Yao, L., Jung, S., Johnson, L. A., and Lamsal, B. (2013). Optimizing protein isolation from defatted and non-defatted *Nannochloropsis* microalgae biomass. *Algal Res.* 2, 145–153. doi:10.1016/j.algal.2013.02.001

Goetz, J., and Koehler, P. (2005). Study of the thermal denaturation of selected proteins of whey and egg by low resolution NMR. *LWT Food Sci. Technol.* 38, 501–512. doi:10.1016/j.lwt.2004.07.009

Gunstone, F. D. (1967). *An Introduction to the Chemistry and Biochemistry of Fatty Acids and Their Glycerides*, 2 Edn. London: Chapman and Hall.

Haque, I., Singh, R., Moosavi-Movahedi, A. A., and Ahmad, F. (2005). Effect of polyol osmolytes on GD, the Gibbs energy of stabilisation of proteins at different pH values. *Biophys. Chem.* 117, 1–12. doi:10.1016/j.bpc.2005.04.004

Hoiczyk, E., and Hansel, A. (2000). Cyanobacterial cell walls: news from an unusual prokaryotic envelope. *J. Bacteriol.* 182, 1191–1199. doi:10.1128/JB.182.5.1191-1199.2000

Jena, U., Das, K. C., and Kastner, J. R. (2011a). Effect of operating conditions of thermochemical liquefaction on biocrude production from *Spirulina platensis*. *Bioresour. Technol.* 102, 6221–6229. doi:10.1016/j.biortech.2011.02.057

Jena, U., Vaidyanathan, N., Chinnasamy, S., and Das, K. (2011b). Evaluation of microalgae cultivation using recovered aqueous co-product from thermochemical liquefaction of algal biomass. *Bioresour. Technol.* 102, 3380–3387. doi:10.1016/j.biortech.2010.09.111

Kritzer, P. (2004). Corrosion in high-temperature and supercritical water and aqueous solutions: a review. *J. Supercrit. Fluids* 29, 1–29. doi:10.1016/S0896-8446(03)00031-7

López Barreiro, D., Prins, W., Ronsse, F., and Brilman, W. (2013). Hydrothermal liquefaction (HTL) of microalgae for biofuel production: state of the art review and future prospects. *Biomass Bioenergy* 53, 113–127. doi:10.1016/j.biombioe.2012.12.029

Lowry, O. H., Rosebrough, N. J., Farr, A. L., and Randall, R. J. (1951). Protein measurement with the Folin phenol reagent. *J. Biol. Chem.* 193, 265–275.

Ma, T., Wang, Q., and Wu, H. (2010). Optimization of extraction conditions for improving solubility of peanut protein concentrates by response surface methodology. *LWT Food Sci. Technol.* 43, 1450–1455. doi:10.1016/j.lwt.2010.03.015

Mendes, R. L., Reis, A. D., Pereira, A. P., Cardoso, M. T., Palavra, A. F., and Coelho, J. P. (2005). Supercritical CO2 extraction of -linolenic acid (GLA) from the cyanobacterium *Arthrospira* (spirulina) maxima: experiments and modeling. *Chem. Eng. J.* 105, 147–151. doi:10.1016/j.cej.2004.10.006

Palinska, K. A., and Krumbein, W. E. (2000). Perforation patterns in the peptidoglycan wall of filamentous cyanobacteria. *J. Phycol.* 36, 139–145. doi:10.1046/j.1529-8817.2000.99040.x

Payne, M. F., and Rippingale, R. J. (2000). Evaluation of diets for culture of the calanoid copepod gladioferens imparipes. *Aquaculture* 187, 85–96. doi:10.1016/S0044-8486(99)00391-9

Piorreck, M., Baasch, K.-H., and Pohl, P. (1984). Biomass production, total protein, chlorophylls, lipids and fatty acids of freshwater green and blue-green algae

under different nitrogen regimes. *Phytochemistry* 23, 207–216. doi:10.1016/S0031-9422(00)80304-0

Pons, A., Richet, C., Robbe, C., Herrmann, A., Timmerman, P., Huet, G., et al. (2003). Sequential GC/MS analysis of sialic acids, monosaccharides, and amino acids of glycoproteins on a single sample as heptafluorobutyrate derivatives. *Biochemistry* 42, 8342–8353. doi:10.1021/bi034250e

Prabakaran, P., and Ravindran, A. D. (2011). A comparative study on effective cell disruption methods for lipid extraction from microalgae. *Lett. Appl. Microbiol.* 53, 150–154. doi:10.1111/j.1472-765X.2011.03082.x

Quanhong, L., and Caili, F. (2005). Application of response surface methodology for extraction optimization of germinant pumpkin seeds protein. *Food Chem.* 92, 701–706. doi:10.1016/j.foodchem.2004.08.042

Safi, C., Charton, M., Pignolet, O., Pontalier, P.-Y., and Vaca-Garcia, C. (2013a). Evaluation of the protein quality of Porphyridium cruentum. *J. Appl. Phycol.* 25, 497–501. doi:10.1007/s10811-012-9883-4

Safi, C., Charton, M., Pignolet, O., Silvestre, F., Vaca-Garcia, C., and Pontalier, P.-Y. (2013b). Influence of microalgae cell wall characteristics on protein extractability and determination of nitrogen-to-protein conversion factors. *J. Appl. Phycol.* 25, 523–529. doi:10.1007/s10811-012-9886-1

Safi, C., Ursu, A. V., Laroche, C., Zebib, B., Merah, O., Pontalier, P.-Y., et al. (2014). Aqueous extraction of proteins from microalgae: effect of different cell disruption methods. *Algal Res.* 3, 61–65. doi:10.1016/j.algal.2013.12.004

Shekharam, K. M., Venkataraman, L. V., and Salimath, P. V. (1987). Carbohydrate composition and characterization of two unusual sugars from the blue green alga *Spirulina platensis*. *Phytochemistry* 26, 2267–2269. doi:10.1016/S0031-9422(00)84698-1

Sluiter, A., Hames, B., Hyman, D., Payne, C., Ruiz, R., Scarlata, C., et al. (2008a). *"Determination of Total Solids in Biomass and Total Dissolved Solids in Liquid Process Samples"*. NREL Technical Report No. NREL/TP-510-42621. Golden, CO: National Renewable Energy Laboratory.

Sluiter, A., Hames, B., Ruiz, R., Scarlata, C., Sluiter, J., and Templeton, D. (2008b). *"Determination of Ash in Biomass"*. NREL Technical Report No. NREL/TP-510-42622. Golden, CO: National Renewable Energy Laboratory.

Spolaore, P., Joannis-Cassan, C., Duran, E., and Isambert, A. (2006). Commercial applications of microalgae. *J. Biosci. Bioeng.* 101, 87–96. doi:10.1263/jbb.101.87

Ursu, A.-V., Marcati, A., Sayd, T., Sante-Lhoutellier, V., Djelveh, G., and Michaud, P. (2014). Extraction, fractionation and functional properties of proteins from the microalgae *Chlorella vulgaris*. *Bioresour. Technol.* 157, 134–139. doi:10.1016/j.biortech.2014.01.071

Valdez, P. J., Tocco, V. J., and Savage, P. E. (2014). A general kinetic model for the hydrothermal liquefaction of microalgae. *Bioresour. Technol.* 163, 123–127. doi:10.1016/j.biortech.2014.04.013

Vollmer, W. (2008). Structural variation in the glycan strands of bacterial peptidoglycan. *FEMS Microbiol. Rev.* 32, 287–306. doi:10.1111/j.1574-6976.2007.00088.x

Wang, X., and Zhang, X. (2012). Optimal extraction and hydrolysis of *Chlorella pyrenoidosa* proteins. *Bioresour. Technol.* 126, 307–313. doi:10.1016/j.biortech.2012.09.059

York, W. S., Darvill, A. G., McNeil, M., Stevenson, T. T., and Albersheim, P. (1986). Isolation and characterization of plant cell walls and cell wall components. *Meth. Enzymol.* 118, 3–40.

Zhang, S., Wang, Z., and Xu, S. (2007). Optimization of the aqueous enzymatic extraction of rapeseed oil and protein hydrolysates. *J. Am. Oil Chem. Soc.* 84, 97–105. doi:10.1007/s11746-006-1004-6

Conflict of Interest Statement: The authors declare that the research was conducted in the absence of any commercial or financial relationships that could be construed as a potential conflict of interest. The Guest Associate Editor S. Kent Hoekman declares that, despite having collaborated with author Keshav C. Das, the review process was handled objectively and no conflict of interest exists.

11

Phenotypic Changes in Transgenic Tobacco Plants Overexpressing Vacuole-Targeted *Thermotoga maritima* BglB Related to Elevated Levels of Liberated Hormones

Quynh Anh Nguyen[1], Dae-Seok Lee[2], Jakyun Jung[1] and Hyeun-Jong Bae[1,2]*

[1] *Department of Bioenergy Science and Technology, Chonnam National University, Gwangju, South Korea,* [2] *Bio-Energy Research Center, Chonnam National University, Gwangju, South Korea*

Edited by:
Robert Henry,
The University of Queensland,
Australia

Reviewed by:
Tianju Chen,
Chinese Academy of Sciences, China
Chiranjeevi Thulluri,
Jawaharlal Nehru Technological
University Hyderabad, India

*Correspondence:
Hyeun-Jong Bae
baehj@chonnam.ac.kr

The hyperthermostable β-glucosidase *BglB* of *Thermotoga maritima* was modified by adding a short C-terminal tetrapeptide (AFVY, which transports phaseolin to the vacuole, to its C-terminal sequence). The modified β-glucosidase *BglB* was transformed into tobacco (*Nicotiana tabacum* L.) plants. We observed a range of significant phenotypic changes in the transgenic plants compared to the wild-type (WT) plants. The transgenic plants had faster stem growth, earlier flowering, enhanced root systems development, an increased biomass biosynthesis rate, and higher salt stress tolerance in young plants compared to WT. In addition, programed cell death was enhanced in mature plants. Furthermore, the C-terminal AFVY tetrapeptide efficiently sorted *T. maritima* BglB into the vacuole, which was maintained in an active form and could perform its glycoside hydrolysis function on hormone conjugates, leading to elevated hormone [abscisic acid (ABA), indole 3-acetic acid (IAA), and cytokinin] levels that likely contributed to the phenotypic changes in the transgenic plants. The elevation of cytokinin led to upregulation of the transcription factor *WUSCHELL*, a homeodomain factor that regulates the development, division, and reproduction of stem cells in the shoot apical meristems. Elevation of IAA led to enhanced root development, and the elevation of ABA contributed to enhanced tolerance to salt stress and programed cell death. These results suggest that overexpressing vacuole-targeted *T. maritima* BglB may have several advantages for molecular farming technology to improve multiple targets, including enhanced production of the β-glucosidase BglB, increased biomass, and shortened developmental stages, that could play pivotal roles in bioenergy and biofuel production.

Keywords: *Thermotoga maritima*, hyperthermostable β-glucosidase BglB, C-terminal AFVY tetrapeptide, vacuole-targeted, hormone conjugates, shoot apical meristem

INTRODUCTION

β-glucosidase is critical for many developmental processes in plants, and the hydrolysis of phytohormone conjugates is one of its most important roles (Schliemann, 1984; Sembdner et al., 1994; Kleczkowski et al., 1995). The rolC gene of the bacterial pathogen *Agrobacterium rhizogenes* encodes β-glucosidase, and results in abnormal development when transformed into plants. In particular, heterologous β-glucosidase can release active forms of phytohormones from their inactive conjugates that consist of glycoside links (Spena et al., 1992; Brzobohaty et al., 1993). Inactive conjugates of each phytohormone can be found abundantly in plant tissues. Their active forms are liberated via β-glucosidase-mediated hydrolysis. Furthermore, many studies have revealed that the inactive forms of phytohormone conjugates act as reversible deactivated storage molecules, and are important for the regulation of physiologically active hormone levels; however, their normal biological functions remain unknown (Staswick, 2009; Piotrowska and Bajguz, 2011).

The vacuole is considered a storage organelle and is an important component of the secretory pathway in plants. Detailed knowledge of the sorting mechanisms, out of and into the vacuole, is lacking (Hall, 2000; Vitale and Hinz, 2005). Previous studies of several lytic enzymes that are specifically targeted to the vacuole (e.g., phaseolin) have revealed some sorting signals (N-terminal or C-terminal polypeptides or internal sequences) that can sort proteins into the vacuole (Frigerio et al., 2001; De Marcos Lousa et al., 2012). Unfortunately, because the internal environment of the vacuole leads to the rapid degradation and hydrolysis of proteins, and other compounds, it has been difficult to determine whether heterologous expressed proteins maintain their functions and features inside the vacuole.

We previously expressed the hyperthermostable β-glucosidase BglB of *T. maritima* in tobacco plants to obtain transgenic plants for application in bioconversion. The optimal temperature and pH of the plant-expressed BglB were 80°C and 4.5, respectively (Jung et al., 2010). Moreover, we also observed some phenotypic modifications, such as longer stems, larger leaves, and shortened developmental stages (Jung et al., 2010, 2013), which we hypothesized may have been due to changes in hormone homeostasis. Therefore, in the present study, we overexpressed heterologous BglB of *T. maritima* in tobacco plants. We targeted the vacuole by insertion of the AFVY tetrapeptide to examine whether BglB maintain its functions of hydrolyzing glycoside bonds to release free hormones from its conjugates, and to determine how such changes in hormones levels may affect the growth and development of transgenic plants. All of the changes in the aboveground or belowground organs in plants can be explained via the development, division, and reproduction of stem cells harbored in the shoot and root apical meristems, which are regulated by the expression of homeodomain genes and hormone levels. For example, in the shoot apical meristem, the transcription factor *WUSCHELL (WUS)* can be upregulated via cytokinin, and a group of dividing cells called the quiescent center (QC) is upregulated by indole 3-acetic acid (IAA) in the root apical meristem (Kerk et al., 2000; Overvoorde et al., 2010; Yadav et al., 2010; Zhao et al., 2010).

MATERIALS AND METHODS

Vector Constructions, Plant Transformation, and Molecular Analysis

For cytosol expression, the full-length sequence of the *T. maritima* β-glucosidase *BglB* gene (Jung et al., 2010) was constructed under control of the 35S promoter, and named Cyt-BglB (CB). For vacuole targeting, *BglB* was modified by replacing its stop codon

with nucleotide sequences encoding the AFVY signal tetrapeptide from the vacuolar storage glycoprotein phaseolin (Frigerio et al., 2001), with a stop codon inserted at the end, and named Vac-BglB (VB). According to previous studies, AFVY tetrapeptide signals are sufficient to target a heterologous protein to the vacuole (Frigerio et al., 2001; Lau et al., 2010). The 35S promoter was also used for vacuole targeting of the recombinant variants. These expression cassettes were then sub-cloned into the modified multiple cloning sites of the binary vector pCambia 2300 (Kim et al., 2010), as shown in **Figure 1A**. *Agrobacterium tumefaciens* strain GV3013 was used for transformation of tobacco (*Nicotiana tabacum* L.) via the leaf-disk method (Helmer et al., 1984). Transformed shoots were selected on solid Murashige–Skoog (MS) medium (Murashige and Skoog, 1962) containing 100 μg/ml kanamycin and 500 μg/ml cefotaxime. Transgenic tobacco plants were grown in a growth chamber under a 16-/8-h light/dark cycle at $25 \pm 3°C$. After the presence of the transgene was confirmed by genomic DNA polymerase chain reaction (PCR), reverse transcription (RT)-PCR, and Western blotting, the T_0 generation of transgenic and wild-type (WT) plants was moved to a greenhouse for development.

Total genomic DNA was isolated from the T_0 generation transgenic plant leaves using genomic DNA extraction buffer [200 mM Tris–HCl, 250 mM NaCl, 25 mM Na$_2$-EDTA, 0.5% sodeum dodecyl sulfate (SDS)]. The concentration of genomic DNA was measured using a NanoDrop spectrophotometer (Thermo Fisher Scientific, USA). To confirm the presence of *BglB*, PCR of genomic DNA was performed using two sets of flanking primers: first FP 5′-GTC GCT CAT CAC GAA ACC GT-3′ and RP 5′-ACT ACA GAG GAA AAG GTG AA-3′ for checking the presence of a 0.7-kb sequence within *BglB* in the CB and VB constructs, and second FP 5′-TAT GCA GGC TCC CAC CCC TT-3′ and RP 5′-GTA TAC GAA TGC ACT ACA GA-3′ for checking the presence of a 0.4-kb sequence within *BglB* and nucleotides sequences of AFVY tetrapeptide in VB constructs. For RT-PCR, total RNA was extracted from the leaf tissues and cDNA was synthesized using avian myeloblastosis virus (AMV) reverse transcriptase (Promega, USA) with random hexamers, and the RT-PCR of *BglB* was performed using first primers set mentioned above. RT-PCR was also performed to determine the expression level of *N. tabacum WUS* (JQ686923.1) after RNA was extracted from the stems of seedlings and cDNA was synthesized, with the specific primers: FP 5′-ATG CAC ATG AGA GGT GTT TG-3′ and RP 5′-TTA GGG GGA ATT AGG AGA TC-3′.

BglB proteins were extracted from the T_0 to T_3 generation transgenic tobacco leaves by grinding the leaf material to a powder in liquid nitrogen and then suspending the powder in protein extraction buffer at pH 8.0 (50 mM Tris–HCl, 5 mM Na$_2$-EDTA, 20 mM Na$_2$S$_2$O$_5 \times$ 5H$_2$O, 100 mM KCl, 5% glycerol, 1% β-mercaptoethanol). Leaf debris was removed by centrifugation at $13,000 \times g$ for 20 min at 4°C. Total soluble protein (TSP) in the supernatants was measured using the Bradford method (Bradford, 1976). Using transfer buffer (39 mM glycine, 48 mM Tris, 10% SDS, 20% methanol), 10 μg of protein was electrophoresed on 12% polyacrylamide gels and transferred to polyvinylidene fluoride membranes (Immobilon-P; Millipore). The membrane was blocked by incubation with 5% skimmed milk (Difco, USA)

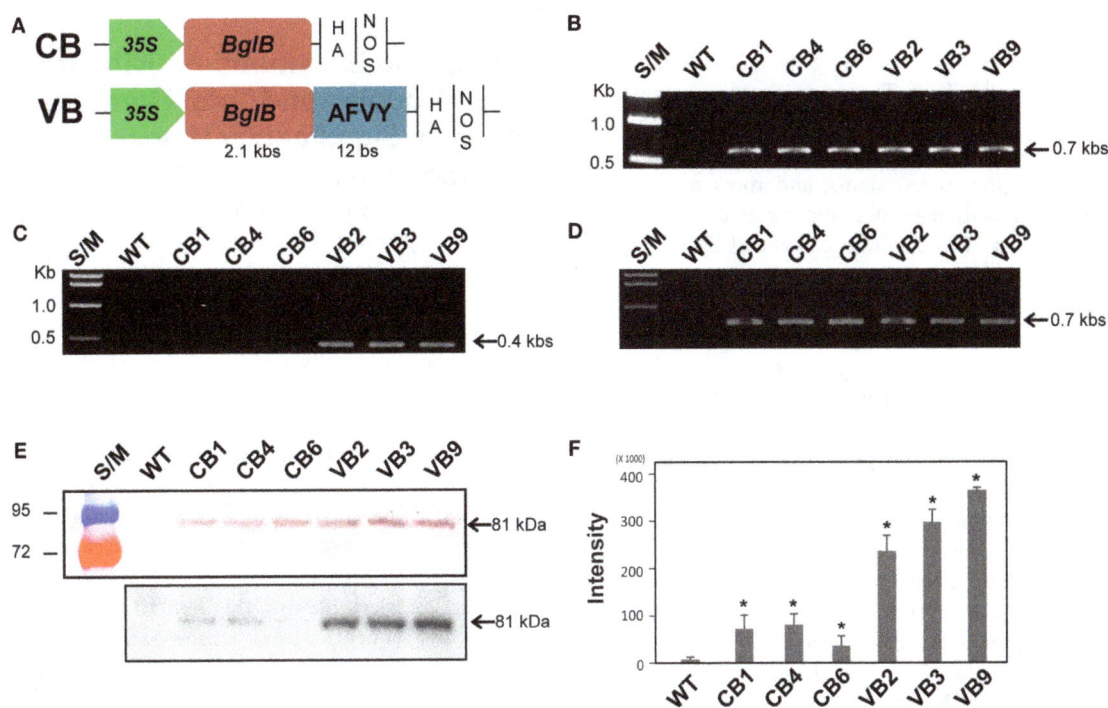

FIGURE 1 | Schematic representation of the BglB expression cassettes and confirmation of heterologously expressed BglB in the transgenic plants.
(A) Vector constructions of CB (cytosol-targeted BglB) and VB (vacuole-targeted BglB, with the 12 bp represent AFVY tetrapeptide). **(B)** Confirmation of the presence of BglB in CB and VB constructs in tobacco genome by PCR genomic DNA. **(C)** Confirmation of the presence of the sequence encoding AFVY tetrapeptide in the C-terminal of BglB of VB transformants by PCR genomic DNA. **(D)** Confirmation of the transcript sequence encoding BglB in CB and VB constructs after mRNA extraction, cDNA synthesis, and RT-PCR. **(E)** Confirmation of the presence of heterologously expressed BglB in the transgenic plants, with BglB present in total soluble protein (TSP) extracted from leaf of CB and VB transgenic plants but no from WT plant (above panel, with 10 μg of TSP was used); and the presence of the heterologous BglB in TSP extracted from isolated vacuole of CB and VB plants but no from those of WT plant (below panel, with 10 μg of TSP from isolated vacuole was used), after concentrated by UFC 710008/Centricon Plus-70 Centrifugal Filter (EM Millipore, USA) by using fluorophore-conjugated secondary antibody anti-rabbit IgG, incubate with substrate within 5 min before transferring to exposure film and keeping within 10 min. **(F)** The intensity of bands exposed in below panel from **(E)**, measured by Adobe Photoshop CS6. *indicates significant differences from the control (WT) ($P < 0.05$).

in phosphate-buffered saline at pH 7.0 (1 mM KH_2PO_4, 10 mM $Na_2HPO_4 \times 12H_2O$, 137 mM NaCl, 2.7 mM KCl), and then incubated with a polyclonal anti-β-glucosidase antibody as the primary antibody. Alkaline phosphatase-conjugated goat anti-rabbit IgG antibody (Promega, USA) was used at a 1:2500 dilution as the secondary antibody. For detecting β-glucosidase *BglB* targeting vacuole, total protein from isolated vacuole was used to conduct western blot with the same as above, except incubating with a fluorophore-conjugated secondary antibody anti-rabbit IgG (H + L) (DyLight TM 680 Conjugate) (red) in a 1:2500 dilution.

Growth Conditions, Sampling, and Phenotypic Observation

After the presence of the transgene was confirmed, T_0 generation transgenic and WT plants were moved to the greenhouse for development. Seeds from the T_0 generation transgenic plants were sprayed in MS medium containing kanamycin and grown in a growth chamber under a 16-/8-h light/dark cycle at $25 \pm 3°C$ to produce the T_1 generation, and the germination day was recorded. After 2 weeks, seedlings from ten lines of the CB and VB transgenic plants were used to determine β-glucosidase activity. A

100 mg sample of grinded powder from the leaf was used to extract TSP (Jung et al., 2010), and after checking TSP by the Bradford method, an amount of extracted protein equivalent to 10 μg of TSP were used to examine β-glucosidase enzymatic activity by using p-nitrophenyl b-D-glucopyranoside (pNPG) as the substrate. One unit of β-glucosidase is defined as the amount of enzyme that released 1 mmol of p-nitrophenol from the pNPG substrate under the assay conditions described below. The assay mixture containing 10 mM pNPG in citrate–phosphate buffer (pH 4.5) was incubated with the enzyme for 30 min at 70°C in a total volume of 1 ml. The reaction was stopped by adding 1M Na_2CO_3, and absorbance was measured at 405 nm. Based on these results, we selected three transgenic lines from each of the CB and VB transgenic plants that showed the highest crude extract β-glucosidase enzymatic activity. Thirty plants from each of the chosen lines were grown in a growth chamber and then moved to 10-l pots in soil–perlite mixtures at $25 \pm 3°C$ under a 16-/8-h light/dark photoperiod and a light intensity of 100 mmol m^{-2} s^{-1} in a greenhouse for further analysis.

The fifth leaf from the tops of three plants of each of the chosen transgenic lines and W plants was harvested at the same time, and after each 20 days, from 30 to 90 day after germination (DAG), stored at −70°C and ground in liquid N_2 to analyze β-glucosidase enzymatic activity.

The phenotypic characteristic of the transgenic and WT plants were also recorded from the T_1 to T_3 generation of transgenic plants, including: stem height, number of leaves, root lengths, number of lateral roots, time from germination to initial flowering, and dry weight (the leaves, stems, and roots of non-sample plants were separately harvested and freeze-dried after harvesting the seeds). The carbohydrate content of each part of the plant was also determined using gas chromatography (Coleman et al., 2009).

To conduct a salt stress tolerance experiment in the growth chamber, after germination, the seedlings were transferred to new MS media (without sucrose) containing 200 mM NaCl, and phenotypic characteristics were measured after 15 DAG. Mature (80 DAG) transgenic and WT plants grown in the greenhouse were used for the salt stress experiment, which the plants were watered with the same amount of 200 mM NaCl within 10 days. The weight of spots (including spot, soil, and plants) was measured before and after 10 days NaCl treatment. Simultaneously, the same position on the leaf (10th leaf from the ground to above) was harvested, ground with liquid N_2, chlorophyll was extracted 90% ethanol, boiled for 5 min and absorbance was measured at an optical density of 620 nm to calculate the chlorophyll concentration (Lichtenthaler and Wellburn, 1983).

Phytohormone Extraction and Measurement

The phytohormones [include abscisic acid (ABA), IAA, and cytokinin] from young leaves or seedlings of CB, VB, and WT plants were extracted with 80% methanol (Oliver et al., 2007) and measured using the Phytodetek competitive enzyme-linked immunosorbent assay (ELISA) kit (Agdia; Elkhardt, IN, USA) the Phytodetek competitive ELISA kits (Agdia). Briefly, young leaves or seedlings of transgenic and WT plants were harvested and ground in N_2 liquid and stored at −70°C for further analysis. One gram of ground powder was mixed with 1 ml 80% methanol and incubated overnight at −4°C. One milliliter (ml) of the supernatant was collected after centrifuging at 13,000 rpm for 10 min to remove debris, and then freeze-dried. The freeze-dried powder was used to measure the levels of each hormone, according to the kit protocols. Each measurement was conducted in triplicate.

Vacuoles Isolation

For purification of vacuole-targeted BglB, transformed protoplasts from young plants (30 DAG) were isolated by hydrolysis with cell-wall hydrolysis enzymes and fractionated by ultracentrifugation according to Mettler et al. (Mettler and Leonard, 1979) and Raikhel et al. (Robert et al., 2007), with some modifications. Due to the requirement of a highly purified of vacuole, the transformed protoplasts were loaded on top of step gradients consisting of 4, 7, 12, and 15% Ficoll, and centrifuged at $97,000 \times g$ for 4 h. The vacuole was isolated in the top layer of the fraction, and then disrupted by sonication before measuring β-glucosidase enzymatic activity with 10 μg TSP.

RESULTS

β-glucosidase Enzymatic Activity from Isolated Vacuoles and Total Hormone Levels were Significantly Higher in Transgenic than in WT Plants

Hormone conjugates, which are found in each class of plant hormones, are mainly localized in the vacuoles of plant. The mechanism controlling their transport across membranes and between plant organs remain unknown (Bajguz and Piotrowska, 2009). To analyze the effects of thermostable *T. maritima* BglB on changes in phytohormone metabolism and the consequences for plant development, we built two constructs of BglB. The CB construct was for ectopic expression of *BglB* in the cytosol, and the VB construct was for expression of vacuole-targeted BglB, under the control of the 35S promoter (**Figure 1A**). In total, 10 and 12 lines of CB and VB transgenic plants, respectively, were confirmed, and three of the lines were used for further analysis after confirmation of the transgenes by genomic DNA PCR, reverse transcription (RT)-PCR, and Western blotting (**Figures 1B–E**). The presence of the nucleotide sequences encoding the AFVY tetrapeptide in the VB construct was confirmed using PCR with a reverse primer specific to the VB construct (**Figure 1C**), and the transcript of the heterologous BglB in the transgenic plants were confirmed by RT-PCR (**Figure 1D**). The existence of BglB in TSP from the CB and VB plants had molecular weights similar to BglB, as mentioned in the previous study (Jung et al., 2010), were detected by western blot (**Figure 1E**, above panel), presented no different between CB and VB plants, but showed significantly higher level of BglB in the isolated vacuoles of VB plant compared to CB plant (**Figure 1E**, below panel), indicated by the higher intensity of the BglB band exposed by the present of BglB in the isolated vacuoles from the VB transgenic than from the CB transgenic and WT plants (**Figure 1F**). These results obviously indicate that VB plants were the highest vacuole-targeted heterologous BglB.

The three best-performing T_0 generation transgenic lines were selected according to their β-glucosidase enzymatic activity, and then self-pollinated. The β-glucosidase enzymatic activity was significantly higher in the transgenic plants, compared to WT plants over three generations (T_1 to T_3; **Figure 2A**). A slight reduction of β-glucosidase enzymatic activity after a few generations was observed, possibly due to factors such as epigenetic silencing mechanisms (Iyer et al., 2000; Matzke et al., 2000). Heterologous *BglB* was also stably expressed in the transgenic plants, as indicated by the pattern of β-glucosidase enzymatic activity during plant development from 30 to 90 days after germination (DAG) of the CB1 and VB9 transgenic plants (**Figure 2B**).

To examine the efficiency of vacuole targeting, we isolated the vacuoles of the WT plants, and the CB and VB transgenic lines, from the T_1 generation. The results showed that the isolated vacuoles of the VB transgenic lines had the highest β-glucosidase enzymatic activity, compared to Cyt-BglB and WT plants (**Figure 2C**). In particular, compared to WT plants, increased β-glucosidase enzymatic activity of 452 and 759% were recorded in the vacuoles of CB1 and VB2 transformants, respectively. These results accompany to above identification (presented

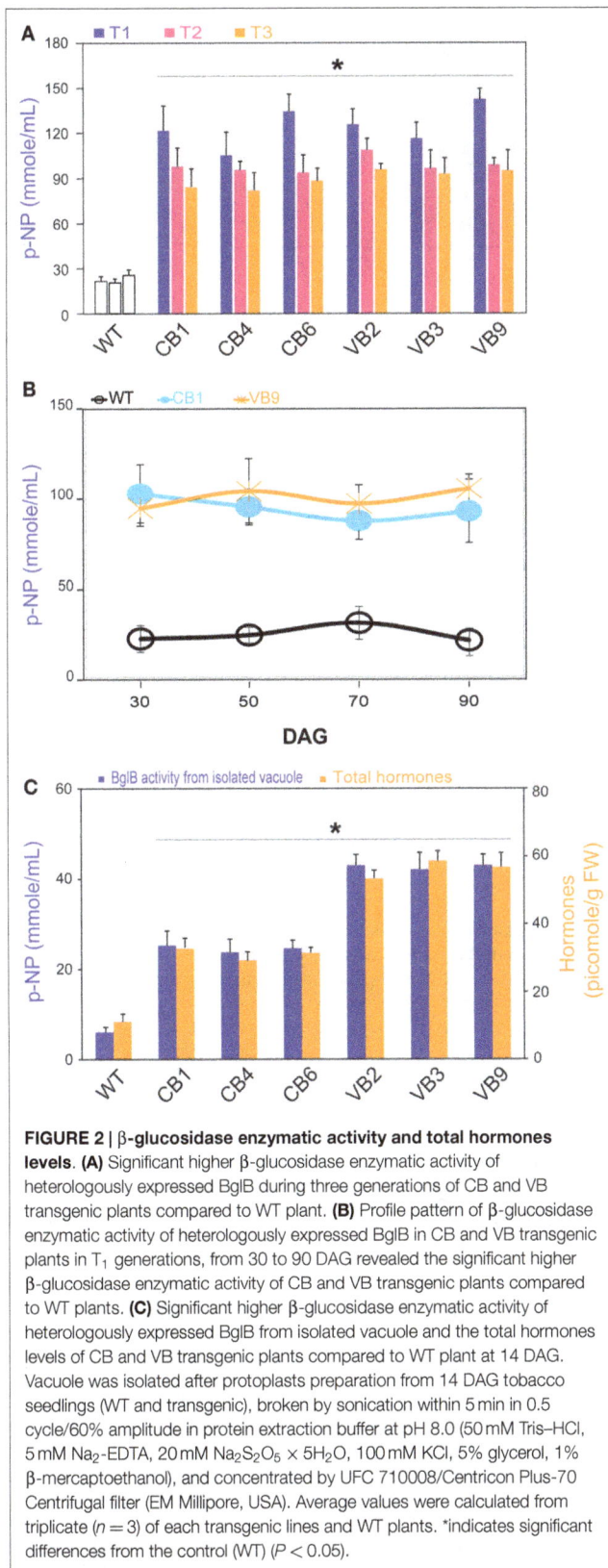

FIGURE 2 | β-glucosidase enzymatic activity and total hormones levels. **(A)** Significant higher β-glucosidase enzymatic activity of heterologously expressed BglB during three generations of CB and VB transgenic plants compared to WT plant. **(B)** Profile pattern of β-glucosidase enzymatic activity of heterologously expressed BglB in CB and VB transgenic plants in T_1 generations, from 30 to 90 DAG revealed the significant higher β-glucosidase enzymatic activity of CB and VB transgenic plants compared to WT plants. **(C)** Significant higher β-glucosidase enzymatic activity of heterologously expressed BglB from isolated vacuole and the total hormones levels of CB and VB transgenic plants compared to WT plant at 14 DAG. Vacuole was isolated after protoplasts preparation from 14 DAG tobacco seedlings (WT and transgenic), broken by sonication within 5 min in 0.5 cycle/60% amplitude in protein extraction buffer at pH 8.0 (50 mM Tris–HCl, 5 mM Na_2-EDTA, 20 mM $Na_2S_2O_5 \times 5H_2O$, 100 mM KCl, 5% glycerol, 1% β-mercaptoethanol), and concentrated by UFC 710008/Centricon Plus-70 Centrifugal filter (EM Millipore, USA). Average values were calculated from triplicate ($n = 3$) of each transgenic lines and WT plants. *indicates significant differences from the control (WT) ($P < 0.05$).

in **Figure 1E**), indicate that the VB constructs which imposed AFVY tetrapeptide effectively targeted β-glucosidase to the vacuole, and that BglB was still active in the vacuole.

Moreover, significantly higher total hormone (including IAA, ABA, and cytokinin) levels were recorded in the transgenic plants compared to the WT plants, based on the ELISA results, with the highest hormone levels obtained from the VB transformants (**Figure 2C**). In particularly, maximum increases of 268 and 463%, when comparing total extracted hormone levels in CB1 and VB3 to WT plants, respectively, were attributed to higher levels of each hormone in the transgenic plants (Figure S1 in Supplementary Material).

Pronounced Phenotypic Changes in the Transgenic Plants

The transgenic CB and VB tobacco plants displayed pronounced phenotypic changes compared to WT plants. Phenotypic characteristics such as stem height, time from germination to initial flowering, and dry weight were proportional to the levels of β-glucosidase enzymatic activity of transgenic and WT plants, suggesting a correlation between the enhancement of β-glucosidase enzymatic activity and these phenotypic changes. In particular, faster development was observed in the transgenic plants than in the WT plants, as indicated by increased stem height, earlier flowering, increased biomass accumulation, and enhanced root system development (**Figure 3A**; Figure S2A in Supplementary Material). Moreover, a shorter time from germination to initial flowering was recorded in the T_1 generation of transgenic plants compared to WT plants. We reported an average of 103.6 and 94.1 DAG in the CB and VB transgenic plants, respectively, compared to 141.7 DAG in WT plants; **Figure 3B**), and similar results were observed for the T_2 and T_3 generations (Figure S2B in Supplementary Material). Higher β-glucosidase enzymatic activity and total hormone levels were recorded at flowering time, with maximum increases in total hormone levels of 222 and 387% for CB1 and VB3 compared to WT plants, respectively (**Figure 3C**; Figure S2C in Supplementary Material), while no significant differences in β-glucosidase enzymatic activity between the CB and VB transgenic plants was observed (**Figure 3C**). These results indicate that more liberated hormones were released in the VB than the CB transgenic plants. After the seeds were harvested, the stem height and dry weight of total biomass accumulation were significantly higher in the mature transgenic plants than in the WT plants, with maximum increases of 133% for stem height (CB1 compared to WT plants) and 124% for total dry weight (VB9 compared to WT plants; **Figures 3D,E**). Similar results were obtained for the T_2 and T_3 generations (Figures S2D,E in Supplementary Material). These results clearly indicated that the increase in liberated hormone levels (particularly IAA and cytokinin) contributed to increased biomass accumulation, despite the shortened growth cycle (earlier flowering after germination) in the transgenic plants. The same phenotypic characteristics were observed in previous studies that targeted β-glucosidase to either general or particular cellular compartments (Jung et al., 2010; Jin et al., 2011). However, despite the significant changes in biomass accumulation and shortened growth cycle, no significant differences in carbohydrate content were observed between the transgenic and WT plants (**Table 1**), indicating that only total biomass accumulation was influenced in the transgenic plants.

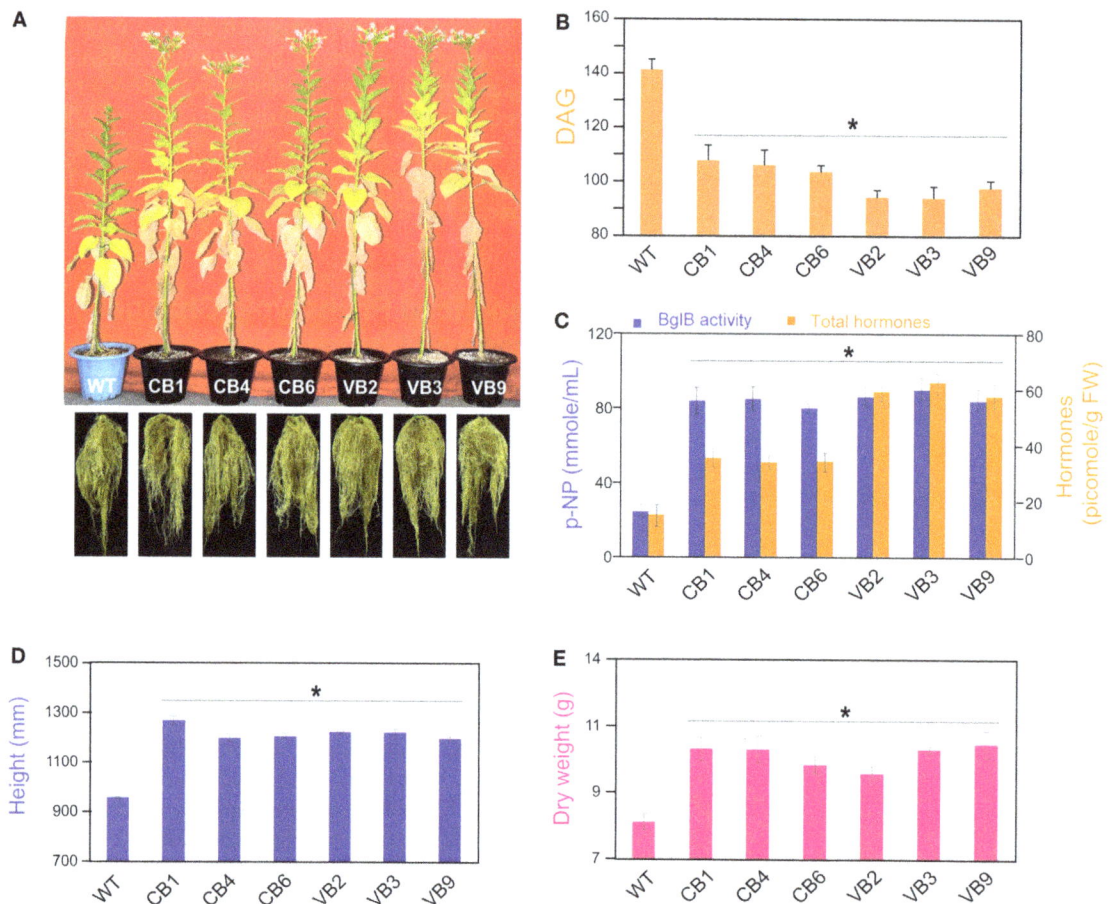

FIGURE 3 | Pronounced phenotypic changes occurred in heterologously expressed BglB transgenic tobacco plants. (A) Faster growth in stem height, shortened growth cycle with earlier flowering, and enhanced root development in the transgenic plants compared to WT plants. (B) Time required from germination to flowering of the transgenic vs. WT plants, showed the shortest time required belong to VB plants. (C) Combination of β-glucosidase enzymatic activity and total hormones levels of the transgenic and WT plants, measured by harvesting the fifth leaf from the top of plant immediately after flowering. (D) Stem height and (E) dry weight of the transgenic and WT plants were calculated after harvesting seeds and freeze-dried. Average values were calculated from data recorded from twenty individuals ($n = 20$) of each transgenic lines and WT plants for (B,D,E), and from triplicate ($n = 3$) of each transgenic lines and WT plants for (C). *indicates significant differences from the control (WT) ($P < 0.05$).

Transgenic Plants Showed Faster Development of the Stem and Roots, Elevation of IAA and Cytokinin Levels, and Upregulation of WUS

Based on the increase in stem height and enhanced roots system development, which appeared to be correlated with increased hormone levels of the transgenic plants, we asked whether the increased levels of cytokinin and auxin would affect the development of stems and roots of transgenic seedlings. Seeds from the T_1 to T_3 generations, and the WT plants, were sprayed with MS media containing kanamycin. Immediately after the seeds germinated, tiny seedlings were transferred to new MS media in a line to compare stem development. Faster development of the transgenic plants was clearly observed, as presented by the larger size of the transgenic plants compared to WT plants (**Figure 4A**). At 15 DAG, along with the increase in β-glucosidase enzymatic activity, IAA and cytokinin levels were significantly higher in

the transgenic plants compared to WT plants, with the highest hormone level obtained in VB transgenic plants (increase in 585% in VB3 compared to WT plants), whereas there was no difference in β-glucosidase enzymatic activity for the CB and VB transgenic plants (**Figure 4B**). These results indicate that more liberated hormones were released from the vacuole in VB than in CB transgenic plants.

Correlation analysis between the development of the stem (height) and root system (root lengths) to cytokinin and auxin levels showed a maximum increase in 200% for stem height corresponded to an increase in 458% in cytokinin level in VB9 compared to WT plants (**Figure 4C**). The maximum increases in roots lengths and IAA level were 186 and 725%, respectively, in VB3 compared to WT plants (**Figure 4D**). Moreover, these results indicate that, despite slight differences in stem height and root lengths between the CB and VB transgenic plants, the increase in liberated IAA and cytokinin levels promoted faster development in the transgenic lines compared to WT plants. We observed

TABLE 1 | Comparison of carbohydrate content in leaves, stem, and roots in WT and vacuole-targeted *T. maritima* BglB transgenic plants.

	(%)	WT	CB1	CB4	CB6	VB2	VB3	VB9
Leaves	Rhamnose	2.3 ± 0.2	2.2 ± 0.3	2.3 ± 0.5	2.6 ± 0.6	2.4 ± 0.3	2.6 ± 0.4	2.6 ± 0.5
	Arabinose	1.4 ± 0.1	1.4 ± 0.2	1.5 ± 0.1	1.5 ± 0.3	1.1 ± 0.1	1.1 ± 0.1	1.1 ± 0.1
	Xylose	1.6 ± 0.2	1.7 ± 0.3	1.6 ± 0.2	1.7 ± 0.0	1.5 ± 0.1	1.6 ± 0.1	1.3 ± 0.0
	Mannose	0.7 ± 0.1	0.9 ± 0.1	1.0 ± 0.2	0.7 ± 0.1	0.7 ± 0.2	0.7 ± 0.3	1.1 ± 0.0
	Galactose	2.1 ± 0.2	2.2 ± 0.4	2.2 ± 0.4	2.2 ± 0.4	2.0 ± 0.3	2.2 ± 0.5	1.9 ± 0.1
	Glucose	45.3 ± 3.5	46.2 ± 4.1	44.6 ± 3.3	44.9 ± 0.8	46.1 ± 3.8	43.9 ± 2.1	45.1 ± 1.9
	Total	53.4 ± 2.8	54.6 ± 3.3	53.2 ± 4.2	53.6 ± 2.6	53.8 ± 4.1	52.1 ± 3.2	53.1 ± 2.5
Stems	Rhamnose	1.8 ± 0.4	1.2 ± 0.1	0.9 ± 0.4	1.2 ± 0.2	1.0 ± 0.2	0.8 ± 0.1	1.2 ± 0.2
	Arabinose	0.7 ± 0.2	0.8 ± 0.0	1.0 ± 0.0	1.1 ± 0.0	1.3 ± 0.0	0.9 ± 0.1	1.1 ± 0.1
	Xylose	5.6 ± 0.7	5.9 ± 1.5	6.2 ± 0.6	7.1 ± 1.1	6.2 ± 0.1	5.8 ± 0.7	5.9 ± 0.9
	Mannose	1.0 ± 0.2	1.3 ± 0.3	1.1 ± 0.3	1.2 ± 0.0	1.1 ± 0.1	1.2 ± 0.1	1.2 ± 0.0
	Galactose	1.2 ± 0.1	1.4 ± 0.2	1.3 ± 0.2	1.2 ± 0.1	1.0 ± 0.1	1.1 ± 0.1	1.0 ± 0.2
	Glucose	44.7 ± 1.2	45.7 ± 6.1	46.2 ± 5.6	47.1 ± 5.4	46.1 ± 2.9	45.5 ± 3.6	43.9 ± 3.2
	Total	55.0 ± 4.2	56.3 ± 6.5	56.7 ± 7.2	58.9 ± 6.5	56.7 ± 3.1	55.3 ± 4.7	54.3 ± 4.8
Roots	Rhamnose	1.3 ± 0.5	1.1 ± 0.0	1.3 ± 0.3	1.0 ± 0.3	1.3 ± 0.1	1.4 ± 0.2	1.3 ± 0.3
	Arabinose	0.8 ± 0.1	0.9 ± 0.1	1.1 ± 0.0	0.8 ± 0.1	1.1 ± 0.1	0.8 ± 0.2	0.9 ± 0.2
	Xylose	6.4 ± 0.5	6.9 ± 1.1	7.1 ± 1.1	6.4 ± 0.3	6.8 ± 0.5	6.6 ± 0.4	6.7 ± 1.9
	Mannose	1.1 ± 0.5	1.3 ± 0.1	1.2 ± 0.0	1.3 ± 0.0	1.2 ± 0.1	1.2 ± 0.0	1.0 ± 0.2
	Galactose	1.2 ± 0.1	1.1 ± 0.3	1.0 ± 0.1	1.1 ± 0.0	1.2 ± 0.2	1.1 ± 0.1	0.8 ± 0.2
	Glucose	46.1 ± 3.8	45.2 ± 2.8	44.8 ± 2.9	44.3 ± 1.5	45.6 ± 3.4	44.8 ± 1.9	46.1 ± 2.8
	Total	56.9 ± 5.1	56.5 ± 3.7	56.5 ± 3.9	54.9 ± 2.4	57.2 ± 4.9	55.9 ± 3.1	56.8 ± 4.8

Average values were calculated from triplicate (n = 3) of each CB and VB transgenic lines and WT plants.

maximum increases in stem height and root lengths of 164 vs. 200%, and 183 vs. 194%, for CB and VB vs. WT plants, respectively. Faster development of the stem and roots was also observed in the T_2 and T_3 generations (Figures S3A,B in Supplementary Material). Furthermore, a higher number of leaves and lateral roots, and greater average fresh weight of 20 young plants, were also observed in the transgenic plants compared to WT plants (Figures S3C–E in Supplementary Material), indicating that the faster development of the transgenic plants, compared to WT plants, was stable after three generations.

Plant stem cells are harbored inside the meristem, which is located in the growing tips of the shoots and roots. The faster development observed in the transgenic plants suggests stronger stimulation of stem cell reproduction, which could then induce changes in plant growth and organogenesis (Murray et al., 2012). The population of stem cells in shoot apical meristems is regulated by expression of the homeodomain gene *WUS*, a transcription factor that can be upregulated by cytokinin level. In the root apical meristem, a group of dividing cells, called the quiescent center (QC) in the root apical meristem, is upregulated by IAA level (Yadav et al., 2010; Zhao et al., 2010). To determine the expression levels of *WUS* for transgenic lines and WT plants, RNA was extracted from the stems of young plants (15 DAG) for cDNA synthesis and RT-PCR. The results showed higher *WUS* expression levels in the transgenic plants compared to WT plants (**Figure 4E**), providing evidence that superior development of the transgenic plants compared to WT plants was due to elevated hormone levels.

Enhanced Resistance to NaCl Stress and Elevation of ABA in Transgenic Plants

Next, we asked whether increased ABA levels in the transgenic plants led to increased tolerance of salt stress, as mentioned in previous studies (Lee et al., 2006; Wang et al., 2011; Han et al.,

2012; Xu et al., 2012). Seeds from the T_1 to T_3 generations were used, and after germination, tiny seedlings were transferred to new MS media containing 200 mM NaCl to examine the response to high NaCl stress. Observations at 15 DAG showed that the transgenic plants were more resistant to high NaCl, as indicated by enhanced development in the transgenic compared to WT seedlings in term of increased root length, number of leaves, number of lateral roots, and fresh weight (**Figure 5A**). As shown in **Figure 5B**, the higher ABA level was clearly related to increased β-glucosidase enzymatic activity in the transgenic seedlings, with the highest ABA levels recorded in the VB transformants (maximum increase in 504% in VB3 compared to WT seedlings). Increased tolerance to high NaCl stress was also displayed by the obviously longer root lengths and greater fresh weight of 100 transgenic seedlings compared to WT seedlings (maximum increase in 271% in root lengths and 256% in fresh weight in the VB2 compared to WT seedlings; **Figures 5C,D**).

Next, to examine the resistance of mature plants to salt stress, transgenic and WT plants grown in the greenhouse were subjected to a salt stress experiment at 80 DAG. As shown in **Figure 5E**, more senescent leaves appeared in the transgenic than WT plants, which may explain the higher rate of weight reduction (8.3% in VB2 compared to 3.4% in WT plants; **Figure 5F**). The appearance of leaves senescence indicated that a faster programed cell death process occurred in the transgenic than WT plants, which was also represented by the higher rate of chlorophyll degradation in transgenic plants (43.2% in VB3 compared to 14.1% in WT plants; **Figure 5G**).

DISCUSSION

Because of its thermostability and transglycosylation properties, the *T. maritima* BglB enzyme is considered to be a useful

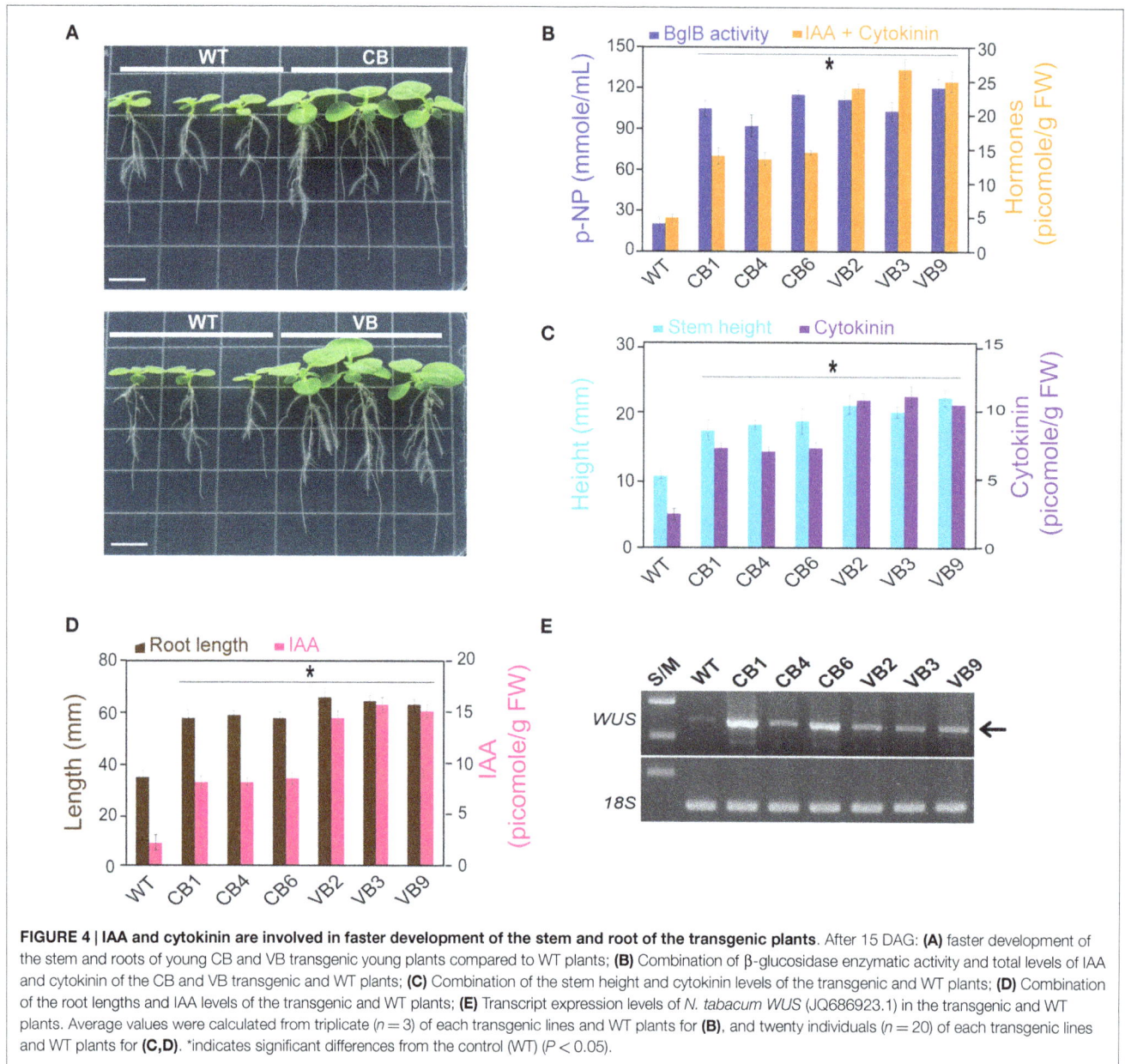

FIGURE 4 | IAA and cytokinin are involved in faster development of the stem and root of the transgenic plants. After 15 DAG: **(A)** faster development of the stem and roots of young CB and VB transgenic young plants compared to WT plants; **(B)** Combination of β-glucosidase enzymatic activity and total levels of IAA and cytokinin of the CB and VB transgenic and WT plants; **(C)** Combination of the stem height and cytokinin levels of the transgenic and WT plants; **(D)** Combination of the root lengths and IAA levels of the transgenic and WT plants; **(E)** Transcript expression levels of *N. tabacum WUS* (JQ686923.1) in the transgenic and WT plants. Average values were calculated from triplicate (*n* = 3) of each transgenic lines and WT plants for **(B)**, and twenty individuals (*n* = 20) of each transgenic lines and WT plants for **(C,D)**. *indicates significant differences from the control (WT) ($P < 0.05$).

catalyst for biotechnological applications (Goyal et al., 2001). According to Jung et al. (2010), transgenic tobacco plants can not only be utilized for the mass production of BglB, but also, the overexpression of heterologous BglB in tobacco has led to changes in phenotypic characteristics (such as larger leaves and taller plants) (Jung et al., 2013). Plants contain their own β-glucosidase genes, and previous studies have demonstrated that the expression of β-glucosidase, including heterologous expression, affects the hydrolysis of hormone conjugates and homeostasis in plants, which in turn control plant development (Schliemann, 1984; Brzobohaty et al., 1993; Dietz et al., 2000; Kiran et al., 2006). In the present study, by observing pronounced phenotypic changes in the *T. maritima* BglB transgenic tobacco compared to WT plants. The transgenic tobacco remained stable

over three offspring generations (**Figures 3–5**). We were encouraged to evaluate the relationship between β-glucosidase enzymatic activity of *T. maritima* BglB and changes in plant hormone levels.

For vacuole targeting, among the three different types of vacuolar sorting signals (N- or C-terminal polypeptides or internal sequences) that have been identified (Jiang and Rogers, 1998; Matsuoka and Neuhaus, 1999), C-terminal polypeptides, such as the C-terminal amino acids AFVY tetrapeptide from phaseolin, are considered be the most efficient (Frigerio et al., 2001; Nausch et al., 2012a,b). However, due to the presence of numerous hydrolytic enzymes in the vacuole of plant cells, it is generally difficult for proteins to maintain their activity inside the vacuole (Boller and Kende, 1979; Marty, 1999). Here, we showed that

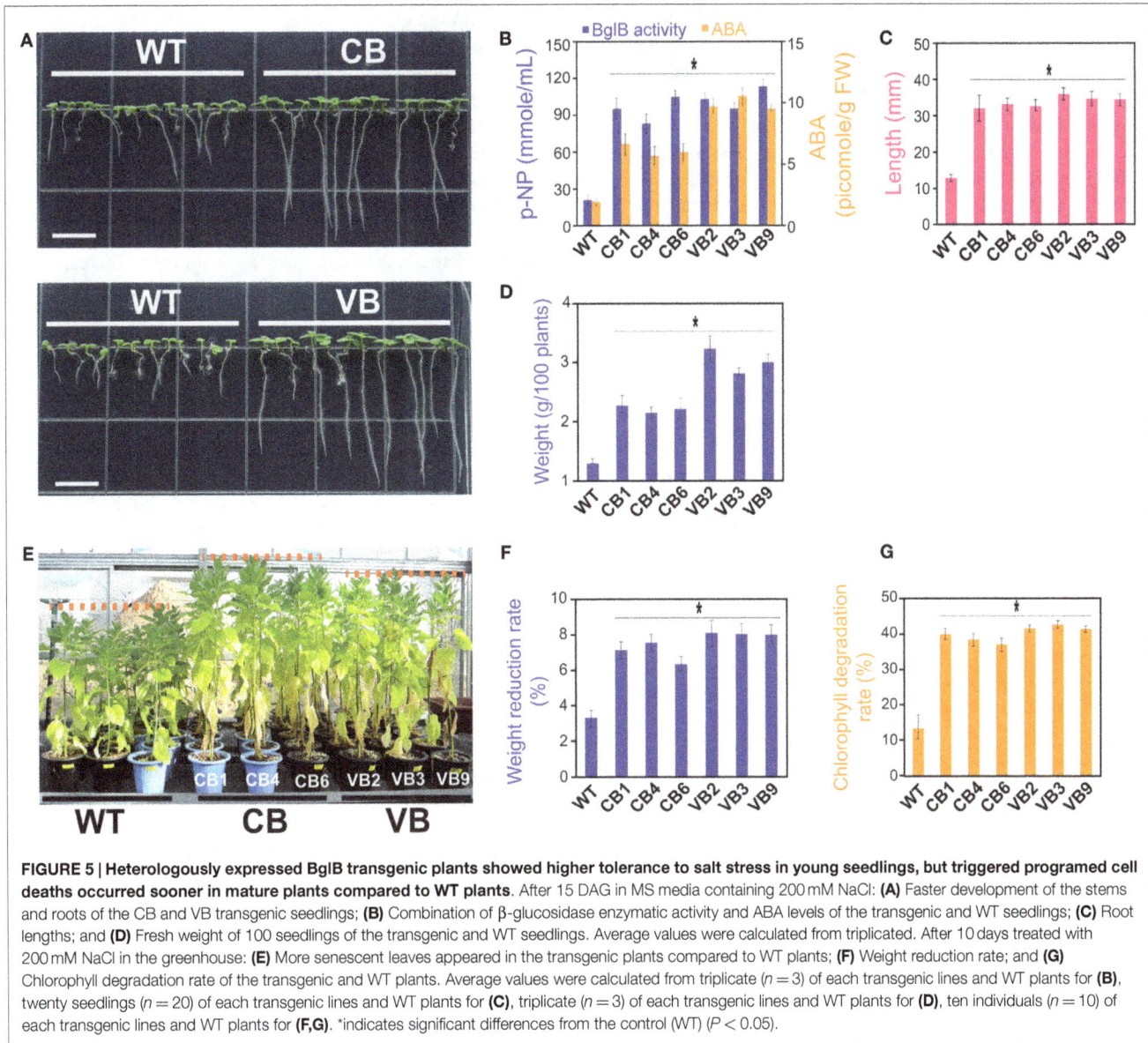

FIGURE 5 | Heterologously expressed BglB transgenic plants showed higher tolerance to salt stress in young seedlings, but triggered programed cell deaths occurred sooner in mature plants compared to WT plants. After 15 DAG in MS media containing 200 mM NaCl: **(A)** Faster development of the stems and roots of the CB and VB transgenic seedlings; **(B)** Combination of β-glucosidase enzymatic activity and ABA levels of the transgenic and WT seedlings; **(C)** Root lengths; and **(D)** Fresh weight of 100 seedlings of the transgenic and WT seedlings. Average values were calculated from triplicated. After 10 days treated with 200 mM NaCl in the greenhouse: **(E)** More senescent leaves appeared in the transgenic plants compared to WT plants; **(F)** Weight reduction rate; and **(G)** Chlorophyll degradation rate of the transgenic and WT plants. Average values were calculated from triplicate ($n = 3$) of each transgenic lines and WT plants for **(B)**, twenty seedlings ($n = 20$) of each transgenic lines and WT plants for **(C)**, triplicate ($n = 3$) of each transgenic lines and WT plants for **(D)**, ten individuals ($n = 10$) of each transgenic lines and WT plants for **(F,G)**. *indicates significant differences from the control (WT) ($P < 0.05$).

the β-glucosidase enzymatic activity of heterologously expressed BglB was significantly higher in the transgenic (both the CB and VB transformants) plant compared to WT plants (**Figure 2A**). The transgenic plants remained stable during the life cycle and durable after three offspring generation (**Figures 2A,B**). These results indicate that the *T. maritima* BglB was effectively expressed in the transgenic tobacco plants.

For the first time, the present of the heterologous BglB and β-glucosidase enzymatic activity assays were conducted after vacuole isolation, which showed that BglB expression was dramatically higher in the VB than CB transgenic plants (**Figures 1E** and **2C**). This result clearly indicated that AFVY tetrapeptide were effective for sorting *T. maritima* BglB into the vacuole, and that its β-glucosidase enzymatic activity was maintained and could tolerate the protein-degrading conditions of the vacuole environment. Therefore, vacuole-targeted *T. maritima* BglB transgenic

plants should be considered candidates for plant molecular farming, where plants are used as bioreactors to produce degrading enzymes for hydrolysis of lignocellulosic material, which is similar to chloroplast-targeted *T. maritima* BglB transgenic plants (Jung et al., 2010, 2013).

Hormone glucoside conjugates, which are mainly stored in the plant vacuole, are considered inactive forms in hormone metabolism, and can be liberated by β-glucosidases, a large group of enzymes that can hydrolyze glucoside ester linkages (Sembdner et al., 1994; Bajguz and Piotrowska, 2009). A wide variety of β-glucosidase enzymes from plants have been proven to be hormone conjugates with hydrolysis capability (Schliemann, 1984; Brzobohaty et al., 1993; Dietz et al., 2000; Kiran et al., 2006; Lee et al., 2006; Yao et al., 2007; Jin et al., 2011). We demonstrated a novel approach in which transformation of BglB, encoding a thermostable β-glucosidase from the bacterium *T. maritima* (Goyal et al., 2001), affected plant hormone levels through hydrolyzation of glucoside ester links in hormone conjugates in the transgenic plants, which seemed to be the result of non-specific activity. For example, previous studies demonstrated that each kind of β-glucosidase likely performs its functions in specific hormone conjugates (Brzobohaty et al., 1993; Dietz et al., 2000). Kiran et al. (2012) reported that Zm-p60.1 is capable of releasing active cytokinin from O- and N-glucosides, and confirmed that the liberated hormones are still in the active state. Knowledge of the transportation mechanism from inside to outside of the vacuole is still lacking (Vitale and Hinz, 2005; De Marcos Lousa et al., 2012). In the present study, significantly higher enzymatic activity, particularly in isolated vacuoles, was accompanied by dramatically higher levels of hormones (IAA, ABA, and cytokinin) in the VB plants compared to CB transgenic plants, with WT plants showing the lowest levels (**Figure 2C**). These results clearly demonstrated that, when greater amounts of BglB were targeted to the vacuole, more liberated hormones were released.

In contrast to the results obtained by Kiran et al. (2012), who found no significant phenotypic changes in vacuole-targeted *Zm-p60.VAL* transgenic plants, our results showed pronounced phenotypic changes in *T. maritima* BglB transgenic plants compared to WT plants. Mature transgenic plants exhibited enhanced development, in terms of faster growth in stem height and a shortened growth cycle, with earlier flowering (**Figure 3**). Young seedlings had increased stem height and longer roots (**Figure 4**), which were accompanied by significantly higher hormones levels that were maintained over three offspring generations of the transgenic plants. These results provide clear evidence that heterologously expressed BglB increases the plant hormones levels, which then influence their phenotypes.

Due to the elevated levels of IAA, ABA, and cytokinin, it is difficult to determine the specific factor that directly contributes to the phenotypic changes in the transgenic plants. Fortunately, previous works can provide clues to trace the cause of such changes. For example, IAA is known to regulate root development (Overvoorde et al., 2010), cytokinin plays pivotal roles in the formation and activity of shoot meristems (Werner et al., 2003; Werner and Schmülling, 2009), and ABA functions in the plant response to dehydrating/salinity stresses and programed cell death (Finkelstein, 2006; Yang et al., 2014). Previous studies have also shown that the reproduction and differentiation of stem cells harbored in the shoot and root apical meristem contribute to development and organogenesis in plants (Williams and Fletcher, 2005; Powell and Lenhard, 2012). Therefore, the taller stem height, longer roots, and earlier flowering observed in the transgenic plants could indicate enhancement of the shoot and root apical meristem in the transgenic plants compared to WT plants. Specifically, the expression level of *WUS*, a transcription factor that regulates the development and division of stem cells in the shoot apical meristem, is upregulated by cytokinin (Kurakawa et al., 2007; Werner and Schmülling, 2009; Zhao et al., 2010), shedding light on the mechanism contributing to the role of cytokinin, which was increased in our transgenic plants, in enhancing the development of the stems and aboveground organs.

Our result showed enhanced development of the root systems (represented by increased roots dry weight, number of lateral roots, and root length), confirming the effect of a larger amount of IAA on the development of root systems in the transgenic plants (**Figures 3** and **4**). ABA mainly functions in the plant's response to dehydration by inducing stomatal opening/closing, and also plays a role in limiting cell division and expansion, decreasing shoot growth and lateral root initiation, and promoting developmental phase changes such as vegetative-to-reproductive transitions (Finkelstein, 2006). In the present study, the increased ABA levels were related to increased salt stress tolerance in young seedlings. The faster chlorophyll degradation and higher rates of weight reduction after treatment with NaCl solution in mature plants revealed that programed cell death was promptly triggered in the transgenic plants for both the VB and CB transformants, compared to WT plants (**Figure 5**). Notably, no significant difference in β-glucosidase enzymatic activity, but significantly higher hormones levels, in the VB transgenic plants compared to CB transgenic plants, were observed, confirming that the hormone conjugates are mainly stored in the vacuole, and more liberated hormones were released from the conjugates in the VB transgenic plants, which contributed to the greater effect on plant development in the VB transgenic plants.

CONCLUSION

After *T. maritima BglB* was first overexpressed and effectively targeted into the vacuole by the addition of AFVY C-terminal tetrapeptides, BglB was still active and functional. The main results emerging from this study are that the hormone (ABA, IAA, and cytokinin) conjugates are mainly stored in the vacuole, and perhaps more importantly, higher levels of hormones liberated from their conjugates via BglB-mediated hydrolysis enhance the growth and development in VB transgenic plants to a greater extent than in CB transgenic plants. Therefore, the use of heterologously overexpressed vacuole-targeted *T. maritima BglB* may be an approach to develop molecular farming technology to achieve multiple targets: increased production of the β-glucosidase BglB, increased biomass accumulation, and shortened of developmental stages. Also this *BglB* vacuole-targeted plant farming system influences of total biomass accumulation and as such may be useful in increasing biomass production for bioenergy and biofuel production.

FUNDING

This work was supported by Priority Centers Program (2010-0020141) through the National Research Foundation of Korea (NRF) funded by the Ministry of Education, Science and technology, and by a grant (S211314L010120) from Forest Science & Technology Projects, Forest Service, Republic of Korea.

REFERENCES

Bajguz, A., and Piotrowska, A. (2009). Conjugates of auxin and cytokinin. *Phytochemistry* 70, 957–969. doi:10.1016/j.phytochem.2009.05.006

Boller, T., and Kende, H. (1979). Hydrolytic enzymes in the central vacuole of plant cells. *Plant Physiol.* 63, 1123–1132. doi:10.1104/pp.63.6.1123

Bradford, M. M. (1976). A rapid and sensitive method for the quantitation of microgram quantities of protein utilizing the principle of protein-dye binding. *Anal. Biochem.* 72, 248–254. doi:10.1016/0003-2697(76)90527-3

Brzobohaty, B., Moore, I., Kristoffersen, P., Bako, L., Campos, N., Schell, J., et al. (1993). Release of active cytokinin by a beta-glucosidase localized to the maize root meristem. *Science* 262, 1051–1054. doi:10.1126/science.8235622

Coleman, H. D., Yan, J., and Mansfield, S. D. (2009). Sucrose synthase affects carbon partitioning to increase cellulose production and altered cell wall ultrastructure. *Proc. Natl. Acad. Sci. U. S. A* 106, 13118–13123. doi:10.1073/pnas.0900188106

De Marcos Lousa, C., Gershlick, D. C., and Denecke, J. (2012). Mechanisms and concepts paving the way towards a complete transport cycle of plant vacuolar sorting receptors. *Plant Cell* 24, 1714–1732. doi:10.1105/tpc.112.095679

Dietz, K. J., Sauter, A., Wichert, K., Messdaghi, D., and Hartung, W. (2000). Extracellular β-glucosidase activity in barley involved in the hydrolysis of ABA glucose conjugate in leaves. *J. Exp. Bot.* 51, 937–944. doi:10.1093/jexbot/51.346.937

Finkelstein, R. R. (2006). Studies of abscisic acid perception finally flower. *Plant Cell* 18, 786–791. doi:10.1105/tpc.106.041129

Frigerio, L., Foresti, O., Felipe, D. H., Neuhaus, J.-M., and Vitale, A. (2001). The C-terminal tetrapeptide of phaseolin is sufficient to target green fluorescent protein to the vacuole. *J. Plant Physiol.* 158, 499–503. doi:10.1078/0176-1617-00362

Goyal, K., Selvakumar, P., and Hayashi, K. (2001). Characterization of a thermostable β-glucosidase (BglB) from *Thermotoga maritima* showing transglycosylation activity. *J. Mol. Catal. B Enzym.* 15, 45–53. doi:10.1016/S1381-1177(01)00003-0

Hall, J. L. (2000). Deepesh N. De.Plant cell vacuoles: an introduction. CSIRO Publishing, Collingwood, 2000. Pp. 288. Price US$ 60.00. ISBN 0643 062548. *J. Exp. Bot.* 51, 2127. doi:10.1093/jexbot/51.353.2127

Han, Y.-J., Cho, K.-C., Hwang, O.-J., Choi, Y.-S., Shin, A.-Y., Hwang, I., et al. (2012). Overexpression of an *Arabidopsis* β-glucosidase gene enhances drought resistance with dwarf phenotype in creeping bentgrass. *Plant Cell Rep.* 31, 1677–1686. doi:10.1007/s00299-012-1280-6

Helmer, G., Casadaban, M., Bevan, M., Kayes, L., and Chilton, M.-D. (1984). A new chimeric gene as a marker for plant transformation: the expression of *Escherichia coli* [beta]-galactosidase in sunflower and tobacco cells. *Nat. Biotech* 2, 520–527. doi:10.1038/nbt0684-520

Iyer, L., Kumpatla, S., Chandrasekharan, M., and Hall, T. (2000). Transgene silencing in monocots. *Plant Mol. Biol.* 43, 323–346. doi:10.1023/A:1006412318311

Jiang, L., and Rogers, J. C. (1998). Integral membrane protein sorting to vacuoles in plant cells: evidence for two pathways. *J. Cell Biol.* 143, 1183–1199. doi:10.1083/jcb.143.5.1183

Jin, S., Kanagaraj, A., Verma, D., Lange, T., and Daniell, H. (2011). Release of hormones from conjugates: chloroplast expression of beta-glucosidase results in elevated phytohormone levels associated with significant increase in biomass and protection from aphids or whiteflies conferred by sucrose esters. *Plant Physiol.* 155, 222–235. doi:10.1104/pp.110.160754

Jung, S., Kim, S., Bae, H., Lim, H.-S., and Bae, H.-J. (2010). Expression of thermostable bacterial β-glucosidase (BglB) in transgenic tobacco plants. *Bioresour. Technol.* 101, 7144–7150. doi:10.1016/j.biortech.2010.03.140

Jung, S., Lee, D.-S., Kim, Y.-O., Joshi, C., and Bae, H.-J. (2013). Improved recombinant cellulase expression in chloroplast of tobacco through promoter engineering and 5' amplification promoting sequence. *Plant Mol. Biol.* 83, 317–328. doi:10.1007/s11103-013-0088-2

Kerk, N. M., Jiang, K., and Feldman, L. J. (2000). Auxin metabolism in the root apical meristem. *Plant Physiol.* 122, 925–932. doi:10.1104/pp.122.3.925

Kim, S., Lee, D.-S., Choi, I., Ahn, S.-J., Kim, Y.-H., and Bae, H.-J. (2010). *Arabidopsis thaliana* Rubisco small subunit transit peptide increases the accumulation of *Thermotoga maritima* endoglucanase Cel5A in chloroplasts of transgenic tobacco plants. *Transgenic Res.* 19, 489–497. doi:10.1007/s11248-009-9330-8

Kiran, N. S., Benkova, E., Rekova, A., Dubova, J., Malbeck, J., Palme, K., et al. (2012). Retargeting a maize beta-glucosidase to the vacuole – evidence from intact plants that zeatin-O-glucoside is stored in the vacuole. *Phytochemistry* 79, 67–77. doi:10.1016/j.phytochem.2012.03.012

Kiran, N. S., Polanska, L., Fohlerova, R., Mazura, P., Valkova, M., Smeral, M., et al. (2006). Ectopic over-expression of the maize beta-glucosidase Zm-p60.1 perturbs cytokinin homeostasis in transgenic tobacco. *J. Exp. Bot.* 57, 985–996. doi:10.1093/jxb/erj084

Kleczkowski, K., Schell, J., and Bandur, R. (1995). Phytohormone conjugates: nature and function. *CRC Crit. Rev. Plant Sci.* 14, 283–298. doi:10.1080/07352689509382361

Kurakawa, T., Ueda, N., Maekawa, M., Kobayashi, K., Kojima, M., Nagato, Y., et al. (2007). Direct control of shoot meristem activity by a cytokinin-activating enzyme. *Nature* 445, 652–655. doi:10.1038/nature05504

Lau, O. S., Ng, D. W., Chan, W. W., Chang, S. P., and Sun, S. S. (2010). Production of the 42-kDa fragment of *Plasmodium falciparum* merozoite surface protein 1, a leading malaria vaccine antigen, in *Arabidopsis thaliana* seeds. *Plant Biotechnol. J.* 8, 994–1004. doi:10.1111/j.1467-7652.2010.00526.x

Lee, K. H., Piao, H. L., Kim, H.-Y., Choi, S. M., Jiang, F., Hartung, W., et al. (2006). Activation of glucosidase via stress-induced polymerization rapidly increases active pools of abscisic acid. *Cell* 126, 1109–1120. doi:10.1016/j.cell.2006.07.034

Lichtenthaler, H., and Wellburn, A. (1983). Determination of total carotenoids and chlorophylls a and b of leaf extracts in different solvents. *Biochem. Soc. Trans.* 11, 591–592. doi:10.1042/bst0110591

Marty, F. (1999). Plant vacuoles. *Plant Cell* 11, 587–599. doi:10.1105/tpc.11.4.587

Matsuoka, K., and Neuhaus, J.-M. (1999). Cis-elements of protein transport to the plant vacuoles. *J. Exp. Bot.* 50, 165–174. doi:10.1093/jxb/50.331.165

Matzke, M. A., Mette, M. F., and Matzke, A. J. M. (2000). Transgene silencing by the host genome defense: implications for the evolution of epigenetic control mechanisms in plants and vertebrates. *Plant Mol. Biol.* 43, 401–415. doi:10.1023/A:1006484806925

Mettler, I. J., and Leonard, R. T. (1979). Isolation and partial characterization of vacuoles from tobacco protoplasts. *Plant Physiol.* 64, 1114–1120. doi:10.1104/pp.64.6.1114

Murashige, T., and Skoog, F. (1962). A revised medium for rapid growth and bio assays with tobacco tissue cultures. *Physiol. Plant.* 15, 473–497. doi:10.1111/j.1399-3054.1962.tb08052.x

Murray, J. A. H., Jones, A., Godin, C., and Traas, J. (2012). Systems analysis of shoot apical meristem growth and development: integrating hormonal and mechanical signaling. *Plant Cell* 24, 3907–3919. doi:10.1105/tpc.112.102194

Nausch, H., Mikschofsky, H., Koslowski, R., Meyer, U., Broer, I., and Huckauf, J. (2012a). High-level transient expression of ER-targeted human interleukin 6 in *Nicotiana benthamiana*. *PLoS ONE* 7:e48938. doi:10.1371/journal.pone.0048938

Nausch, H., Mischofsky, H., Koslowski, R., Meyer, U., Broer, I., and Huckauf, J. (2012b). Expression and subcellular targeting of human complement factor C5a in *Nicotiana* species. *PLoS ONE* 7:e53023. doi:10.1371/journal.pone.0053023

Oliver, S. N., Dennis, E. S., and Dolferus, R. (2007). ABA regulates apoplastic sugar transport and is a potential signal for cold-induced pollen sterility in rice. *Plant Cell Physiol.* 48, 1319–1330. doi:10.1093/pcp/pcm100

Overvoorde, P., Fukaki, H., and Beeckman, T. (2010). Auxin control of root development. *Cold Spring Harb. Perspect. Biol.* 2, a001537. doi:10.1101/cshperspect.a001537

Piotrowska, A., and Bajguz, A. (2011). Conjugates of abscisic acid, brassinosteroids, ethylene, gibberellins, and jasmonates. *Phytochemistry* 72, 2097–2112. doi:10.1016/j.phytochem.2011.08.012

Powell, A. E., and Lenhard, M. (2012). Control of organ size in plants. *Curr. Biol.* 22, R360–R367. doi:10.1016/j.cub.2012.02.010

Robert, S., Zouhar, J., Carter, C., and Raikhel, N. (2007). Isolation of intact vacuoles from *Arabidopsis rosette* leaf-derived protoplasts. *Nat. Protoc.* 2, 259–262. doi:10.1038/nprot.2007.26

Schliemann, W. (1984). Hydrolysis of conjugated gibberellins by β-glucosidases from dwarf rice (Oryza sativa L. cv. «Tan-ginbozu»). *J. Plant Physiol.* 116, 123–132. doi:10.1016/S0176-1617(84)80069-3

Sembdner, G., Atzorn, R., and Schneider, G. (1994). Plant hormone conjugation. *Plant Mol. Biol.* 26, 1459–1481. doi:10.1007/BF00016485

Spena, A., Estruch, J. J., Prinsen, E., Nacken, W., Van Onckelen, H., and Sommer, H. (1992). Anther-specific expression of the rolB gene of *Agrobacterium rhizogenes* increases IAA content in anthers and alters anther development and whole flower growth. *Theor. Appl. Genet.* 84, 520–527. doi:10.1007/BF00224147

Staswick, P. (2009). Plant hormone conjugation: a signal decision. *Plant Signal. Behav.* 4, 757–759. doi:10.1104/pp.109.138529

Vitale, A., and Hinz, G. (2005). Sorting of proteins to storage vacuoles: how many mechanisms? *Trends Plant Sci.* 10, 316–323. doi:10.1016/j.tplants.2005.05.001

Wang, P., Liu, H., Hua, H., Wang, L., and Song, C.-P. (2011). A vacuole localized β-glucosidase contributes to drought tolerance in *Arabidopsis*. *Chin. Sci. Bull.* 56, 3538–3546. doi:10.1007/s11434-011-4802-7

Werner, T., Motyka, V., Laucou, V., Smets, R., Van Onckelen, H., and Schmülling, T. (2003). Cytokinin-deficient transgenic *Arabidopsis* plants show multiple developmental alterations indicating opposite functions of cytokinins in the regulation of shoot and root meristem activity. *Plant Cell* 15, 2532–2550. doi: 10.1105/tpc.014928

Werner, T., and Schmülling, T. (2009). Cytokinin action in plant development. *Curr. Opin. Plant Biol.* 12, 527–538. doi:10.1016/j.pbi.2009.07.002

Williams, L., and Fletcher, J. C. (2005). Stem cell regulation in the *Arabidopsis* shoot apical meristem. *Curr. Opin. Plant Biol.* 8, 582–586. doi:10.1016/j.pbi.2005.09.010

Xu, Z. Y., Lee, K. H., Dong, T., Jeong, J. C., Jin, J. B., Kanno, Y., et al. (2012). A vacuolar beta-glucosidase homolog that possesses glucose-conjugated abscisic acid hydrolyzing activity plays an important role in osmotic stress responses in *Arabidopsis*. *Plant Cell* 24, 2184–2199. doi:10.1105/tpc.112.095935

Yadav, R. K., Tavakkoli, M., and Reddy, G. V. (2010). WUSCHEL mediates stem cell homeostasis by regulating stem cell number and patterns of cell division and differentiation of stem cell progenitors. *Development* 137, 3581–3589. doi: 10.1242/dev.054973

Yang, J., Worley, E., and Udvardi, M. (2014). A NAP-AAO3 regulatory module promotes chlorophyll degradation via ABA biosynthesis in *Arabidopsis* leaves. *Plant Cell* 26, 4862–4874. doi:10.1105/tpc.114.133769

Yao, J., Huot, B., Foune, C., Doddapaneni, H., and Enyedi, A. (2007). Expression of a β-glucosidase gene results in increased accumulation of salicylic acid in transgenic *Nicotiana tabacum* cv. Xanthi-nc NN genotype. *Plant Cell Rep.* 26, 291–301. doi:10.1007/s00299-006-0212-8

Zhao, Z., Andersen, S. U., Ljung, K., Dolezal, K., Miotk, A., Schultheiss, S. J., et al. (2010). Hormonal control of the shoot stem-cell niche. *Nature* 465, 1089–1092. doi:10.1038/nature09126

Conflict of Interest Statement: The authors declare that the research was conducted in the absence of any commercial or financial relationships that could be construed as a potential conflict of interest.

12

Comparison of Chemical Composition and Energy Property of Torrefied Switchgrass and Corn Stover

Jaya Shankar Tumuluru*

Idaho National Laboratory, Idaho Falls, ID, USA

In the present study, 6-mm ground corn stover and switchgrass were torrefied in temperatures ranging from 180 to 270°C for 15- to 120-min residence time. Thermogravimetric analyzer was used to do the torrefaction studies. At a torrefaction temperature of 270°C and a 30-min residence time, the weight loss increased to >45%. At 180°C and 120 min, there was about 56 and 73% of moisture loss in the corn stover and switchgrass; further increasing the temperature to 270°C and 120 min resulted in about 78.8–88.18% moisture loss in both the feedstock. Additionally, at these temperatures, there was a significant decrease in the volatile content and increase in the fixed carbon content, and the ash content for both the biomasses tested. The ultimate composition like carbon content increased and hydrogen content decreased with increase in the torrefaction temperature and time. At 270°C and 120-min residence time, the carbon content observed was 56.63 and 58.04% and hydrogen content observed was 2.74 and 3.14%. Nitrogen and sulfur content measured at 270°C and 120 min were 0.98, 0.8, 0.076, and 0.07% for both the corn stover and switchgrass. The hydrogen/carbon and oxygen/carbon ratios calculated decreased to the lowest values of 0.59 and 0.64, and 0.71 and 0.76 for both biomasses. The van Krevelen diagram drawn for corn stover and switchgrass torrefied at 270°C indicated that H/C and O/C values are closer to coals like Illinois Basis and Powder River Basin. In the present study, the maximum higher heating value that was observed by corn stover and switchgrass was 21.51 and 21.53 MJ/kg at 270°C and a 120-min residence time. From these results, it can be concluded that corn stover and switchgrass, after torrefaction, shows consistent proximate, ultimate, and energy properties.

Keywords: corn stover, switchgrass, torrefaction temperature and time, chemical composition, energy property, mathematical model

Edited by:
Cherng-Yuan Lin,
National Taiwan Ocean University,
Taiwan

Reviewed by:
Shanmugaprakash Muthusamy,
Kumaraguru College of Technology,
India
Jaime Puna,
Instituto Superior Engenharia Lisboa,
Portugal

*Correspondence:
Jaya Shankar Tumuluru
jayashankar.tumuluru@inl.gov

INTRODUCTION

There is growing concern to reduce the use of fossil fuels due to the greenhouse gas emissions, which have a direct impact on the global warming temperatures. This has led researchers to explore alternative renewable energy sources. Biomass is considered a potential resource as it is considered carbon neutral because it is still part of the carbon cycle. According to the Bioenergy Technology office at the U.S. Department of Energy (DOE), biomass feedstock is defined as any renewable biological material that can be used directly as a fuel or converted to another form of fuel, such as ethanol, butanol, biodiesel, and other hydrocarbon fuels. Some examples of the biomass feedstocks that are widely used for bioenergy applications are corn starch, sugarcane juice, and crop residues (e.g., corn stover, sugarcane bagasse, purpose-grown grass crops, and woody plants). Corn stover is the largest quantity of biomass residue in the United States, and 120 million tons are available for biofuels production (U.S. Department of Energy, 2011). Corn stover is the leaf, husk, and cob remaining in the field after the harvest of cereal grain. The residue is the stalk of the leaves of maize (*Zea mays ssp. mays L.*) plants left in a field after harvest. Stover makes up about half of the yield of a crop and is similar to straw. Switchgrass (*Panicum virgatum*) is a warm-season perennial grass native to North America. It is a dedicated perennial crop and is considered a viable energy crop that could significantly increase the amount of biomass available for conversion to biofuel (Simmons et al., 2008). U.S. DOE in the 1990s selected switchgrass as a potential bioenergy feedstock due to adaptation to different growing conditions (Newman et al., 2014), and is successfully used for biopower application.

According to Nhuchhen et al. (2014), the lignocellulosic biomass, in spite of all its positive attributes, has different short-comings like structural heterogeneity, non-uniform physical properties, low-energy density, hygroscopic nature, and low bulk density. These challenges in terms of physical (lower mass density, high-moisture content, irregular size and shape, and hydrophilic in nature), chemical (low carbon and high hydrogen, oxygen, and high volatiles), and energy properties [high hydrogen/carbon (H/C) and oxygen/carbon (O/C), and lower heating values] limit the use of woody, herbaceous, and agricultural straws and other biomasses for energy application. These limitations create difficulties in transportation, handling, storage, and conversion processes (Arias et al., 2008; Medic et al., 2011; Phanphanich and Mani, 2011; Uemura et al., 2011; Wannapeera et al., 2011). Tumuluru et al. (2012) in their review indicated that raw biomass physical properties and chemical composition does not make them suitable for co-firing higher percentages with coal. These authors stated that boiler inefficiency, due to higher moisture and volatiles and lower energy content of the biomass fuels as compared to coal, is a major limitation to cofire higher percentages with coal.

To overcome the biomass challenges in terms of physical, chemical, and energy properties, a torrefaction process was developed. Torrefaction is a thermal pretreatment method where the biomass is thermally pretreated in a temperature range of 200–300°C for about 30 min in absence of air at atmospheric pressure (Tumuluru et al., 2011). Torrefaction makes biomass (a) brittle product making it easier to grind (better particle size and shape), (b) changes the chemical composition (removing the moisture and low-energy content of volatiles), and (c) increases the net energy content of the biomass. According to Lu and Chen (2014), biomass torrefaction helps to retain most of its energy and simultaneously loses its hygroscopic properties. Sarkar et al. (2014) and Yang et al. (2014) have successfully used torrefied switchgrass for gasification and pyrolysis application. Nhuchhen et al. (2014), in their review on torrefaction of biomass, suggested that the process makes biomass suitable for co-firing applications and can be promoted as an alternative to charcoal. They also stated that the torrefaction increases its hydrophobicity, grindability, and energy density making it more suitable for thermochemical applications like gasification and pyrolysis.

The common biomass reactions during torrefaction are dehydration, devolatilization, depolymerization, and carbonization. During initial heating at a temperature about 180°C, the dehydration reactions remove most of the moisture in the biomass. Furthermore, heating biomass in the temperature range of 180–270°C promotes the reactions like depolymerization, devolatilization, and carbonization of hemicellulose, cellulose, and lignin. For torrefaction, process temperatures over 270°C can lead to extensive devolatilization of the biomass due to the initiation of the pyrolysis process (Tumuluru et al., 2011). Also during these reactions, biomass loses some of the lipophilic compounds that make it hydrophobic.

The torrefaction process is influenced by process parameters like temperature, heating rate, absence of oxygen, residence time, ambient pressure, and the properties of feedstock like moisture content and particle size. Torrefaction temperature is typically between 200 and 300°C and the residence time is adjusted to produce a brittle, hydrophobic, and high-calorific value product. Typically, biomass is pre-dried to <10% moisture content prior to torrefaction. The particle size of the biomass will influence the reaction mechanisms, kinetics, duration, and its specific heating rate of the process. Several researchers have worked on the torrefaction of agricultural and woody biomass using a thermogravimetric analyzer (TGA) (Bridgeman et al., 2008; Chen and Kuo, 2010; Repellin et al., 2010). These researchers have successfully used TGA to estimate the effects of particle size, temperature, and moisture to better understand the torrefaction kinetics, chemical, proximate, and energy properties of both woody and herbaceous biomass.

The focus of this work is to understand the effect of the torrefaction time and temperature on proximate and ultimate composition and energy property of corn stover and switchgrass using a TGA. The specific objectives for this research is to (a) understand the effect of torrefaction temperature (180–270°C) and its residence time (15–120 min) on the weight loss, proximate composition (moisture, ash, volatiles, and fixed carbon), ultimate composition (hydrogen, carbon, sulfur, nitrogen, and oxygen), and energy property (higher heating value), (b) understand the significance of torrefaction temperature and time on weight loss, chemical composition and higher heating value, and (c) develop mathematical models for weight loss, proximate and ultimate composition, and higher heating value experimental data.

MATERIALS AND METHODS

Corn stover and switchgrass samples were used in the present torrefaction experiments. These biomass samples were harvested from farms in Iowa and Nebraska in the form of bales. It was initially ground to bigger particle sizes using a 50.8-mm screen with a Vermeer HG200 grinder (Vermeer Corporation–Agriculture, Pella, IA, USA). The ground material was evaluated for moisture content and stored in sealed plastic containers that were maintained at about 4°C until it was further size-reduced to 6 mm.

Thermogravimetric Analyzer

Torrefaction was performed on the LECO TGA701 (see **Figure 1**) in a batch procedure. Experiments were conducted in a temperature range of 180–270°C and a residence time of 15–120 min.

Biomass sample preparation includes grinding corn stover and switchgrass to 6 mm using a Retsch splitter. These samples were double-bagged and stored in air-tight containers and used for torrefaction studies. A method file was also developed to carry out the torrefaction experiments using TGA (Bridgeman et al., 2008). This method file includes the steps for drying and torrefaction, and its associated cooling steps.

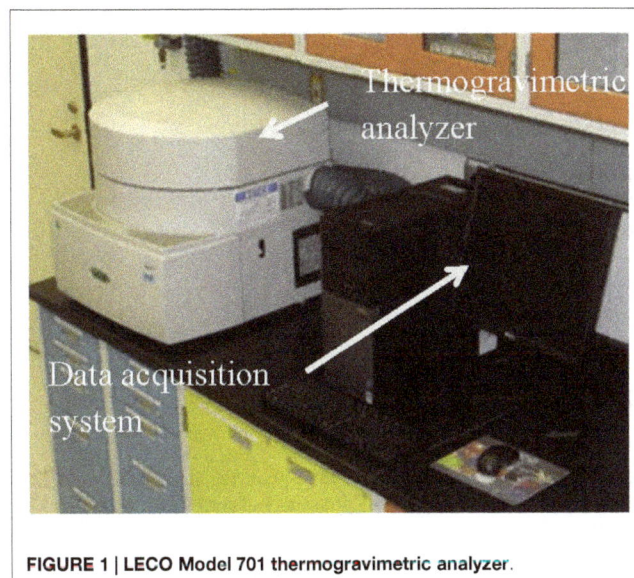

FIGURE 1 | LECO Model 701 thermogravimetric analyzer.

TABLE 1 | Methods followed for measurement of chemical and energy properties of corn stover and switchgrass biomass samples.

	Chemical composition	
	Proximate	
1	Moisture	ASTM International D3173 (2008)
2	Ash	ASTM International D3174 (2002)
3	Volatiles	ASTM International D3175 (2007)
4	Fixed carbon	Fixed carbon by difference method
	Ultimate composition	
1	Carbon	ASTM International D3178 (2002)
2	Hydrogen	ASTM International D3178 (2002)
3	Nitrogen	ASTM International D3179 (2002)
4	Sulfur	ASTM International D3177 (2002)
5	Oxygen	Oxygen by difference method
6	H/C ratio	H/C: Number of hydrogen atoms/number of Carbon atoms = (%H/1)/(%C/12)
7	O/C ratio	O/C: Number of oxygen atoms/number of Carbon atoms = (%O/8)/(%C/12)
8	HHV (higher heating value)	ASTM International D5865 (2010)

For the present study, the heating rate selected was 10°C/min. LECO instrumentation contains an easy-to-follow menu, driven by Microsoft™ Windows®-based software, which allows its analysis method. Temperature, temperature ramp rate, and atmosphere are selectable at each step. Analysis methods can also be entered to evaluate the moisture, volatiles, and ash content. American Society for Testing and Materials (ASTM) standard methods for estimating the chemical composition was used to measure the proximate and ultimate composition of the torrefied biomass (**Table 1**). **Table 2** shows the experimental design used for conducting the torrefaction tests.

Data Analysis

The experimental data on weight loss, proximate and ultimate composition, and higher heating value obtained at different torrefaction temperatures (180°, 230°, and 270°C) and different residence times (15, 30, 60, and 120 min) for both corn stover and switchgrass was used to draw the bar plots, develop multiple regression equations (Eq. 1), and understand the significance of the torrefaction process variables with respect to torrefied material properties studied. Statistica (Version 9) statistical software was used to develop the multiple regression models and analysis of variance (ANOVA) for the experimental data. Coefficient of determination was used to understand the model fit and the ANOVA was used to understand the significance of the process variables with respect to the weight loss, proximate, ultimate composition, and higher heating value.

$$y = b_0 + b_1 x_1 + b_2 x_2 \qquad (1)$$

where b_0, b_1, and b_2 are equation constants; x_1 and x_2 are torrefaction temperature (°C) and torrefaction residence time (min), respectively; and y is dependent variable (weight loss, proximate, ultimate composition, and higher heating value).

RESULTS AND DISCUSSION

The proximate and ultimate composition of raw corn stover and switchgrass are indicated in **Table 3**. Ash content of corn stover is slightly higher compared to switchgrass, and volatiles are slightly higher in corn stover when compared to switchgrass. The carbon content of the corn stover and switchgrass were in the range of 42–43%. The heating values of both feedstocks were close. The oxygen content of the biomass feedstocks are in the range of 40–41%. The H/C and O/C ratios calculated are slightly higher for corn stover (1.64 and 1.38) compared to switchgrass (1.52 and 1.45). **Table 4** indicates the ultimate composition and H/C and O/C ratios of Appalachian, Powder River Basin, Illinois Basin, and North Dakota lignite coals. The Appalachian coal has the highest carbon and lowest oxygen (66.93 and 7.55), and North Dakota lignite has the lowest carbon and highest oxygen values (31.8 and 26.35). The H/C and O/C ratios calculated for

TABLE 2 | Experimental design for torrefaction experiments using TGA.

Process	Temperatures (°C)	Residence time (min)	Particle size (mm)	Heating rate (°C/min)	Feedstocks
Torrrefaction	180, 230, and 270	15, 30, 60, and 120	6	10	Corn stover and Switchgrass

TABLE 3 | Proximate, ultimate, and energy property of raw corn stover and switchgrass.

Chemical composition		Feedstock	
Proximate		**Corn stover**	**Switchgrass**
1	Moisture	4.01	6.01
2	Ash	5.13	4.01
3	Volatiles	75.63	73.32
4	Fixed carbon	15.23	16.66
Ultimate composition and higher heating value			
1	Carbon	43.92	42.08
2	Hydrogen	6.01	5.44
3	Nitrogen	0.42	0.36
4	Sulfur	0.07	0.05
5	Oxygen	40.44	41.38
6	H/C	1.64	1.52
7	O/C	1.38	1.45
8	HHV (higher heating value)	17.31	17.36

TABLE 4 | Ultimate composition and H/C and O/C ratio of different coals (source: Tillman et al., 2009; Tumuluru et al., 2012).

Coals	Hydrogen	Carbon	Oxygen	H/C	O/C
Coal Appalachian	4.43	66.93	7.55	0.79	0.17
Illinois Basin	4.77	60.68	13.61	0.94	0.33
Powder River Basin	3.55	51.89	12.77	0.82	0.37
North Dakota Lignite	4.51	31.8	26.35	1.70	1.24

the coals indicated that Appalachian coals has the lowest values of 0.79 and 0.17, and North Dakota lignite has the highest values of 1.7 and 1.34.

Analysis of Variance and Mathematical Models

The experimental data was analyzed further to understand the significance of the process variables, and to develop multiple regression models for weight loss, proximate, ultimate composition, and energy data. **Table 5** indicates the significance of the process variables, and **Table 6** shows the multiple regression equations fitted for the experimental data. For the weight loss for both corn stover and switchgrass, the torrefaction temperature was found to be significant at $P < 0.001$ and the torrefaction residence time at $P < 0.05$. In the case of moisture content, volatiles, and fixed carbon, torrefaction temperature was found to be significant at $P < 0.001$ for both switchgrass and corn stover. Torrefaction residence time was found to be significant for switchgrass at $P < 0.05$ for moisture content, volatiles, and fixed carbon at $P < 0.01$; whereas for corn stover, it was found to be non-significant. In case of ash content, torrefaction temperature was found to be significant at $P < 0.001$, and residence time was found to be non-significant for both corn stover and switchgrass. Ultimate composition, hydrogen, oxygen, and carbon content were influenced by the torrefaction

TABLE 5 | Significance of the torrefaction temperature and time with respect to proximate, ultimate, and energy property.

Proximate composition	Corn stover		Switchgrass	
	Process variables		**Process variables**	
	Torrefaction temperature (x_1)	**Torrefaction time (x_2)**	**Torrefaction temperature (x_1)**	**Torrefaction time (x_2)**
Moisture content (%, w.b.)	(−)***	(−)*	(−)***	(−)*
Ash (%)	(+)***	ns	(+)*	ns
Volatiles (%)	(−)***	ns	(−)***	(−)*
Fixed carbon (%)	(+)***	ns	(+)***	(+)**
Ultimate composition and higher heating value				
Hydrogen (%)	(−)***	(−)*	(−)***	(−)*
Carbon (%)	(+)***	ns	(+)***	ns
Oxygen (%)	(−)***	ns	(−)***	ns
H/C ratio	(−)***	(−)*	(−)***	ns
O/C ratio	(−)***	ns	(−)***	ns
Higher heating value (MJ/kg)	(+)***	ns	(+)***	ns
Weight loss (%)	(+)***	(+)*	(+)***	(+)*

*$p < 0.05$; **$p < 0.01$; ***$p < 0.001$.*

TABLE 6 | Multiple regression equations for weight loss, proximate and ultimate composition, and energy properties of corn stover and switchgrass.

	Multiple regression equation	(R^2)
Corn stover		
Proximate composition		
Moisture content (%, w.b.)	$y = 4.45 − 0.011x_1 − 0.0057x_2$	0.81
Ash (%)	$y = −6.12 + 0.060x_1 + 0.013x_2$	0.75
Volatiles (%)	$y = 138.22 − 0.32x_1 − 0.057x_2$	0.88
Fixed carbon (%)	$y = 36.55 + 0.28x_1 + 0.050x_2$	0.91
Ultimate composition, weight loss and higher heating value		
Hydrogen (%)	$y = 10.89 − 0.025x_1 − 0.0073x_2$	0.92
Carbon (%)	$y = 24.76 + 0.108x_1 + 0.0098x_2$	0.90
Oxygen (%)	$y = 66.41 − 0.133x_1 − 0.014x_2$	0.83
H/C ratio	$y = 3.19 − 0.0084x_1 − 0.0018x_2$	0.95
O/C ratio	$y = 2.516 − 0.0062x_1 − 0.00052x_2$	0.89
Higher heating value (MJ/kg)	$y = 12.55 + 0.030x_1 + 0.0051x_2$	0.89
Weight loss (%)	$y = −82.10 + 0.474x_1 + 0.076x_2$	0.94
Switchgrass		
Proximate composition		
Moisture content (%, w.b.)	$y = − 4.18 − 0.0107x_1 − 0.0047x_2$	0.76
Ash (%)	$y = − 1.01 + 0.024x_1 + 0.0151x_2$	0.62
Volatiles (%)	$y = 111.30 − 0.20x_1 − 0.070x_2$	0.90
Fixed carbon (%)	$y = −14.47 + 0.187x_1 + 0.060x_2$	0.91
Ultimate composition, weight loss and higher heating value		
Hydrogen (%)	$y = 8.33 − 0.0138x_1 − 0.00636x_2$	0.75
Carbon (%)	$y = 27.12 + 0.099x_1 + 0.017x_2$	0.85
Oxygen (%)	$y = 61.15 − 0.100x_1 − 0.019x_2$	0.85
H/C ratio	$y = 2.48 − 0.0053x_1 − 0.0017x_2$	0.83
O/C ratio	$y = 2.31 − 0.0050x_1 − 0.00086x_2$	0.88
Higher heating value (MJ/kg)	$y = 12.41 + 0.031x_1 + 0.0046x_2$	0.87
Weight loss (%)	$y = −73.26 + 0.41x_1 + 0.090x_2$	0.88

x_1 = torrefaction temperature (degree Celsius) and x_2 = torrefaction time (min).

temperature at $P < 0.001$ for both corn stover and switchgrass. Torrefaction residence time was found to be significant only for hydrogen content at $P < 0.05$; whereas for carbon, it was found to be non-significant. For the H/C ratio and O/C ratio, the torrefaction temperature was found to be significant at $P < 0.001$; however, the residence time was significant only for corn stover at $P < 0.05$. In the case of the higher heating value, only the torrefaction temperature was found to be statistically significant at $P < 0.05$. The regression equation developed for the experimental data has adequately fitted based on the coefficient of determination values. Also, all the equations were found to be statistically significant at $P < 0.001$. These equations can help to predict the weight loss, proximate and ultimate composition, higher heating values, and H/C and O/C ratios of corn stover and switchgrass at different torrefaction temperatures and residence times studied in this paper.

Weight Loss

The weight loss for both corn stover and switchgrass at different torrefaction temperatures and residence times are indicated in **Figure 2**. At 180°C, and 15–120 min residence time, the weight loss was in the range of 8.1–9.3% for switchgrass and 9.1–10.04% for corn stover. Increasing the torrefaction temperature and residence time increased the weight loss in both corn stover and switchgrass. The maximum weight loss observed for corn stover at this temperature was about 33% and about 26% for switchgrass. At 270°C and 30 min, the weight loss observed for corn stover and switchgrass was 42.63 and 51.26%, and at 120 min the weight loss values increased to final values of 54.93–58.29%.

Proximate Composition

The loss of moisture was significant at all the torrefaction temperature and residence times. At lower temperature of 180°C and 30 min residence times, the loss of moisture was about 26% for corn stover and 54% for switchgrass compared to its original value. Furthermore, increasing the residence time to 120 min reduced the final moisture content in both corn stover and switchgrass to about 1.67 and 1.62% – a reduction of 56 and 73%

compared with their original value. The decrease in moisture content of the switchgrass and corn stover had increased with the increase in torrefaction temperature and residence time. At 270°C and 120-min residence time, a maximum moisture reduction of 78.8% for corn stover and 88.18% for switchgrass was observed (**Figure 3**). Tumuluru et al. (2011) in their review indicated that lower temperatures of <200°C contributes to the loss of moisture resulting from the dehydration reactions; whereas, at temperatures of >200°C, the loss of moisture is attributed to the complicated interactions of biomass components to temperature and residence time. Phanphanich and Mani (2011) in their studies indicated that torrefying the biomass at temperature >200°C results in improved grinding characteristics. The main reason for this behavior is due to loss of moisture and other low energy volatiles, which makes biomass brittle and easy to grind. **Figure 4** shows the photos of corn stover and switchgrass torrefied at 180°, 230°, and 270°C for 120-min residence time. It is very clear from the figure that the color changes significantly with the torrefaction temperature, which can be a good indicator for the degree of torrefaction. According to Tumuluru et al. (2011), a temperature regime of 150–200°C, also called the reactive drying range, initiates the breakage of hydrogen and carbon bonds and results in a structural deformity that does not result in a significant change in color of the biomass. The same authors indicated that at a temperature range of 200–300°C, also called as destructive drying, results in the disruption of most of the inter- and intramolecular hydrogen bonds and C–C and C–O bonds and emits condensable and non-condensable gases, which result in darkening of the biomass. At these temperatures, cell structure is completely destroyed as the biomass loses its fibrous nature, becomes brittle, and is easier to grind (Bergman and Kiel, 2005).

During torrefaction, the ash components in the biomass do not change. All ash components of biomass are still present in torrefied biomass. The change in the ash content is more relative to the change in the original biomass components. As the biomass loses some of the moisture and volatiles during the process, the ash content is more of a relative increase with respect to the original components. In the present study, the initial ash content

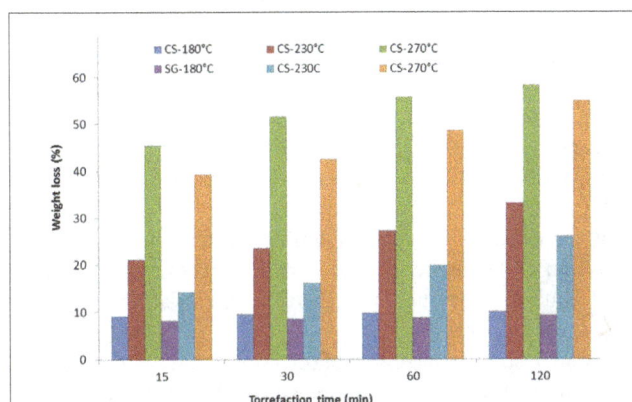

FIGURE 2 | Weight loss in corn stover and switchgrass at different torrefaction temperatures and times.

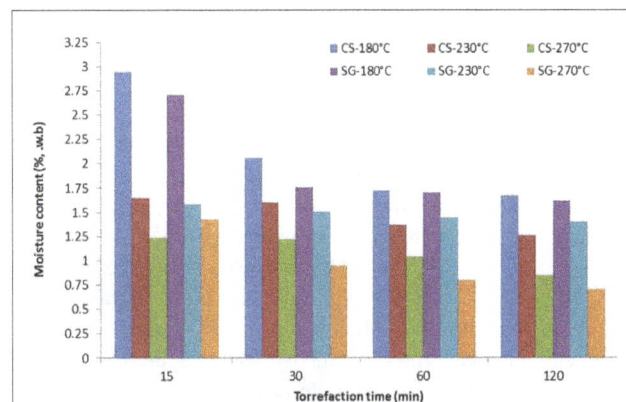

FIGURE 3 | Torrefaction temperature and time effect on moisture content of the switchgrass and corn stover.

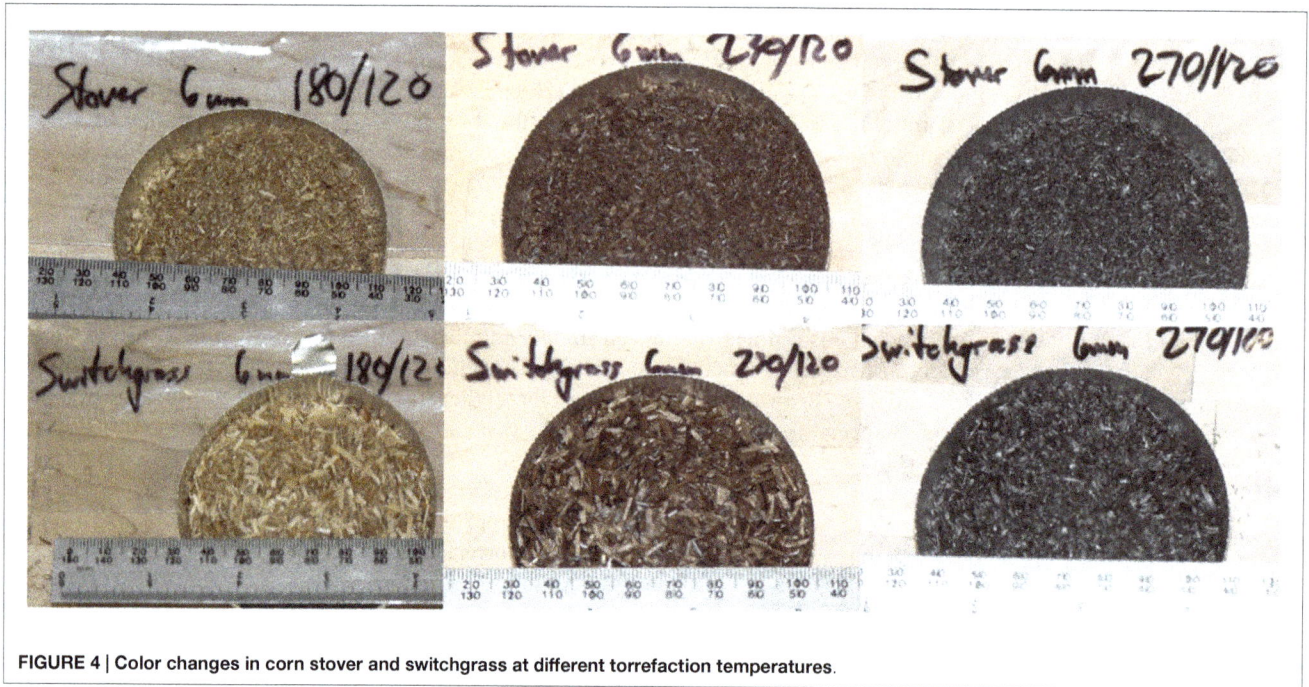

FIGURE 4 | Color changes in corn stover and switchgrass at different torrefaction temperatures.

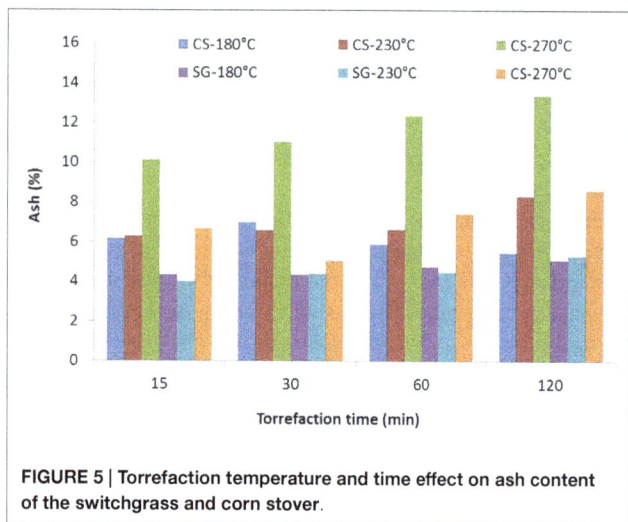

FIGURE 5 | Torrefaction temperature and time effect on ash content of the switchgrass and corn stover.

of corn stover and switchgrass was 5.13 and 4.01%. At 180°C and 15-min residence time, the percent increase in the ash content was about 20.07% for corn stover, and in the case of switchgrass, it increased by about 8.22%. Increasing the residence time to 30 min did increase the ash content to 6.98 and 4.35%. This increase is marginal from 15–30 min for both the feedstocks. **Figure 5** clearly indicates that the ash content in both the biomasses tested increased with the increasing of the torrefaction temperature and residence time. At 230° and 270°C torrefaction temperature and 15-min residence time, the ash content of both the biomasses tested increased. According to Poudel and Oh (2012) and Chen et al. (2014), during torrefaction the volatile content in the biomass decreases leading to an increase in ash content.

Initial volatile content of corn stover was 75.63% and the switchgrass was 73.32%. At 180°C and 15-min residence time, the volatile content decreased to 74.03% (a decrease of about 2% from the original value for corn stover). In the case of switchgrass, the decrease was from 73.32 to 71.39% (a decrease of about 2.6% from the original value). The loss of volatiles increased with the increase in torrefaction temperature and residence time. At 230°C and between 15- and 120-min residence time, the decrease was in the range of 61.23–69.42%; however, at 270°C the decrease in the volatile content was further decreased and was in the range of 50.27–58.78% (**Figure 6**). The study indicated that the volatile losses are higher at higher torrefaction temperature and residence time. At torrefaction temperature of 200–300°C, weight loss in the biomass is mainly due to loss of moisture and hemicellulose and lignin decomposition. Xylan-based hemicellulose generally decomposes around 250–280°C. Lignin decomposition proceeds more slowly, but will gradually increase starting at about 200°C. At these temperatures, the disruption of most inter- and intramolecular hydrogen bonds and C–C and C–O bonds will result in a formation of hydrophilic extractives, carboxylic acids, alcohols, aldehydes, ether, and gases like CO, CO_2, and CH_4 (Tumuluru et al., 2011). The common reactions at these temperatures are limited devolatilization and carbonization of the hemicellulose. At temperatures of >250°C, the hemicellulose decomposes extensively into volatiles and a char-like solid product.

The initial fixed carbon content of corn stover and switchgrass is 15.23 and 16.66%. At lower temperatures the change in the fixed carbon is not significant, where at higher temperature of 270°C the change in the fixed carbon is more significant. At 180°C torrefaction temperature and 15-min residence time, the change in the fixed carbon is marginal for corn stover (16.86%);

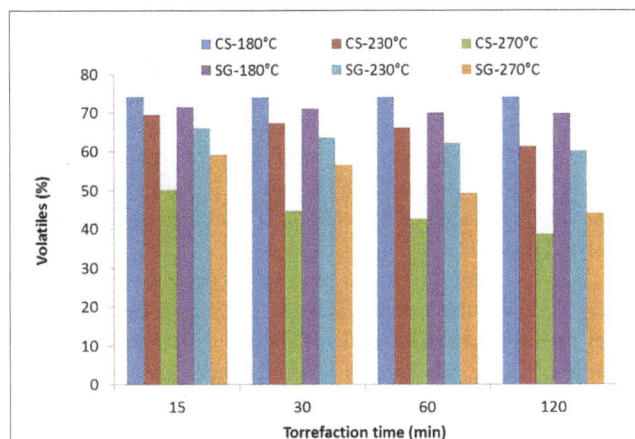

FIGURE 6 | Torrefaction temperature and time effect on volatile content of the switchgrass and corn stover.

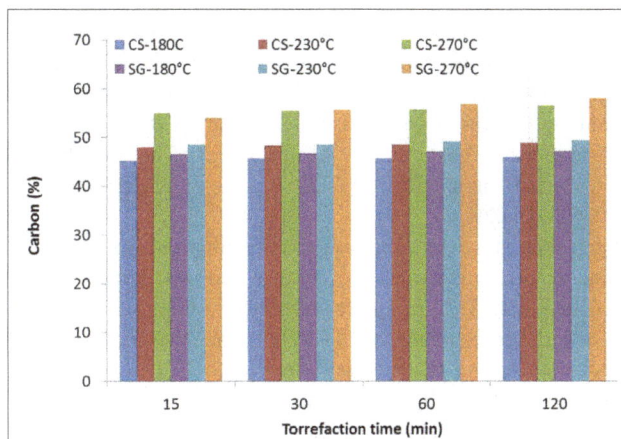

FIGURE 8 | Torrefaction temperature and time effect on elemental carbon of the switchgrass and corn stover.

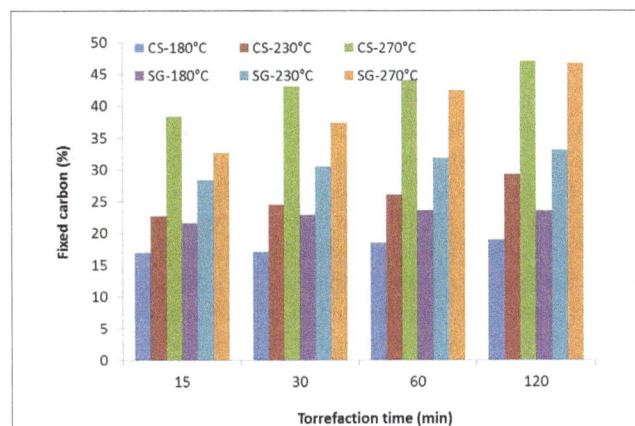

FIGURE 7 | Torrefaction temperature and time effect on fixed carbon of the switchgrass and corn stover.

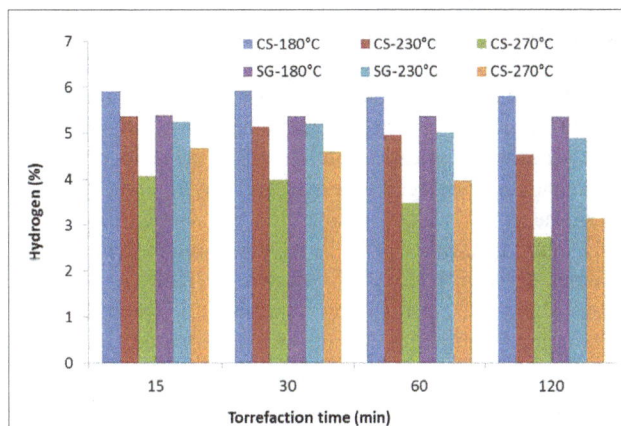

FIGURE 9 | Torrefaction temperature and time effect on hydrogen content of the switchgrass and corn stover.

however, for switchgrass the fixed carbon increased to 21.56%. Increasing the residence time to 30–120 min at 180°C torrefaction temperature increased the fixed carbon to 18.95% for corn stover, and for switchgrass the fixed carbon increased to 23.49% (**Figure 7**). At 230 and 270°C, the increase in fixed carbon content was significant compared with its original value. At 270°C and 15 min, the fixed carbon content of both the biomasses tested almost doubled (28.35 and 32.71%).

Ultimate Composition

The initial carbon content of corn stover and switchgrass is about 43.92 and 42.08%. The increased carbon content at 180°C at 15-min residence time is 45.26% for corn stover and 46.56% for switchgrass. Furthermore, increasing the residence time to 120 min increased the carbon content to a final value of 46.03% for corn stover and 47.28% for switchgrass (**Figure 8**). By increasing the torrefaction temperature to 230 and 270°C at 15-min residence time, the carbon content values observed from both corn stover and switchgrass were 47.92 and 48.53%, and 54.92

and 53.94%. Increasing the residence time further to 120 min also increased the carbon content, but marginally.

The initial hydrogen content of the corn stover and switchgrass is about 6.01 and 5.44%. The hydrogen content of corn stover and switchgrass at 180°C and 15-min residence time decreased to 5.9 and 5.39%. By further increasing the residence to 120 min, the decrease was marginal (5.8 and 5.35%). Increasing the torrefaction temperature to 230°C and 15 min, the hydrogen content of corn stover and switchgrass samples were found to be 5.14 and 5.20%, but increasing the residence time to 120 min decreased the hydrogen content of the samples to 4.53 and 4.89% (**Figure 9**). At 270°C and 15 min, the hydrogen content of corn stover and switchgrass were found to be 4.07 and 4.68%; however, at 30, 60, and 120 min, the hydrogen content of the samples observed was about 3.99 and 4.6; 3.48 and 3.97; and 2.74 and 3.14% for corn stover and switchgrass.

The initial nitrogen content of the corn stover and switchgrass was 0.42 and 0.36%. At 180°C and 120 min, the nitrogen content increased to 0.50% for corn stover; however, for switchgrass, the nitrogen content increased to 0.38%. Increasing the torrefaction temperature to 270°C and 120 min, the nitrogen content observed in the case of corn stover was 0.98%, while in the case of switchgrass it was about 0.8%. The initial sulfur content of corn stover and switchgrass was found to be 0.07 and 0.05%. There is not a significant change in the sulfur content of both of the biomasses tested, but at 270°C and 120 min the observed values were 0.076 and 0.07%.

Oxygen content is determined based on the difference method. The initial oxygen content of the corn stover and switchgrass was about 40.44 and 41.38%. At a lower torrefaction temperature of 180°C and smaller residence time of 15 min, the oxygen content of the corn stover and switchgrass decreased to 39.99 and 41% (**Figure 10**). Additionally, increasing the residence time to 120 min did not bring much change in the oxygen content of the samples. At 230°C and 15-min residence time, the observed oxygen content of the corn stover and switchgrass samples was 38.53 and 39.53%, which is also marginal. Torrefying the switchgrass and corn stover at higher temperatures of 270°C did have an impact on the oxygen content of both corn stover and switchgrass. At 270°C and 15 min residence time, the observed oxygen content of the corn stover and switchgrass samples was 29.67 and 33.29%. Furthermore, increasing the residence time to 120 min still reduced the oxygen content of the samples to 26.41 and 29.5%. The results indicated that the torrefaction temperature had a more-significant effect on the oxygen content compared to residence time.

The H/C and O/C ratios of the corn stover and switchgrass raw samples are calculated and indicated in **Table 1**. Corn stover has higher H/C ratio when compared to switchgrass, and switchgrass has higher O/C ratio when compared to corn stover (H/C and O/C ratios of corn stover showed 1.64 and 1.52 and switchgrass showed 1.38 and 1.45). At a lower torrefaction temperature of 180°C and 15 min, the observed H/C ratio for corn stover was 1.56 and switchgrass was 1.38. Increasing the torrefaction residence time to 120 min reduced the H/C ratio marginally (corn stover 1.51 and switchgrass 1.35). Increasing the torrefaction temperature to 230°C at 15 min residence

time, reduced the H/C ratio to 1.27 for corn stover and 1.28 for switchgrass (**Figure 11**). Further increasing the residence time to 120 min at the same torrefaction temperature of 230°C, the H/C ratio values observed were 1.11 for corn stover and 1.18 for switchgrass. At 270°C torrefaction temperature and 15 min residence time, the observed H/C ratio value for corn stover was 0.88 and switchgrass was 1.04, though at 120 min residence time H/C values reduced to 0.59 and 0.64 for corn stover and switchgrass (**Figure 11**).

With the O/C ratio, a similar trend was observed, where increasing the torrefaction temperature and residence decreased the values for both corn stover and switchgrass. At a torrefaction temperature of 180°C and 15-min residence time, the observed values were 1.32 for both corn stover and switchgrass. Increasing the residence time to 120 min did change the values marginally (1.31 corn stover and 1.29 switchgrass). At the other torrefaction temperatures of 230 and 270°C and a 15-min residence time, the observed O/C ratio values for corn stover and switchgrass observed were 1.20, 1.22 and 0.81, 0.92. At 120 min residence time at 230 and 270°C, O/C values reduced to 1.13, 1.16 and 0.71, 0.76 (**Figure 12**). It is clear from the data that both torrefaction temperature and residence time had significant effect on the H/C ratio and O/C ratios.

The van Krevelen diagram, which is drawn for O/C and H/C ratio, was drawn for corn stover and switchgrass and was compared to different grades of the coal (**Figure 13**). The lower ratio of H/C to O/C in coal is mainly due to lower oxygen content and higher carbon content when compared to biomass. Torrefaction of switchgrass and corn stover helped to lower the oxygen and hydrogen content of the biomass, and made it comparable to different forms of coal. It is clear from this diagram that lower H/C ratio and O/C ratios can be produced at higher torrefaction temperatures and residence times. Torrefying both switchgrass and corn stover at 230°C at different residence times resulted in H/C and O/C ratio closer to North Dakota lignite coal. Additionally, increasing the temperature to 270°C has moved the torrefied switchgrass and corn stover closer to other coal forms like Illinois Basin and Powder River Basin.

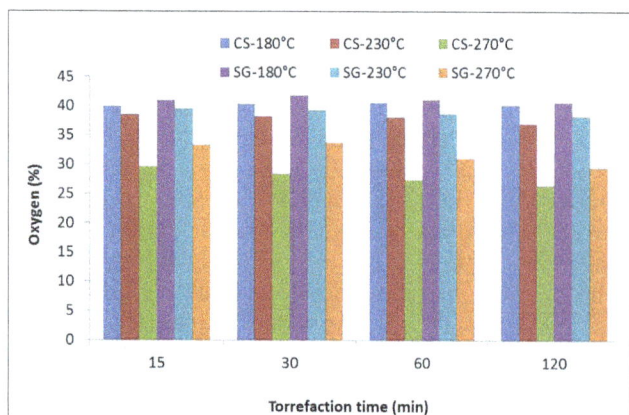

FIGURE 10 | Torrefaction temperature and time effect on oxygen content of the switchgrass and corn stover.

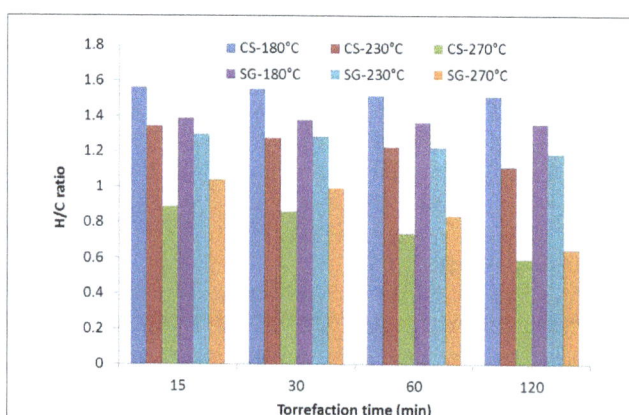

FIGURE 11 | Torrefaction temperature and time effect on H/C ratio of the switchgrass and corn stover.

Higher Heating Value

The calorific value was measured for both raw and torrefied corn stover and switchgrass samples. The higher heating values of raw corn stover and switchgrass observed were 17.31 and 17.36 MJ/kg. At a torrefaction temperature of 180°C and 15-min residence times, the observed higher heating values were 18.43 and 18.06 MJ/kg (**Figure 13**). Increasing the residence to 120 min increased the higher heating values marginally. At 230°C torrefaction temperature and 15-min residence time, the higher heating values observed were 18.77 for corn stover and 18.75 MJ/kg for switchgrass. At 120-min residence time, the higher heating value of corn stover was 19.74 and switchgrass was about 19.34 MJ/kg. At a torrefaction temperature of 270°C and 15-min residence time, the higher heating values observed 20.55 for corn stover and 21.0 MJ/kg for switchgrass. At the same temperature, increasing

the residence time to 120 min increased the higher heating values to 21.51 for corn stover and 21.53 MJ/kg for switchgrass (**Figure 14**).

DISCUSSION

In the present study, both the switchgrass and corn stover lost moisture at different torrefaction temperatures and residence times tested. Tumuluru et al. (2011) and other researchers indicated that at temperatures of 150–200°C the loss of moisture is due to dehydration reactions. The dehydration reactions mainly result in loss of unbound moisture. According to Bridgeman et al. (2008) and other researchers, increasing the torrefaction temperature to 230 and 270°C, the loss of moisture can be attributed to drying and depolymerization of hemicellulose. The loss of volatiles at the temperature of 180°C is marginal, also at this temperature thermal devolatilization and depolymerization reactions do not initiate. At 230 and 270°C, the loss of volatiles is more significant, which might be due to depolymerization of hemicellulose, cellulose, and lignin (Tumuluru et al., 2012). Torrefaction temperatures (200–300°C) of biomass will affect the ash content in the biomass. The increase of ash content observed, in the present study, is more relative to a decrease in the biomass components when compared with the original biomass components. In case of ultimate composition, the carbon content of the biomass increased with increases in torrefaction temperature and residence time. In the present study, the increase in elemental carbon is higher at 270°C when compared with 230 and 180°C temperatures. According to Tumuluru et al. (2011) and other researchers, at a temperature of <250°C the decarbonizing reactions are more limited, but at a temperature of >250°C the biomass undergoes extensive devolatilization resulting with more volatiles loss and an increase in carbon content. The decrease in

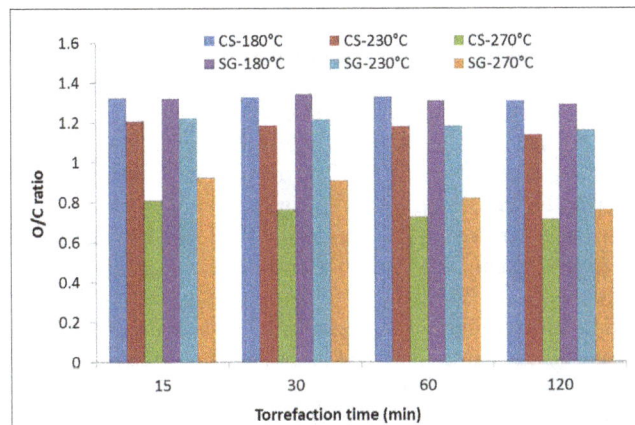

FIGURE 12 | Torrefaction temperature and time effect on O/C ratio of the switchgrass and corn stover.

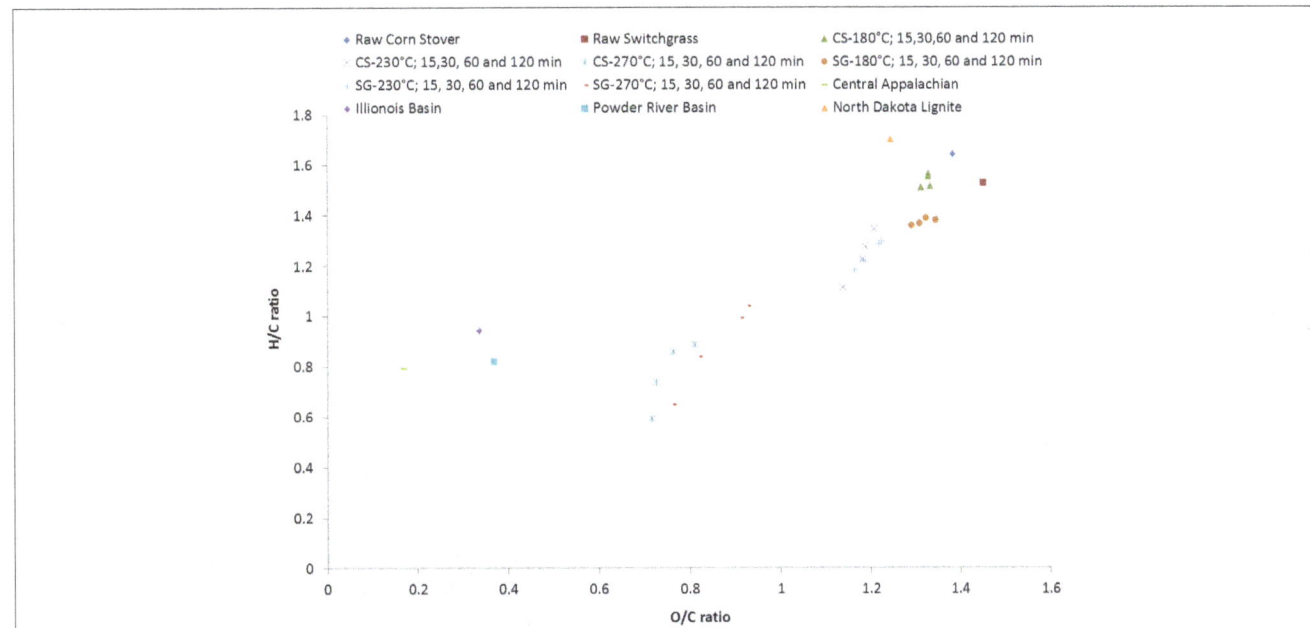

FIGURE 13 | van Krevelen diagram for corn stover, switchgrass, and coals.

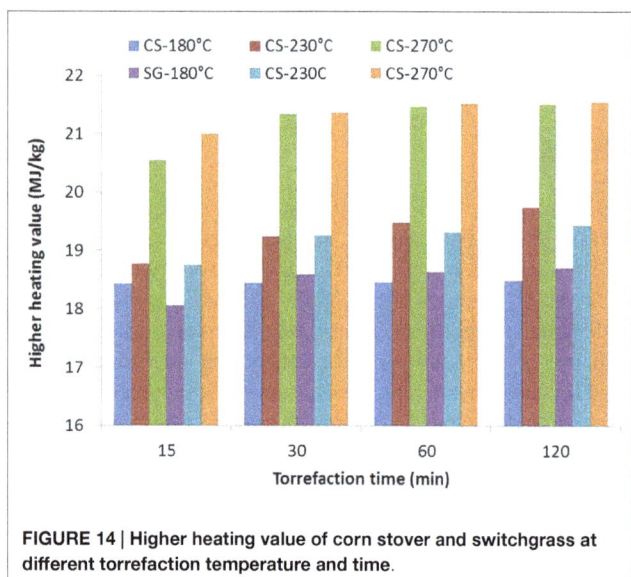

FIGURE 14 | Higher heating value of corn stover and switchgrass at different torrefaction temperature and time.

hydrogen and oxygen content at a lower temperature of 180°C was marginal; whereas, at higher torrefaction temperatures of 230°C and 270°C, the decrease in hydrogen and oxygen is more significant. The major reasons for the decrease in hydrogen and oxygen content at a higher torrefaction temperature are due to formation of water, carbon monoxide, and carbon dioxide. In the present study, it indicates that the nitrogen content increased slightly, while the sulfur content did not change much. The increase in nitrogen and sulfur is more a relative and is due to the decrease of oxygen content. H/C and O/C ratios that were calculated were marginal at the temperature of 180° compared to 230 and 270°C. This observation has corroborated with the findings of the other researchers (Poudel and Oh, 2012; Tumuluru et al., 2012b,c, and Tumuluru et al., 2011). Lower O/C ratio observed at higher torrefaction temperature can be due to the generation of volatiles rich in oxygen, such as CO, CO_2, and H_2O. Lower H/C ratios observed at higher torrefaction temperature can be due to the formation of hydrocarbons, such as CH_4 and C_2H_6. In general, fuels with less H/C and O/C ratios resulted in less smoke and water vapor formation and less energy loss during combustion and gasification processes. The van Krevelen diagram drawn for torrefied corn stover and switchgrass at a torrefaction temperature of 270°C at different torrefaction residence times, indicated that the both corn stover and switchgrass moved closer to higher quality coals like Power River Basin and Illinois Basis. The lower H/C and O/C ratios of these fossil fuels is mainly due low oxygen content and hydrogen content, which makes them suitable for power generation and gasification applications. Torrefaction of switchgrass and corn stove at 270°C and different residence times lowered H/C and O/C ratios to <1 making them more suitable for biopower generation. Also, based on this study it can be concluded that torrefaction residence times of >30 min may not be necessary, as most of the changes in the biomass proximate, ultimate and higher heating value occur at ≤30 min of the residence time. Additionally in the present study, the higher heating value

measured increased with higher torrefaction temperatures. Maximum heating values were observed at 270°C for both corn stover and switchgrass. The increase in higher heating values at higher torrefaction temperature can be due to the loss of lower energy content volatiles, resulting in net increases in the energy content of the torrefied biomass. Nevertheless, torrefaction of corn stover and switchgrass resulted in consistent proximate, ultimate, and energy properties making it more suitable for co-firing applications. The TGA data obtained will be further used to understand the mechanism that results in proximate and ultimate changes in biomass during torrefaction, kinetics of the biomass torrefaction process, and textural changes in torrefied biomass using a scanning electron microscope.

CONCLUSION

Torrefaction of corn stover and switchgrass has resulted in improved chemical and energy properties. Torrefaction temperature had more significant effect on the chemical and energy property compared to residence time. Weight loss observed during torrefaction at 180°C was about 10%, but at 270°C it increased to >45% for both corn stover and switchgrass. At 180°C and 120 min, about 78.8% in corn stover moisture and 88.18% in switchgrass moisture were lost. Loss of moisture at this temperature is mainly due to dehydration reactions. Increasing the temperatures to 230 and 270°C caused significant changes in proximate and ultimate composition. The relative ash content significantly increased to about 10.1 in the corn stover and 6.66% in the switchgrass at 270°C and 15 min. At this temperature, the volatiles decreased to about 50.27 in the corn stover and 52.03% in the switchgrass. In case of ultimate composition, the carbon content increased to about 54.92 in the corn stover and 53.94% in the switchgrass, and hydrogen content decreased to about 4.07 and 4.68%. Furthermore, the nitrogen and sulfur content observed in corn stover and switchgrass were 0.98 and 0.8%, and 0.076 and 0.07%. The oxygen content of the corn stover and switchgrass observed at 270°C and 120 min was about 26.41 and 29.5%. The H/C and O/C ratios also decreased with increasing torrefaction temperature. The van Krevelen diagram drawn for H/C and O/C ratios at 270°C and 15–30 min residence time is closer to some of the coals like Illinois Basis and Powder River Basin. Maximum higher heating values observed for corn stover and switchgrass were 21.51 and 21.53 MJ/kg at 270°C and 120 min. From the present study, it can be concluded that torrefaction resulted in improved and consistent chemical composition and energy properties of corn stover and switchgrass.

ACKNOWLEDGMENTS

This work was supported by the Department of Energy, Office of Energy Efficiency and Renewable Energy under the Department of Energy Idaho Operations Office Contract DE-AC07-05ID14517. Accordingly, the publisher, by accepting the article for publication, acknowledges that the U.S. government retains a non-exclusive, paid-up, irrevocable, worldwide license to publish or reproduce the published form of this manuscript, or allow others to do so, for U.S. government purposes.

REFERENCES

Arias, B., Pevida, C., Fermoso, J., Plaza, M. G., Rubiera, F., and Pis, J. J. (2008). Influence of torrefaction on the grindability and reactivity of woody biomass. *Fuel Process Technol.* 89, 169–175. doi:10.1016/j.fuproc.2007.09.002

ASTM International D3173. (2008). *Standard Test Methods for Moisture in the Analysis Sample of Coal and Coke", Last Modified 2015.* Available from: http://www.astm.org/Standards/D3173.htm

ASTM International D3174. (2002). *Standard Test Methods for Ash Analysis of Coal and Coke", Last Modified 2015.* Available from: https://edis.ifas.ufl.edu/pdffiles/AG/AG29600.pdf.

ASTM International D3175. (2007). *Standard Test Methods for Volatile Matter in the Analysis Sample of Coal and Coke" Last Modified 2015.* Available from: http://www.astm.org/Standards/D3175.htm

ASTM International D3177. (2002). *Standard Test Methods for Total Sulfur in the Analysis Sample of Coal and Coke" Last Modified 2015.* Available from: http://www.astm.org/Standards/D3177.htm

ASTM International D3178. (2002). *Standard Test Methods for Carbon and Hydrogen in the Analysis Sample of Coal and Coke" Last Modified 2015.* Available from: http://www.astm.org/Standards/D5373.htm

ASTM International D3179. (2002). *Standard Test Methods for Nitrogen in the Analysis Sample of Coal and Coke" Last Modified 2015.* Available from: http://www.astm.org/Standards/D3179.htm

ASTM International D5865. (2010). *Standard Test Methods for Gross Calorific Value of Coal and Coke" Last Modified 2015.* Available from: http://www.astm.org/Standards/D5865.htm

Bergman, P. C. A., and Kiel, J. H. A. (2005). "Torrefaction for biomass upgrading," in Proceedings of the 14th European Biomass Conference & Exhibition (Paris), 17–21. Available at: http://www.energy.ca.gov/2009_energypolicy/documents/2009-04-21_workshop/comments/Torrefaction_for_Biomass_Upgrading_TN-51257.PDF

Bridgeman, T. G., Jones, J. M., Shield, I., and Williams, P. T. (2008). Torrefaction of reed canary grass, wheat straw and willow to enhance solid fuel qualities and combustion properties. *Fuel.* 87, 844–856. doi:10.1016/j.fuel.2007.05.041

Chen, D., Zhou, J., Zhang, Q., Zhu, X., and Lu, Q. (2014). Upgrading of rice husk by torrefaction and its influence on the fuel properties. *Bioresour.* 9, 5893–5905.

Chen, W. H., and Kuo, P. C. (2010). A study on torrefaction of various biomass materials and its impact on lignocellulosic structure simulated by a thermogravimetry. *Energy* 35, 2580–2586. doi:10.1016/j.energy.2010.02.054

Lu, J. J., and Chen, W. H. (2014). Product yields and characteristics of corncob waste under various torrefaction atmospheres. *Energies* 7, 13–27. doi:10.3390/en7010013

Medic, D., Darr, M., Shah, A., Potter, B., and Zimmerman, J. (2011). Effects of torrefaction process parameters on biomass feedstock upgrading. *Fuel* 91, 147–154. doi:10.1016/j.fuel.2011.07.019

Newman, Y., Williams, M. J., Helsel, Z., and Vendramini, J. (2014). *Production of Biofuel Crops in Florida: Switchgrass.* Gainesville, FL: IFAS Extension University of Florida, SSAGR291.

Nhuchhen, D. R., Basu, P., and Acharya, B. (2014). A comprehensive review on biomass torrefaction. *Interantinla J. Renew. Energy ad Biofuels* 2014, 1–55. doi:10.5171/2014.506376

Phanphanich, M., and Mani, S. (2011). Impact of torrefaction on the grindability and fuel characteristics of forest biomass. *Bioresour. Technol.* 102, 1246–1253. doi:10.1016/j.biortech.2010.08.028

Poudel, J., and Oh, S. C. (2012). A kinetic analysis of wood degradation in supercritical alcohols. *Ind. Eng. Chem. Res.* 51, 4509–4514. doi:10.1021/ie200496b

Repellin, V., Govin, A., Rolland, M., and Guyonnet, R. (2010). Energy requirement for fine grinding of torrefied wood. *Biomass Bioenergy* 34, 923–930. doi:10.1016/j.biombioe.2010.01.039

Sarkar, M., Kumar, A., Tumuluru, J. S., Patil, K. N., and Bellmer, D. D. (2014). Gasification performance of switchgrass pretreated with torrefaction and densification. *Appl. Energy* 127, 194–201. doi:10.1016/j.apenergy.2014.04.027

Simmons, B. A., Loque, D., and Blanch, H. W. (2008). Next-generation biomass feedstocks for biofuel production. *Gen. Biol.* 9, 242. doi:10.1186/gb-2008-9-12-242

Tillman, D. A., Duong, D. N. B., Miller, B. G., and Bradley, L. C. (2009). "Combustion effects of biomass co-firing in coal-fired boiler," in Proceedings of Power-Gen International (Presentation) (Las Vegas, NV), December 8–10 (2009).

Tumuluru, J. S., Boardman, R. D., and Wright, C. T. (2012c). Response surface analysis of elemental composition and energy properties of corn stover during torrefaction. *J. Biobased Mater. bioenergy.* 6, 25–35. doi:10.1166/jbmb.2012.1187

Tumuluru, J. S., Boardman, R. D., Wright, C. T., and Hess, J. R. (2012b). Some chemical compositional changes in *Miscanthus* and white oak sawdust samples during torrefaction. *Energies* 5, 3928–3947. doi:10.3390/en5103928

Tumuluru, J. S., Hess, J. R., Boardman, R. D., Wright, C. T., and Westover, T. L. (2012). Formulation, pretreatment, and densification options to improve biomass specifications for co-firing high percentages with coal. *Ind. Biotechnol.* 8, 113–132. doi:10.1089/ind.2012.0004

Tumuluru, J. S., Shahab, S., Hess, J. R., Wright, C. T., and Boardman, R. D. (2011). A review on biomass torrefaction process and product properties for energy applications. *Ind. Biotechnol.* 7, 384–401. doi:10.1089/ind.2011.7.384

U.S. Department of Energy. (2011). *U.S. Billion-Ton Update: Biomass Supply for a Bioenergy and Bioproducts Industry.* R. D. Perlack and B. J. Stokes (Leads), ORNL/TM-2011/224. Oak Ridge, TN: Oak ridge National Laboratory. p. 227.

Uemura, Y., Omar, W. N., Tsutsui, T., and Yusup, S. B. (2011). Torrefaction of oil palm wastes. *Fuel* 90, 2585–2591. doi:10.1016/j.fuel.2011.03.021

Wannapeera, J., Fungtammasan, B., and Worasuwannarak, N. (2011). Effects of temperature and holding time during torrefaction on the pyrolysis behaviors of woody biomass. *J. Anal. Appl. Pyrolysis* 92, 99–105. doi:10.1016/j.jaap.2011.04.010

Yang, Z., Sarkar, M., Kumar, A., Tumuluru, J. S., and Huhnke, R. L. (2014). Effects of torrefaction and densification on switchgrass pyrolysis products. *Bioresour. Technol.* 174, 266–273. doi:10.1016/j.biortech.2014.10.032

13

Relationships between biomass composition and liquid products formed via pyrolysis

Fan Lin[1†], Christopher L. Waters[2†], Richard G. Mallinson[2], Lance L. Lobban[2] and Laura E. Bartley[1]*

[1] Department of Microbiology and Plant Biology, University of Oklahoma, Norman, OK, USA, [2] School of Chemical, Biological, and Materials Engineering, University of Oklahoma, Norman, OK, USA

Edited by:
Jason Lupoi,
Joint BioEnergy Institute, USA;
University of Queensland, Australia

Reviewed by:
Suyin Gan,
The University of Nottingham
Malaysia Campus, Malaysia
Xu Fang,
Shandong University, China

*Correspondence:
Laura E. Bartley
lbartley@ou.edu

†Fan Lin and Christopher L. Waters
have contributed equally to this work.

Thermal conversion of biomass is a rapid, low-cost way to produce a dense liquid product, known as bio-oil, that can be refined to transportation fuels. However, utilization of bio-oil is challenging due to its chemical complexity, acidity, and instability – all results of the intricate nature of biomass. A clear understanding of how biomass properties impact yield and composition of thermal products will provide guidance to optimize both biomass and conditions for thermal conversion. To aid elucidation of these associations, we first describe biomass polymers, including phenolics, polysaccharides, acetyl groups, and inorganic ions, and the chemical interactions among them. We then discuss evidence for three roles (i.e., models) for biomass components in the formation of liquid pyrolysis products: (1) as direct sources, (2) as catalysts, and (3) as indirect factors whereby chemical interactions among components and/or cell wall structural features impact thermal conversion products. We highlight associations that might be utilized to optimize biomass content prior to pyrolysis, though a more detailed characterization is required to understand indirect effects. In combination with high-throughput biomass characterization techniques, this knowledge will enable identification of biomass particularly suited for biofuel production and can also guide genetic engineering of bioenergy crops to improve biomass features.

Keywords: thermochemical conversion, plant biomass, bio-oil, lignin, polysaccharides, cell wall, fast pyrolysis, minerals

Biomass can be a renewable and sustainable source of transportation fuels not associated with fossil CO_2 release. Numerous studies highlight the advantages of displacing petroleum fuels with industrial production of liquid fuels from thermochemical conversion of biomass (Bridgwater et al., 1999; Perlack et al., 2005; Mohan et al., 2006; NSF, 2008). Thermochemical conversion entails heating of biomass in an anoxic environment; condensation of organic liquid products, known as bio-oil; and subsequent treatment of the products with catalysts to create liquid fuels, i.e., refined bio-oil, similar to petroleum-derived gasoline or diesel. This is in contrast to biochemical conversion, which utilizes enzymes to release sugars followed by microbial production of ethanol or other fuel molecules (Somerville, 2007; Youngs and Somerville, 2012). Relative to biochemical approaches, thermal conversion has the potential to make use of all carbon (C)-containing biomass components, would allow society to retain existing infrastructure associated with liquid hydrocarbon fuels, and, due to the rapidity of the process, may reduce production costs by permitting scalability and distribution of production (Huber et al., 2006; Mettler et al., 2012). For both thermochemical and biochemical biofuels, lowering processing costs and improving fuel yields per hectare are major engineering challenges that hinder economic viability. Thermochemical fuel production also faces challenges related to maintaining a high C-yield while obtaining a fungible fuel. We posit that this latter challenge might be addressed by understanding the relationships between biomass composition and bio-oil components and using this information to alter biomass through genetic, chemical, or thermal means.

THERMAL CONVERSION CHALLENGES

Two types of pyrolysis have been developed: fast pyrolysis and slow pyrolysis. Slow pyrolysis is usually performed over several hours and has a high solid yield, and as such has little relevance for liquid fuels production. Fast pyrolysis, however, is typically performed quickly, in seconds, at temperatures between 400 and 600°C and decomposes most of the solid biomass into a volatile mixture of various organic molecules, water, and CO/CO_2. Pyrolysis oil or bio-oil constitutes the condensable portion of this vapor. Non-condensable components (primarily CO_2 and CO) and a mineral-rich solid (char) are other product classes that will not be addressed here, except in that they detract from the overall C-yield of raw and refined bio-oil. Bio-oil comprises water (15–30%) plus compounds from several chemical families including the following (**Table 1**): organic acids, light (C1–C3) oxygenates, furan and furan derivatives, phenolic species with various methyl and methoxy substituents, pyrones, and sugar derivatives like levoglucosan (Faix et al., 1991a,b). Bio-oil's chemically complex nature prohibits its direct use in combustion applications or petroleum refining. The reasons for this include low heating value; ignition difficulty; high chemical reactivity, which results in oligomerization and polymerization over time and upon heating, prohibiting distillative separation (Oasmaa and Czernik, 1999; Demirbas, 2011; Patwardhan et al., 2011a); immiscibility with petroleum; and high corrosivity (Oasmaa and Czernik, 1999). Many of these features are associated with the high oxygen content of biomass and the resulting bio-oil, relative to fossil fuels.

In order to obtain desirable fuel properties and allow integration with the existing transportation fuels infrastructure (gasoline and diesel engines), the bio-oil must be chemically converted to reduce the undesirable characteristics mentioned above. Catalytic upgrading is typically used to refine bio-oil, improving its stability and making it an acceptable liquid fuel. The simplest method is hydrotreating or hydrodeoxygenation, which removes oxygen via catalytic hydrogenation (Furimsky, 2000), decreasing both the chemical reactivity and corrosivity. However, this process converts any C1–C5 oxygenates, representing as much as half of the carbon in bio-oil, to C1–C5 hydrocarbons that are too volatile for liquid fuels (Resasco, 2011). Another straightforward approach is to "crack" the pyrolysis vapors using acidic zeolite catalysts into light olefins and aromatic hydrocarbons (primarily benzene, toluene, and o/m/p-xylene) (Bridgwater, 1994; Carlson et al., 2008, 2009). This approach is appealing because of the lack of an external H_2 requirement and the simplicity of the product streams. Furthermore, since zeolite cracking is widely used in traditional petroleum refining/valorization (Wan et al., 2015), other advantages are the product compatibility with existing refinery infrastructure and the maturity of the process (Wan et al., 2015). However, zeolite cracking is crippled by poor usable carbon yield due to the high amounts of coke, CO, and CO_2 formed during the catalytic process (Carlson et al., 2008) and the concomitant rapid catalyst deactivation. Additionally, further catalytic oligomerization and reforming for olefins and aromatics, respectively, are needed to make these products suitable for addition to refinery fuel product streams, increasing the process costs and further reducing overall carbon yield.

More advanced strategies propose to use reactions such as ketonization, condensation, alkylation, and others to retain a higher fraction of the biomass carbon into liquid fuel-range molecules (Zhu et al., 2011; Zapata et al., 2012; Pham et al., 2013; Gonzalez-Borja and Resasco, 2015). However, catalytic upgrading of any one family of compounds (e.g., light oxygenates) typically requires a catalyst and reaction conditions different than those required for another family of compounds (e.g., substituted phenolics). Moreover, catalysts used for upgrading one family of compounds may be ill-suited for other families, either facilitating undesirable reactions (breaking C–C bonds unnecessarily or increasing H:C ratios above the 2:1 optimum) or undergoing rapid deactivation due to reactions with other non-targeted bio-oil oxygenates. These upgrading challenges suggest the desirability of thermal conversion producing more selective product streams, i.e., each stream comprising fewer families of chemical compounds. Developing such thermal conversion processes would be aided by clearer knowledge of the relationship between biomass composition and thermal conversion products.

BIOMASS COMPOSITION AND CHEMICAL STRUCTURES

Recent reviews have addressed the general relationships between biomass composition and thermal products, such as

TABLE 1 | The percentage ranges and categories of major bio-oil components.

Category	Major components	Wet weight (%)
Light oxygenates	Glycolaldehyde, acetol	3–26
Organic acids	Acetic acid, formic acid, propanoic acid	2–27
Aldehydes	Acetaldehyde, formaldehyde, ethanedial	3–18
Sugars	1,6-Anhydroglucose (levoglucosan)	5–14
Phenols	Phenol, catechol (di-OH benzene), methyl phenol, dimethyl phenol	3–13
Guaiacols	Isoeugenol, eugenol, 4-methylguaiacol	3–15
Furans	Furfurol, hydroxymethyl furfural, furfural	2–11
Syringols	2,6-Dimethoxy phenol, syringaldehyde, propyl syringol	2–9
Ketones	Acetone	4–6
Alcohols	Methanol, ethylene glycol, ethanol	2–6
Esters	Methyl formate, butyrolactone, methylfuranone	<1–3

Source: Huber et al. (2006).

increasing the content of phenolics relative to carbohydrates to reduce the oxygen content of bio-oil (Tanger et al., 2013). Here, we provide a more detailed description of the chemical structure and interactions among major cell wall components to aid in understanding more subtle relationships between biomass and bio-oil content. Biomass consists of cell walls that establish the structure of the plant and, to a lesser extent, non-structural components (**Table 2**). Cell walls determine the shape of leaves and stems and the cells that compose them and consist of cellulose, hemicellulose, lignin, as well as structural proteins and wall-associated mineral components (O'Neill and York, 2009; Vogel et al., 2011; Tanger et al., 2013). Non-structural components include sugars, proteins, and additional minerals (O'Neill and York, 2009; Vogel et al., 2011; Tanger et al., 2013). For example, in switchgrass, an important potential bioenergy crop, dry biomass consists of ~70% cell walls, 9% intrinsic water, 8% minerals, 6% proteins, and 5% non-structural sugars (Vogel et al., 2011). The relative fractions of different components, chemical linkages within and between polymers, and cellular patterning vary among plant species, organs, developmental stages, and growth conditions (Adler et al., 2006; El-Nashaar et al., 2009; Singh et al., 2012; Zhao et al., 2012). Here, we review the components of secondary cell walls, which are formed as plant growth ceases, as they constitute the majority of plant biomass (Pauly and Keegstra, 2008), and then discuss evidence for interactions among components. **Table 2** lists the different major and minor components of biomass and the broad ranges of their representation within biomass for biofuel conversion.

TABLE 2 | The variation of biomass components among vascular plants including grasses, softwoods, and hardwoods.

Biomass component	Dry weight (%)[a]
Cellulose	15–49[b]
Hemicellulose	12–50[b,c,d]
Xylan	5–50[c]
Mixed-linkage glucan	0–5[c,e]
Xyloglucan	Minor[c]
Mannan (and galactoglucomannan)	0–30[c,f]
Soluble (mainly sucrose)	9–67[b,g]
Pectin	<0.1[h]
Lignin	6–28[b]
Ferulic acid and p-coumaric acid	<1.5[h]
Protein	4–5[b]
Ash (mainly silicate)	0.4–14.4[b]
Intrinsic moisture	11–34[i]

[a]Percent mass composition of secondary cell walls.
[b]Pauly and Keegstra (2008).
[c]Scheller and Ulvskov (2010).
[d]As the highest percentage of xylan in Scheller and Ulvskov (2010) is higher than the highest percentage of hemicellulose in Pauly and Keegstra (2008), the highest percentage of hemicellulose is set to the highest percentage of xylan.
[e]MLG is only abundant in grasses. The maximum percentage of MLG we are aware of is that of the mature rice stem after flowering (Vega-Sanchez et al., 2012).
[f]Galactoglucomannan is only abundant in gymnosperm woods. Dicots and grasses possess <8% of mannan and galactoglucomannan (Scheller and Ulvskov, 2010).
[g]The high abundance of solubles is only for sorghum biomass. Other plants usually have less than 15% soluble content (Pauly and Keegstra, 2008).
[h]Vogel (2008).
[i]McKendry (2002).

Figure 1(1–3) shows the chemical structures and atom numbering of the most abundant cell wall monomeric species.

Cellulose and hemicellulose represent 15–49% and 12–50% of biomass by dry weight, respectively (Pauly and Keegstra, 2008; Vogel, 2008; Zhao et al., 2012). Cellulose is an unbranched homopolymer of >500 β-(1,4)-linked glucose units. In plant cell walls, cellulose is primarily in the form of crystalline microfibrils consisting of approximately 36 hydrogen-bonded cellulose chains, but also has amorphous regions (Somerville, 2006; Newman et al., 2013).

Hemicelluloses are typically branched polysaccharides substituted with various sugars and acyl groups. As discussed further in the Section "Evidence Relating Biomass Content and Bio-oil Composition," the different sugar composition and linkages of hemicelluloses influence thermal products (Shafizadeh et al., 1972; Mante et al., 2014). The structure and composition of hemicellulosic polysaccharides differ depending on plant species classification, i.e., taxonomy. Major taxonomic divisions with relevance to bioenergy production are grasses, such as switchgrass and wheat; woody dicots, i.e., hardwoods, such as poplar; and woody gymnosperms, i.e., softwoods, such as pine. The most abundant grass hemicelluloses are mixed-linkage glucan (MLG) and glucuronoarabinoxylan (GAX) (Scheller and Ulvskov, 2010; Vega-Sanchez et al., 2013); the hemicelluloses of hardwood are primarily composed of glucuronoxylans (GX) but also contain a small amount of galactomannans (GM) (Pauly and Keegstra, 2008); and softwood hemicelluloses are largely galactoglucomannan (GGM) and GAXs (Scheller and Ulvskov, 2010). MLG is an unbranched glucose polymer similar to cellulose but containing both β-(1-3)- and β-(1-4)-linkages (Vega-Sanchez et al., 2012). MLG is nearly unique to the order Poales, which includes the grasses, but has also been found in horsetail (*Equisetum*). Its abundance in mature tissues and secondary cell walls has recently been recognized (Vega-Sanchez et al., 2013). Xylans consist of a β-(1-4)-linked xylose backbone with various substitutions. GXs are xylans substituted mostly by glucuronic acid and 4-O-methyl glucuronic acid through α-(1-2)-linkages. GAXs are not only substituted by glucuronic acid but also substituted by arabinofuranoses at the O-3, which can be further substituted by the phenylpropanoid acids, to form feruloyl- and p-coumaryl esters linked at the O-5 (Scheller and Ulvskov, 2010). Acetyl groups are often attached to the O-3 of backbone xyloses but also attach to the O-2. Unlike xylans, which mainly consist of pentoses, mannans consist of hexoses like mannose, glucose, and galactose. GM and GGM have a β-(1-4)-linked backbone with mannose or a combination of glucose and mannose, respectively. Both GM and GGM can be acetylated and substituted by α-(1-6)-linked galactoses (Scheller and Ulvskov, 2010; Rodriguez-Gacio Mdel et al., 2012; Pauly et al., 2013). Relatively depleted in secondary walls, but rich in growing primary walls of dicot species, xyloglucan and pectins are two other polysaccharides in cell walls. Xyloglucan consists of β-(1-4)-linked glucose residues, modified by xylose and other sugar residues; and pectin is another branched or unbranched polymer that is rich in galacturonic acid, rhamnose, galactose, and several other monosaccharide residues (Somerville et al., 2004; Scheller and Ulvskov, 2010).

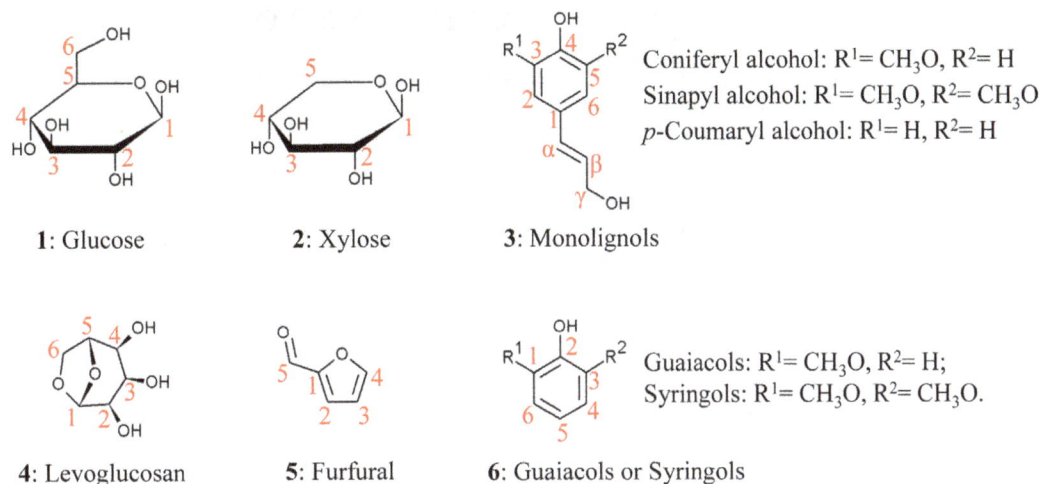

FIGURE 1 | Chemical structure of the major basic units of biomass polymers and related products.

Lignin is a cross-linked, heteropolyphenol mainly assembled from three monolignols – sinapyl (S), coniferyl (G), and p-coumaryl (H) alcohols. As waste products are often selected as biofuel feedstocks, it is also relevant to note that lignin derived from other monolignols such as caffeyl alcohol and 5-hydroxyconiferyl have been found in the seedcoat of both monocots and dicots (Chen et al., 2012, 2013). Lignin structural heterogeneity and various types of incorporated groups can lead to a variety of different depolymerization reactions during pyrolysis (Kawamoto et al., 2007). Often traceable to the corresponding bio-oil components, the three major lignin units differ in the degree of methoxylation of their carbon ring. S-units are methoxylated at both O-3 and O-5 ring positions; G-unit have one methoxy group at the O-3 position; and H-units lack ring methoxy groups (3, Figure 1) (Boerjan et al., 2003). Lignin units undergo oxidative coupling in the cell wall to form many types of dimers, including β–O–4, β–5, β–β, 5–5, 5–O–4, and β–1, leaving other atoms free to further polymerize, which significantly increases the structural heterogeneity of lignin. Lignin units can also be esterified with p-coumaryl, p-hydroxybenzoyl, and acetyl groups, primarily at the γ position of terminal units (Petrik et al., 2014; Lu et al., 2015). Lignin compositions and the acylation groups vary among plant clades (Boerjan et al., 2003). Woody dicot lignins have G- and S-units and trace amount of H-units. Poplar wood, for example, has a G:S:H ratio of 55:45:1 (Vanholme et al., 2013). The lignin of many hardwoods is acylated by p-hydroxybenzoates (Lu et al., 2015) and acetyl groups in low amounts (Sarkanen et al., 1967). Biomass from other species, such as palms and kenaf, possess a high degree of lignin acetylation (Lu and Ralph, 2002). Grass lignins also contain G- and S-units with slightly higher amount of H-units than woody dicots. Wheat straw, for example, has a G:S:H ratio of 64:30:6 (Bule et al., 2013). Grass lignin possesses high levels of p-coumarate esters (Hatfield et al., 2008) and can also be etherified by tricin and ferulic acid (Ralph et al., 1995; Lan et al., 2015), as discussed further below. Woody gymnosperm

lignins are different from angiosperm lignins, being primarily composed of G-units and a lower amount of H-units (Boerjan et al., 2003).

Biomass also contains inorganic elements including Ca, K, Si, Mg, Al, S, Fe, P, Cl, and Na and some trace elements (<0.1%) such as Mn and Ti, according to ash analysis, formed by oxidation of biomass at 575°C (Masia et al., 2007; Vassilev et al., 2010). As with other biomass components, the abundance of mineral elements varies among species. In general, compared with grass biomass, woody biomasses contain less ash, Cl, K, N, S, and Si, but more Ca (Vassilev et al., 2010).

Plant biomass components do not accumulate independently of each other, though their relationships are still an active area of research (Dick-Perez et al., 2011; Tan et al., 2013; Mikkelsen et al., 2015). Biomass component amounts can correlate because they are physically bound to each other through covalent and non-covalent bonds or because they accumulate in the same plant organ or stage of plant development, though a physical interaction may not exist. Because the abundance of some biomass components is correlated, the thermal products from one biomass component may also correlate with other components. For example, the abundance of cellulose correlates with the abundance of lignin in five different biomass sources (Pearson's correlation coefficient = 0.83) and lignin-derived thermal products correlate with cellulosic glucose (Mante et al., 2014). Many mineral elements are also correlated with each other, for example, N, S, and Cl; Si, Al, Fe, Na, and Ti; Ca, Mg, and Mn; K, P, S, and Cl (Vassilev et al., 2010). Numerous interactions between lignins and hemicelluloses and among hemicelluloses have been observed. Among the best-studied examples, GAXs of grasses and other recently evolved monocot species covalently link to lignin through ether bonds with ferulate esters on arabinose moieties of arabinoxylan (Bunzel et al., 2004). In poplar and spruce wood, NMR results indicate that lignin and carbohydrates are directly bonded through several types of ether linkages (Yuan et al., 2011; Du et al., 2014). The data provide evidence for ether bonds between lignin and C1, C5, and C6 atoms of pentoses and hexoses (Yuan et al., 2011). Generally, xylan is the most closely associated polysaccharide to lignin, and NMR studies have also clearly identified lignin–glucuronic acid ester bonds (Yuan et al., 2011). Also, MLGs closely coat low-substituted xylan regions, likely via non-covalent interactions (Carpita et al., 2001; Kozlova et al., 2014). Furthermore, some components can also affect the distribution of other components. For example, rice plants that overexpress an enzyme that cleaves MLG exhibit reduced MLG and have an altered distribution profile of Si though maintain the same total amount of Si (Kido et al., 2015). In sum, mounting evidence supports covalent and non-covalent interactions among cell wall polymers and components; however, these connections have been difficult to study with questions persisting related to how different cell wall preparations and manipulation may alter observations.

MODELS FOR RELATIONSHIPS BETWEEN BIOMASS COMPONENTS AND BIO-OIL PRODUCT COMPOSITION

Reaction pathways of individual biomass components to formation of thermal products have been described (Collard and Blin, 2014). However, the pyrolysis literature suggests that biomass components tend to have more complex effects on bio-oil yield and product composition than simply their quantity. Here, we introduce three possible "models" of how biomass components may influence the yield or composition of thermal products, and in Section "Evidence Relating Biomass Content and Bio-oil Composition," we discuss evidence supporting each of them. **Figure 2** provides schematic representations of the following models:

Model 1: Biomass components are the direct sources of thermal products. Components are converted to products through depolymerization and secondary reactions such as cracking, i.e., splitting, and recombination (**Figure 2A**).
Model 2: Components or their derived products act as catalysts that accelerate thermal reactions of other components, altering product yields and ratios (**Figure 2B**).
Model 3: Chemical interactions or structural relationships among cell wall components alter bio-oil composition and/or yield (**Figure 2C**). This "indirect" model applies when variation in a biomass component alters the yield of a chemically unrelated product in a manner not easily explained by a catalytic effect. Chemical interactions that alter products may either be covalent or non-covalent chemical bonds between cell wall components. Structural relationships refer to correlations between components, often minor ones, and physical features of the biomass. For example, the abundance of a cell wall component may be indicative of the structure of the plant material, such as biomass bulk density differences caused by different leaf to stem ratios, but do not reflect chemical bonding between components. As of the preparation of this review, very little evidence addresses how biological correlations effect bio-oil products, so the discussion focuses on potential chemical interactions.

EVIDENCE RELATING BIOMASS CONTENT AND BIO-OIL COMPOSITION

Evidence in the literature for the three models described above is presented in **Table 3** and discussed below. In the reviewed experiments, relationships between biomass components and pyrolysis products have been identified by varying the starting biomass, either through experimentation on purified components, via naturally occurring variation among different biomass sources, or via pretreatment of the biomass. Most studies included in this discussion report the chemical products derived from pyrolysis of biomass or biomass components. Studies that only reported weight losses or elemental balances were not considered. The two dominant techniques present in this corpus of literature are either pyrolysis-gas chromatography/mass spectroscopy, where pyrolysis vapors from microgram- to milligram-scale samples are directly transported to a GC for analysis, or pyrolysis in a gram- to kilogram-scale reactor system followed by condensation of the vapors and subsequent chromatographic analysis of the liquid.

Model 1: Direct Products of Cellulose, Hemicellulose, and Lignin

Thermal breakdown of purified cellulose, hemicellulose, and lignin has been relatively well studied. Levoglucosan, a six-carbon 1,6-anhyrosugar (see **Figure 1**), was identified as the main product of cellulose pyrolysis nearly a century ago (Pictet and Sarasin, 1918). Levoglucosan is formed alongside other smaller decomposition products, with maximum levoglucosan production occurring at 500°C (Shafizadeh et al., 1979). Minor products of cellulose pyrolysis are dominated by other anhydrosugars that retain all six carbons of glucose, such as 1,6-anhydroglucofuranose and 5-hydroxymethyl furfural, but also smaller molecules, like furfural (**5**, **Figure 1**), formic acid, and glycolaldehyde, among others (Patwardhan et al., 2011b).

As with cellulose, hemicellulose pyrolysis products depend mostly on the number of carbons in the monosaccharide residues of the starting polymer (Shafizadeh et al., 1972). Pentoses and hexoses produce similar light C1–C3 oxygenates but differ in the types and selectivities (i.e., relative ratios) of heavier C4–C6 products. Consistent with expectations, pyrolysis of monosaccharides reveals that hexoses can form more unique compounds than pentoses, including pyranic species; additionally, pentoses yield more lighter fragmentation products than hexoses and only trace amounts of C6 and higher products (Raisanen et al., 2003).

Lignin thermal degradation products generally retain the characteristic ring decoration of the monolignols from which they originate (**3**, **Figure 1**). For example, syringol derivatives are bio-oil products derived from S-lignin units and guaiacols are products derived from G-lignin units (**6**, **Figure 1**). The derivative groups possess 1–3 carbons and/or oxygenate moieties at the fourth position (**6**, **Figure 1**). Consistent with expectations, softwood lignins yield almost exclusively guaiacyl derivatives, while hardwood lignins yield both guaiacyl and syringyl derivatives. Grasses yield not only guaiacyl, syringyl, and *p*-hydroxyphenyl derivatives but also vinylphenol, propenyl-phenols, and *p*-hydroxybenzaldehyde that are not produced during pyrolysis of

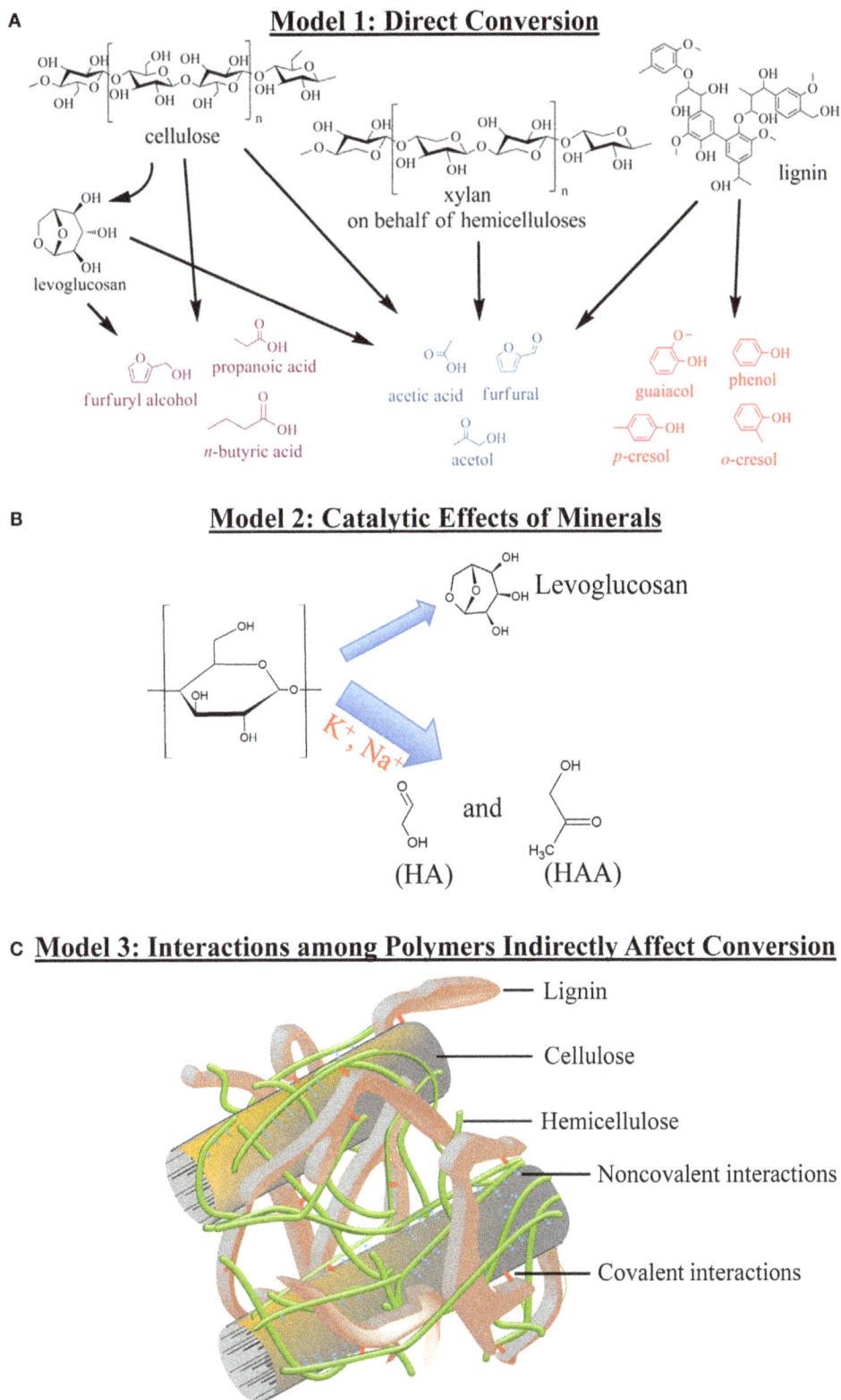

FIGURE 2 | Three models of how biomass components and their interactions affect the formation of thermal products. HA, hydroxyacetone; HAA, hydroxyacetaldehyde. **(A)** Direct conversion. **(B)** Catalytic effect of minerals. **(C)** Interactions among polymers indirectly affect conversion. **(A)** and **(C)** were adapted and modified with permission (Vanholme et al., 2010; Zhang et al., 2013).

TABLE 3 | Possible models of how biomass components affect the composition and yield of liquid thermal products.

Possible model	Biomass component	Variation of input material (independent variable)	Effect on products (dependent variable)	Sample type	Thermal conversion condition	Reference
1	Hemicellulose	Arabinose, xylose, mannose, arabinitol	Arabinose and xylose produce similar products but slightly different yields and product ratios. Mannose gives more decomposition products than arabinose and xylose and a unique product, 5-hydroxymethyl furfural	Powdered high-purity monosaccharides (>99%)	500 and 550°C, 10 s	Raisanen et al. (2003)
1	Hemicellulose or non-structural polysaccharides	Variation in sugar composition	Fermented grain samples produce less acetic acid, furfural, and acetone than hull and straw	Barley straw, hull and yeast-fermented grain	Fluidized bed, 500°C	Mullen et al. (2010)
1	Lignin	Lignin variation among hardwoods, softwoods, and grasses	Hardwood lignin produces guaiacyl and syringyl derivatives. Softwood lignin produces mostly guaiacyl derivatives but no hydroxyphenyl or syringyl compounds. Grass lignin uniquely produces vinylphenol, propenyl-phenols, and p-hydroxybenzaldehyde	Solvent extracted lignin from milled spruce, beech, aspen, and bamboo	510°C, 10 s	Saizjimenez and Deleeuw (1986)
1	Lignin	Natural variation and variation by hydrogen peroxide treatment in Klason lignin	Samples with high Klason lignin produce more 4-vinylguaiacol. Klason lignin positively correlates with 4-vinylguaiacol	Hybrid maize, bm1 mutant, bm3 mutant, and switchgrass biomass treated by different [H₂O₂] to remove lignin	650°C, 20 s	Li et al. (2012)
1	Lignin	Natural variation in lignin abundance	High-lignin endoscarp biomasses produce more phenolic compounds, like phenol, 2-methoxyphenol, 2-methylphenol, 2-methoxy-4-methylphenol, and 4-ethyl-2-methoxyphenol, compared to switchgrass biomass	Walnut, olive, coconut husks, and peach drupe endocarp biomass and switchgrass biomass	650°C, 20 s	Mendu et al. (2011)
1	Lignin	Natural variation in lignin composition or structure	Compared with endocarp lignin, switchgrass lignin produces more acetic acid, toluene, furfural, and 4-methylphenol, and lower or undetectable amounts of 4-ethyl-2-methoxyphenol, 2-methoxy-4-vinylphenol and 2-methoxy-4-(2-propenyl)-phenol. Compared with walnut and olive endocarp lignin, coconut endocarp lignin produces strikingly more phenol and less 2-methoxy-4-methylphenol. Coconut shell lignin produces unique compounds among the analyzed feedstocks, such as 2,6-dimethoxyphenol, 2-methoxy-4-(2-propenyl)-phenol and vanillin, but less 2-methoxy-4-(1-propenyl)-phenol compared to walnut shell and olive lignin	Lignin extracted from various endocarps and switchgrass	650°C, 20 s	
1	Lignin interunit linkages (lignin dimer)	Variation of lignin interunit linkages, including β-O-4' ethers, pinoresinols, phenylcoumaran, and dibenzodioxocins, among lignin-carbohydrate complexes with different compositions	Glucomannan-associated lignin produces more guaiacol than xylan-associated lignin and glucan-associated lignin	Lignin-carbohydrate complexes extracted from spruce wood	500°C, 1 min	Mendu et al. (2011)
2	Mineral content	Variation of K⁺, Mg²⁺, and Ca²⁺ by demineralization and impregnation	Samples with more K⁺ produce more glycolaldehyde, acetic acid, acetol, butanedial, guaiacol, syringol, and 4-vinylsyringol, but less levoglucosan, furans, and pyrans. Samples with more Mg²⁺ produce less glycolaldehyde, levoglucosan, and 3-methoxycatechol. Samples with more Ca²⁺ produce more pyrans and cyclopentenes but less levoglucosan and 3-methoxycatechol	Poplar wood powder demineralized with hydrofluoric acid and impregnated with KCl, MgCl₂, and CaCl₂	550°C, 10 s	Eom et al. (2012)

(Continued)

TABLE 3 | Continued

Possible model	Biomass component	Variation of input material (independent variable)	Effect on products (dependent variable)	Sample type	Thermal conversion condition	Reference
2	Mineral content	Increase of minerals by adding 1% and 5% NaCl	Samples with more NaCl added produce more furans, acids, ketones, and phenols than the samples with no NaCl added	Rice straw and bamboo with NaCl	400–900°C, 4 min	Lou et al. (2013)
2	Mineral content	Decrease minerals by acid wash	Cellulose with the acid-washed char produces more levoglucosan but less formic acid, furane-type derivatives, and light oxygenates, than cellulose with unwashed char	Cellulose with acid-washed and unwashed red oak char	400°C, 15 s	Ronsse et al. (2012)
2	Mineral content	Decrease of K^+ and Na^+ by washing	Washed samples yield more levoglucosan, hydroxyacetaldehyde, and char than unwashed samples	Washed and unwashed lolium, festuca, willow, and switchgrass biomass	500°C, 10 s	Fahmi et al. (2007)
2	Mineral content	Increase mineral content by impregnation with NaCl, KCl, $MgCl_2$, $CaCl_2$, $Ca(OH)_2$, $Ca(NO_3)_2$, $CaCO_3$, and $CaHPO_4$	Impregnated samples produce less levoglucosan, but more glycolaldehyde and acetol, compared to untreated cellulose	Microcrystalline powdered cellulose impregnated with salt or switchgrass ash solution	500°C, 30 s and a range from 350 to 600°C with 50°C increments	Patwardhan et al. (2010)
2	Mineral content	Natural variation in ash	Ash negatively correlates with hydroxyacetaldehyde and phenolic compounds, such as trimethoxybenzene, syringol, 4-allyl-2,6-dimethoxyphenol, syringaldehyde, and 3,5-dimethoxy-4-hydroxycinnamaldehyde	Poplar, willow, switchgrass, hot-water extracted sugar maple, and debarked sugar maple	550°C, 20 s	Mante et al. (2014)
2	Mineral content	Increase mineral content by adding $CaCl_2$, NaCl, KCl, K_2SO_4, $KHCO_3$ individually to washed wood	Washed wood with calcium salt added produces more $C_6H_8O_2$ compound compared to washed wood with no additive. Samples with sodium salt and potassium salt added produce less 2-hydroxy-butanedial, 4-hydroxy-5,6-dihydro-(2H)-pyran-2-one, and levoglucosan, but more tetrahydro-4-hydroxy-pyran-2-one, compared to washed samples	Hornbeam wood washed and mixed with salts	280°C, 20 min, 100–500°C	Muller-Hagedorn et al. (2003)
2	Mineral content	Increase mineral content by adding 1% NaCl	Xylans with NaCl added yield more glycolaldehyde but less 1,4-anhydro-α-D-xylopyranose, 1,5-anhydro-4-deoxypent-1-en-3-ulose and char	Synthesized xylan	280°C, 30 min	Ponder and Richards (1991)
2	Mineral content	Replace the cations in biomass with Na^+ by ion-exchange or increase mineral content by adding Na_2SO_4 or $NaHCO_3$	Sodium-containing pulps produce less polysaccharide-derived products and similar amounts of lignin-derived products compared to untreated pulps	Spruce pulp, treated with ion exchange and salt impregnation	620°C, 2 s	Kleen and Gellerstedt (1995)
3	Intercomponent linkages	Biomass with hemicellulose selectively removed, compared with a physical mixtures of lignin and cellulose, and separate pyrolysis of lignin and cellulose	The native grass cellulose-lignin samples produce more C1 to C3 products and furan, but less pyrans and levoglucosan	Corn stover, red oak, pine, and switchgrass depleted of hemicellulose and compared with different extracted component mixtures	500°C	Zhang et al. (2015)

(Continued)

TABLE 3 | Continued

Possible model	Biomass component	Variation of input material (independent variable)	Effect on products (dependent variable)	Sample type	Thermal conversion condition	Reference
1 and 2 or 3	Acetyl content	Natural variation in acetyl content among five species	Acetyl content positively correlates with acetic acid, but also methyl pyruvate and 2-furanone	Poplar, willow, switchgrass, hot-water extracted sugar maple, and debarked sugar maple	550°C, 20 s	Mante et al. (2014)
1 and 3	Acetyl content	Variation of acetylation between acetylxylan and xylan	The acetylxylan produces more acetic acid, furan, and acetone, but less 2-furfural and acetaldehyde, than xylan	Xylans extracted from cotton wood	500°C	Shafizadeh et al. (1972)
1 and 3	Cellulose	Natural variation in glucans, mostly cellulose, among five species	Glucans positively correlate with levoglucosan and hydroxymethyl furfural, but also phenolic compounds like syringaldehyde, vanillin, and 3,5-dimethoxy-4-hydroxycinnamaldehyde	Poplar, willow, switchgrass, hot-water extracted sugar maple, and debarked sugar maple	550°C, 20 s	Mante et al. (2014)
1 and 3	Lignin	Natural variation in lignin among five species	Lignin positively correlates with hydroxyacetaldehyde and phenolic compounds, trimethoxybenzene; syringol; 4-allyl-2,6-dimethoxyphenol; syringaldehyde; and 3,5-dimethoxy-4-hydroxycinnamaldehyde	Poplar, willow, switchgrass, hot-water extracted sugar maple, and debarked sugar maple	550°C, 20 s	Mante et al. (2014)
1 and 3	Xylan	Natural variation in xylan among five species	Xylan positively correlates with hydroxyacetone, but negatively correlates with hydroxymethyl furfural and syringaldehyde	Poplar, willow, switchgrass, hot-water extracted sugar maple, and debarked sugar maple	550°C, 20 s	Mante et al. (2014)
2 and 3	Mineral content or other structure	Variation in weakly associated and strongly associated K, Na, Mg, and Ca by washing with deionized water or nitric acid	Acid-washed samples with less mineral content produce bio-oil with a greater fraction of water-insoluble content and produce more levoglucosan and sugar compounds but less monophenols like phenol, 2-methyl-phenol, syringol compared to unwashed samples	Eucalyptus wood	Fluidized bed, 500°C	Mourant et al. (2011)
3	Water	Natural variation in moisture content (0–20%)	Samples with greater moisture content produce more char and gas but less water compared to lower moisture samples	Pine wood	Fluidized bed, 480°C, 2 s	Westerhof et al. (2007)
3	Water	Natural variation in moisture content (2–55%)	Samples with greater moisture content produce more condensible products, but less char than dry samples	Norway spruce	500°C, 30 min	Burhenne et al. (2013)
3	Water	Natural variation in moisture content (5–15%)	Samples with greater moisture content produce less levoglucosan at 450 and 500°C than samples with lower moisture content. Other products are also significantly affected by moisture content, but trends depend on pyrolysis temperature	Switchgrass	Fluidized bed, 450, 500, and 550°C	He et al. (2009)
3	Intercomponent linkages	Variation in polysaccharide amounts in lignin–carbohydrate complexes by enzyme treatment	Samples with lignin-associated polysaccharides removed produce more coniferyl alcohol than untreated samples	Lignin–carbohydrate complexes extracted from spruce wood	500°C, 1 min	Du et al. (2014)

softwood and hardwood (Saizjimenez and Deleeuw, 1986; Mante et al., 2014) and are likely derived from ferulate and coumarate esters (Penning et al., 2014b). Phenol derivatives are the large majority of the products formed from lignin pyrolysis; aromatic hydrocarbons and some furan derivatives are also detectable, but at very low amounts that might represent lignin sample contaminants (Saizjimenez and Deleeuw, 1986). Lignins from spruce wood with different dimer compositions also show different product distributions, including variations in the yield of major products like guaiacol (Du et al., 2014). This suggests that bonds between lignin units and the lignin structure determined by those bonds may impact pyrolysis as well.

Model 2: Secondary Reactions Catalyzed by Inorganic Components

The biopolymers that make up the majority of the biomass by weight are established as the primary source of bio-oil products formed during thermal degradation. However, secondary reactions occur during the pyrolysis process involving other components present within the biomass (Ponder and Richards, 1991; Kleen and Gellerstedt, 1995; Muller-Hagedorn et al., 2003; Fahmi et al., 2007; Patwardhan et al., 2010; Ronsse et al., 2012; Lou et al., 2013; Mante et al., 2014). As products form, they can interact with catalytic minerals in the residual solid. For example, levoglucosan has been shown to react on minerals present in the residual char from pyrolysis of biomass. The products formed include levoglucosenone, furan derivatives, and lighter oxygenates such as acetic acid, acetone, and acetol. Demineralization prohibits the formation of these products (Fahmi et al., 2007; Ronsse et al., 2012).

Different inorganics are responsible for different kinds of secondary reactions. In general, the presence of metal cations enhances the homolytic cleavage of pyranose ring bonds over the heterolytic cleavage of glycosidic linkages, leading to the increased formation of light oxygenate decomposition products at the expense of levoglucosan formation. While Na^+, K^+, Mg^{2+}, and Ca^{2+} all catalyze levoglucosan decomposition, the effects of group 1 (alkali metals) and group 2 (alkaline) elements differ. Increased Na^+ and K^+ alkali metal loading increased formic acid, glycolaldehyde, and acetol more than similar amounts of the alkaline metals, Mg^{2+} and Ca^{2+}, though more furfural is produced with increasing concentrations of Mg^{2+} and Ca^{2+}. Additionally, the alkali metals reduce levoglucosan production at very low thresholds. This suggests that Na^+ and K^+ ultimately promote cracking reactions while Mg^{2+} and Ca^{2+} promote dehydration reactions (Muller-Hagedorn et al., 2003; Patwardhan et al., 2010; Eom et al., 2012).

Model 3: Interactions and Linkages Between Primary Components

While the first two models address the direct conversion of biopolymer organic components to related bio-oil products and their further reaction catalyzed by biomass inorganics, the third addresses compositional and structural relationships among cell wall components and their impact on products. Interactions between polysaccharides and lignin have been shown to alter pyrolysis products (Du et al., 2014; Zhang et al., 2015). The cellulose–lignin interaction can lead to a decrease in levoglucosan

yield and an increase in light (C1–C3) compounds, especially glycolaldehyde and furans. Based on the nature of the small products, Zhang et al. (2015) hypothesized that the cellulose–lignin interaction occupies the C6 position, disfavoring glycosidic bond cleavage that is required for the formation of levoglucosan and favoring light compound and furan formation through ring scission, rearrangement, and dehydration reactions. The strength of this effect on pyrolysis products is most pronounced in grasses, followed by softwood and then hardwood, possibly due to the increased prevalence of covalent bonds between cellulose and lignin in grass cell walls (Jin et al., 2006; Zhou et al., 2010). Hemicellulose–lignin interactions, especially the xylan–lignin interaction revealed in NMR experiments (Yuan et al., 2011), may also affect pyrolysis. Indeed, enzymatic removal of hemicelluloses from lignin–carbohydrate complexes increased coniferyl alcohol yields (Du et al., 2014).

An example of a compositional feature that may impact product distribution is the degree of acetylation of the biopolymers. As mentioned, acetyl groups decorate hemicellulose side chains and are also present in the lignin. The increased abundance of these groups in biomass correlates with increasing yields of acetic acid, methyl pyruvate, acetone, and furan; additionally, this acetylation correlates with decreasing yields of furfural and acetaldehyde (Shafizadeh et al., 1972; Mante et al., 2014). While the acetic acid and perhaps the methyl pyruvate can be explained by the direct production of these compounds upon pyrolytic decomposition (Model 1), the nature of the relationship between acetate and the furanic and other 4-carbon species has not been clearly defined. The production of the 4-carbon species may be due to an indirect effect (Model 3) or may be the result of catalytic reaction of acetate with itself (Model 2).

Several investigations (Westerhof et al., 2007; He et al., 2009; Burhenne et al., 2013) suggest that feedstock moisture content can also play a role in the yield and product distribution of the organic fraction of the bio-oil. As previously discussed, the presence of water in bio-oil prohibits its direct use and creates challenges to catalytic valorization. For these reasons, biomass is typically subjected to drying prior to pyrolysis, which both reduces the required energy of the pyrolysis step and limits the water in the liquid condensate to water produced by decomposition reactions. However, the degree to which the feedstock moisture content should be eliminated is still under investigation. Burhenne et al. (2013) found that higher feedstock moisture content led to slightly lower char and gas yields upon pyrolysis with minimal changes to the elemental composition of the char. However, this is in disagreement with Westerhof et al. (2007) who observed slightly higher char yields with increasing moisture content. The water weight fraction distribution of the feedstocks in the two studies were quite different, 2.4–55.4% in Burhenne et al. versus 0–20% in Westerhof et al. Beyond impacts to the yields, He et al. (2009) studied the change in selectivity to the organic fraction produced upon pyrolysis of switchgrass with 5, 10, and 15% feedstock moisture contents. The authors found that at 500°C, the lowest moisture content feedstock produced the highest amounts of levoglucosan and acetic acid. The authors note that while significant differences in pyrolysis products were observed, they could not identify clear trends in their data. Among these studies, the observable but sometimes contradictory or unclear trends suggest that the feedstock moisture content may have multiple impacts on the pyrolysis process, possibly related to the physical location of the water in biomass.

In addition to compositional factors, morphological factors also influence the bio-oil product distributions. Biomass undergoing thermal decomposition retains its morphology even in harsh thermal treatment regimes (Pohlmann et al., 2014). Biomass is a poor conductor of heat (conductivity <0.1 W/m K) (Bridgwater et al., 1999), and large temperature gradients occur in heated biomass particles (Bryden et al., 2002). Most reactor systems for thermal degradation require size reduction of biomass particles; as an example, fluidized beds require particle sizes no larger than 2 mm (Bridgwater et al., 1999) to ensure rapid reaction. These particle sizes are larger than the tissue structures present in biomass. While the overall tissue and cellular morphology remain intact, micropore formation and shrinkage during the reaction process can occur in a non-uniform manner throughout the biomass (Davidsson and Pettersson, 2002; Pohlmann et al., 2014). Piskorz and colleagues observed decreasing liquid yields with increasing particle size, attributed to increasing incidence of secondary reactions with in wood particles (Scott and Piskorz, 1984). The principles of internal and external diffusion and the impacts of tortuosity, surface area, and diffusion path lengths are all fundamental to catalytic reaction engineering, and in the case of thermal biomass conversion, these important parameters are all dictated by the reacting feedstock (Fogler, 2006). Some evidence supports the notion that different plant developmental stages, which are related to the ratio of leaves to stems and biomass density, result in different pyrolysis products. For example, switchgrass harvested at later times during the growing season produced increased yields of condensable products, relative to that from younger, leafier material (Boateng et al., 2006), though compositional and developmental differences of the starting material were not carefully assessed.

CONCLUSION

Years of research have led to understanding of the direct pyrolysis conversion pathways of the major monomeric and polymeric constituents of biomass (Model 1, **Table 2**). The observation that these constituents often represent minor components in raw bio-oil (**Table 1**) highlights the importance of catalytic degradation (Model 2) and possibly indirect effects (Model 3) on pyrolysis products. The latter model is only recently receiving attention as knowledge of cell wall structures and analytical repertoires blossom (Mante et al., 2014; Zhang et al., 2015). Detailed examination of the relationships between components and products is still sparse, with the biological literature providing detailed characterization of cell wall components, while the engineering literature analyzes the chemical components, or often just total yields, of different pyrolysis fractions. We would argue that further investigations on the relationships between biomass components and thermal products will allow improvement of thermal product "quality." Short of attaining (or improving on) petroleum fuel-like properties, even the criteria for a high-quality thermal product remain unclear. As discussed, this is, in part, because methods

for upgrading are so dependent on bio-oil composition. Thus, methods that economically separate and/or simplify the different product streams, while still maintaining C–C bonds and overall C-content, are more likely to be amenable to catalytic upgrading.

Greater and more systematic analysis of biomass composition and pyrolysis products within species that show significant compositional variation will aid in better understanding biomass–bio-oil relationships. Much of the existing literature relies on comparisons of thermal degradation products across diverse taxonomic groups that vary greatly in cell wall composition beyond the biomass components measured (**Table 3**). An analysis of more subtle compositional differences, in which compositional factors are varied across different samples, may aid in refining biomass–bio-oil relationships. For example, genetic mutants that vary in only one component relative to near isogenic, unmutated "wild-type" plants can directly address relationships between starting components and products (Li et al., 2012). In addition to genetically determined compositional differences, biomass composition also depends on growth conditions and developmental stage, which relates to harvest time. Taken together, the scale of the problem points to the value of developing high-throughput methods to help identify species and genotypes that are most suitable for production of specific thermal products and to guide the optimization of genetic stocks and growth condition for bioenergy crops. Methods available to identify such "high-quality" biomass include near-infrared reflectance spectroscopy (Vogel et al., 2011), Fourier transform near-infrared spectroscopy (Liu et al., 2010), and pyrolysis molecular beam mass spectrometry, at least for lignin components (Sykes et al., 2009; Penning et al., 2014b). In general, these methods can be trained, either rationally or in a model-independent manner, to detect spectroscopic or molecular signatures in biological materials with linear or non-linear relationships to thermal products.

Besides selecting or breeding for natural variation in biomass composition (Wegrzyn et al., 2010; Penning et al., 2014a), it is also possible to genetically modify biomass composition (Bartley and Ronald, 2009). Most simply, genetic engineering of bioenergy plants can be achieved by modifying the plant's genome to (1) express genes from other organisms, (2) increase expression of native genes, or (3) reduce expression of native genes. More complex schemes are also possible, in which expression patterns of genes are altered through synthetic biology approaches that recombine various genetic elements (Yang et al., 2013). The most common method for plant genetic engineering co-opts the molecular machinery of a bacterial pathogen that introduces genes into plant chromosomes to facilitate its pathogenesis.

Genetic engineering to improve bio-oil production would aim to increase biomass components that enhance the yield of favored products and/or to decrease components that produce disfavored products or interfere with upgrading strategies. Advances in understanding cell wall biosynthesis, including genes responsible for synthesizing the major polymer classes (Bonawitz and Chapple, 2010; Scheller and Ulvskov, 2010; Pauly et al., 2013) and covalent interactions among them (Chiniquy et al., 2012; Bartley et al., 2013; Schultink et al., 2015); regulation of expression of the cell wall biosynthesis genes (Zhao and Dixon, 2011); and metal ion transport proteins that determine the abundance and distribution of plant mineral content (Ma et al., 2006; Yamaji and Ma, 2009; Zhong and Ye, 2015), lay the foundation for genetically engineering bioenergy crop cell wall content and structure. For example, lignin is an important target for genetic engineering for pyrolysis since the major lignin-derived products have a lower O:C ratio, a higher energy value, and are more stable than sugar-derived products (Tanger et al., 2013; Mante et al., 2014). Some important genes that participate in or regulate lignin synthesis have already been modified in energy crops without major interference with plant biomass yield (Baxter et al., 2014, 2015; reviewed in Bartley et al., 2014). However, current genetic engineering strategies are focused on developing low lignin biomass for saccharification and biochemical conversion to fuels. Therefore, more work is required to develop biomass with high-lignin content for thermal conversion. Producing corrosive acetic acid in bio-oil (Mante et al., 2014), acetyl groups on cell wall polymers are another potential target for genetic engineering of "pyrolysis crops." Three enzyme classes, including the reduced wall acetylation (RWA) proteins, Trichome birefringence-like (TBL) and Altered Xyloglucan (AXY) proteins acetylate cell wall polysaccharides (Lee et al., 2011; Xiong et al., 2013; Schultink et al., 2015). A mutant of the dicot reference plant, *Arabidopsis thaliana*, which lacks expression of all four RWA genes, shows a 40% reduction in secondary wall-associated acetyl groups (Lee et al., 2011). Reducing expression of this family of genes in bioenergy crops may help to solve the problems caused by acetic acid in bio-oil produced from such plants.

Pretreatments such as washing/leaching and torrefaction are another class of strategies to improve biomass quality by changing biomass composition (Zheng et al., 2013; Banks et al., 2014).

For example, by washing biomass with detergent (Triton) or acid to remove minerals, the yield of bio-oil is increased and reaction water content is reduced (Banks et al., 2014). Coupling biochemical conversion of biomass, which depletes the polysaccharide fraction, with pyrolysis of the resulting residue, or bagasse, is another avenue to explore further (Islam et al., 2010; Cunha et al., 2011). Torrefaction is a low-temperature (200–400°C) thermal pretreatment that decomposes hemicellulose and may segregate disfavored products such as water and acid into intermediate streams before the next stage of pyrolysis (Zheng et al., 2013). More efficient torrefaction may be achieved by changing the composition or chemical structure of hemicellulose through genetic methods to further separate the decomposition temperatures of hemicellulose from lignin and cellulose. By identifying and studying the roles of key biomass components during thermal conversion, it will be possible to maximize the economic and environmental benefits of plant biomass-derived biofuels in the future.

AUTHOR CONTRIBUTIONS

FL and CW drafted and revised the manuscript. LB, LL, and RM revised the manuscript.

ACKNOWLEDGMENTS

This material is based upon work supported by the National Institute of Food and Agriculture, U.S. Department of Agriculture, under award number 2010-38502-21836, through the South Central Sun Grant Program.

REFERENCES

Adler, P. R., Sanderson, M. A., Boateng, A. A., Weimer, P. I., and Jung, H. J. G. (2006). Biomass yield and biofuel quality of switchgrass harvested in fall or spring. *Agron. J.* 98, 1518–1525. doi:10.2134/agronj2005.0351

Banks, S. W., Nowakowski, D. J., and Bridgwater, A. V. (2014). Fast pyrolysis processing of surfactant washed *Miscanthus*. *Fuel Process. Technol.* 128, 94–103. doi:10.1016/j.fuproc.2014.07.005

Bartley, L. E., Peck, M. L., Kim, S. R., Ebert, B., Manisseri, C., Chiniquy, D. M., et al. (2013). Overexpression of a BAHD acyltransferase, *OsAt10*, alters rice cell wall hydroxycinnamic acid content and saccharification. *Plant Physiol.* 161, 1615–1633. doi:10.1104/pp.112.208694

Bartley, L. E., and Ronald, P. C. (2009). Plant and microbial research seeks biofuel production from lignocellulose. *Calif. Agric.* 63, 178–184. doi:10.3733/ca.v063n04p178

Bartley, L. E., Tao, X., Zhang, C., Nguyen, H., and Zhou, J. (2014). "Switchgrass biomass content, synthesis, and biochemical conversion to biofuels," in *Compendium of Bioenergy Plants*, eds Luo H., and Wu Y. (Boca Raton, FL: Science), 109–169.

Baxter, H. L., Mazarei, M., Labbe, N., Kline, L. M., Cheng, Q., Windham, M. T., et al. (2014). Two-year field analysis of reduced recalcitrance transgenic switchgrass. *Plant Biotechnol. J.* 12, 914–924. doi:10.1111/pbi.12195

Baxter, H. L., Poovaiah, C. R., Yee, K. L., Mazarei, M., Rodriguez, M., Thompson, O. A., et al. (2015). Field evaluation of transgenic switchgrass plants overexpressing *PvMYB4* for reduced biomass recalcitrance. *Bioenerg. Res.* 8, 910–921. doi:10.1007/s12155-014-9570-1

Boateng, A. A., Hicks, K. B., and Vogel, K. P. (2006). Pyrolysis of switchgrass (*Panicum virgatum*) harvested at several stages of maturity. *J. Anal. Appl. Pyrolysis* 75, 55–64. doi:10.1016/j.jaap.2005.03.005

Boerjan, W., Ralph, J., and Baucher, M. (2003). Lignin biosynthesis. *Annu. Rev. Plant Biol.* 54, 519–546. doi:10.1146/annurev.arplant.54.031902.134938

Bonawitz, N. D., and Chapple, C. (2010). The genetics of lignin biosynthesis: connecting genotype to phenotype. *Ann. Rev. Genet.* 44, 337–363. doi:10.1146/annurev-genet-102209-163508

Bridgwater, A. V. (1994). Catalysis in thermal biomass conversion. *Appl. Catal. A Gen.* 116, 5–47. doi:10.1016/0926-860x(94)80278-5

Bridgwater, A. V., Meier, D., and Radlein, D. (1999). An overview of fast pyrolysis of biomass. *Org. Geochem.* 30, 1479–1493. doi:10.1016/S0146-6380(99)00120-5

Bryden, K. M., Ragland, K. W., and Rutland, C. J. (2002). Modeling thermally thick pyrolysis of wood. *Biomass Bioenergy* 22, 41–53. doi:10.1016/S0961-9534(01)00060-5

Bule, M. V., Gao, A. H., Hiscox, B., and Chen, S. (2013). Structural modification of lignin and characterization of pretreated wheat straw by ozonation. *J. Agric. Food Chem.* 61, 3916–3925. doi:10.1021/jf4001988

Bunzel, M., Ralph, J., Lu, F., Hatfield, R. D., and Steinhart, H. (2004). Lignins and ferulate-coniferyl alcohol cross-coupling products in cereal grains. *J. Agric. Food Chem.* 52, 6496–6502. doi:10.1021/jf040204p

Burhenne, L., Damiani, M., and Aicher, T. (2013). Effect of feedstock water content and pyrolysis temperature on the structure and reactivity of spruce wood char produced in fixed bed pyrolysis. *Fuel* 107, 836–847. doi:10.1016/j.fuel.2013.01.033

Carlson, T. R., Tompsett, G. A., Conner, W. C., and Huber, G. W. (2009). Aromatic production from catalytic fast pyrolysis of biomass-derived feedstocks. *Top. Catal.* 52, 241–252. doi:10.1007/s11244-008-9160-6

Carlson, T. R., Vispute, T. R., and Huber, G. W. (2008). Green gasoline by catalytic fast pyrolysis of solid biomass derived compounds. *ChemSusChem* 1, 397–400. doi:10.1002/cssc.200800018

Carpita, N. C., Defernez, M., Findlay, K., Wells, B., Shoue, D. A., Catchpole, G., et al. (2001). Cell wall architecture of the elongating maize coleoptile. *Plant Physiol.* 127, 551–565. doi:10.1104/pp.010146

Chen, F., Tobimatsu, Y., Havkin-Frenkel, D., Dixon, R. A., and Ralph, J. (2012). A polymer of caffeyl alcohol in plant seeds. *Proc. Natl. Acad. Sci. U.S.A.* 109, 1772–1777. doi:10.1073/pnas.1120992109

Chen, F., Tobimatsu, Y., Jackson, L., Nakashima, J., Ralph, J., and Dixon, R. A. (2013). Novel seed coat lignins in the Cactaceae: structure, distribution and implications for the evolution of lignin diversity. *Plant J.* 73, 201–211. doi:10.1111/tpj.12012

Chiniquy, D., Sharma, V., Schultink, A., Baidoo, E. E., Rautengarten, C., Cheng, K., et al. (2012). XAX1 from glycosyltransferase family 61 mediates xylosyltransfer to rice xylan. *Proc. Natl. Acad. Sci. U.S.A.* 109, 17117–17122. doi:10.1073/pnas.1202079109

Collard, F. X., and Blin, J. (2014). A review on pyrolysis of biomass constituents: mechanisms and composition of the products obtained from the conversion of cellulose, hemicelluloses and lignin. *Renew. Sustain. Energ. Rev.* 38, 594–608. doi:10.1016/j.rser.2014.06.013

Cunha, J. A., Pereira, M. M., Valente, L. M. M., De La Piscina, P. R., Homs, N., and Santos, M. R. L. (2011). Waste biomass to liquids: low temperature conversion of sugarcane bagasse to bio-oil. The effect of combined hydrolysis treatments. *Biomass Bioenergy* 35, 2106–2116. doi:10.1016/j.biombioe.2011.02.019

Davidsson, K. O., and Pettersson, J. B. C. (2002). Birch wood particle shrinkage during rapid pyrolysis. *Fuel* 81, 263–270. doi:10.1016/S0016-2361(01)00169-7

Demirbas, A. (2011). Competitive liquid biofuels from biomass. *Appl. Energy* 88, 17–28. doi:10.1016/j.apenergy.2010.07.016

Dick-Perez, M., Zhang, Y., Hayes, J., Salazar, A., Zabotina, O. A., and Hong, M. (2011). Structure and interactions of plant cell-wall polysaccharides by two- and three-dimensional magic-angle-spinning solid-state NMR. *Biochemistry* 50, 989–1000. doi:10.1021/bi101795q

Du, X., Perez-Boada, M., Fernandez, C., Rencoret, J., Del Rio, J. C., Jimenez-Barbero, J., et al. (2014). Analysis of lignin-carbohydrate and lignin-lignin linkages after hydrolase treatment of xylan-lignin, glucomannan-lignin and glucan-lignin complexes from spruce wood. *Planta* 239, 1079–1090. doi:10.1007/s00425-014-2037-y

El-Nashaar, H. M., Banowetz, G. M., Griffith, S. M., Casler, M. D., and Vogel, K. P. (2009). Genotypic variability in mineral composition of switchgrass. *Bioresour. Technol.* 100, 1809–1814. doi:10.1016/j.biortech.2008.09.058

Eom, I. Y., Kim, J. Y., Kim, T. S., Lee, S. M., Choi, D., Choi, I. G., et al. (2012). Effect of essential inorganic metals on primary thermal degradation of lignocellulosic biomass. *Bioresour. Technol.* 104, 687–694. doi:10.1016/j.biortech.2011.10.035

Fahmi, R., Bridgwater, A. V., Darvell, L. I., Jones, J. M., Yates, N., Thain, S., et al. (2007). The effect of alkali metals on combustion and pyrolysis of *Lolium* and *Festuca* grasses, switchgrass and willow. *Fuel* 86, 1560–1569. doi:10.1016/j.fuel.2006.11.030

Faix, O., Fortmann, I., Bremer, J., and Meier, D. (1991a). Thermal-degradation products of wood – a collection of electron-impact (EI) mass-spectra of polysaccharide derived products. *Holz Roh Werkst.* 49, 299–304. doi:10.1007/Bf02663795

Faix, O., Fortmann, I., Bremer, J., and Meier, D. (1991b). Thermal-degradation products of wood – gas-chromatographic separation and mass-spectrometric characterization of polysaccharide derived products. *Holz Roh Werkst.* 49, 213–219. doi:10.1007/Bf02613278

Fogler, H. S. (2006). *Elements of Chemical Reaction Engineering*. Upper Saddle River, NJ: Prentice Hall PTR.

Furimsky, E. (2000). Catalytic hydrodeoxygenation. *Appl. Catal. A Gen.* 199, 147–190. doi:10.1016/S0926-860x(99)00555-4

Gonzalez-Borja, M. A., and Resasco, D. E. (2015). Reaction pathways in the liquid phase alkylation of biomass-derived phenolic compounds. *AIChE J.* 61, 598–609. doi:10.1002/aic.14658

Hatfield, R. D., Marita, J. M., and Frost, K. (2008). Characterization of *p*-coumarate accumulation, *p*-coumaroyl transferase, and cell wall changes during the development of corn stems. *J. Sci. Food Agric.* 88, 2529–2537. doi:10.1002/jsfa.3376

He, R., Ye, X. P., English, B. C., and Satrio, J. A. (2009). Influence of pyrolysis condition on switchgrass bio-oil yield and physicochemical properties. *Bioresour. Technol.* 100, 5305–5311. doi:10.1016/j.biortech.2009.02.069

Huber, G. W., Iborra, S., and Corma, A. (2006). Synthesis of transportation fuels from biomass: chemistry, catalysts, and engineering. *Chem. Rev.* 106, 4044–4098. doi:10.1021/cr068360d

Islam, M. R., Parveen, M., and Haniu, H. (2010). Properties of sugarcane waste-derived bio-oils obtained by fixed-bed fire-tube heating pyrolysis. *Bioresour. Technol.* 101, 4162–4168. doi:10.1016/j.biortech.2009.12.137

Jin, Z., Katsumata, K. S., Lam, T. B., and Iiyama, K. (2006). Covalent linkages between cellulose and lignin in cell walls of coniferous and nonconiferous woods. *Biopolymers* 83, 103–110. doi:10.1002/bip.20533

Kawamoto, H., Horigoshi, S., and Saka, S. (2007). Pyrolysis reactions of various lignin model dimers. *J. Wood Sci.* 53, 168–174. doi:10.1007/s10086-006-0834-z

Kido, N., Yokoyama, R., Yamamoto, T., Furukawa, J., Iwai, H., Satoh, S., et al. (2015). The matrix polysaccharide (1;3,1;4)-beta-D-glucan is involved in silicon-dependent strengthening of rice cell wall. *Plant Cell Physiol.* 56, 268–276. doi:10.1093/pcp/pcu162

Kleen, M., and Gellerstedt, G. (1995). Influence of inorganic species on the formation of polysaccharide and lignin degradation products in the analytical pyrolysis of pulps. *J. Anal. Appl. Pyrolysis* 35, 15–41. doi:10.1016/0165-2370(95)00893-J

Kozlova, L. V., Ageeva, M. V., Ibragimova, N. N., and Gorshkova, T. A. (2014). Arrangement of mixed-linkage glucan and glucuronoarabinoxylan in the cell walls of growing maize roots. *Ann. Bot.* 114, 1135–1145. doi:10.1093/aob/mcu125

Lan, W., Lu, F., Regner, M., Zhu, Y., Rencoret, J., Ralph, S. A., et al. (2015). Tricin, a flavonoid monomer in monocot lignification. *Plant Physiol.* 167, 1284–1295. doi:10.1104/pp.114.253757

Lee, C., Teng, Q., Zhong, R., and Ye, Z. H. (2011). The four *Arabidopsis* reduced wall acetylation genes are expressed in secondary wall-containing cells and required for the acetylation of xylan. *Plant Cell Physiol.* 52, 1289–1301. doi:10.1093/pcp/pcr075

Li, M., Foster, C., Kelkar, S., Pu, Y., Holmes, D., Ragauskas, A., et al. (2012). Structural characterization of alkaline hydrogen peroxide pretreated grasses exhibiting diverse lignin phenotypes. *Biotechnol. Biofuels* 5, 38. doi:10.1186/1754-6834-5-38

Liu, L., Ye, X. P., Womac, A. R., and Sokhansanj, S. (2010). Variability of biomass chemical composition and rapid analysis using FT-NIR techniques. *Carbohydr. Polym.* 81, 820–829. doi:10.1016/j.carbpol.2010.03.058

Lou, R., Wu, S. B., Lv, G. J., and Zhang, A. L. (2013). Factors related to minerals and ingredients influencing the distribution of pyrolysates derived from herbaceous biomass. *Bioresources* 8, 1345–1360. doi:10.15376/biores.8.1.1345-1360

Lu, F., Karlen, S. D., Regner, M., Kim, H., Ralph, S. A., Sun, R.-C., et al. (2015). Naturally *p*-hydroxybenzoylated lignins in palms. *Bioenergy Res.* 8, 934–952. doi:10.1007/s12155-015-9583-4

Lu, F. C., and Ralph, J. (2002). Preliminary evidence for sinapyl acetate as a lignin monomer in kenaf. *Chem. Commun.* 90–91. doi:10.1039/b109876d

Ma, J. F., Tamai, K., Yamaji, N., Mitani, N., Konishi, S., Katsuhara, M., et al. (2006). A silicon transporter in rice. *Nature* 440, 688–691. doi:10.1038/nature04590

Mante, O. D., Babu, S. P., and Amidon, T. E. (2014). A comprehensive study on relating cell-wall components of lignocellulosic biomass to oxygenated species formed during pyrolysis. *J. Anal. Appl. Pyrolysis* 108, 56–67. doi:10.1016/j.jaap.2014.05.016

Masia, A. A. T., Buhre, B. J. P., Gupta, R. P., and Wall, T. F. (2007). Characterising ash of biomass and waste. *Fuel Process. Technol.* 88, 1071–1081. doi:10.1016/j.fuproc.2007.06.011

McKendry, P. (2002). Energy production from biomass (part 1): overview of biomass. *Bioresour. Technol.* 83, 37–46. doi:10.1016/S0960-8524(01)00118-3

Mendu, V., Harman-Ware, A. E., Crocker, M., Jae, J., Stork, J., Morton, S. III, et al. (2011). Identification and thermochemical analysis of high-lignin feedstocks for biofuel and biochemical production. *Biotechnol. Biofuels* 4, 43. doi:10.1186/1754-6834-4-43

Mettler, M. S., Vlachos, D. G., and Dauenhauer, P. J. (2012). Top ten fundamental challenges of biomass pyrolysis for biofuels. *Energy Environ. Sci.* 5, 7797–7809. doi:10.1039/C2EE21679E

Mikkelsen, D., Flanagan, B. M., Wilson, S. M., Bacic, A., and Gidley, M. J. (2015). Interactions of arabinoxylan and (1,3)(1,4)-beta-glucan with cellulose networks. *Biomacromolecules* 16, 1232–1239. doi:10.1021/acs.biomac.5b00009

Mohan, D., Pittman, C. U., and Steele, P. H. (2006). Pyrolysis of wood/biomass for bio-oil: a critical review. *Energy Fuel* 20, 848–889. doi:10.1021/ef0502397

Mourant, D., Wang, Z. H., He, M., Wang, X. S., Garcia-Perez, M., Ling, K. C., et al. (2011). Mallee wood fast pyrolysis: effects of alkali and alkaline earth metallic species on the yield and composition of bio-oil. *Fuel* 90, 2915–2922. doi:10.1016/j.fuel.2011.04.033

Mullen, C. A., Boateng, A. A., Hicks, K. B., Goldberg, N. M., and Moreau, R. A. (2010). Analysis and comparison of bio-oil produced by fast pyrolysis from three barley biomass/byproduct streams. *Energy Fuel* 24, 699–706. doi:10.1021/ef900912s

Muller-Hagedorn, M., Bockhorn, H., Krebs, L., and Muller, U. (2003). A comparative kinetic study on the pyrolysis of three different wood species. *J. Anal. Appl. Pyrolysis* 6, 231–249. doi:10.1016/S0165-2370(03)00065-2

Newman, R. H., Hill, S. J., and Harris, P. J. (2013). Wide-angle x-ray scattering and solid-state nuclear magnetic resonance data combined to test models for cellulose microfibrils in mung bean cell walls. *Plant Physiol.* 163, 1558–1567. doi:10.1104/pp.113.228262

NSF. (2008). "Selective thermal processing of cellulosic biomass and lignin" in *Breaking the Chemical and Engineering Barriers to Lignocellulosic Biofuels: Next Generation Hydrocarbon Biorefineries.* ed. Huber G. W. (Washington, DC: University of Massachusetts, Amherst), 30–47.

Oasmaa, A., and Czernik, S. (1999). Fuel oil quality of biomass pyrolysis oils state of the art for the end users. *Energy Fuel* 13, 914–921. doi:10.1021/ef980272b

O'Neill, M. A., and York, W. S. (2009). "The plant cell wall," in *Annual Plant Reviews.* ed. Rose J. K. C. (Blackwell: Wiley), 1–44.

Patwardhan, P. R., Brown, R. C., and Shanks, B. H. (2011a). Understanding the fast pyrolysis of lignin. *ChemSusChem* 4, 1629–1636. doi:10.1002/cssc.201100133

Patwardhan, P. R., Dalluge, D. L., Shanks, B. H., and Brown, R. C. (2011b). Distinguishing primary and secondary reactions of cellulose pyrolysis. *Bioresour. Technol.* 102, 5265–5269. doi:10.1016/j.biortech.2011.02.018

Patwardhan, P. R., Satrio, J. A., Brown, R. C., and Shanks, B. H. (2010). Influence of inorganic salts on the primary pyrolysis products of cellulose. *Bioresour. Technol.* 101, 4646–4655. doi:10.1016/j.biortech.2010.01.112

Pauly, M., Gille, S., Liu, L., Mansoori, N., De Souza, A., Schultink, A., et al. (2013). Hemicellulose biosynthesis. *Planta* 238, 627–642. doi:10.1007/s00425-013-1921-1

Pauly, M., and Keegstra, K. (2008). Cell-wall carbohydrates and their modification as a resource for biofuels. *Plant J.* 54, 559–568. doi:10.1111/j.1365-313X.2008.03463.x

Penning, B. W., Sykes, R. W., Babcock, N. C., Dugard, C. K., Held, M. A., Klimek, J. F., et al. (2014a). Genetic determinants for enzymatic digestion of lignocellulosic biomass are independent of those for lignin abundance in a maize recombinant inbred population. *Plant Physiol.* 165, 1475–1487. doi:10.1104/pp.114.242446

Penning, B. W., Sykes, R. W., Babcock, N. C., Dugard, C. K., Klimek, J. F., Gamblin, D., et al. (2014b). Validation of PyMBMS as a high-throughput screen for lignin abundance in lignocellulosic biomass of grasses. *Bioenerg. Res.* 7, 899–908. doi:10.1007/s12155-014-9410-3

Perlack, R. D., Wright, L. L., Turhollow, A., Graham, R. L., Stokes, B. J., and Erbach, D. C. (2005). *Biomass as Feedstock for a Bioenergy and Bioproducts Industry: The Technical Feasibility of a Billion-Ton Annual Supply.* Oak Ridge, TN: Oak Ridge National Laboratory.

Petrik, D. L., Karlen, S. D., Cass, C. L., Padmakshan, D., Lu, F., Liu, S., et al. (2014). p-Coumaroyl-CoA:monolignol transferase (PMT) acts specifically in the lignin biosynthetic pathway in Brachypodium distachyon. *Plant J.* 77, 713–726. doi:10.1111/tpj.12420

Pham, T. N., Sooknoi, T., Crossley, S. P., and Resasco, D. E. (2013). Ketonization of carboxylic acids: mechanisms, catalysts, and implications for biomass conversion. *ACS Catal.* 3, 2456–2473. doi:10.1021/cs400501h

Pictet, A., and Sarasin, J. (1918). Sur la distillation de la cellulose et de l'amidon sous pression réduite. *Helv. Chim. Acta* 1, 87–96. doi:10.1002/hlca.19180010109

Pohlmann, J. G., Osorio, E., Vilela, A. C. F., Diez, M. A., and Borrego, A. G. (2014). Integrating physicochemical information to follow the transformations of biomass upon torrefaction and low-temperature carbonization. *Fuel* 131, 17–27. doi:10.1016/j.fuel.2014.04.067

Ponder, G. R., and Richards, G. N. (1991). Mechanisms of pyrolysis of polysaccharides. 4. Thermal synthesis and pyrolysis of a xylan. *Carbohydr. Res.* 218, 143–155. doi:10.1016/0008-6215(91)84093-T

Raisanen, U., Pitkanen, I., Halttunen, H., and Hurtta, M. (2003). Formation of the main degradation compounds from arabinose, xylose, mannose and arabinitol during pyrolysis. *J. Therm. Anal. Calorim.* 72, 481–488. doi:10.1023/A:1024557011975

Ralph, J., Grabber, J. H., and Hatfield, R. D. (1995). Lignin-ferulate cross-links in grasses – active incorporation of ferulate polysaccharide esters into ryegrass lignins. *Carbohydr. Res.* 275, 167–178. doi:10.1016/0008-6215(95)00237-N

Resasco, D. E. (2011). What should we demand from the catalysts responsible for upgrading biomass pyrolysis oil? *J. Phys. Chem. Lett.* 2, 2294–2295. doi:10.1021/jz201135x

Rodriguez-Gacio Mdel, C., Iglesias-Fernandez, R., Carbonero, P., and Matilla, A. J. (2012). Softening-up mannan-rich cell walls. *J. Exp. Bot.* 63, 3976–3988. doi:10.1093/jxb/ers096

Ronsse, F., Bai, X. L., Prins, W., and Brown, R. C. (2012). Secondary reactions of levoglucosan and char in the fast pyrolysis of cellulose. *Environ. Prog. Sustain.* 31, 256–260. doi:10.1002/ep.11633

Saizjimenez, C., and Deleeuw, J. W. (1986). Lignin pyrolysis products – their structures and their significance as biomarkers. *Org. Geochem.* 10, 869–876. doi:10.1016/S0146-6380(86)80024-9

Sarkanen, K. V., Chang, H., and Allan, G. G. (1967). Species variation in lignins. 3. Hardwood lignins. *Tappi* 50, 587.

Scheller, H. V., and Ulvskov, P. (2010). Hemicelluloses. *Annu. Rev. Plant Biol.* 61, 263–289. doi:10.1146/annurev-arplant-042809-112315

Schultink, A., Naylor, D., Dama, M., and Pauly, M. (2015). The role of the plant-specific ALTERED XYLOGLUCAN9 protein in Arabidopsis cell wall polysaccharide O-acetylation. *Plant Physiol.* 167, 1271–U1243. doi:10.1104/pp.114.256479

Scott, D. S., and Piskorz, J. (1984). The continuous flash pyrolysis of biomass. *Can. J. Chem. Eng.* 62, 404–412. doi:10.1016/j.biortech.2012.11.114

Shafizadeh, F., Mcginnis, G. D., and Philpot, C. W. (1972). Thermal degradation of xylan and related model compounds. *Carbohydr. Res.* 25, 23–33. doi:10.1016/S0008-6215(00)82742-1

Shafizadeh, F., Furneaux, R. H., Cochran, T. G., Scholl, J. P., and Sakai, Y. (1979). Production of levoglucosan and glucose from pyrolysis of cellulosic materials. *J. Appl. Polym. Sci.* 23, 3525–3539. doi:10.1002/app.1979.070231209

Singh, M. P., Erickson, J. E., Sollenberger, L. E., Woodard, K. R., Vendramini, J. M. B., and Fedenko, J. R. (2012). Mineral composition and biomass partitioning of sweet sorghum grown for bioenergy in the southeastern USA. *Biomass Bioenergy* 47, 1–8. doi:10.1016/j.biombioe.2012.10.022

Somerville, C. (2006). Cellulose synthesis in higher plants. *Annu. Rev. Cell Dev. Biol.* 22, 53–78. doi:10.1146/annurev.cellbio.22.022206.160206

Somerville, C. (2007). Biofuels. *Curr. Biol.* 17, R115–R119. doi:10.1016/j.cub.2007.01.010

Somerville, C., Bauer, S., Brininstool, G., Facette, M., Hamann, T., Milne, J., et al. (2004). Toward a systems approach to understanding plant cell walls. *Science* 306, 2206–2211. doi:10.1126/science.1102765

Sykes, R., Yung, M., Novaes, E., Kirst, M., Peter, G., and Davis, M. (2009). "High-throughput screening of plant cell-wall composition using pyrolysis molecular beam mass spectroscopy," in *Biofuels,* ed. Mielenz J. R. (New York City: Humana Press), 169–183.

Tan, L., Eberhard, S., Pattathil, S., Warder, C., Glushka, J., Yuan, C., et al. (2013). An Arabidopsis cell wall proteoglycan consists of pectin and arabinoxylan covalently linked to an arabinogalactan protein. *Plant Cell* 25, 270–287. doi:10.1105/tpc.112.107334

Tanger, P., Field, J. L., Jahn, C. E., Defoort, M. W., and Leach, J. E. (2013). Biomass for thermochemical conversion: targets and challenges. *Front. Plant Sci.* 4:218. doi:10.3389/fpls.2013.00218

Vanholme, B., Cesarino, I., Goeminne, G., Kim, H., Marroni, F., Van Acker, R., et al. (2013). Breeding with rare defective alleles (BRDA): a natural Populus nigra HCT mutant with modified lignin as a case study. *New Phytol.* 198, 765–776. doi:10.1111/nph.12179

Vanholme, R., Van Acker, R., and Boerjan, W. (2010). Potential of Arabidopsis systems biology to advance the biofuel field. *Trends Biotechnol.* 28, 543–547. doi:10.1016/j.tibtech.2010.07.008

Vassilev, S. V., Baxter, D., Andersen, L. K., and Vassileva, C. G. (2010). An overview of the chemical composition of biomass. *Fuel* 89, 913–933. doi:10.1016/j.fuel.2009.10.022

Vega-Sanchez, M. E., Verhertbruggen, Y., Christensen, U., Chen, X., Sharma, V., Varanasi, P., et al. (2012). Loss of cellulose synthase-like F6 function affects mixed-linkage glucan deposition, cell wall mechanical properties, and defense responses in vegetative tissues of rice. *Plant Physiol.* 159, 56–69. doi:10.1104/pp.112.195495

Vega-Sanchez, M. E., Verhertbruggen, Y., Scheller, H. V., and Ronald, P. C. (2013). Abundance of mixed linkage glucan in mature tissues and secondary cell walls of grasses. *Plant Signal. Behav.* 8, e23143. doi:10.4161/psb.23143

Vogel, J. (2008). Unique aspects of the grass cell wall. *Curr. Opin. Plant Biol.* 11, 301–307. doi:10.1016/j.pbi.2008.03.002

Vogel, K. P., Dien, B. S., Jung, H. G., Casler, M. D., Masterson, S. D., and Mitchell, R. B. (2011). Quantifying actual and theoretical ethanol yields for switchgrass strains using NIRS analyses. *Bioenergy Res.* 4, 96–110. doi:10.1007/s12155-010-9104-4

Wan, S. L., Waters, C., Stevens, A., Gumidyala, A., Jentoft, R., Lobban, L., et al. (2015). Decoupling HZSM-5 catalyst activity from deactivation during upgrading of pyrolysis oil vapors. *ChemSusChem* 8, 552–559. doi:10.1002/cssc.201402861

Wegrzyn, J. L., Eckert, A. J., Choi, M., Lee, J. M., Stanton, B. J., Sykes, R., et al. (2010). Association genetics of traits controlling lignin and cellulose biosynthesis in black cottonwood (*Populus trichocarpa*, Salicaceae) secondary xylem. *New Phytol.* 188, 515–532. doi:10.1111/j.1469-8137.2010.03415.x

Westerhof, R. J. M., Kuipers, N. J. M., Kersten, S. R. A., and Van Swaaij, W. P. M. (2007). Controlling the water content of biomass fast pyrolysis oil. *Ind. Eng. Chem. Res.* 46, 9238–9247. doi:10.1021/ie070684k

Xiong, G. Y., Cheng, K., and Pauly, M. (2013). Xylan O-acetylation impacts xylem development and enzymatic recalcitrance as indicated by the *Arabidopsis* mutant tbl29. *Mol. Plant.* 6, 1373–1375. doi:10.1093/mp/sst014

Yamaji, N., and Ma, J. F. (2009). A transporter at the node responsible for intervascular transfer of silicon in rice. *Plant Cell* 21, 2878–2883. doi:10.1105/tpc.109.069831

Yang, F., Mitra, P., Zhang, L., Prak, L., Verhertbruggen, Y., Kim, J. S., et al. (2013). Engineering secondary cell wall deposition in plants. *Plant Biotechnol. J.* 11, 325–335. doi:10.1111/Pbi.12016

Youngs, H., and Somerville, C. (2012). Growing better biofuel crops. *Scientist* 26, 46–52. doi:10.2134/jeq2013.05.0171

Yuan, T. Q., Sun, S. N., Xu, F., and Sun, R. C. (2011). Characterization of lignin structures and lignin-carbohydrate complex (LCC) linkages by quantitative 13C and 2D HSQC NMR spectroscopy. *J. Agric. Food Chem.* 59, 10604–10614. doi:10.1021/jf2031549

Zapata, P. A., Faria, J., Ruiz, M. P., and Resasco, D. E. (2012). Condensation/hydrogenation of biomass-derived oxygenates in water/oil emulsions stabilized by nanohybrid catalysts. *Top. Catal.* 55, 38–52. doi:10.1007/s11244-012-9768-4

Zhang, J., Choi, Y. S., Yoo, C. G., Kim, T. H., Brown, R. C., and Shanks, B. H. (2015). Cellulose-hemicellulose and cellulose-lignin interactions during fast pyrolysis. *ACS Sustain. Chem. Eng.* 3, 293–301. doi:10.1021/sc500664h

Zhang, X. S., Yang, G. X., Jiang, H., Liu, W. J., and Ding, H. S. (2013). Mass production of chemicals from biomass-derived oil by directly atmospheric distillation coupled with co-pyrolysis. *Sci. Rep.* 3, 1120. doi:10.1038/srep01120

Zhao, Q., and Dixon, R. A. (2011). Transcriptional networks for lignin biosynthesis: more complex than we thought? *Trends Plant Sci.* 16, 227–233. doi:10.1016/j.tplants.2010.12.005

Zhao, Y. L., Steinberger, Y., Shi, M., Han, L. P., and Xie, G. H. (2012). Changes in stem composition and harvested produce of sweet sorghum during the period from maturity to a sequence of delayed harvest dates. *Biomass Bioenergy* 39, 261–273. doi:10.1016/j.biombioe.2012.01.020

Zheng, A., Zhao, Z., Chang, S., Huang, Z., Wang, X., He, F., et al. (2013). Effect of torrefaction on structure and fast pyrolysis behavior of corncobs. *Bioresour. Technol.* 128, 370–377. doi:10.1016/j.biortech.2012.10.067

Zhong, R., and Ye, Z. H. (2015). Secondary cell walls: biosynthesis, patterned deposition and transcriptional regulation. *Plant Cell Physiol.* 56, 195–214. doi:10.1093/pcp/pcu140

Zhou, Y., Stuart-Williams, H., Farquhar, G. D., and Hocart, C. H. (2010). The use of natural abundance stable isotopic ratios to indicate the presence of oxygen-containing chemical linkages between cellulose and lignin in plant cell walls. *Phytochemistry* 71, 982–993. doi:10.1016/j.phytochem.2010.03.001

Zhu, X. L., Lobban, L. L., Mallinson, R. G., and Resasco, D. E. (2011). Bifunctional transalkylation and hydrodeoxygenation of anisole over a Pt/HBeta catalyst. *J. Catal.* 281, 21–29. doi:10.1016/j.jcat.2011.03.030

Conflict of Interest Statement: The authors declare that the research was conducted in the absence of any commercial or financial relationships that could be construed as a potential conflict of interest.

14

Non-invasive rapid harvest time determination of oil-producing microalgae cultivations for biodiesel production by using chlorophyll fluorescence

Yaqin Qiao[1,2], Junfeng Rong[3], Hui Chen[1], Chenliu He[1] and Qiang Wang[1]*

[1] *Key Laboratory of Algal Biology, Institute of Hydrobiology, Chinese Academy of Sciences, Wuhan, China,* [2] *University of Chinese Academy of Sciences, Beijing, China,* [3] *SINOPEC Research Institute of Petroleum Processing, Beijing, China*

Edited by:
P. C. Abhilash,
Banaras Hindu University, India

Reviewed by:
Jason Ryan Hattrick-Simpers,
University of South Carolina, USA
Shanmugaprakash Muthusamy,
Kumaraguru College of
Technology, India

*Correspondence:
Qiang Wang,
Key Laboratory of Algal Biology,
Institute of Hydrobiology, Chinese
Academy of Sciences, 7 South
Donghu Road, Hubei Province,
Wuhan 430072, China
wangqiang@ihb.ac.cn

For the large-scale cultivation of microalgae for biodiesel production, one of the key problems is the determination of the optimum time for algal harvest when algae cells are saturated with neutral lipids. In this study, a method to determine the optimum harvest time in oil-producing microalgal cultivations by measuring the maximum photochemical efficiency of photosystem II, also called Fv/Fm, was established. When oil-producing *Chlorella* strains were cultivated and then treated with nitrogen starvation, it not only stimulated neutral lipid accumulation, but also affected the photosynthesis system, with the neutral lipid contents in all four algae strains – *Chlorella sorokiniana* C1, *Chlorella* sp. C2, *C. sorokiniana* C3, and *C. sorokiniana* C7 – correlating negatively with the Fv/Fm values. Thus, for the given oil-producing algae, in which a significant relationship between the neutral lipid content and Fv/Fm value under nutrient stress can be established, the optimum harvest time can be determined by measuring the value of Fv/Fm. It is hoped that this method can provide an efficient way to determine the harvest time rapidly and expediently in large-scale oil-producing microalgae cultivations for biodiesel production.

Keywords: chlorophyll fluorescence, harvest time, oil-producing microalgae, neutral lipid, photosynthesis

Introduction

The limiting supply, increasing cost, and ever increasing environmental pollution and health problems from conventional fossil fuels is placing an increasing demand on biodiesel to replace some fossil fuel usage (Hill et al., 2006; Hu et al., 2006; Hu et al., 2008; Amaro et al., 2011). With this in mind, microalgae have attracted considerable attention in recent years as potential sources of renewable fuel, due to their fast growth rate, high adaptability to environment conditions, and their no competition with crops for arable land and potable water (Santos et al., 2011). Microalgae have been promising feedstock for biodiesel production (Zhang et al., 2014b). The green microalgae genus *Chlorella* (*Chlorophyta*), which are capable of photoautotrophic, mixotrophic, and heterotrophic growth with high biomass accumulation, appears to contain good candidate strains for biodiesel production (Petkov and Garcia, 2007).

The cell metabolism of microalgae varies with the change in growth conditions. Nitrogen (N) starvation is one of the most effective environmental stresses and stimulates the accumulation of lipids in many microalgae (Illman et al., 2000). The general principle is that the lack of N limits protein biosynthesis, and the fixed carbon from photosynthesis will then channeled into high energy density compounds such as triglycerides and/or starch (Scott et al., 2010). With limited N supply, *Chlorella* could accumulate doubled or even tripled amount of lipid (Converti et al., 2009; Widjaja et al., 2009). However, even though cultivation with N limiting media results in an increase in the lipid content on a per cell weight basis, it also lowers the total biomass productivity significantly, resulting in decreased lipid productivity. A two-stage cultivation strategy has then been proposed by Zhu et al. (2014) to avoid this problem, in which a full-strength medium was used in the first stage to promote algal cell growth and to accumulate biomass, then the cells were subjected to N-starvation conditions in the second stage to trigger the target product accumulation (Zhu et al., 2014).

For the application of microalgae in biodiesel production, there remain many problems that need to be solved, and development of an effective method to ascertain the optimum time for algal harvest to obtain high neutral lipid productivity is one of the most important challenges. The chlorophyll-fluorescence measurement has been utilized for decades for non-invasive analyses of stress-induced perturbations to photosynthesis (Schreiber, 1978; Conroy et al., 1986). The maximum quantum efficiency of photosystem II (PSII, Fv/Fm) measured under dark-adapted conditions represents the theoretical capacity for light energy absorbed by PSII to be utilized in photosynthesis. The Fv/Fm value, which remains relatively constant under normal conditions, reflects the potential quantum efficiency of PSII and is used as one of the most sensitive indicators of the photosynthetic performance, which decreases when exposed to stress (Maxwell and Johnson, 2000; Oxborough, 2004). The Fv/Fm value was got with no extra agent add to the algae suspension and no harm to the algae influencing the downstream apply of microalgae. And the measurement of Fv/Fm was performed in action light that is the optimal light condition for algae growth for a few minutes and only no more one second saturation pulse. So Fv/Fm value was rapid and no-invasive parameter to characterize microalgae.

In this study, it was found that the lipid content increased with the decreasing Fv/Fm in the green algae *Chlorella* sp. C2 treated by N starvation, which is consistent with the reports in our previous study (Zhang et al., 2013). On this basis, four oil-producing green algae strains – *Chlorella sorokiniana* C1, *Chlorella* sp. C2, *C. sorokiniana* C3, and *C. sorokiniana* C7 – were cultivated and treated under N-starvation conditions to investigate the relationship between neutral lipid accumulation and the Fv/Fm values. A significant negative correlation between the neutral lipid contents and the Fv/Fm values was found for all four algal strains. Then, a method to determine the optimum harvest time in oil-producing microalgae cultivation by measuring Fv/Fm only was established and validated in lab-scale cultivation. This could provide a rapid and inexpensive way to determine the harvest time in large-scale oil-producing microalgae cultivation for biodiesel production.

Materials and Methods

Growth Conditions and N-Starvation Treatment

The N-sufficient medium (N+) used was full-strength BG11 medium (Tran et al., 2014). The N-deficient medium (N−) was BG11 without $NaNO_3$. Alga strains *C. sorokiniana* C1, *Chlorella* sp. C2, *C. sorokiniana* C3 and *C. sorokiniana* C7 in the exponential phase were each inoculated with an initial OD_{700} of 0.05 into a 1 l flask containing 500 ml BG11 medium at controlled temperature of 25°C, under continuous illumination of white fluorescence light at 70 μmol m^{-2} s^{-1}, and bubbled continuously with filtered air. For the N-starvation treatment, the cells were harvested during the middle exponential growth phase (OD_{700} approximately 0.8, about 1.1×10^7 cells ml^{-1}) and centrifuged 3 min at 3,000 g at room temperature (about 25°C). The pellet was then washed and resuspended in the N-medium to OD_{700} 0.8. Then, cultivation of the cultures was continued with the same light, temperature, and air supply as the N-sufficient medium.

Fluorescence Microscope Analysis

Microscopic visualization of algae cells were carried out using a confocal scanner (Zeiss LSM 710 NLO) and fluorescence microscope (OLYMPUS system microscope BX53, Japan) as described previously (Zhang et al., 2013). The lipid bodies were stained with the fluorescent dye, Bodipy 505/515 (Invitrogen Molecular Probes, Carlsbad, CA, USA) with a final labeling concentration of 1 mM, according to Cooper et al. (2010). Confocal laser scanning microscopy (CLSM) analysis was carried out as previously described in Zhang et al. (2013, 2014a). Fluorescence microscope analysis of the green bodipy fluorescence was excited at 488 nm and detected at 505–515 nm. The red auto-fluorescence of the chloroplasts was simultaneously detected at 650–700 nm.

Thin Layer Chromatography Analysis of Lipid

The cell lipids were extracted and thin layer chromatography (TLC) analyzed according to Reiser and Somerville (1997) with minor modifications as described by Zhang et al. (2013).

Flow Cytometry Analysis

Flow cytometry (FCM) analysis was used to visualize the florescence of lipid in large number cells level. Five hundred microliters of cells (1.1×107 cells ml^{-1}) were stained with Bodipy 505/515 at 37°C for 30 min under dark condition and then analyzed by FACS Aria Flow Cytometer (Becton Dickinson, San Jose, CA, USA) equipped with a laser emitting at 488 nm and an optical filter FL1 (530/30 nm). And Flow Jo software (Tree Star, San Carlos, CA, USA) was used to analysis the collected data from Flow Cytometer.

Pigments Quantification

Total pigments were extracted with 100% methanol. One milliliter of cells (1.1×107 cells ml^{-1}) was collected and centrifuged at 3,000 g for 3 min at room temperature. The pellet was resuspended in 1 ml of 100% methanol for 12 h. And the pigment concentrations were determined spectrophotometrically and calculated according to the formula developed by Lichtenthaler (1987) as chlorophyll a (Chl a) (microgram per milliliter) = 16.72

$A_{665.2} - 29.16\ A_{652.4}$, chlorophyll b (Chl b) (microgram per milliliter) = $34.09\ A_{652.4} - 15.28\ A_{665.2}$, total chlorophylls (Chl $a + b$) (microgram per milliliter) = $1.44\ A_{665.2} + 24.93\ A_{652.4}$, total carotenoids (Car) (microgram per milliliter) = $(1000\ A_{470} - 21.63$ Chl a $- 104.96$ Chl b)$/221$.

Chl Fluorescence Analysis

A Dual-PAM-100 Chl fluorometer (Walz, Germany) was used for all Chl fluorescence measurements. Cells were fully dark-adapted for 15 min before the measurement of Initial (Fo) and maximum Chl fluorescence level. The maximum quantum yields of PSII electron transport was calculated as Fv/Fm = (Fm − Fo)/Fm according to Genty et al. (1989).

Photosynthetic Steady-State Oxygen Evolution and Dark Respiration Rates Measurement

Photosynthetic oxygen evolution and dark respiration rates were measured using a Clark-type oxygen electrode (Oxylab 2, Hansatech, UK) at 20°C as described by Zhang et al. (2013). The dark respiration rates were measured with $10\ \mu M$ NaHCO3 in cell suspensions (2 ml) under dark condition and the oxygen evolution rates were measured with cell suspensions (2 ml) illuminated at a quantum flux density of 600 μmol m^{-2} s^{-1}. The collected data were the rate of oxygen changes (μmol O$_2$ min^{-1}). The chlorophyll (Chl a) contents were determined as described in above 2.5 pigments quantification. And the result of oxygen evolution and dark respiration rates were expressed with the rate of oxygen released (oxygen evolution μmol O$_2$ mg^{-1} Chla h^{-1}) and consumption (respiration rate μmol O$_2$ mg^{-1}Chla h^{-1}).

Assessment of Lipid Peroxidation and ROS Scavenging Enzyme Activity Assays

The malondialdehyde (MDA) level, and the enzyme activities of catalase (CAT), peroxidase (POD), and superoxide dismutase (SOD) were measured according to Shi et al. (2009) as described by Zhang et al. (2013).

Lab-Scale (3L) Cultivation

The lab-scale cultivation was performed in a 3 l photobioreactor (Zhang et al., 2014a). The four algal strains were inoculated with an initial OD$_{700}$ of 0.05 into the photobioreactor flask containing 2 l BG11 medium at controlled temperature of 25°C, under continuous illumination of white fluorescence light at 70 μmol m^{-2} s^{-1}, and bubbled continuously with filtered air. Cells reached their mid logarithmic growth phase (about 0.8 at OD$_{700}$) were harvested by centrifugation for 3 min at 3,000 g at room temperature. The algae pellets were then washed and resuspended in N-medium to OD$_{700}$ 0.8. Then, cultivation of the cultures continued under the same growth conditions.

Statistical Analysis

Each result shown represents at least the mean of three independent biological replicates. The statistical analysis of the collected data was made using SPSS-13. The correlation between the variables was analyzed with Pearson's correlation.

Results

N Starvation Stimulates Neutral Lipid Accumulation in *Chlorella* sp. C2

To understand the regulation of neutral lipid accumulation in *Chlorella* sp. C2, the cell neutral lipid levels at various time intervals of N depletion were extracted and examined using TLC. As shown in **Figure 1A**, neutral lipid accumulation in both N− (**Figure 1A**, lanes 1, 2) and N+ medium (**Figure 1A**, lanes 5, 6) were undetectable during the first 2 days of cultivation, which came into sight after 2 days in the cells cultured in N-medium (**Figure 1A**, lane 3), and significant accumulation could be observed after 8 days of N starvation (**Figure 1A**, lane 4). While as contrast, only a trace amount of neutral lipids could be detected in N repletion medium grown cells (**Figure 1A**, lane 8) even after 8 days.

A CLSM analysis was used to visualize the neutral lipid accumulation of the algal cells cultured in N-medium at different stages. In accordance with the TLC (**Figure 1A**) analysis, **Figure 1B** shows that no green Bodipy 505/515 fluorescence could be detected at 0 day (**Figure 1B**) and 0.5 day (**Figure 1B**) after N-treatment, while a weak green fluorescence was first detected after 2 days (**Figure 1B**) and a strong green fluorescence signal was observed after 8 days of treatment (**Figure 1B**). Moreover, to further characterize the neutral lipid accumulation during N starvation at statistical level, a group of Chlorella sp. C2 cells (>10,000) at different stages were analyzed using FCM. Compared with day 0, the Bodipy fluorescence intensity of the

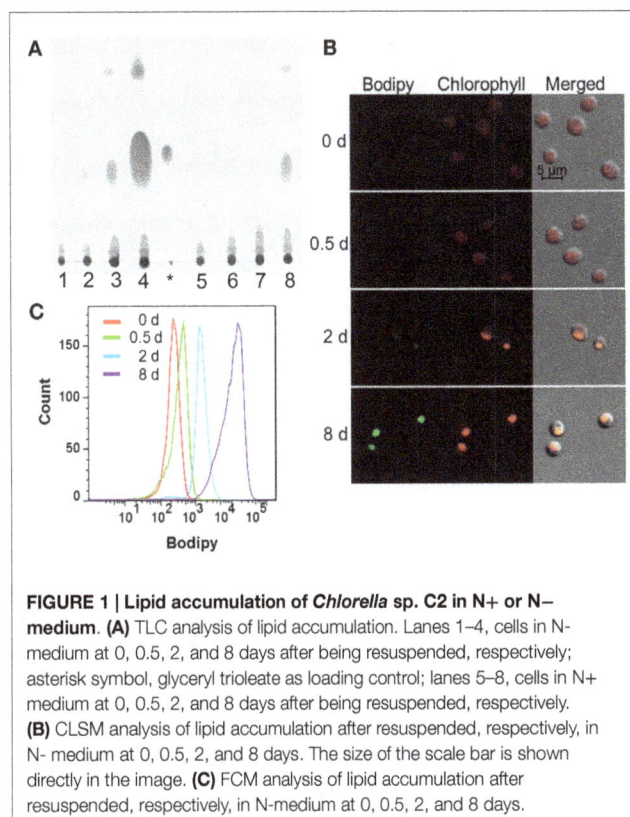

FIGURE 1 | Lipid accumulation of *Chlorella* sp. C2 in N+ or N− medium. (A) TLC analysis of lipid accumulation. Lanes 1–4, cells in N-medium at 0, 0.5, 2, and 8 days after being resuspended, respectively; asterisk symbol, glyceryl trioleate as loading control; lanes 5–8, cells in N+ medium at 0, 0.5, 2, and 8 days after being resuspended, respectively. **(B)** CLSM analysis of lipid accumulation after resuspended, respectively, in N- medium at 0, 0.5, 2, and 8 days. The size of the scale bar is shown directly in the image. **(C)** FCM analysis of lipid accumulation after resuspended, respectively, in N-medium at 0, 0.5, 2, and 8 days.

cell populations at days 0.5, 2, and 8 of N starvation showing a constant increase, indicating constantly increasing levels of neutral lipid (**Figure 1C**). Thus, as in *Chlorella* sp. C3 (Zhang et al., 2013), days 0, 0.5, 2, and 8 of N- treatment could be defined as the control stage (Cs), pre-oil droplet formation stage (PDFs), oil droplet formation stage (ODFs), and late-oil droplet formation stage (LDFs), respectively. These were also the key stages in the further tests.

N starvation Damages Photosynthesis System

Nitrate is one of the most important elements contributing to algae growth, so its depletion may dramatically change the physiological activity. As a stressor, the N starvation will not only induce lipid accumulation but also lead to the depression of photosynthesis. As shown in **Figure 2A**, all the pigment contents, including Chl *a*, Chl *b*, Chl *a* + *b*, and carotenoid (Car), had a significant decrease before the ODFs and then decreased continuously during neutral lipid accumulation in N-medium.

In photosynthetic organisms, nutrient stresses could be generally detected with a decrease in the Fv/Fm value, as the Fv/Fm value in healthy cells has been reported to be constant (Conroy et al., 1986; Genty et al., 1989; Maxwell and Johnson, 2000), but decreases under various stresses (Conroy et al., 1986; Lu et al., 1999; Lee et al., 2013; Zhang et al., 2013). **Figure 2B** shows the Fv/Fm value declined linearly during

oil droplet formation, suggesting that the algae cell was under environmental stress.

Rates of steady-state photosynthetic oxygen evolution and dark respiration were examined to further understand the variation in photosynthesis of *Chlorella* sp. C2 during N-induced oil droplet formation. Compared with the untreated cells, both the photosynthetic oxygen evolution rate and the respiration rate of the N starved cells reduced significantly during oil droplet formation (**Figure 2C**), indicating that severe damage to the photosynthetic apparatus, as well as to the respiratory apparatus, had occurred.

Most stresses in oxygenic photosynthetic organisms would ultimately lead to oxidative stress (Elstner, 1991). Lipid peroxidation level, the most commonly accepted indicator of oxidative stress (Apel and Hirt, 2004; Zhang et al., 2013), could be estimated by measuring the formation of MDA in cells. During oxidative stress, ROS, including $^{\bullet}O_2^-$, $^1O_2$, and H_2O_2, levels increase (Apel and Hirt, 2004), and the antioxidant enzyme, including SOD, POD, and CAT, will be induced or activated in cells (Ali et al., 2005). As shown in **Figure 3**, N starvation induced an increase in the MDA level in the cells. The relative activities of SOD and CAT dramatically increased at the PDFs and dropped gradually at the ODFs and LDFs. By contrast, the relative activities of POD declined slightly at the PDFs, and then increased at the ODFs and LDFs. Therefore, this suggests that POD, CAT, and SOD

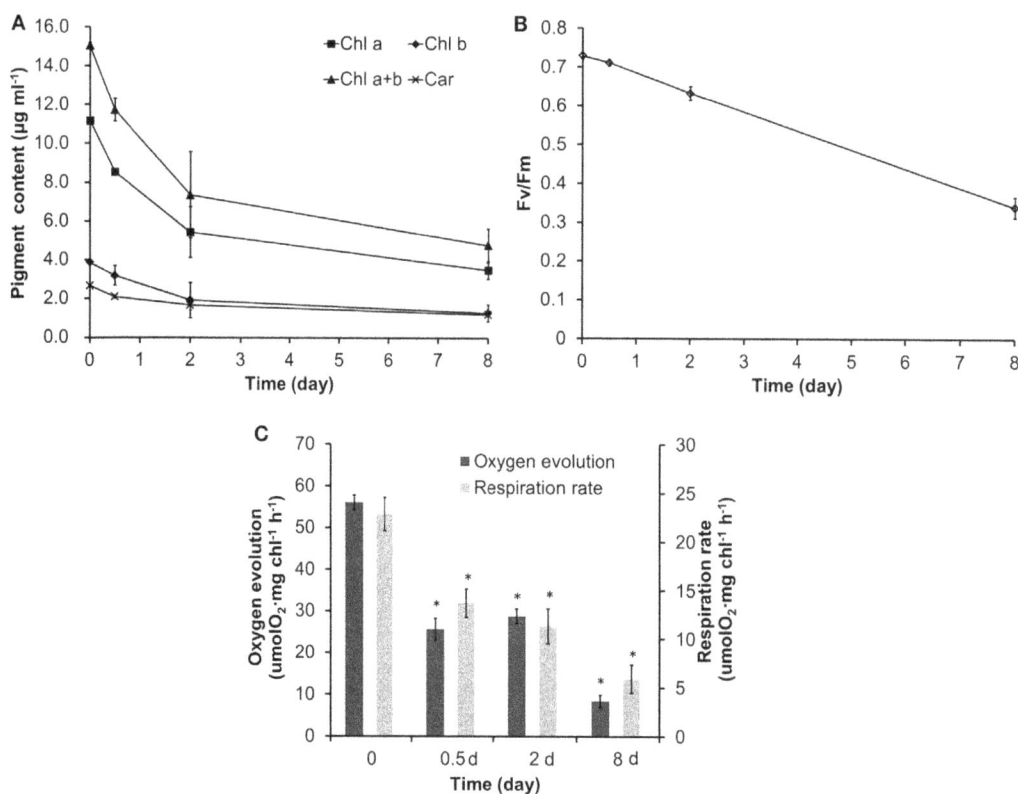

FIGURE 2 | The variation of photosynthetic physiology in *Chlorella* sp. C2 during lipid droplet formation. (A) Pigment content. (B) The maximum photochemical efficiency of PSII (Fv/Fm). (C) Steady-state oxygen evolution and in dark respiration. All data points in the current and following figures represent the means of three replicated studies in each independent culture, with the SD of the means (*$p < 0.05$; **$p < 0.01$).

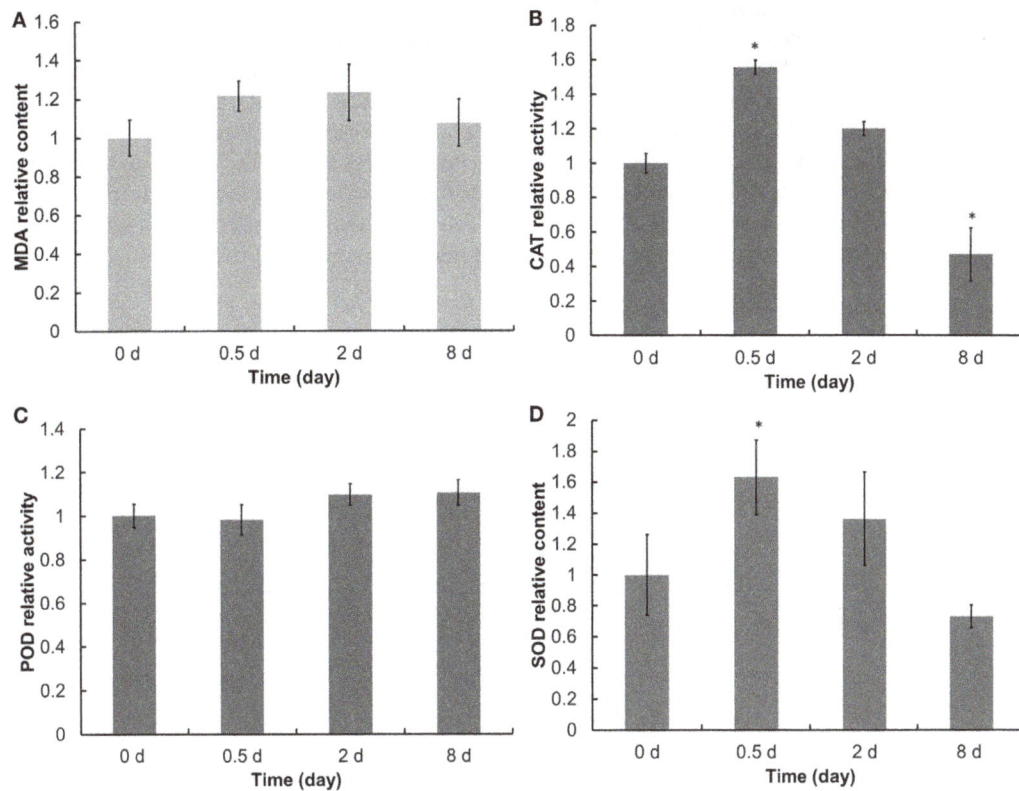

FIGURE 3 | Lipid peroxidation level and antioxidant enzymes activities of *Chlorella* sp. C2 during OD formation. (A–D) represents the MDA content, CAT, POD, and SOD activities, respectively.

play important roles in scavenging ROS and reducing MDA in *Chlorella* sp. C2 cells under N starvation at different stages (**Figure 3**).

The above results suggest that both the neutral lipid accumulation and the damage to the photosynthetic system might be caused by N starvation-induced oxidative stress. Furthermore, as the most sensitive indicator for stress response under unfavorable conditions, the Fv/Fm value decreased linearly (**Figure 2B**) along with the increasing triacylglycerol (TAG) accumulation (**Figure 1**) under N starvation. So it is predicted that the Fv/Fm value may be an excellent indicator for identifying the level of N-induced TAG accumulation and could provide a non-invasive and quick tools of determining the timing for microalgae harvest if a significant relationship existed between the Fv/Fm value and the TAG content.

Fv/Fm Value Can be Used to Estimate Neutral Lipid Levels in the Four Oil-Production Algae

Besides *Chlorella* sp. C2, three further oil-producing green algae strains – *C. sorokiniana* C1, *C. sorokiniana* C3, and *C. sorokiniana* C7 – were selected to identify the relationship between the Fv/Fm value and the neutral lipid level. During N-starvation treatment, the neutral lipid levels in cells of all four strains rose gradually with the prolongation of N-depletion stress (**Figure 4**, green fluorescence). However, Chl auto-fluorescence (**Figure 4** red

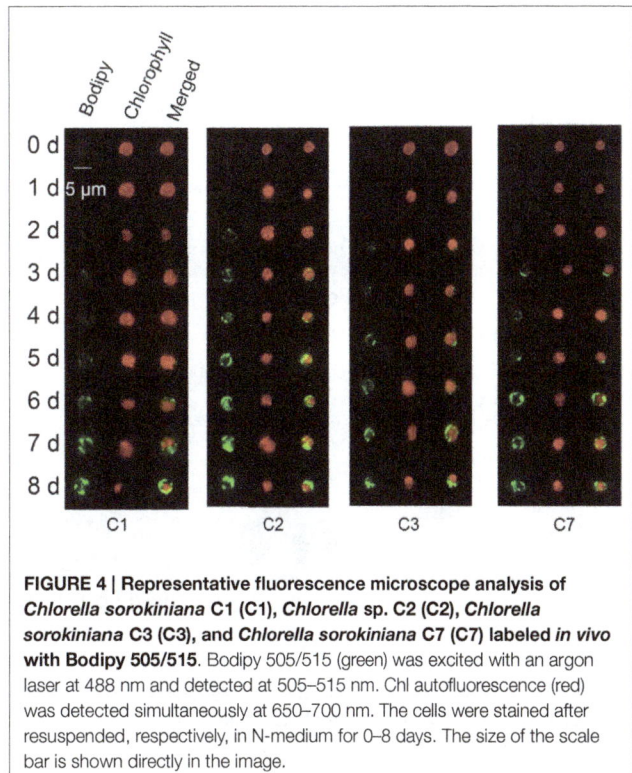

FIGURE 4 | Representative fluorescence microscope analysis of *Chlorella sorokiniana* C1 (C1), *Chlorella* sp. C2 (C2), *Chlorella sorokiniana* C3 (C3), and *Chlorella sorokiniana* C7 (C7) labeled *in vivo* with Bodipy 505/515. Bodipy 505/515 (green) was excited with an argon laser at 488 nm and detected at 505–515 nm. Chl autofluorescence (red) was detected simultaneously at 650–700 nm. The cells were stained after resuspended, respectively, in N-medium for 0–8 days. The size of the scale bar is shown directly in the image.

fluorescence) signals became more heterogeneous as the chloroplast shapes became abnormal. Moreover, in accordance with the fluorescence microscope results (**Figure 4**), the TLC results also showed that increasing neutral lipid levels were detected with time (**Figure 5A**). The relative contents of neutral lipid were measured using Image J (v1.41, NIH), and it was seen that the contents had significantly positive association with time under N starvation in all four algae (**Figure 5B**, $r > 0.9$, $p < 0.01$, correlation test by SPSS-13). As show in **Figure 6**, the values of Fv/Fm in all four algae declined linearly during N starvation, and

the Fv/Fm values were significant negatively correlated with time (correlation test, $r < -0.9$, $p < 0.01$). Notably, during N starvation, the significant negative linear correlations appeared between the Fv/Fm values and relative lipid content in all four algae (**Figure 7**, correlate test, $r > 0.9$, $p < 0.01$). Therefore, for these four oil-rich algae, in which we have established the relationship between the neutral lipid content and the Fv/Fm values, the neutral lipid level under N starvation can be got indirectly by measuring the Fv/Fm value. This can therefore be used to determine the optimal harvest time of microalgae for lipid production.

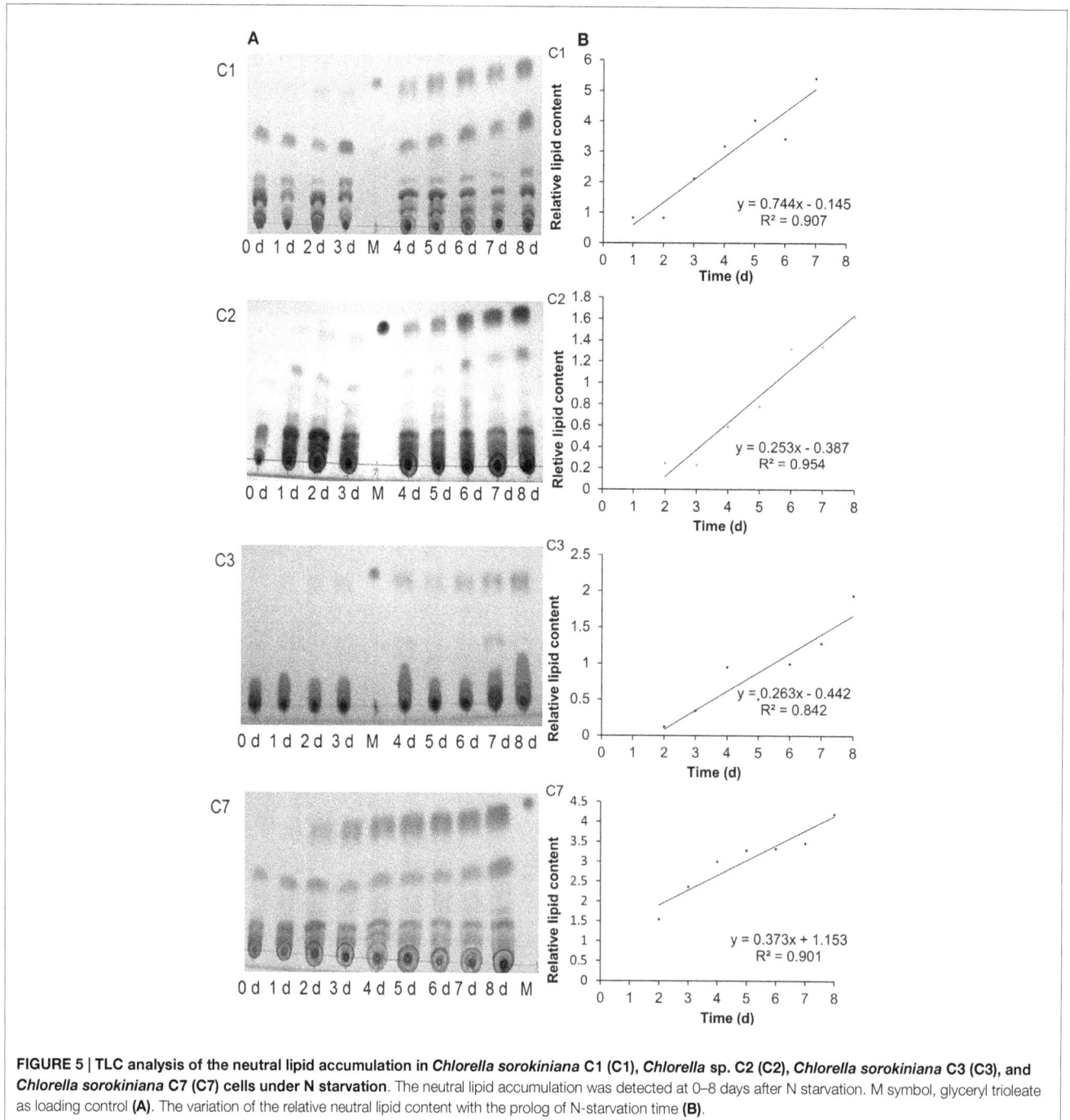

FIGURE 5 | TLC analysis of the neutral lipid accumulation in *Chlorella sorokiniana* **C1 (C1),** *Chlorella* **sp. C2 (C2),** *Chlorella sorokiniana* **C3 (C3), and** *Chlorella sorokiniana* **C7 (C7) cells under N starvation**. The neutral lipid accumulation was detected at 0–8 days after N starvation. M symbol, glyceryl trioleate as loading control **(A)**. The variation of the relative neutral lipid content with the prolog of N-starvation time **(B)**.

Lab-Scale Experiments Demonstrated the Application of the Fv/Fm Index for Determining the Harvest Time in a Given Oil-Producing Algae Cultivation

To further test the practicability of using the Fv/Fm value for determining the harvest time in a given oil-producing algae, lab-scale experiments using four algal strains were performed in 3 l bioreactors. During the N-starvation time, the algae were collected at random time points to detect the neutral lipid accumulation and the Fv/Fm values. Similar to our previous results, the accumulation of neutral lipids increased with the decreasing Fv/Fm values in all four algae in the 3 l bioreactors (**Figure 8**). This indicated that the Fv/Fm value could be an indicator for determining the neutral lipid level in a given oil-producing algae cultivated in a 3 l or even larger scale bioreactor. Furthermore, for a given oil-rich microalgae, this indicates that if the relationship between the neutral lipid content and the Fv/Fm value under nutrient stress can be established, the optimum harvest time for lipid production in cultivations under nutrient stress can be determined by measuring the Fv/Fm value.

Discussion

One of the challenges of microalgae biodiesel production is the measurement of lipid content to determine the optimum harvest time obtaining high levels of lipids. In order to determine the lipid accumulation level in microalgae cells, lipid extraction and detection are required using traditional technology. TLC and the gravimetric method are the two most commonly used methods for lipid-content detection (Bligh and Dyer, 1959; Reiser and Somerville, 1997). However, the lipid extraction and detection steps used are complicated, need many organic reagents, resulting in environmental toxicity, and are laboring intensive and expensive. Recently, fluorescent probes such as the lipophilic probes Nile red (Cooksey et al., 1987) and BODIPY 505/515

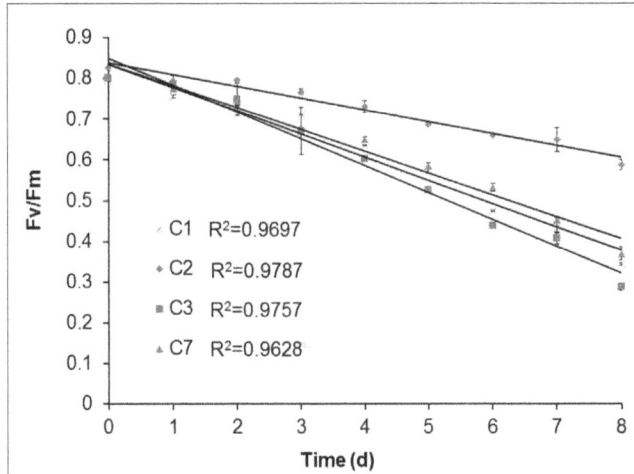

FIGURE 6 | The variation of maximum photochemical efficiency of PSII (Fv/Fm) of *Chlorella sorokiniana* C1 (C1), *Chlorella* sp. C2 (C2), *Chlorella sorokiniana* C3 (C3), *Chlorella sorokiniana* C7 (C7) with the prolog of N-starvation time.

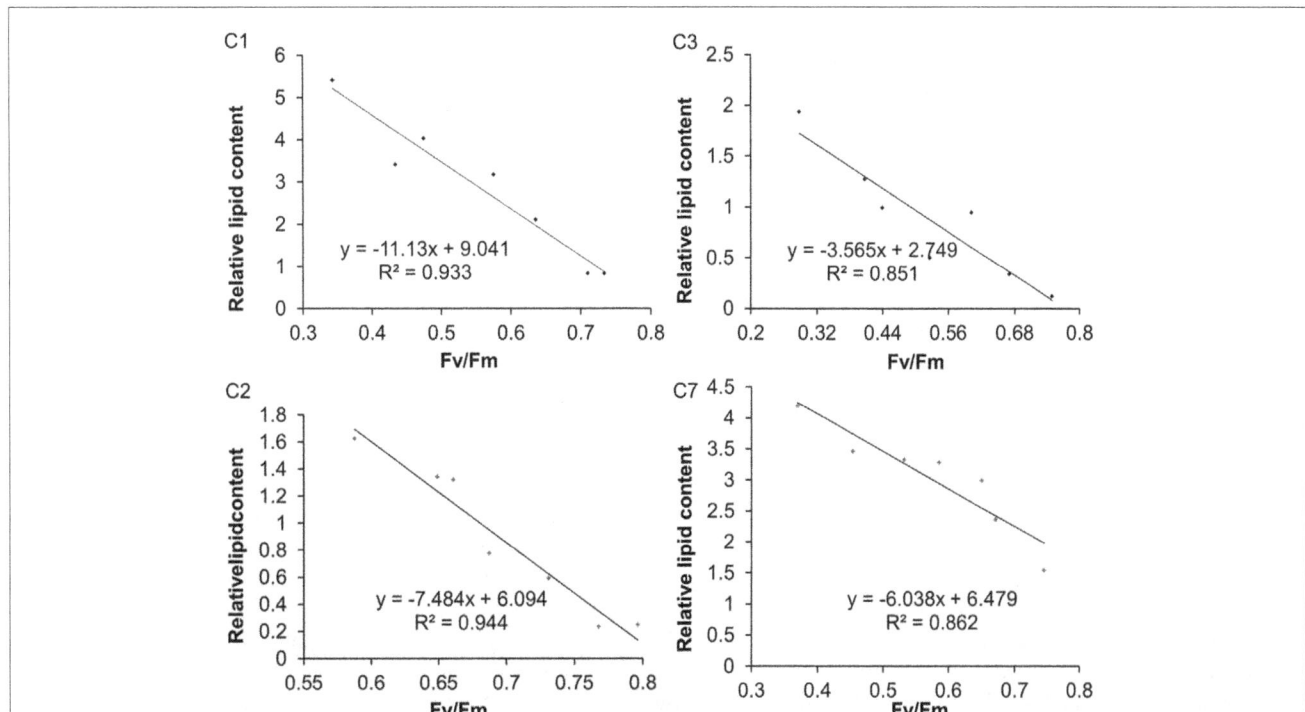

FIGURE 7 | The correlation between Fv/Fm and relative neutral lipid content in *Chlorella sorokiniana* C1 (C1), *Chlorella* sp. C2 (C2), *Chlorella sorokiniana* C3 (C3), and *Chlorella sorokiniana* C7 (C7) under N starvation.

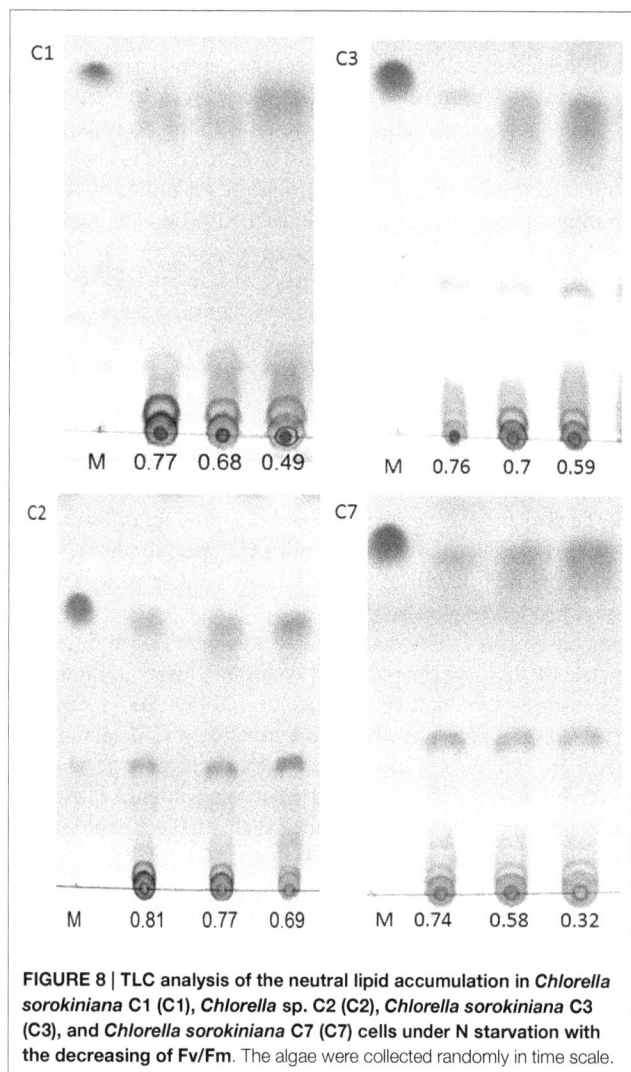

FIGURE 8 | TLC analysis of the neutral lipid accumulation in *Chlorella sorokiniana* C1 (C1), *Chlorella* sp. C2 (C2), *Chlorella sorokiniana* C3 (C3), and *Chlorella sorokiniana* C7 (C7) cells under N starvation with the decreasing of Fv/Fm. The algae were collected randomly in time scale.

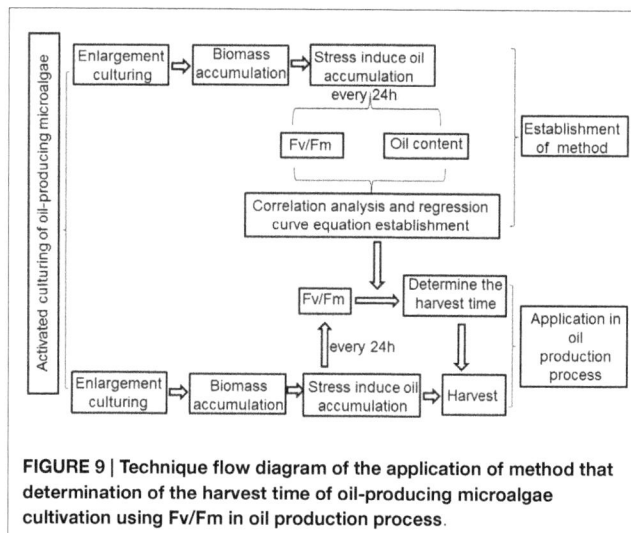

FIGURE 9 | Technique flow diagram of the application of method that determination of the harvest time of oil-producing microalgae cultivation using Fv/Fm in oil production process.

(Cooper et al., 2010), which can measure the neutral lipid level in intact cells without lipid extraction, have been used to estimate lipid accumulation. However, there are also some disadvantages of these florescent probes, namely, the relatively high cost of time and money for staining and detecting the fluorescent probe, and the potential errors caused by the different permeability of the fluorescent probe into diverse microalgae cells. Therefore, a simple and low-cost method is needed to establish the lipid level to then determine the optimum harvest time rapidly and expediently in oil-producing microalgae cultivations. In our study, we established a new method by measuring the chlorophyll-fluorescence parameter Fv/Fm.

Under N-starvation condition, microalgae cells preferentially degraded nitrate containing macromolecules, resulting in a decrease of total nitrogen content as well as the accumulation of excess carbon in the form of lipids (Dawes, 1976). In this article, the lipid content was linearly increased after 2 days of N-starvation treatment (**Figures 1** and **5**). N depletion also induced the decrease of Fv/Fm values, which may be a consequence of reduced

photosynthetic pigment (**Figures 2** and **4**). Previous studies also reported the decrease in Fv/Fm values under osmotic, light, and nutrient stress (Lu et al., 1999; Beardall et al., 2001). Under N-starvation condition, the neutral lipid accumulation increasing and Fv/Fm decreasing occurs simultaneously. So we predicted there may be some relationship between lipid increasing and Fv/Fm decreasing. The result shows that the lipid increasing was negatively correlative with Fv/Fm decreasing. Therefore, Fv/Fm may be a tool to determine the lipid accumulation for microalgae under N starvation. A previous research reported the use of PAM fluorometry to measure the biosynthesis of neutral lipid under nutrient stress (White et al., 2011). So the measurement of Fv/Fm for the determination of harvest time is a feasible method. The Fv/Fm value was obtained from specimens in the dark-adapted state, and the measurement could be completed within a few seconds using a single saturating pulse. So this procedure is simple, rapid, non-invasive, low-cost, and highly appropriate for large-scale application.

Even though Fv/Fm can be a parameter to determine lipid content, there were some interesting reports. A study using the Fv/Fm value as a screening tool for oil-rich mutant microalgae reported that under stress conditions, for different algae, algae with high remaining Fv/Fm values also have a high total lipid content, thus presenting a positive correlation ($R^2 = 0.906$) between the lipid content and the Fv/Fm value (Huangfu et al., 2013). Similar phenomenon was also observed in naturally occurring microalgae strains (Pan et al., 2011). In our study, the Fv/Fm values were significantly negatively correlated with the relative lipid content (**Figure 7**, $r < -0.9$, $p < 0.01$), and the result was verified in a 3 l scale culture by testing samples at random time points (**Figure 8**). The difference of our negative correlation and the previous positive correlation report may be because that the previous report focused on the comparison of different algae, while we concentrate on the comparison of the given algae at different stress time. Under stress condition, the Fv/Fm value decreased and the lipid content increased. However, for different algae or mutants under same stress, the alga with higher remaining

Fv/Fm value behaved a higher toleration to the stress and has a higher potential photosynthesis for fixing more carbon and supplying more energy and carbon source for lipid production. So the remaining Fv/Fm was positive correlation with lipid content. As for our study, Fv/Fm value is the most sensitive indicator for stress response under unfavorable conditions, and neutral lipid accumulation coupled with N starvation-induced oxidative stress (Zhang et al., 2013). With the decreasing of Fv/Fm under N starvation, the lipid content continuously accumulated to saturated point with a slower and slower rate. Therefore, for the given algae, comparison with the stress time, the Fv/Fm value was negative correlation with lipid content. Thus, we predicted that for a given oil-producing algae under stress conditions, there is a negative relationship between the Fv/Fm value and the TAG content that can be used to estimate the TAG content and thus determine the optimum harvest time for lipid production.

In summary, we have established a simple and convenient method to determine the harvest time of microalgae under stress conditions for lipid production. For the technological process shown in **Figure 9**, for a given oil-rich microalgae, we can establish the relationship between the lipid content and the Fv/Fm value, thus obtaining a range for the Fv/Fm value around the lipid saturation point. When cultivating algae on a large scale for oil production, we can determine the optimum harvest time by measuring the Fv/Fm value and referring it to the established relationship.

Author Contributions

YQ was responsible for study conception and design, data collection and analysis, manuscript writing, and final approval of the manuscript; JR for study conception and design, data collection and analysis, and final approval of the manuscript; HC for data collection and analysis, manuscript writing, and final approval of the manuscript; CH for data analysis and final approval of the manuscript; and QW for conception and design, critical revision and manuscript writing, and final approval of the manuscript. All authors read and approved the final manuscript.

Acknowledgments

This work was supported jointly by the National Program on Key Basic Research Project (2012CB224803, 2011CB200902), the National Natural Science Foundation of China (31300030, 31270094), the Natural Science Foundation of Hubei Province of China (2013CFA109), Sinopec (S213049), and the Knowledge Innovation Program of the Chinese Academy of Sciences (Y35E05).

References

Ali, M. B., Yu, K. W., Hahn, E. J., and Paek, K. Y. (2005). Differential responses of anti-oxidants enzymes, lipoxygenase activity, ascorbate content and the production of saponins in tissue cultured root of mountain *Panax ginseng* C.A. Mayer and *Panax quinquefolium* L. in bioreactor subjected to methyl jasmonate stress. *Plant Sci.* 169, 83–92. doi:10.1016/j.plantsci.2005.02.027

Amaro, H. M., Guedes, A. C., and Malcata, F. X. (2011). Advances and perspectives in using microalgae to produce biodiesel. *Appl. Energy* 88, 3402–3410. doi:10.1016/j.apenergy.2010.12.014

Apel, K., and Hirt, H. (2004). Reactive oxygen species: metabolism, oxidative stress, and signal transduction. *Annu. Rev. Plant Biol.* 55, 373–399. doi:10.1146/annurev.arplant.55.031903.141701

Beardall, J., Young, E., and Roberts, S. (2001). Approaches for determining phytoplankton nutrient limitation. *Aquat. Sci.* 63, 44–69. doi:10.1007/PL00001344

Bligh, E. G., and Dyer, W. J. (1959). A rapid method of total lipid extraction and purification. *Can. J. Biochem. Physiol.* 37, 911–917. doi:10.1139/o59-099

Conroy, J. P., Smillie, R. M., Kuppers, M., Bevege, D. I., and Barlow, E. W. (1986). Chlorophyll a fluorescence and photosynthetic and growth responses of *Pinus radiata* in using microalgae to phosphorus deficiency, drought stress, and high CO2. *Plant Physiol.* 81, 423–429. doi:10.1104/pp.81.2.423

Converti, A., Casazza, A. A., Ortiz, E. Y., Perego, P., and Del Borghi, M. (2009). Effect of temperature and nitrogen concentration on the growth and lipid content of *Nannochloropsis oculata* and *Chlorella vulgaris* for biodiesel production. *J. Chem. Eng. Process.* 48, 1146–1151. doi:10.1016/j.cep.2009.03.006

Cooksey, K. E., Guckert, J. B., Williams, S. A., and Callis, P. R. (1987). Fluorometric-determination of the neutral lipid-content of microalgal cells using nile red. *J. Microbiol. Methods* 6, 333–345. doi:10.1016/0167-7012(87)90019-4

Cooper, M. S., Hardin, W. R., Petersen, T. W., and Cattolico, R. A. (2010). Visualizing "green oil" in live algal cells. *J. Biosci. Bioeng.* 109, 198–201. doi:10.1016/j.jbiosc.2009.08.004

Dawes, E. (1976). "Endogenous metabolism and the survival of starved prokaryotes," in *The Survival of Vegetative Microbes*, eds Gray T.R.G. and Postgate J.R. (Cambridge: Cambridge University Press), 19–53.

Elstner, E. F. (1991). "Mechanisms of oxygen activation in different compartments of plant cells," in *Active Oxygen/Oxidative Stress in Plant Metabolism*, eds Pell E. J. and Steffen K. L. (Rockville, MD: American Society of Plant Physiologists), 13–25.

Genty, B., Briantais, J. M., and Baker, N. R. (1989). The relationship between the quantum yield of photosynthetic electron-transport and quenching of chlorophyll fluorescence. *Biochim. Biophys. Acta* 990, 87–92. doi:10.1016/S0304-4165(89)80016-9

Hill, J., Nelson, E., Tilman, D., Polasky, S., and Tiffany, D. (2006). Environmental, economic, and energetic costs and benefits of biodiesel and ethanol biofuels. *Proc. Natl. Acad. Sci. U.S.A.* 103, 11206–11210. doi:10.1073/pnas.0604600103

Hu, Q., Sommerfeld, M., Jarvis, E., Ghirardi, M., Posewitz, M., Seibert, M., et al. (2008). Microalgal triacylglycerols as feedstocks for biofuel production: perspectives and advances. *Plant J.* 54, 621–639. doi:10.1111/j.1365-313X.2008.03492.x

Hu, Q., Zhang, C., and Sommerfeld, M. (2006). Biodiesel from algae: lessons learned over the past 60 years and future perspectives. *PSA Abstracts. J. Phycol.* 42, 12–12. doi:10.1111/j.1529-8817.2006.20064201.x

Huangfu, J. Q., Liu, J., Sun, Z., Wang, M. F., Jiang, Y., Chen, Z. Y., et al. (2013). Antiaging effects of astaxanthin-rich alga *Haematococcus pluvialis* on fruit flies under oxidative stress. *J. Agric. Food Chem.* 61, 7800–7804. doi:10.1021/jf402224w

Illman, A. M., Scragg, A. H., and Shales, S. W. (2000). Increase in *Chlorella* strains calorific values when grown in low nitrogen medium. *Enzyme Microb. Technol.* 27, 631–635. doi:10.1016/S0141-0229(00)00266-0

Lee, C. W., Lim, J. H., and Heng, P. L. (2013). Investigating the spatial distribution of phototrophic picoplankton in a tropical estuary. *Environ. Monit. Assess.* 185, 9697–9704. doi:10.1007/s10661-013-3283-3

Lichtenthaler, H. K. (1987). "Chlorophylls and carotenoids: pigments of photosynthetic biomembranes," in *Methods in Enzymology*, ed. Lester Packer R. D. (San Diego: Academic Press), 350–382.

Lu, C. M., Torzillo, G., and Vonshak, A. (1999). Kinetic response of photosystem II photochemistry in the cyanobacterium Spirulina platensis to high salinity is characterized by two distinct phases. *Aust. J. Plant Physiol.* 26, 283–292. doi:10.1071/PP98119

Maxwell, K., and Johnson, G. N. (2000). Chlorophyll fluorescence–a practical guide. *J. Exp. Bot.* 51, 659–668. doi:10.1093/jexbot/51.345.659

Oxborough, K. (2004). Imaging of chlorophyll a fluorescence: theoretical and practical aspects of an emerging technique for the monitoring of photosynthetic performance. *J. Exp. Bot.* 55, 1195–1205. doi:10.1093/Jxb/Erh145

Pan, Y. Y., Wang, S. T., Chuang, L. T., Chang, Y. W., and Chen, C. N. (2011). Isolation of thermo-tolerant and high lipid content green microalgae: oil accumulation is predominantly controlled by photosystem efficiency during stress treatments in *Desmodesmus*. *Bioresour. Technol.* 102, 10510–10517. doi:10.1016/j.biortech.2011.08.091

Petkov, G., and Garcia, G. (2007). Which are fatty acids of the green alga *Chlorella*? *Biochem. Syst. Ecol.* 35, 281–285. doi:10.1016/j.bse.2006.10.017

Reiser, S., and Somerville, C. (1997). Isolation of mutants of *Acinetobacter calcoaceticus* deficient in wax ester synthesis and complementation of one mutation with a gene encoding a fatty acyl coenzyme a reductase. *J. Bacteriol.* 179, 2969–2975.

Santos, C. A., Ferreira, M. E., da Silva, T. L., Gouveia, L., Novais, J. M., and Reis, A. (2011). A symbiotic gas exchange between bioreactors enhances microalgal biomass and lipid productivities: taking advantage of complementary nutritional modes. *J. Ind. Microbiol. Biotechnol.* 38, 909–917. doi:10.1007/s10295-010-0860-0

Schreiber, U. (1978). Chlorophyll fluorescence assay for ozone injury in intact plants. *Plant Physiol.* 61, 80–84. doi:10.1104/pp.61.1.80

Scott, S. A., Davey, M. P., Dennis, J. S., Horst, I., Howe, C. J., Lea-Smith, D. J., et al. (2010). Biodiesel from algae: challenges and prospects. *Curr. Opin. Biotechnol.* 21, 277–286. doi:10.1016/j.copbio.2010.03.005

Shi, S., Tang, D., and Liu, Y. (2009). Effects of an algicidal bacterium *Pseudomonas mendocina* on the growth and antioxidant system of *Aphanizomenon* flosaquae. *Curr. Microbiol.* 59, 107–112. doi:10.1007/s00284-009-9404-0

Tran, T. H., Govin, A., Guyonnet, R., Grosseau, P., Lors, C., Damidot, D., et al. (2014). Influence of the intrinsic characteristics of mortars on their biofouling by pigmented organisms: comparison between laboratory and field-scale experiments. *Int. Biodeterior. Biodegradation* 86, 334–342. doi:10.1016/j.ibiod.2013.10.005

White, S., Anandraj, A., and Bux, F. (2011). PAM fluorometry as a tool to assess microalgal nutrient stress and monitor cellular neutral lipids. *Bioresour. Technol.* 102, 1675–1682. doi:10.1016/j.biortech.2010.09.097

Widjaja, A., Chien, C.-C., and Ju, Y.-H. (2009). Study of increasing lipid production from fresh water microalgae *Chlorella vulgaris*. *J. Taiwan Inst. Chem. Eng.* 40, 13–20. doi:10.1016/j.jtice.2008.07.007

Zhang, X., Chen, H., Chen, W., Qiao, Y., He, C., and Wang, Q. (2014a). Evaluation of an oil-producing green alga *Chlorella* sp. C2 for biological DeNOx of industrial flue gases. *Environ. Sci. Technol.* 48, 10497–10504. doi:10.1021/es5013824

Zhang, X., Rong, J., Chen, H., He, C., and Wang, Q. (2014b). Current status and outlook in the application of microalgae in biodiesel production and environmental protection. *Front. Energy Res.* 2:32. doi:10.3389/fenrg.2014.00032

Zhang, Y. M., Chen, H., He, C. L., and Wang, Q. (2013). Nitrogen starvation induced oxidative stress in an oil-producing green alga *Chlorella sorokiniana* C3. *PLoS ONE* 8:e69225. doi:10.1371/journal.pone.0069225

Zhu, S., Wang, Y., Huang, W., Xu, J., Wang, Z., Xu, J., et al. (2014). Enhanced accumulation of carbohydrate and starch in *Chlorella zofingiensis* induced by nitrogen starvation. *Appl. Biochem. Biotechnol.* 174, 2435–2445. doi:10.1007/s12010-014-1183-9

Conflict of Interest Statement: The authors declare that the research was conducted in the absence of any commercial or financial relationships that could be construed as a potential conflict of interest.

15

Stability and Activity of Doped Transition Metal Zeolites in the Hydrothermal Processing

Thomas François Robin*, Andrew B. Ross, Amanda R. Lea-Langton and Jenny M. Jones

School of Chemical and Process Engineering, University of Leeds, Leeds, UK

This study investigates the stability and activity of HZSM-5 doped with metals such as molybdenum, nickel, copper, and iron under hydrothermal conditions used for the direct liquefaction of microalgae. Catalysts have been prepared by ion-exchange techniques, and MoZSM-5 was also prepared by wet incipient impregnation for comparison. Hydrothermal liquefaction is considered a potential route to convert microalgae into a sustainable fuel. One of the drawbacks of this process is that the bio-crude produced contains significant levels of nitrogen and oxygen compounds that have an impact on the physical and chemical properties of the fuel. Heterogeneous catalysts have been shown to improve the quality of the bio-crude by reducing nitrogen and oxygen contents. Zeolites, such as HZSM-5, are strong candidates due to their low cost compared to noble metal catalysts, but their stability and activity under hydrothermal conditions are not well understood. The stability of the catalysts has been determined under hydrothermal conditions at 350°C. Catalysts have been characterized before and after treatment using X-ray diffraction, BET physisorption, and scanning transmission electronic microscopy. Metal leaching was determined by the analysis of the water phase following the hydrothermal treatment. The inserted cation following ion-exchange can influence the physical properties of HZSM-5, for example, molybdenum improves the crystallinity of the zeolite. In general, metal-doped zeolites were relatively stable in subcritical water. The activity of the catalysts for processing lipids, protein, and microalgae has been assessed. Four feedstocks were selected: sunflower oil, soya proteins, *Chlorella*, and *Pseudochoricystis ellipsoidea*. The catalysts exhibited greater activity toward converting lipids, for example, MoZSM-5 enhanced the formation of aromatic compounds. NiZSM-5 and CuZSM-5 were observed to be more efficient for deoxygenation.

Keywords: subcritical, HZSM5, transition metals, liquefaction, biomass

Edited by:
Luca Fiori,
University of Trento, Italy

Reviewed by:
Zhidan Liu,
China Agricultural University, China
Uwe Schröder,
Technische Universität Braunschweig,
Germany

*Correspondence:
Thomas François Robin
thomas.cognac@gmail.com

INTRODUCTION

Hydrothermal liquefaction has been accepted as a sustainable and efficient technique to convert biomass into bio-crude oil, for the reason that wet biomass feedstock such as microalgae can be processed without prior drying (Peterson et al., 2008). An advantage of water under hydrothermal conditions is that its chemical and physical proprieties as a solvent can be tuned in relation to its temperature and pressure. The solubility of non-polar molecules, for example, increases with temperature as the polarity of water decreases from 78 F/m at room temperature to 14 F/m at 350°C (Peterson et al., 2008). A disadvantage is that the bio-crude produced by the process contains a significant content of heteroatoms including oxygen (from approximately 10 to 20 wt.%) and nitrogen (from approximately 5 to 11 wt.%), which reduces the heating value of the fuel and produces NO_x when combusted (Biller et al., 2011). Furthermore, heteroatoms such as oxygen can have an effect on the long-term storage and the quality of the fuel.

Heterogeneous catalysts have been used during *in situ* and *ex situ* liquefaction of microalgae in order to improve the quality of the fuels (Duan and Savage, 2010; Biller et al., 2011; Duan et al., 2013). Noble metal catalysts such as Pd/C and Pt/C have been selected to upgrade lipids and bio-crude from microalgae (Fu et al., 2010; Duan et al., 2013). Good rates of deoxygenation and high stability after several regeneration cycles were achieved with these catalysts. These results are encouraging, but lower-cost catalysts would improve the competitiveness of hydrothermal liquefaction compared to fossil fuel.

Zeolites such as HZSM-5 are strong candidates because of their high availability in the petrochemical industry and their acidic and microporous structures (Flanigen et al., 1991). However, a disadvantage is that this catalyst needs to be regularly regenerated because of the coke produced during the cracking of crude oils (Corma et al., 2007). Moreover, steam conditions induced a structural change with the migration of aluminum out of the catalyst framework called "dealumination." This effect is to reduce the Brønsted acidic sites and lower the catalytic activity (de Lucas et al., 1997b; Gayubo et al., 2004a,b). The "non-framework" aluminum can be reinserted into the framework by treating the dealuminated HZSM5 under harsh conditions by reflux in a solution of concentrated hydrochloric acid (HCl) for 120 h at 100°C (Sano et al., 2000).

In order to improve the activity and stability, HZSM-5 can be doped using different transition metals. HZSM-5 was shown to be more robust under steam conditions and less coke was produced when nickel was incorporated by a wet impregnation method (from 0.5 to 3 wt.% loading) during the production of hydrocarbon from bio-ethanol (Gayubo et al., 2010). The ammonia temperature-programmed desorption (TPD) analysis detected that nickel enhanced the concentration of Lewis acidic sites with a reduction in Brønsted acidic sites (Gayubo et al., 2010). Molybdenum was impregnated, and due to the high acidic sites it was demonstrated to produce aromatic compounds from methane in the process called "dehydroaromatization" (Song et al., 2006, 2007). Metal-doped HZSM-5 can be prepared with other techniques such as ion-exchange. The addition of iron II to HZSM-5 allowed to convert ammonia into nitrogen and CuZSM5 which was used to convert NO_x gases (Sato et al., 1991; Long and Yang, 2000).

HZSM-5 has been demonstrated to be a more robust catalyst compared to the zeolite HY in hot compressed water (from 150 to 200°C) and supercritical water at 400°C (Ravenelle et al., 2010; Mo and Savage, 2014). A mild change in the structure detected by the X-ray diffraction (XRD) analysis of the silica bonding was observed at 400°C. In addition, HZSM-5 appeared to be more robust after several regeneration cycles by oxidation of the catalyst with a consistent activity to produce hydrocarbons from palmitic acid. More recently, a more acidic zeolite with a low ratio of Si/Al

(or the SiO_2/Al_2O_3 ratio) was more selective to produce aromatic compounds in supercritical water, nevertheless, a higher Si/Al ratio zeolite was more stable under the same condition (Mo et al., 2015). For the reason that the propriety of the water are different at 350°C compared to 400°C, it would be interesting in study the stability of HZSM-5 at 350°C.

Zeolites including HZSM-5 and HY have already been used to process biomass, bio-crude oil, and fatty acids in hydrothermal liquefaction (Duan and Savage, 2010; Yang et al., 2011; Bai et al., 2014; Mo and Savage, 2014). HZSM-5 has been demonstrated to be a good catalyst to produce paraffinic oil with a high carbon content from the upgrading of a microalgal pretreated crude oil (Mo and Savage, 2014). Overall, HZSM-5 was more selective compared to other zeolites such as Hβ and HY to produce aromatic compounds from fatty acids at higher temperatures (Mo et al., 2015). Pretreating a bio-crude oil with the same catalyst reduced the nitrogen content, although the production of coke (28 wt.%) was more significant above 400°C compared to the other catalysts which represents a drawback (Bai et al., 2014).

The aim of this study is to investigate whether HZSM-5 catalysts retain the same chemical properties and stability under subcritical hydrothermal conditions. This study characterizes the metal-doped zeolites with molybdenum, nickel, copper, and iron cations produced using the ion-exchange method. Subsequently, their stabilities are investigated in subcritical water (350°C). The changes of physical or chemical properties are observed using techniques such as X-ray crystallography, surface area (BET), and scanning transmission electronic microscopy (STEM). Moreover, two methods (ion-exchange and wet incipient wetness) were investigated to prepare MoZSM-5; molybdenum was selected because of previous results from the literatures with good activities and to investigate the impact on the stability and the activity from different methods of preparation (Song et al., 2006, 2007). The activity of HZSM-5, MoZSM-5, CuZSM-5, FeZSM-5, and NiZSM-5 was investigated with soya proteins and two microalgae with different compositions: *Pseudochoricystis ellipsoidea* and *Chlorella*. The bio-crude composition was analyzed by elemental analysis, gas chromatography–mass spectrometry (GC–MS), and gel permeation chromatography (GPC). Finally, the deoxygenation and denitrogenation of the processed bio-crude oil, using the catalysts prepared by ion-exchange, were investigated.

MATERIALS AND METHODS

Materials

The microalga strain *P. ellipsoidea* was supplied by Denso Corporation and has been used in other work (Biller et al., 2013). Sunflower oil, *Chlorella vulgaris* and soy protein were obtained from commercial sources. NH_4-ZSM-5 was supplied by Alfa Aesar, the metal salts were supplied by Sigma Aldrich.

Preparation and Characterization of Metal-Doped HZSM-5

The zeolite in the form of NH_4-ZSM5 was calcined for 3 h at 550°C under a constant flow of air (50 ml/min) to yield HZSM5. The Si/Al ratio was 27 (4.9 wt.%, for the aluminum while silica 82.2 wt.%

of silica) where the silica content was determined by colorimetry and the aluminum content by atomic absorption spectroscopy (AAS). According to the method by Long and Yang (2000), the metal ion-exchanged catalysts were prepared as follows: 20 g of NH₄-ZSM5 was mixed with a 0.05 M solution of metal salts [copper acetate Cu(acac)₂, iron nitrate Fe(NO₃)₃, nickel acetate, ammonium molybdate] under constant stirring for 24 h. The solid was washed and filtered under Buchner, then dried at 110°C overnight. Finally, the doped zeolite was calcined at 550°C for 5 h. The chemical composition of metal ion-exchanged zeolites were analyzed by AAS (with an average coefficient of variance of 3.7%) after acid digestion and were determined as containing the following metal loadings: FeZSM5 0.3 wt.%, MoZSM5 0.1 wt.%, CuZSM-5 1.2 wt.%, and NiZSM-5 0.1 wt.%. Two batches of NiZSM-5 were produced, where the metal loading had a coefficient of variance of 10.3%.

For the preparation of the impregnated sample of molybdenum (5 wt.%), 0.9 g of ammonium molybdate tetrahydrate [(NH₄)₆Mo₇O₂₄ 4H₂O] was dissolved into 1.5 ml of water (pore of the HZSM-5 is 0.15 ml/g) and poured in 10 g of HZSM-5. The slurry was crushed, mixed well, and finally dried overnight (more than 12 h) at 105°C. The final step was the calcination of the solid at 550°C for 4 h under a constant flow of air at 50 ml/min. For the impregnated molybdenum sample, 4.5 wt.% of metal was measured by AAS.

The preparation of the kaolin-HZSM-5 pellet was carried out as follows: 2.1 g of kaolin clay and 1.0 g binder polyvinyl alcohol (PVA) with a molecular weight of 88,000 g/mol were mixed in 20 ml of water for 30 min. Subsequently, 22 ml of concentrated HCl (35 vol.%) was carefully added drop-wise for 1 h by stirring, forming a foam. Subsequently, a suspension of 2 g of HZSM-5 in 20 ml of water was added into the slurry under constant mixing for 30 min. The water was evaporated on a hot plate for 6 h, and the residue was further dried in an oven overnight. The solid was calcined at 550°C for 1 h, once it had cooled down. The solid was washed with 45 ml of water and 0.094 g of ammonium chloride (NH₄Cl) to remove any trace of sodium. The mixture was stirred under a mild heating at 60°C for 30 min. The residue was retrieved by filtration and subsequently calcined at 550°C for 3 h.

The XRD measurements were performed using a BRUKER-binary V3 machine and used to determine the crystallinity and phases present in the fresh and the processed sample; HZSM-5 was mixed with a standard solution of alumina corundum as a reference for quantification with a ratio of 3:1. Indeed, intensity depends on different parameters such as the moisture and the presence of amorphous material; with the mixture, this parameter is reduced. The sum of the intensity between 20° and 25° is chosen as this peak is influenced by the Si/Al (Pollack et al., 1984) and 35° and 40° for the alumina. Crystallinity is estimated using Eq. 1, where the ratio of the fresh sample with the used one was taken, I is the increment of the measurement.

$$\% \text{ Crystallinity} = \frac{\dfrac{\Sigma \text{ intensity fresh}[20-25] \times I_a}{\Sigma \text{ intensity fresh}[35-40]}}{\dfrac{\Sigma \text{ intensity HTL}[20-25] \times I_b}{\Sigma \text{ intensity HTL}[35-40]}} \quad (1)$$

The BET surface area and the pore size [Barrett, Joyner, and Halenda (BJH) method] were determined using a Quantachrome 2200e model. Approximately, 0.1 g of sample was degassed under vacuum at 300°C. The adsorption–desorption was carried out at 77 K (−196°C) using 42 points. The surface area of the calcined ZSM5 support was found to be 388 m²/g.

Scanning transmission electronic microscope is conducted using an FEI Tecnai F20 field emission gun (FEG)-TEM operated at 200 kV and equipped with a Gatan Orius SC600A CCD camera and an Oxford Instruments 80-mm X-Max SDD detector. STEM samples were prepared by dispersing powders in isopropanol, with a drop placed on a holey carbon-coated copper grid (Agar Scientific).

Hydrothermal Processing

The hydrothermal processing of different biomasses was performed in an unstirred batch reactor (77 ml, Parr, USA). The processing of biomass was carried out as follows: 3 g of biomass (sunflower oil, soy proteins, *C. vulgaris, P. ellipsoidea*), 0.5 g of catalyst, and 27 ml of deionized water were introduced into the reactor, which were subsequently pressurized with 2 bar nitrogen. Most of the experiments were carried out in duplicate with an average coefficient of variance of 6.5%. The reactor was heated with a heating rate of approximately 9°C/min. The reactor was held at the final temperature of 350°C for 1 h. Pressures achieved were recorded approximately from 140 to 160 bars.

For the regeneration experiments, pellet HZSM-5 was recycled four times in the presence of sunflower oil as follows: 3 g of sunflower oil, 0.5 g of HZSM-5, and 27 ml of water were added into the reactor and heated to 350°C for 1 h. Pellets were prepared by using a press and bringing a pressure of 1 t using HZSM5 of a size of 100–80 mesh. The first experiment was performed six times, for the second cycle, three experiments were carried out, subsequently two and for the last cycle only one. At the end, the catalyst was washed with water and dichloromethane (DCM), and finally it was dried and recycled for the next cycle.

Sample Workup and Analysis

Once the reactor was cooled and opened, it was first rinsed with 50 ml of deionized water and subsequently by 50 ml of DCM. The two phases were mixed together in a separating funnel and allowed to separate. The organic phase (DCM) was separated and filtered using a 1-PS filter and left to evaporate at room temperature for approximately 2 days to obtain the mass yield of crude oil. The aqueous phase was filtered under vacuum using a Buchner filter (filter grade 5) in order to collect the residue and the catalyst. Finally, the aqueous phase was diluted to 1 l. Yields, including the bio-crude and residue (Datasheet S1 in Supplementary Material), were calculated according to the mass of the oil and residue collected as dry weight; the gaseous yield (Datasheet S1 in Supplementary Material) was calculated according to the previous studies (Biller and Ross, 2011; Biller et al., 2011). The residue phase was measured by the weight of the residue collected on filters of the organic and aqueous phases minus the weight of the catalyst. The aqueous phase yield was calculated by the difference of the other phases (bio-crude, gaseous, and residue). **Figure 1** describes the general procedure of the workup.

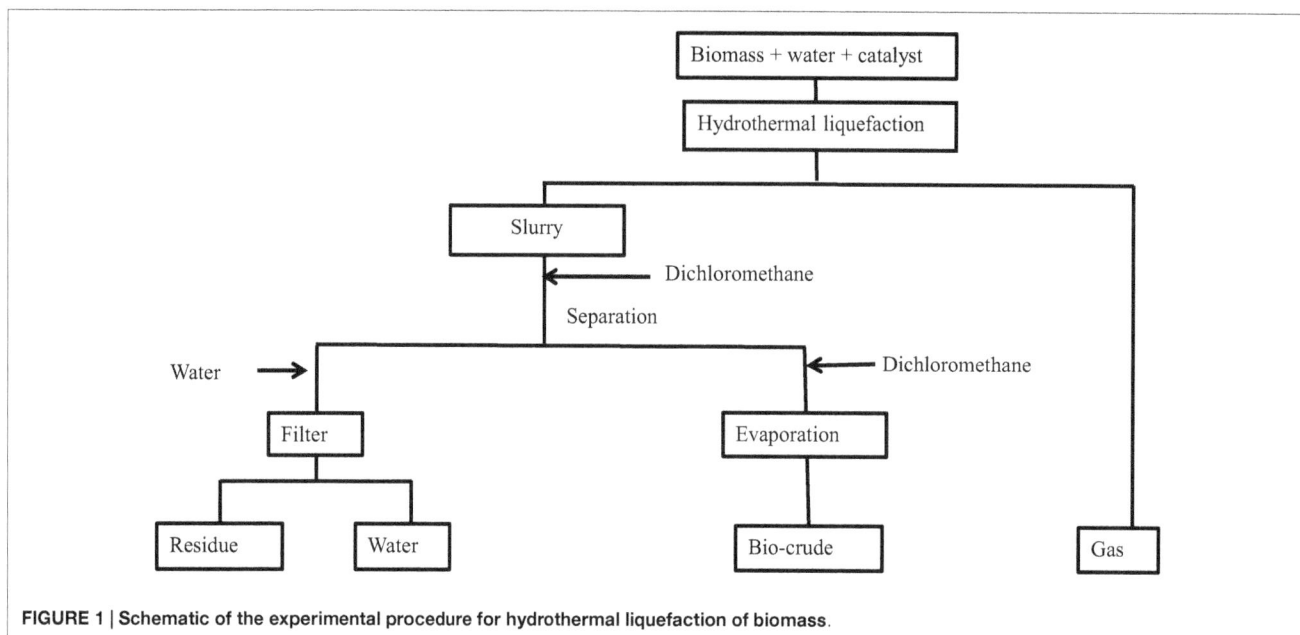

FIGURE 1 | Schematic of the experimental procedure for hydrothermal liquefaction of biomass.

Following evaporation, a fraction of the bio-crude oil was redissolved in DCM to obtain a concentration of approximately 10 wt.% and analyzed by GC–MS using an Agilent 5975B inert MSD. Separation of the products was achieved using an RTx 1701 60-m capillary column, 0.25 id, 0.25 μm film thickness using the following temperature program: 40°C, hold time 2 min, ramped to 280°C at 60°C/min, hold time 10 min with a split ratio of 1:10. The column head pressure was 30 psi at 40°C.

Gel permeation chromatography was performed using a Perkin Elmer Series 200 HPLC instrument fitted with a refractive index detector. The column used was a Varian PGel 30 cm length with a diameter of 7.5 mm and a particle size of 100 Å, the flow rate of mobile phase was 1 ml/min in stabilized tetrahydrofuran (THF). The sample was measured in a 1.5-ml vial, a concentration of 10 wt.% using THF as solvent. The instrument was calibrated using different molecular weight polystyrene polymer standards (poly laboratories). The percent fraction of different weight was determined by the integration of curve by the software Origin 8 at different molecular size ranges: the first fraction was integrated between 8.5 and 11 min (0–200 g/mol), the second fraction from 8 to 8.5 min (200–600 g/mol), third fraction from 7.5 to 8 min (600–1000 g/mol), and the last fraction from 6 to 7.5 min (superior to 1000 g/mol). The ash content of the raw material (1.0 g) was measured in crucible with a muffle furnace for 3 h at 550°C.

For the elemental analysis (CHNS) of the bio-crude oil, the oil was filtered using a 1-PS grade filter to remove any residue or moisture. Subsequently, the oils were dissolved in a DCM/methanol (10 vol.%) solution with a ratio of oil/solvent, 1:4, and placed into smooth wall tin capsules. Solvents were evaporated before the analysis using CE instruments Flash EA 1112 series elemental analyzer. The high heating value (HHV) with a unit of megajoule per kilogram was calculated using the Dulong formula, as shown in Eq. 2 (Corbitt, 1999); C, H, O, and S represent the

weight percentage of the carbon, hydrogen, oxygen, and sulfur, respectively, measured by the elemental analysis. The energy recovery was calculated using Eq. 3, where $m_{bio\text{-}crude}$ and $m_{raw\ sample}$ are the masses in grams as dry basis; HHV were calculated using the previous equation.

$$HHV = 0.338 \times C + 1.428 \times \left(H - \frac{O}{8} \right) + 0.095 \times S \qquad (2)$$

$$\%\ Energy\ recovery = \frac{HHV_{bio\text{-}crude} \times m_{bio\text{-}crude}}{HHV_{raw\ sample} \times m_{raw\ sample}} \times 100 \qquad (3)$$

The determination of coking during the recycling step was carried out using the Netzsch STA (simultaneous thermal analyzers) from a previous study (Ortega et al., 1997). During the first step, impurities were removed by pyrolyzing approximately 20 mg of samples in a constant flow of helium at 100 ml/min with a heating rate of 10°C/min and holding at 500°C for 30 min. This following program was used to stabilize the coke. In the second step, combustion was performed; the amount of coke was measured from 300 to 550°C with a heating rate of 10°C/min in a gaseous mixture of 12.5 vol.% O_2/He, and this temperature is held for 30 min. Combustion gases were determined using an online FT-IR. The wavelength of carbon dioxide (CO_2) measured is 2362 cm^{-1} and for carbon monoxide (CO) 2175 cm^{-1}, and water (H_2O) 3400 cm^{-1}. The STA was calibrated by melting pure metals. At least five points are used to create these calibration curves. Each point corresponds to data obtained from the melting of a different metal. The metals were selected to more or less cover the range of temperatures for the measurements [i.e., from In (indium) (156.6°C) to Au (gold) (1064.4°C)]. In addition, a buoyancy (aka background, correction, and calibration) correction is also done under the same conditions as the experiments (i.e., gases, flow rates, heating rates, and duration of isothermal steps).

Equation 4 is used to calculate the H/C ratio of the coke. n_{H_2O} represents the molar amount of water calculated from the infrared spectrum; n_{CO} and n_{CO_2} are also determined from the absorbance of CO and CO_2. % C represents the total carbon content for both gases. The spectrometer was previously calibrated with calcium oxalate using the following program, in helium gas with a heating rate of 10°C/min to 900°C.

$$\frac{H}{C} = \frac{\dfrac{\% \text{ H in water} \times n_{H_2O}}{2}}{\dfrac{\left(n_{CO} + n_{CO_2}\right) \times \% \text{ C}}{12}} \tag{4}$$

The concentration of CO_2, CO, and water are calculated in order to know the H/C of the coke. The equation is deduced from the calibration of the water in Eq. 5 where x is the mass of water and y is the absorbance. The equation for CO is in Eq. 6 and CO_2 is in Eq. 7.

$$y = 0.6363x + 0.2291 \tag{5}$$

$$y = 0.2147x + 0.2703 \tag{6}$$

$$y = 3.4127x + 4.4405 \tag{7}$$

RESULTS

The Stability of HZSM5 Under Subcritical Condition

Characterization of the Catalyst

The first section presents the experimental results regarding the initial composition and the hydrothermal stability of metal-doped catalysts at 350°C.

The method of ion-exchange is the most efficient with copper with 1.2 wt.%. The metal loading found for the ion-exchanged metal are lower compared to the literature obtained by the same conditions with 1.6, 0.8, and 4.3 wt.% for iron, nickel, and copper, respectively (Long and Yang, 2000).

The metal loading was more significant with molybdenum using the wet incipient impregnation method compared to the ion-exchange. The loading of metal probably depends on the size of the element to enter through the pore. **Figure 2** shows two examples of STEM results for two catalysts **Figure 2A** FeZSM-5 and **Figure 2B** MoZSM-5, where (i) is the STEM images, (ii) EDX, and (iii) HAADF imaging (high-angle annular dark field), where the bright spot represents heavy molecular weight metals. The square corresponds to the place, where the EDX measurement was carried out. The copper found in the EDX scan was from the grid. However, it was difficult to obtain clear pictures with this material, because it began to degrade under the electron beam after a few minutes hence then the structure starts to deteriorate.

The image suggests that iron formed clusters at the edge of the particle with a size of approximately 16 nm, the presence of iron is confirmed by EDX although this technique does not allow the quantification of metals. **Figure 2B** proposes that molybdenum is located inside the pores as observed with the image (iii) forming a bright spot of approximately 1 nm. Nickel (STEM not shown)

probably forms a thin layer as the EDX identifies nickel without seeing any difference, implying that no brighter spot was observed by HAADF technique. A lattice network with a size of 1.6 nm for the unprocessed HZSM-5 was measured from the STEM images in Figure S1 in Supplementary Material.

Stability

The method of subcritical water liquefaction involves harsh conditions with high pressure between 140 and 150 bar at 350°C. Therefore, catalysts should be robust under these circumstances. Metal leachates from the reactor are measured by ICP-OES, a blank experiment is carried out and the main cations detected are iron, calcium, and nickel with a concentration of 1.1, 1.0, and 2.7 mg/l, respectively.

Generally, the dealumination is low with approximately 0.2 wt.% leaching of the HZSM-5. As said earlier, the migration of aluminum was possible out of the framework (de Lucas et al., 1997b). The loss of silica is more significant with approximately 20.0 wt.%. This result is in accordance with the previous study (Ravenelle et al., 2010), where the hydrolysis of the siloxane bond (Si–O–Si) was enhanced in hot compressed water (150–200°C); these phenomena do have an impact on the reactivity for HZSM-5 as discussed in the following section. The leaching of the doped metal at 350°C is high with 6.2 wt.% for FeZSM-5, 7.9 wt.% for CuZSM-5, 10.0 wt.% with NiZSM-5, and 34.0 wt.% with MoZSM-5. These results are counterintuitive with the STEM analysis as described in the previous section, as iron is located outside the particle which is more subject to sintering, while molybdenum is found in the pores. Yet, these leachate ions could have a catalytic impact during the liquefaction of biomass. For MoZSM-5 between the impregnated and ion-exchanged samples, a similar value is measured (35.0 wt.%). Molybdenum might have reacted with the silica leachate as the aqueous phase is blue after the stability test.

The loss of aluminum and silica is lower than the doped metal, especially with FeZSM-5 and MoZSM-5 with 11.2 and 7.2 wt.% of leachate of silicon. Thus, these metals could have an impact on the strength of the silica framework. The STEM imaging of the samples treated in subcritical water at 350°C suggests that the lattice interstice widens to 2.3 nm. For the processed MoZSM-5, the EDX shows no indication of the presence of molybdenum, but the sample is degraded under the beam for a long time and it is difficult to be certain. An agglomeration of iron (shown in Figure S2 in Supplementary Material) is observed in some concentrated part, whereas elsewhere no iron has been detected. Some traces of nickel (shown in Figure S3 in Supplementary Material) are detected in the processed sample, although the size is too small to be seen by STEM HAADF, the EDX show traces of nickel.

The Effect of the Surface Area and Pore Sizes

The pore size of HZSM-5 is located in the majority of the microporous region below 1.8 Å (radius) calculated by the Horváth–Kawazoe (HK) method. **Table 1** compares the surface area (m^2/g) and the cumulative pore volume (ml/g) of the fresh and the used metal-doped HZSM-5. The BJH method allowed the calculation of the cumulative volume in the micro- and mesoporous. This method predicts the pressure when condensation happens inside the cylindrical pore.

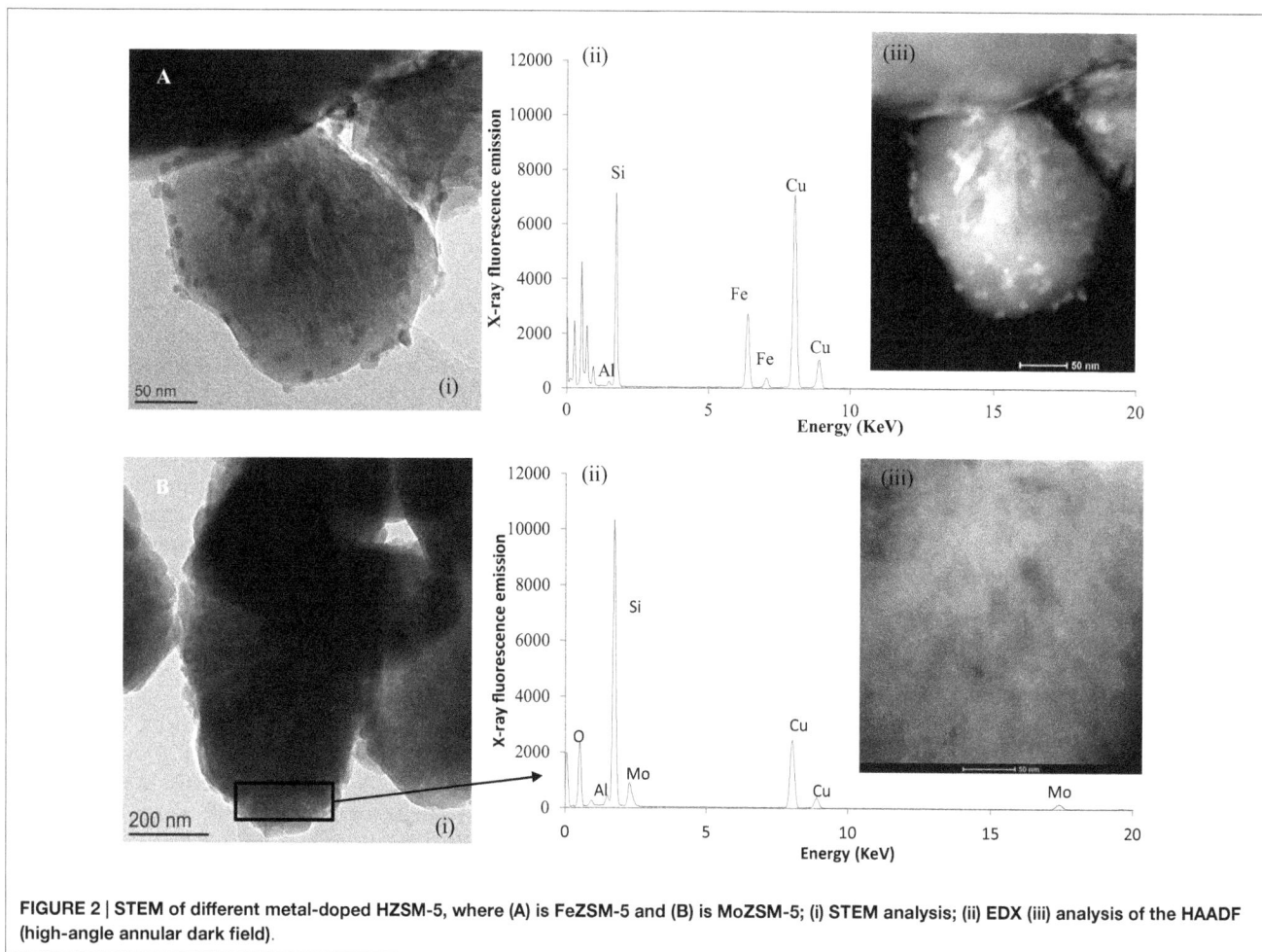

FIGURE 2 | STEM of different metal-doped HZSM-5, where (A) is FeZSM-5 and (B) is MoZSM-5; (i) STEM analysis; (ii) EDX (iii) analysis of the HAADF (high-angle annular dark field).

TABLE 1 | BET values calculated for fresh and used doped HZSM-5; pore volume and desorption calculated by the BJH methods.

Catalysts	Surface area (m²/g)	Cumulative pore volume (ml/g)	Surface area after hydrothermal run (m²/g)	Cumulative hydrothermal pore volume (ml/g)
HZSM-5	382	0.09	141	0.05
CuZSM-5	384	0.09	285	0.05
FeZSM-5	399	0.09	52	0.03
MoZSM-5 ion-exchanged	435	0.1	250	0.07
NiZSM-5	370	0.1	219	0.1
MoZSM-5 impregnated	321	0.1	184	0.1
Kaolin-pellet HZSM-5	174	0.09	200	0.4

During the hydrothermal treatment using HZSM-5, the average pore size remains unchanged, that is, 1.8 Å. The volume of the mesoporous, measured by the BJH method, is 0.05 ml/g. The difference between the fresh and doped samples is probably caused by the metal blocking the pores. Copper increases the cumulative

pore volume compared to HZSM-5. Impregnated HZSM-5 with molybdenum has a lower surface area than HZSM-5 doped by ion-exchange. In the literature, a surface area of 324 m²/g for the plain HZSM-5 was measured which was lower than the value during this study (Idem et al., 1997). A surface area of 334 m²/g and a pore volume of 0.11 ml/g were determined for impregnating HZSM-5 with nickel (0.5 wt.%) (Gayubo et al., 2010). The surface area of the pellet was lower compared to the pure material as the binder (the clay kaolin) had a very low surface area of approximately 15 m²/g.

Under subcritical condition, abrasion of the catalyst was observed with the loss of surface area. Indeed, a 25% of loss was calculated between the fresh and the used sample at 350°C. The reduction was more significant with FeZSM-5 with 87% of loss (between the two values). Under steam, iron enhanced the migration of aluminum from the framework and reduced the dispersion of the metal; the same phenomena might result in the loss of surface area (Pieterse et al., 2007). The observation in the STEM, described in the previous section, confirmed that iron formed cluster during this condition. In the literature, no change in pore volume was noticeable in the microporous and mesoporous in hot compressed water (150–200°C) and in supercritical water as well (Ravenelle et al., 2010; Mo et al., 2015).

TABLE 2 | Peak area ratio of mixture 25% HZSM-5–75% γAl₂O₃ for the determination of the percentage of crystallinity.

	HZSM-5	FeZSM-5	MoZSM-5 exchanged	CuZSM-5	NiZSM-5	MoZSM-5 impregnated
% Crystallinity compared to the fresh catalyst	75.5	77.4	86.2	72.4	79.2	75.6

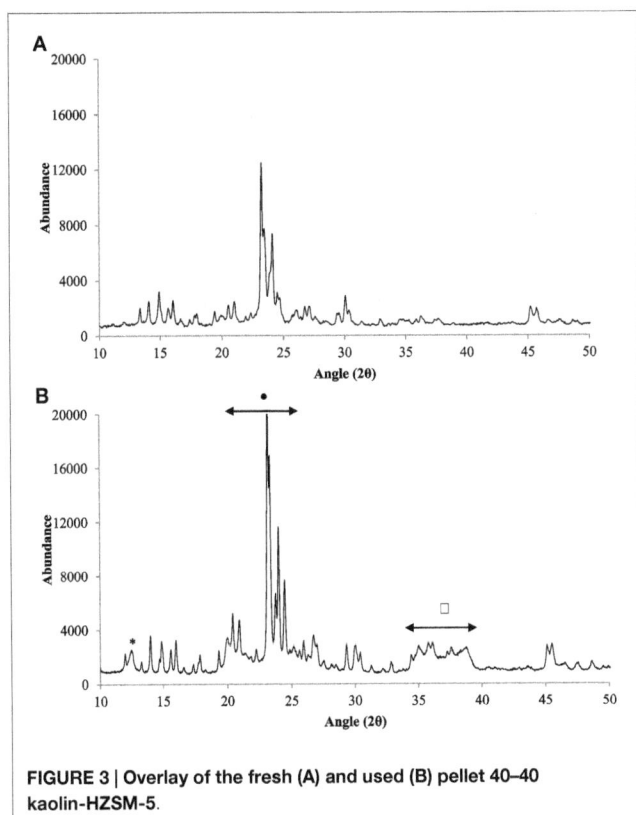

FIGURE 3 | Overlay of the fresh (A) and used (B) pellet 40–40 kaolin-HZSM-5.

TABLE 3 | Yield (wt.%) of the hydrothermal processing of sunflower oil at 350°C for 1 h in a 77 ml bomb reactor recycling HZSM5 four times.

Experiment	Bio-crude yield (wt.%)	Coke formed (wt.%)
HTL of sunflower with pellet HZSM5 cycle 1	82.4	2.6
HTL of sunflower with pellet HZSM5 cycle 2	92.0	3.9
HTL of sunflower with pellet HZSM5 cycle 3	84.5	4.2
HTL of sunflower with pellet HZSM5 cycle 4	91.5	3.4

For the majority of samples, with the exception of the kaolin-pellet HZSM-5, a decrease in cumulative pore volume (mesoporous) is apparent probably caused by the destruction of pores during the elevated pressure. Nevertheless, the BJH desorption profile indicates that new pores have been created in the region of 20–30 Å for HZSM-5 and CuZSM-5, from 80 to 110 Å for FeZSM-5 and in the region of 120 Å for MoZSM-5. Interestingly, the surface area of the kaolin-pellet increases the surface area to 200 m²/g, which implies that the binder and the clay enhance the stability toward abrasion. Moreover, the pore volume is larger than the fresh catalyst. To conclude, subcritical water has an impact on the physical aspect of HZSM-5, yet binder could enhance the stability of the catalyst.

Crystallinity of HZSM-5

HZSM-5 has high crystallinity and the metal loading was too small to observe a peak on the XRD spectrum. **Table 2** lists the crystallinity value calculated using Eq. 1.

Doped metal with the exception of CuZSM-5 enhanced the percentage of the crystallinity after the stability test compared to the plain HZSM-5. Yet, molybdenum prepared by ion-exchange had the highest effect compared to the other samples. Generally,

impregnated samples have a lower crystallinity with 75.6 for molybdenum compared to ion-exchange metal-doped HZSM-5. Even though FeZSM-5 has a low surface area, its crystallinity is higher than HZSM-5. Copper contains slightly more amorphous material than HZSM-5.

Figure 3 represents the X-ray spectrum overlay between the fresh and the samples treated in subcritical water for the kaolin-pellet HZSM-5. The main difference with the used sample is the presence of new peak at approximately 12° (marking with *) and the intensity is larger between two regions of 21–25° (•) and 35–40° (⊔). The change in intensity could be caused by the presence of moisture and amorphous material (explaining why previously zeolite was mixed with a reference material). This pellet is mixed with kaolin clay that was an amorphous material.

Two new peaks appeared after two regeneration tests in supercritical water with no effect, though on the activity as these peaks disappeared after calcination (Mo and Savage, 2014). No change of structure was noticed under low temperature (Ravenelle et al., 2010).

To summarize, these metals have an impact on enhancing the crystallinity or for some the surface area. Impregnated doped metals with a high loading do not have an effect on the physical property of HZSM-5. The next step was to test this catalyst with real biomass.

Regeneration of HZSM-5

The regeneration and the stability of pellet HZSM-5 (compressed pellet) are investigated in this section. Regeneration experiment has already been carried out (Mo et al., 2015) in supercritical water; however, in this study, a different insight brings to the analysis of the composition of the coke. **Table 3** includes the bio-crude yield measured after the HTL reaction with a coefficient of variance of 4.6% and the weight percent of coke which are calculated from the thermogravimetric curve and weight loss (from 300 to 900°C). During this process, according to the details in the study (Ortega et al., 1997), the first stage heated the catalyst to 500°C under high flow of helium to age the coke and subsequently the coke was combusted from 300 to 900°C.

The initial size of the pellet was approximately 1 mm of diameter, yet after each experiment, the size of the particles was reduced due to the abrasion; this is the reason that cycle 1 is carried out six times, in order to have sufficient catalyst (0.5 g) for the last cycle. The bio-crude yield increases from step 1 to 2 and subsequently the bio-crude yield is leveled out with cycle 3 (regarding the error margin) and cycle 4. The yield of the gaseous yield is 4.2 wt.% for the cycle 1 and for the next cycles (from cycle 2 to 4) it is approximately 3.8 wt.%. Yet, the yield is higher than the powder HZSM-5 with 68.0 wt.% (experiment, as discussed in the next section) indicating that the cracking is less efficient with compressed pellet as the surface area of contact was higher than the powder. In other words, lipids could not access the acidic site. The coke content increases throughout cycles 1–3 and decreases for the last regeneration 4. Between each stage, the catalyst is dried, but not calcined. Compared to the pyrolysis, the formation of coke is up to 6 wt.%, whereas under subcritical condition the coking was lower (de Lucas et al., 1997a). The gas emission is analyzed by the FT-IR, where the absorption of the CO, CO_2, and water are detected during the combustion stage. Thanks to a calibration, the mass is calculated and presented in **Table 4**.

The H/C ratio of the coke decreased after the second cycle. The previous study proposed that the reduction in the H/C ratio was caused by the dealumination and the increase in the formation of coke. Moreover, a low H/C ratio indicates the presence of aromatic compounds in the coke (de Lucas et al., 1997a). Thus, HZSM-5 started to degrade after the third cycle. During the first step with pyrolysis, a large amount of methane is produced implying that even with the washing with DCM, some aliphatic compounds are trapped inside the structure of HZSM-5. Subsequently, during the combustion stage mainly CO and CO_2 are formed.

A reduction in the surface area is caused by the abrasion of the pellet, and the saturation of the pores by coke is observed. The surface area levels out between the cycles 3 and 4. The same observation was drawn previously with a reduction in surface area and the loss of cumulative total pore volume (Mo et al., 2015).

Despite some physical change during the regeneration experiment, no major change is observed in the bio-crude yield. Thus, HZSM-5 is stable during the recycling process. The stability of the pellet (kaolin-HZSM-5) should be investigated as it showed good resistance to abrasion and also the ion-exchanged catalysts. Even though, impregnated metal-doped catalysts have a larger metal doped than ion-exchange, the latter show a higher stability and these catalysts therefore are selected during the processing of the different biomass.

TABLE 4 | Mass of water, carbon monoxide, and carbon dioxide determined by the FT-IR and the H/C ratio for a sample of mass of approximately 20 mg.

	Pellet cycle1	Pellet cycle2	Pellet cycle 3
Mass of water (mg)	4.5	7.7	7.4
Mass of carbon dioxide (mg)	17.6	47.9	44.0
Mass of carbon monoxide (mg)	1.0	1.8	1.9
Carbon wt.%	40.3	43.3	43.1
Hydrogen wt.%	2.3	1.5	1.5
H/C ratio	1.1	1.5	0.7

Processing of Biomass with Metal-Doped Zeolites

This section gives a brief overview of the processing of sunflower oil, soy protein, and two microalgae *Chlorella* and the stressed *P. ellipsoidea* regarding the composition of the bio-crude oil. The stressed *P. ellipsoidea* contains a significant lipid content (67.0 wt.%) and low ash content (less than 1.0 wt.%). The lipid fraction was increased by starving the microalga for nitrogen nutrients as described in a previous study (Satoh et al., 2010). Soy protein is 90.0 wt.% pure with 3.5 wt.% of ash. *Chlorella* contains high protein content (55.0 wt.%) and low lipids (25.0 wt.%) and high ash content with 10.9 wt.%. Sunflower oil is refined and contains mainly linoleic acid and oleic acid as fatty acids. **Table 5** summarizes the main results, including the bio-crude yield (as dried basis), the elemental analysis, and the energy efficiency. The average coefficient of variance is approximately 6.5% for the bio-crude yield; 0.9–2.5% for the nitrogen content, and for the carbon, hydrogen, and sulfur content are listed with a coefficient of variance of approximately 2.0%. In this study, results regarding the bio-crude composition are detailed; however, the initial biomass is degraded within different phases, including the aqueous, residue, and gaseous phases. The mass yield of these phases is included in Table S1 in Supplementary Material.

Sunflower Oil

The bio-crude yield (**Table 5**) was more significant for non-catalytic run compared to the run with metal-doped HZSM-5 implying that these catalysts enhanced the cracking of sunflower oil. For the processing of sunflower oil, no residue was produced. The mass of the gaseous yield was more significant for NiZSM-5, as nickel is known to promote gasification (Stucki et al., 2009). The mass of the aqueous phase is also more significant with the catalytic run from 22.8 to 36.4 wt.%.

Between the two catalysts prepared by impregnation and ion-exchange with molybdenum, higher activity was achieved with MoZSM-5 prepared with metal exchanged. HZSM-5 in the form of powder is more active with lower bio-crude yield and a higher fraction of aromatic compound produced. In general, there is a trade-off as the most robust catalyst has lower activity compared to the ion-exchanged catalysts.

The elemental analysis, in **Table 5**, indicates that HZSM-5 and MoZSM-5 enhance the deoxygenation for sunflower oil with low oxygen content with 7.7 and 8.8 wt.%, respectively, compared to the run without any catalyst. Furthermore, these bio-crude oils have a close heating value close to the petroleum oil with 42–44 MJ/kg (Demirbaş, 1998). Higher decarboxylation from palmitic acid was observed in the previous study (Mo and Savage, 2014), since more catalysts and higher temperatures with supercritical water conditions were used.

The molecular weight distribution for different catalysts measured by GPC is included in Figure S4A in Supplementary Material. HZSM-5 contains in majority "long-chain" or fatty acid materials and the lowest fraction of oligomers materials compared to the other catalysts. Oligomers are produced from the cross-linked reaction between the linoleic and oleic acids. MoZSM-5 contains the highest fraction of lower molecular weight materials which could include aliphatic hydrocarbons

TABLE 5 | Elemental composition and the bio-crude yield as received of the bio-crude oils using the different catalysts in water and the % of energy recovery.

	N wt.%	C wt.%	H wt.%	S wt.%	O wt.%	Dulong HHV (MJ/kg)	Bio-crude yield wt.%	% Energy recovery
Sunflower oil								
Raw	–	72.9	11.1		16.0	37.6	–	–
No catalyst	–	76.8	11.5	–	11.7	40.3	86.0	92.3
HZSM-5	–	80.4	11.9	–	7.7	42.8	68.0	72.0
FeZSM-5	–	75.5	11.4	–	13.1	39.5	58.0	65.4
CuZSM-5	–	74.4	11.6	–	13.9	39.3	73.0	76.1
MoZSM-5	–	80	11.6	–	8.4	42.1	60.0	67.1
NiZSM-5	–	77.2	11.7	–	11.1	40.8	55.0	50.0
Soy protein								
Raw	14.2	50.6	7.4	0.4	27.4	22.8	–	–
No catalyst	6.1	72.8	9.2	0.4	11.5	35.7	17.5	65.6
HZSM-5	7.5	71.5	9.7	0.6	10.7	36.1	10.8	28.9
FeZSM-5	6.6	70.5	9.5	0.7	12.8	35.2	15.1	19.8
CuZSM-5	7.8	79.1	10.5	0.7	1.9	41.5	14.0	30.3
MoZSM-5	5.7	67.6	9.3	0.5	16.9	33.2	15.1	15.1
NiZSM-5	7.3	75.7	9.9	0.9	6.3	38.7	15.0	30.3
Chlorella								
Raw	10.3	49.4	7.4	0.5	21.5	21.5	–	–
No catalyst	5.9	74.9	9.3	0.6	9.4	36.9	28.9	50.3
HZSM-5	5.5	75.0	9.3	0.4	9.7	37.0	18.3	36.5
FeZSM-5	5.2	73.4	8.8	0.8	11.8	35.4	29.6	48.9
CuZSM-5	6.7	80	10.1	0.4	2.8	41.0	18.7	42.4
MoZSM-5	6.2	82.5	10.5	0.4	0.5	42.8	21.6	51.1
NiZSM-5	6.5	75.5	9.7	0.6	7.7	38.0	28.1	38.5
Stressed P. ellipsoidea								
Raw	2.9	63.4	9.5	–	30.6	30.6	–	–
No catalyst	1.0	70.5	10.3	–	18.2	35.3	49.3	56.9
HZSM-5	2.1	78.0	11.7	–	8.1	41.7	25.5	35.8
FeZSM-5	1.6	75.2	11.3	–	11.9	39.4	34.2	33.9
CuZSM-5	2.2	81.5	11.9	–	4.5	43.7	22.8	33.7
MoZSM-5	2.2	80.4	11.2	–	6.2	42.1	34.2	46.2
NiZSM-5	1.5	74.6	11.0	–	12.9	38.6	33.5	47.7

and aromatic compounds. The fraction of oligomers is more significant with CuZSM-5.

MoZSM-5 is more selective to produce aromatic compounds as these molecules identified with the GC–MS in the bio-crude oil: 1,3-ethyl-dimethyl-benzene, 1-methyl-butyl-benzene, 1,4-methyl-2-(2-methyl-propyl)-benzene, and decahydronaphthalene. It explains why more "low molecular weight materials" are observed for this catalyst. Aromatic compounds are produced from the Diels–Alder reaction (Benson et al., 2009). The bio-crude oil for HZSM-5 contains more alkenes compounds such as 8-heptadecene, 3-ethyl-2-pentene, and 4-undecene. Oxygenated compounds were also identified with a lower fraction compared to the other catalysts and the non-catalytic run. Oleic acid was the most abundant, followed by Z,Z-10,12-hexadecadienal which is produced from the degradation of linoleic acid.

Previous studies (Mo and Savage, 2014; Mo et al., 2015) demonstrated the formation of benzene, toluene, and xylene (BTX) compounds from fatty acids using HZSM-5 in supercritical water. Nevertheless, a high fraction of coke and gas was obtained. In this study, only 2.5 wt.% of coke was produced and only 2 bar of gas was produced which represent approximately 5 wt.% of mass yield. Furthermore, a ratio catalyst, feedstock of 1:1, was

used which is not economically viable compared to 10 wt.% used in this study.

Therefore, HZSM-5, MoZSM-5, and NiZSM-5 have a good activity in processing sunflower oil into alkenes and aromatic compounds. CuZSM-5 enhanced the formation of oligomer materials in the bio-crude oil. Therefore, in this case, MoZSM-5 would be the most suitable catalyst to process sunflower oil to produce a bio-crude with less oxygen content.

Soy Proteins

At the opposite of the processing of sunflower oil, the majority of the soy protein at 350°C is degraded within the aqueous phase with approximately 73.1 wt.% of yield. In opposition to the previous section, low bio-crude yield (17.8 wt.%) was produced, and the formation of residue is more important compared with the processing of sunflower oil. Furthermore, the addition of metal-doped HZSM-5 has a minor impact on enhancing the bio-crude yield compared to sunflower oil. The lower bio-crude yield with catalyst could be explained that some compounds could be absorbed within the zeolite pores. MoZSM-5 was more selective to enhance the formation of gas with 21.5 wt.% (result in the Datasheet S1 in Supplementary Material).

Elemental values (**Table 5**) indicate that the nitrogen content is significant in the bio-crude oil. A reduction in nitrogen is only achieved using MoZSM-5 (5.1 wt.% with a coefficient of variance of 0.9%) compared to the non-catalytic run. Higher nitrogen content is measured with the other catalysts (with a coefficient of variance of 2.5%). Interestingly, CuZSM-5 and NiZSM-5 with 1.9 and 6.3 wt.% oxygen content, respectively, enhanced the deoxygenation capacity.

The formation of heavy molecular weight materials (with a molecular weight higher than 1000 g/mol) was enhanced with CuZSM-5 (Figure S4B in Supplementary Material). One previous study (Imai et al., 1999) suggests that CuCl₂ enhances the oligomerization of glycine forming long peptide chain in hydrothermal vents in the bottom of the ocean. It could indicate here that the metal leachate present in the aqueous phase could enhance the condensation of small molecules into heavy molecular weight materials. The drawback is that the processed water produced during the HTL of microalgae should be recycled back for the cultivation of microalgae nevertheless metals such as copper and nickel inhibit their growth (Stucki et al., 2009).

The majority of compounds identified in the bio-crude oil by GC–MS are nitrogen heterocyclic molecules, for example, pyrazine, diisopropylpiperazine-2,5-dione, pyrrole, 1-butyl-2-pyrrolidinone, 1-(1-oxo-9,12-octadecadienyl)-pyrrolidine, and 3-methyl-1*H*-indole. The disadvantage of these molecules is that they are relatively stable.

For the catalytic run, more complex structures were identified with GC–MS within the bio-crude oil which include *N*-(1-methyl-2-propynyl)-benzenamine, 1-(2-phenylethyl)-pyrimidine-2,4,6trione, 5,10-diethoxy-2,3,7,8-tetrahydro-dipyrrolopyrazine, and hexahydro-3-(phenylmethyl)-pyrrolopyrazine-1,4-dione. The presence of L-leucine-*N*-cyclopropylcarbonyl-1-methyl ester 9*H*-pyridoindole is one example of compounds produced from the condensation of several monomers which explains why with CuZSM-5 the fraction of oligomers and heavy molecular weight materials are more significant.

To conclude, metal-doped zeolites have lower impact on improving the formation of bio-crude yield. Nevertheless, MoZSM-5 reduced the nitrogen content and CuZSM-5 and NiZSM-5 enhanced the deoxygenation of the bio-crude oil. The drawback of CuZSM-5 is that more oligomers and heavy molecular weight materials are being formed. Therefore, the most efficient catalyst to process soy protein would be NiZSM-5 with a good conversion toward deoxygenation and the production of low molecular weight materials.

Microalgae

Microalgae are composed of various elements, including proteins, lipids, carbohydrates, and inorganic salts; it is the reason why the bio-crude oil has a complex composition. The initial content of the microalgae determines the bio-crude yield as follows: lipids > proteins > carbohydrates (Biller et al., 2011). Therefore, it explains why the higher bio-crude yield was achieved at 350°C for the stressed *P. ellipsoidea* with 49.3 wt.% compared to *Chlorella* with 28.9 wt.% containing a higher protein content.

Similar to the catalytic processing of soy protein and sunflower oil, the bio-crude yields are lower compared to the non-catalytic run, this is especially observed with the stressed *P. ellipsoidea*. In general, not only FeZSM-5 but also NiZSM-5 (with lower extent) enhance the bio-crude yield compared to the other catalysts, especially with *Chlorella* (29.6 wt.%) which is higher than the non-catalytic run. Similarly with soy protein, the formation of gas was more significant with MoZSM-5 particularly with *Chlorella* with approximately 49.3 wt.%. Molybdenum is also known to promote gasification (Elliott et al., 2006).

As previously earlier, for the stressed *P. ellipsoidea*, the nitrogen content values (from 2.1 to 1.5 wt.% with a coefficient of variance of 2.0 wt.%) are particularly higher compared to the non-catalytic run (1.0 wt.% with a coefficient of variance of 2.5 wt.%). For *Chlorella*, HZSM-5 and MoZSM-5 reduce slightly the nitrogen content within the bio-crude oil, nevertheless as previously discussed the catalyst has minor impact on the nitrogen content. A slight increase in nitrogen content within the bio-crude using a "zeolite" was observed compared to the non-catalytic run for the processing of *Nannochloropsis* (Duan and Savage, 2010).

More significant results were observed with deoxygenation especially using MoZSM-5 and *Chlorella* with an oxygen content of 0.5 wt.%. Similar oxygen content values were already obtained pretreating a bio-crude oil with HZSM-5 at higher temperature and time at 500°C and 4 h and 5 wt.% loading or using 50 wt.% for half an hour at the same temperature (Li and Savage, 2013). Therefore, it is an encouraging result implying that lower harsh conditions could be used with less catalyst loading. The most efficient energy recovery is determined using NiZSM-5 with a good deoxygenation.

The GPC results of the processing of *Chlorella* for FeZSM-5 and NiZSM-5 (Figure S4C in Supplementary Material) exhibit larger fractions of oligomers and "heavy molecular weight" materials. Conversely, CuZSM-5 and HZSM-5 result in increased yields of lower molecular weight materials. The presence of catalyst did not have a significant impact on the level of "long-chain" materials which remains constant. For the processing of the stressed *P. ellipsoidea* (Figure S4D in Supplementary Material), the "long-chain" material is the highest fraction since this alga contains 67.0 wt.% dry weight lipids. MoZSM-5, in water, enhances the formation of "heavy molecular weight" materials.

The bio-crude oil is composed of nitrogen heterocyclic compounds similar to the bio-crude from soy protein processing. **Figure 4** shows two examples of GC–MS chromatogram, where **Figure 4A** shows the non-catalytic run and **Figure 4B** shows the MoZSM-5.

The chromatogram (B) contains less compared to the non-catalytic run. Amide fatty acids and hydrocarbon molecules are also identified because of the initial presence of lipids. The processing with the metal-doped zeolite, for example, with MoZSM-5 and *Chlorella* produces lower oxygenated compounds including 2-butylhexanoic acid and 3-pentyloxycarbonylpropyl ester with more hydrocarbons such as 3-octadecene and hexadecane. Complex nitrogen compounds are also identified, such as 1-ethyl-2-undecylimidazole in the bio-crude oil of CuZSM-5. It is also observed that the presence of amide fatty acids is reduced

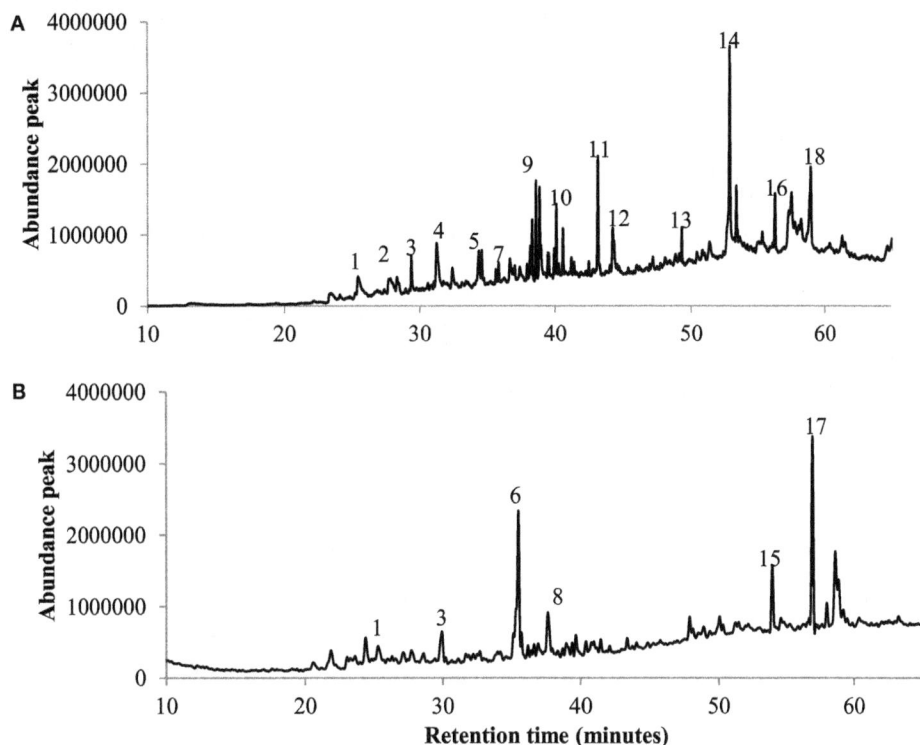

FIGURE 4 | GC–MS chromatogram of the bio-crude oils at 350°C processing *Chlorella*, where (A) shows the non-catalytic run and (B) shows the MoZSM-5. (1) 25.4 min 4-methyl-phenol; (2) 27.8 min 3-ethyl-phenol; (3) 29.4 min hexadecane; (4) 31.7 min 1-butyl-2-pyrrohdinone; (5) 34.3 min 3-methyl-indole; (6) 35.4 min 3-octadecene; (7) 36.6 min benzonitrile; (8) 37.6 min 2,3-dihydro-1-methyl-1H-pyrrole; (9) 38.3 min 3,7,1 l,15-tetramethyl-2-hexadecene; (10) 40.5 min 7-hexadecyn-1-ol; (11) 43.1 min isophytol; (12) 44.1 min ethyl-phenyl-piperidine; (13) 49.1 min 1-nonadecene; (14) 52.8 min hexadecademide; (15) 53.8 min 9-octadecenamide; (16) 53.6 min (5-amino-thiadiazol-2-yl)-propyl-3H-benzooxazol-2-one; (17) 56.3 min 2-butylhexanoic acid, 3-pentyloxycarbonylpropyl ester; (18) 58.8 min octanoic acid, morpholide.

compared to the other catalysts as copper could enhance the hydrolysis of amides (Dzumakaev et al., 1986). For the stressed *P. ellipsoidea*, more alkene and amides were identified within the bio-crude oil. Lower heterocyclic compounds are identified, such as pyrrole.

Overall, an efficient deoxygenation is achieved for MoZSM-5 with *Chlorella*; CuZSM-5 and NiZSM-5 generally enhanced the formation of bio-crude oil with lower oxygen content. Even though a slight reduction in nitrogen content was observed for MoZSM-5, on the whole, metal-doped zeolite had low impact on the denitrogenation. In subcritical water, the activity of catalyst depends on the biomass processed with less side reaction, for example, MoZSM-5 would be more efficient with sunflower oil. A good deoxygenation is achieved with CuZSM-5 but when high protein content is processed, although the formation of oligomers and heavy molecular weight materials is more significant.

DISCUSSION

Comparing pyrolysis condition, HZSM-5 has a lower activity converting biomass feedstock (Idem et al., 1997; Thangalazhy-Gopakumar et al., 2012), nevertheless in subcritical water, HZSM-5 is more selective to produce long-chain materials, which make more suitable for jet fuel or kerosene-like fuel and produce less coke.

The low metal loading on HZSM-5 had a significant impact on the stability and the processing of different biomass feedstock. Nevertheless, the method to insert the metal should be improved in order to increase the metal loading first, and second to reduce the leaching during subcritical water. For example, the stability and also the performance of doped metal HZSM-5 could be improved further using phosphorus (from phosphoric acid) (Xue et al., 2007). Bimetallic doping should be investigated to enhance the catalytic activity of HZSM-5, for example, the combination of molybdenum and copper could reduce the deactivation and increase the stability of the catalyst. Moreover, SAPO catalysts seem to be a good candidate to upgrade biomass due to the high stability under hydrothermal condition and wider pore sizes (Mees et al., 2003).

In this study, doped metal HZSM-5 demonstrated to have a lower impact on the denitrogenation. In literature, harsh conditions (temperature higher to 500°C and using hydrogen) are required to lower nitrogen content, although approximately 2.0 wt.% of nitrogen content are still present in the upgrading bio-crude (Duan et al., 2013; Bai et al., 2014). Therefore, one of the solutions would be to employ zeolites as molecular sieves to reduce the nitrogen content. Doped HY with copper or zeolite-A was previously used as an absorbent for heterocyclic nitrogen compounds. Even with low activity, these zeolites could be used in higher proportion to reduce the amount of heteroatoms (nitrogen and oxygen) in the bio-crude oils (Liu et al., 2008; Xia et al., 2015). Nitrogen compounds found in the bio-crude of soy protein, for example, including 1,3-bis(2-phenylethyl) pyrimidine-2,4,6(1H,3H,5H)-trione, could be extracted and used as high-value molecules which can be used as drug therapy (Xia et al., 2015).

Finally, as in this study, catalysts reduced the formation of bio-crude yield to some extent. It would imply that more raw feed-stock biomasses are necessary to obtain sufficient oil to produce a high-quality fuel; it is the reason why it could be economically more viable to use a microalgal strain containing a high lipid content such as with *P. ellipsoidea*.

CONCLUSION

Subcritical water conditions did have some influence on the physical properties due to abrasion, but less influence on the chemical structure of HZSM-5. Less leaching occurred for nickel compared to molybdenum. The doped metal HZSM-5 had enhanced stability, for example, molybdenum with HZSM-5 improved crystallinity, and copper improved the surface area. Finally, ion-exchange HZSM-5 seemed to produce a more stable form than the impregnated catalyst. HZSM-5 was processed with sunflower oil four times. It was found that the coke amount increased after each test, resulting in the decreased surface area. In general, the bio-crude yield was lower adding the catalyst compared to the non-catalytic run. CuZSM-5 and NiZSM-5 had

good catalytic activities to reduce the oxygen content during the processing of soy protein and microalgae. MoZSM-5 enhanced the formation of aromatic compounds from sunflower oil. An encouraging result was observed with MoZSM-5 and *Chlorella*, as the oil had a lower oxygen content. Some further investigations are required to improve the stability and activity of metal zeolites for the purpose of upgrading biomass in subcritical water.

AUTHOR CONTRIBUTIONS

I hereby TR declare that I carried out all the experiments and measurements except for the STEM measurement by Dr. Nicole Hondow. I have written the manuscript. Dr. AL-L helped me in the analysis of my sample with the XRD and the STEM, she also proofreaded my manuscript. Dr. AR was my main supervisor and supervised the experiments. Prof. JJ was my second supervisor and she helped me in the planification of experiments.

ACKNOWLEDGMENTS

The authors would like to thank the Engineering and Physical Sciences Research Council for financial support (EP/I014365/1). In addition, the authors are grateful not only to Simon Lloyd and Dr. Adrian Cunliff for technical support but also to Dr. Tim Comyn for the XRD, Dr. Leilani Darvell for the STA-FTIR, and Dr. Nicole Hondow for STEM measurement.

REFERENCES

Bai, X., Duan, P., Xu, Y., Zhang, A., and Savage, P. E. (2014). Hydrothermal catalytic processing of pretreated algal oil: a catalyst screening study. *Fuel* 120, 141–149. doi:10.1016/j.fuel.2013.12.012

Benson, T. J., Hernandez, R., French, W. T., Alley, E. G., and Holmes, W. E. (2009). Elucidation of the catalytic cracking pathway for unsaturated mono-, di-, and triacylglycerides on solid acid catalysts. *J. Mol. Catal. A Chem.* 303, 117–123. doi:10.1016/j.molcata.2009.01.008

Biller, P., Friedman, C., and Ross, A. B. (2013). Hydrothermal microwave processing of microalgae as a pre-treatment and extraction technique for bio-fuels and bio-products. *Bioresour. Technol.* 136, 188–195. doi:10.1016/j.biortech.2013.02.088

Biller, P., Riley, R., and Ross, A. B. (2011). Catalytic hydrothermal processing of microalgae: decomposition and upgrading of lipids. *Bioresour. Technol.* 102, 4841–4848. doi:10.1016/j.biortech.2010.12.113

Biller, P., and Ross, A. B. (2011). Potential yields and properties of oil from the hydrothermal liquefaction of microalgae with different biochemical content. *Bioresour. Technol.* 102, 215–225. doi:10.1016/j.biortech.2010.06.028

Corbitt, R. A. (1999). *Standard Handbook of Environmental Engineering*. Washington, DC: McGraw-Hill.

Corma, A., Huber, G., Sauvanaud, L., and Oconnor, P. (2007). Processing biomass-derived oxygenates in the oil refinery: catalytic cracking (FCC) reaction pathways and role of catalyst. *J. Catal.* 247, 307–327. doi:10.1016/j.jcat.2007.01.023

de Lucas, A., Canizares, P., Duran, A., and Carrero, A. (1997a). Coke formation, location, nature and regeneration on dealuminated HZSM-5 type zeolites. *Appl. Catal. A. Gen.* 156, 299–317. doi:10.1016/S0926-860X(97)00045-8

de Lucas, A., Canizares, P., Durán, A., and Carrero, A. (1997b). Dealumination of HZSM-5 zeolites: effect of steaming on acidity and aromatization activity. *Appl. Catal. A Gen.* 154, 221–240. doi:10.1016/s0926-860x(96)00367-5

Demirbaş, A. (1998). Fuel properties and calculation of higher heating values of vegetable oils. *Fuel* 77, 1117–1120. doi:10.1016/S0016-2361(97)00289-5

Duan, P., Bai, X., Xu, Y., Zhang, A., Wang, F., Zhang, L., et al. (2013). Catalytic upgrading of crude algal oil using platinum/gamma alumina in supercritical water. *Fuel* 109, 225–233. doi:10.1016/j.fuel.2012.12.074

Duan, P., and Savage, P. E. (2010). Hydrothermal liquefaction of a microalga with heterogeneous catalysts. *Ind. Eng. Chem. Res.* 50, 52–61. doi:10.1021/ie100758s

Dzumakaev, K. K., Kagarlitskii, A., and Fedolyak, G. (1986). Kinetics of benzonitrile hydration on a skeletal copper catalyst. *React. Kinet. Catal. Lett.* 30, 289–295. doi:10.1007/BF02064305

Elliott, D. C., Hart, T. R., and Neuenschwander, G. G. (2006). Chemical processing in high-pressure aqueous environments. 8. Improved catalysts for hydrothermal gasification. *Ind. Eng. Chem. Res.* 45, 3776–3781. doi:10.1021/ie060031o

Flanigen, E. M., Jansen, J. C., and Van Bekkum, H. (1991). *Introduction to Zeolite Science and Practice (Studies in Surface Science and Catalysis)*. Vol. 58, Elsevier Science.

Fu, J., Lu, X., and Savage, P. E. (2010). Catalytic hydrothermal deoxygenation of palmitic acid. *Energy Environ. Sci.* 3, 311–317. doi:10.1039/b923198f

Gayubo, A. G., Aguayo, A. T., Atutxa, A., Prieto, R., and Bilbao, J. (2004a). Deactivation of a HZSM-5 zeolite catalyst in the transformation of the aqueous fraction of biomass pyrolysis oil into hydrocarbons. *Energy Fuels* 18, 1640–1647. doi:10.1021/ef040027u

Gayubo, A. G., Aguayo, A. T., Atutxa, A., Prieto, R., and Bilbao, J. (2004b). Role of reaction-medium water on the acidity deterioration of a HZSM-5 zeolite. *Ind. Eng. Chem. Res.* 43, 5042–5048. doi:10.1021/ie0306630

Gayubo, A. G., Alonso, A., Valle, B., Aguayo, A. T., Olazar, M., and Bilbao, J. (2010). Hydrothermal stability of HZSM-5 catalysts modified with Ni for the transformation of bioethanol into hydrocarbons. *Fuel* 89, 3365–3372. doi:10.1016/j.fuel.2010.03.002

Idem, R. O., Katikaneni, S. P. R., and Bakhshi, N. N. (1997). Catalytic conversion of canola oil to fuels and chemicals: roles of catalyst acidity, basicity and shape selectivity on product distribution. *Fuel Process. Technol.* 51, 101–125. doi:10.1016/s0378-3820(96)01085-5

Imai, E.-I., Honda, H., Hatori, K., Brack, A., and Matsuno, K. (1999). Elongation of oligopeptides in a simulated submarine hydrothermal system. *Science* 283, 831–833. doi:10.1126/science.283.5403.831

Li, Z., and Savage, P. E. (2013). Feedstocks for fuels and chemicals from algae: treatment of crude bio-oil over HZSM-5. *Algal Res.* 2, 154–163. doi:10.1016/j.algal.2013.01.003

Liu, D., Gui, J., and Sun, Z. (2008). Adsorption structures of heterocyclic nitrogen compounds over Cu (I) Y zeolite: a first principle study on mechanism of the denitrogenation and the effect of nitrogen compounds on adsorptive desulfurization. *J. Mol. Catal. A Chem.* 291, 17–21. doi:10.1016/j.molcata.2008.05.014

Long, R. Q., and Yang, R. T. (2000). Superior ion-exchanged ZSM-5 catalysts for selective catalytic oxidation of ammonia to nitrogen. *Chem. Commun.* 17, 1651–1652. doi:10.1039/b004957n

Mees, F., Martens, L., Janssen, M., Verberckmoes, A., and Vansant, E. (2003). Improvement of the hydrothermal stability of SAPO-34. *Chem. Commun.* 1, 44–45. doi:10.1039/b210337k

Mo, N., and Savage, P. E. (2014). Hydrothermal catalytic cracking of fatty acids with HZSM-5. *ACS Sustain Chem Eng* 2, 88–94. doi:10.1021/sc400368n

Mo, N., Tandar, W., and Savage, P. E. (2015). Aromatics from saturated and unsaturated fatty acids via zeolite catalysis in supercritical water. *J. Supercrit. Fluids* 102, 73–79. doi:10.1016/j.supflu.2015.03.018

Ortega, J. M., Gayubo, A. G., Aguayo, A. T., Benito, P. L., and Bilbao, J. (1997). Role of coke characteristics in the regeneration of a catalyst for the MTG process. *Ind. Eng. Chem. Res.* 36, 60–66. doi:10.1021/ie9507336

Peterson, A. A., Vogel, F., Lachance, R. P., Froling, M., Antal, J. M. J., and Tester, J. W. (2008). Thermochemical biofuel production in hydrothermal media: a review of sub- and supercritical water technologies. *Energy Environ. Sci.* 1, 32–65. doi:10.1039/b810100k

Pieterse, J. A. Z., Pirngruber, G. D., Van Bokhoven, J. A., and Booneveld, S. (2007). Hydrothermal stability of Fe-ZSM-5 and Fe-BEA prepared by wet ion-exchange for N2O decomposition. *Appl. Catal. B Environ.* 71, 16–22. doi:10.1016/j.apcatb.2006.08.011

Pollack, S., Adkins, J., Wetzel, E., and Newbury, D. (1984). SiO2Al2O3 ratios of ZSM – 5 crystals measured by electron microprobe and X-ray diffraction. *Zeolites* 4, 181–187. doi:10.1016/0144-2449(84)90058-7

Ravenelle, R. M., Schüßler, F., D'amico, A., Danilina, N., Van Bokhoven, J. A., Lercher, J. A., et al. (2010). Stability of zeolites in hot liquid water. *J. Phys. Chem. C* 114, 19582–19595. doi:10.1021/jp104639e

Sano, T., Uno, Y., Wang, Z. B., Ahn, C. H., and Soga, K. (2000). Erratum to: realumination of dealuminated HZSM-5 zeolites by acid treatment and their catalytic properties [Microporous and Mesoporous Materials 31 (1999) 89–95]. *Microporous Mesoporous Mater.* 34, 348. doi:10.1016/s1387-1811(99)00284-x

Sato, S., Yu-U, Y., Yahiro, H., Mizuno, N., and Iwamoto, M. (1991). Cu-ZSM-5 zeolite as highly active catalyst for removal of nitrogen monoxide from emission of diesel engines. *Appl. Catal.* 70, L1–L5. doi:10.1016/S0166-9834(00)84146-9

Satoh, A., Kato, M., Yamato, K., Ishibashi, M., Sekiguchi, H., Kurano, N., et al. (2010). Characterization of the lipid accumulation in a new microalgal species, *Pseudochoricystis ellipsoidea* (Trebouxiophyceae). *J. Jpn. Inst. Energy* 89, 909–913. doi:10.3775/jie.89.909

Song, Y., Sun, C., Shen, W., and Lin, L. (2006). Hydrothermal post-synthesis of HZSM-5 zeolite to enhance the coke-resistance of Mo/HZSM-5 catalyst for methane dehydroaromatization. *Catal. Lett.* 109, 21–24. doi:10.1007/s10562-006-0066-2

Song, Y., Sun, C., Shen, W., and Lin, L. (2007). Hydrothermal post-synthesis of HZSM-5 zeolite to enhance the coke-resistance of Mo/HZSM-5 catalyst for methane dehydroaromatization reaction: reconstruction of pore structure and modification of acidity. *Appl. Catal. A Gen.* 317, 266–274. doi:10.1016/j.apcata.2006.10.037

Stucki, S., Vogel, F., Ludwig, C., Haiduc, A. G., and Brandenberger, M. (2009). Catalytic gasification of algae in supercritical water for biofuel production and carbon capture. *Energy Environ. Sci.* 2, 535–541. doi:10.1039/b819874h

Thangalazhy-Gopakumar, S., Adhikari, S., Chattanathan, S. A., and Gupta, R. B. (2012). Catalytic pyrolysis of green algae for hydrocarbon production using H+ ZSM-5 catalyst. *Bioresour. Technol.* 118, 150–157. doi:10.1016/j.biortech.2012.05.080

Xia, R., Na, D., Zhang, Y., Baoming, L., Zhidan, L., and Haifeng, L. (2015). Nitrogen and phosphorous adsorption from post-hydrothermal liquefaction wastewater using three types of zeolites. *Int. J. Agric. Biol. Eng.* 8, 86–95. doi:10.3965/j.ijabe.20150805.1561

Xue, N., Chen, X., Nie, L., Guo, X., Ding, W., Chen, Y., et al. (2007). Understanding the enhancement of catalytic performance for olefin cracking: hydrothermally stable acids in P/HZSM-5. *J. Catal.* 248, 20–28. doi:10.1016/j.jcat.2007.02.022

Yang, C., Jia, L., Chen, C., Liu, G., and Fang, W. (2011). Bio-oil from hydro-liquefaction of *Dunaliella salina* over Ni/REHY catalyst. *Bioresour. Technol.* 102, 4580–4584. doi:10.1016/j.biortech.2010.12.111

Conflict of Interest Statement: The authors declare that the research was conducted in the absence of any commercial or financial relationships that could be construed as a potential conflict of interest.

16

Optimization of alkaline and dilute acid pretreatment of agave bagasse by response surface methodology

Abimael I. Ávila-Lara[1], Jesus N. Camberos-Flores[1], Jorge A. Mendoza-Pérez[2], Sarah R. Messina-Fernández[3], Claudia E. Saldaña-Duran[3], Edgar I. Jimenez-Ruiz[4], Leticia M. Sánchez-Herrera[4] and Jose A. Pérez-Pimienta[1]*

[1] Department of Chemical Engineering, Universidad Autónoma de Nayarit, Tepic, Mexico, [2] Department of Engineering in Environmental Systems, Instituto Politécnico Nacional, Mexico City, Mexico, [3] Cuerpo Académico de Sustentabilidad Energética, Universidad Autónoma de Nayarit, Tepic, Mexico, [4] Food Technology Unit, Universidad Autónoma de Nayarit, Tepic, Mexico

Edited by:
Robert Henry,
The University of Queensland,
Australia

Reviewed by:
Jian Xu,
Chinese Academy of Sciences, China
Maria Gonzalez Alriols,
University of the Basque Country,
Spain

*Correspondence:
Jose A. Pérez-Pimienta,
Department of Chemical Engineering,
Universidad Autónoma de Nayarit,
Ciudad de la Cultura "Amado Nervo"
S/N, Tepic, Nayarit 63155, Mexico
japerez@uan.edu.mx

Utilization of lignocellulosic materials for the production of value-added chemicals or biofuels generally requires a pretreatment process to overcome the recalcitrance of the plant biomass for further enzymatic hydrolysis and fermentation stages. Two of the most employed pretreatment processes are the ones that used dilute acid (DA) and alkaline (AL) catalyst providing specific effects on the physicochemical structure of the biomass, such as high xylan and lignin removal for DA and AL, respectively. Another important effect that need to be studied is the use of a high solids pretreatment ($\geq$15%) since offers many advantaged over lower solids loadings, including increased sugar and ethanol concentrations (in combination with a high solids saccharification), which will be reflected in lower capital costs; however, this data is currently limited. In this study, several variables, such as catalyst loading, retention time, and solids loading, were studied using response surface methodology (RSM) based on a factorial central composite design of DA and AL pretreatment on agave bagasse using a range of solids from 3 to 30% (w/w) to obtain optimal process conditions for each pretreatment. Subsequently enzymatic hydrolysis was performed using Novozymes Cellic CTec2 and HTec2 presented as total reducing sugar (TRS) yield. Pretreated biomass was characterized by wet-chemistry techniques and selected samples were analyzed by calorimetric techniques, and scanning electron/confocal fluorescent microscopy. RSM was also used to optimize the pretreatment conditions for maximum TRS yield. The optimum conditions were determined for AL pretreatment: 1.87% NaOH concentration, 50.3 min and 13.1% solids loading, whereas DA pretreatment: 2.1% acid concentration, 33.8 min and 8.5% solids loading.

Keywords: agave bagasse, high solids, biomass pretreatment, optimization, characterization

Introduction

Lignocellulosic biomass is the most abundant renewable carbohydrate source in the world and it is proposed to dominate the biofuel production in the future (Avci et al., 2013). Mainly composed by cellulose, hemicellulose, and lignin, their organization and interaction between these polymeric structures, the plant cell wall is naturally recalcitrant to biological degradation (da Costa Sousa et al., 2009). A pretreatment step is fundamental to alter the structure of cellulosic biomass to make cellulose more accessible to the enzymes that convert the carbohydrate polymers into fermentable sugars (Mosier et al., 2005).

Many options exist for pretreatment of biomass, increase saccharification efficiency and improve the yields of monomerics sugars; the leading examples use liquid catalysts, such as sulfuric acid, ammonia, ionic liquid, or water, which penetrate the cell wall and alter its chemistry and ultrastructure (Dadi et al., 2006; Chundawat et al., 2011).

Recently, agave bagasse (AGB) byproduct of the Tequila industry that represent 40% of the harvested plant, with an annual generation in Mexico of about $1.12 \, kg \times 10^8 \, kg$ has been studied for biomass conversion using different pretreatment approaches, such as ionic liquid (Perez-Pimienta et al., 2013) and organosolv (Caspeta et al., 2014). Moreover, AGB was also been used with acid and enzymatic hydrolysis followed by a fermentation step using a native microorganism (Pichia caribbica UM-5) obtaining ~57% of theoretical ethanol (w/w) (Saucedo-Luna et al., 2011) or for the production of n-butanol and ethanol from different Agave species (Mielenz et al., 2015).

Dilute acid (DA) and alkaline (AL; NaOH) are among the most extensively studied biomass pretreatments in different feedstocks, such as grasses, agricultural residues, and woods (Kumar et al., 2009; Xu et al., 2010; Sathitsuksanoh et al., 2013; Zhang et al., 2014). The mode of action of the DA pretreatment typically use sulfuric acid that removes hemicellulose in a great extent improving the enzyme accessibility to cellulose which its effectiveness depends on the acid concentration and temperature applied during the process, however, if severe conditions are applied several degradation products are formed, mainly furfural, 5-hydroxymethylfurfural, phenolic acids and aldehydes, levulinic acid, and other aliphatic acids, which can inhibit both, enzymatic hydrolysis and fermentation (Mosier et al., 2005; da Costa Sousa et al., 2009). On the other hand, ALs pretreatment uses AL catalyst, such as sodium hydroxide, which are effective depending on the lignin content on the biomass, increasing cellulose digestibility through lignin solublization/removal, exhibiting minor cellulose and hemicellulose solubilization than acid or hydrothermal processes (Avira et al., 2010).

In recent years, the need to investigate the use of high solids loading ($\geq 15\%$) in biomass pretreatment has increase hence offers many advantaged over lower solids loadings, including increased sugar and ethanol concentrations, which will be reflected in lower capital costs (Modenbach and Nokes, 2012; Li et al., 2013); however, this data is currently limited for DA and AL pretreatments in AGB (Hernández-Salas et al., 2009; Saucedo-Luna et al., 2011).

In the present manuscript, optimization of DA and AL pretreatment strategies for conversion of AGB to sugars using a central composite design (CCD) for response surface methodology (RSM) was studied. The objective of this study was to identify the optimum process conditions for the selected operating variables namely catalyst concentration, retention time, and solid loading for the maximum production of fermentable sugars. Furthermore, the untreated and selected samples from both pretreatments were characterized by calorimetric techniques (TGA), fluorescence and

energy dispersive X-ray spectroscopy (EDS), and scanning electron microscopy (SEM).

Materials and Methods

The biomass used in this study was obtained from Destilería Rubio, a Tequila plant from western Mexico. The AGB was harvested in August 2014. The biomass was milled with a Thomas-Wiley Mini Mill fitted with a 40-mesh screen (Model 3383-L10 Arthur H. Thomas Co., Philadelphia, PA, USA) and stored at 4°C in a sealed plastic bag. Cellic® CTec2 (Cellulase complex for degradation of cellulose) and HTec2 (Endoxylanase with high specificity toward soluble hemicellulose) were a gift from Novozymes (Davis, CA, USA).

Experimental Design

Optimization of processing conditions for fermentable sugars recovery was studied using a factorial CCD of RSM. The independent variables were catalyst concentration, residence time, and solids loading. The experimental data were fit using Eq. 1, a low-order polynomial equation to evaluate the effect of each independent variable to the response, which was later analyzed to obtain the optimum process conditions (Tan et al., 2011). In this study, a polynomial quadratic equation was employed as follows:

$$y = \beta_0 + \sum_{i=1}^{3} \beta_i X_i + \sum_{i=1}^{3} \beta_{ii} X_i^2 + \sum_{i=1}^{3} \sum_{j=1}^{3} \beta_{ii} X_i X_j \quad (1)$$

where y is the response, X_i and X_j are independent variables, β_0 is the constant coefficient, β_i is the ith linear coefficient, β_{ii} is the quadratic coefficient, and β_{ij} is the ijth interaction coefficient. CCD consists of 2^k factorial points, $2k$ axial points ($\pm \alpha$), and six central points, where k is the number of independent variables. Each of the variables were investigated at five coded levels ($-\alpha$, -1, 0, 1, α), as listed in **Table 1**, and the complete experimental design matrix for this study is shown in **Table 2**. For each pretreatment (DA and AL), a total of 20 experiments per pretreatment were carried out, including eight per factorial design, six for axial points and six repetitions at the central point.

Alkaline Pretreatment

A NaOH solution at a specific concentration were placed in a serum bottle and mixed with AGB using a glass rod, forming

---

TABLE 1 | Levels of the pretreatment condition variables tested in the CCD.

Variable	Unit	Coding	Coded level				
			$-\alpha$ [a]	-1	0	1	$+\alpha$ [a]
Catalyst concentration	% (w/w)	A	0.15	0.73	1.58	2.42	3.00
Residence time	Min	B	15.00	30.20	52.50	74.80	90.00
Solids loading	% (w/w)	C	3.00	8.47	16.50	24.53	30.00

[a] *(axial distance)* $= \sqrt[4]{N}$, where N is the number of experiments of the factorial design. In this case, 1.6818.

TABLE 2 | Experimental design matrix of CCD and corresponding results (sugars and solids recovery).

Run	Experimental variables			Solids recovery (%)		TRS yield (mg/g biomass)	
	Catalyst concentration, A (%, w/w)	Retention time, B (min)	Solids loading, C (%, w/w)	AL	DA	AL	DA
1	1.58	15.00	16.50	70.3	71.9	506.1	419.3
2	2.42	30.20	8.47	74.3	58.5	476.9	391.3
3	0.73	30.20	8.47	75.5	83.0	447.5	410.9
4	2.42	30.20	24.53	72.3	66.8	468.2	371.6
5	0.73	30.20	24.53	84.3	82.7	415.6	385.7
6	0.15	52.50	16.50	86.2	80.8	360.3	339.7
7	1.58	52.50	3.00	65.5	57.6	513.6	457.2
8	3.00	52.50	16.50	60.7	58.3	457.9	364.5
9	1.58	52.50	30.00	80.6	68.7	421.0	399.7
10	2.42	74.80	8.47	63.4	54.4	460.5	360.5
11	0.73	74.80	8.47	72.4	77.3	460.1	428.0
12	2.42	74.80	24.53	74.6	65.3	453.7	353.7
13	0.73	74.80	24.53	87.6	86.1	437.2	409.0
14	1.58	90.00	16.50	72.8	70.4	521.5	397.9
15	1.58	52.50	16.50	76.2	68.6	532.8	394.4
16	1.58	52.50	16.50	76.7	71.6	517.1	415.2
17	1.58	52.50	16.50	78.2	68.8	497.2	443.8
18	1.58	52.50	16.50	79.3	70.6	521.3	431.5
19	1.58	52.50	16.50	79.9	69.5	502.7	439.4
20	1.58	52.50	16.50	77.8	70.4	511.6	435.9
Untreated	–	–	–	100		135.1	

AL, alkaline pretreatment; DA, dilute acid pretreatment.

a slurry at with a precise biomass concentration and the pretreatment was performed in autoclave conditions (121°C and ~15 psi) during the appropriate time according to **Table 1** (Xu et al., 2010). Pretreated biomass was recovered by filtration and washed with 400 mL of distilled water to remove excess alkali and dissolved byproducts. All experiments were conducted in triplicate.

Dilute Acid Pretreatment

The DA pretreatment with H_2SO_4 was conducted using the appropriate acid concentration and solids loading referred to **Table 1** at 130°C and 20 psi in an autoclave for a specific time (Sathitsuksanoh et al., 2013). After DA, the hydrolyzate was separated by filtration and the pretreated AGB was washed with 400 mL of distilled water prior to enzymatic hydrolysis. All experiments were conducted in triplicate.

Scanning Electron Microscopy

The morphology of untreated and selected pretreated AGB solids was analyzed using a high resolution SEM by a JEOL JSM-7800F equipment. The representative images were acquired with a 1 kV accelerating voltage and analysis using 20 kV.

Confocal Fluorescent Microscopy

The confocal fluorescent microscope images of untreated and selected pretreated AGB samples were taken using a Carl Zeiss LSM 710 NLO with two laser sources (405 and 633 nm). To demonstrate the microstructure based on the distribution of lignin (autofluorescence) and cellulose, all samples were labeled with Calcofluor white stain (0.1%) for 5 min, subsequently were

washed four times using distilled water and allowed to dry in the dark until analysis under the confocal microscope.

DSC and TGA Analysis

A differential scanning calorimeter (Pyris 1) from Perkin Elmer was employed with an argon atmosphere in the range of 50–450°C, at 10°C/min ramp. DSC curves were obtained with 3.3 mg. The TGA curves were obtained using around 3.8 mg of AGB as initial sample mass. The samples was tested in a SETARAM thermal analysis instrument, with temperature range of 50–800°C and heating rate of 10°C/min in argon atmosphere. Untreated and selected pretreated samples were measured by DSC and TGA.

Biomass Porosimetry

Nitrogen porosimetry (ASAP 2406) from Mca-Micromeritics was employed to measure the surface area, pore volume and pore size distribution of the untreated and selected pretreated AGB with the following methods from ASTM: ASTM D-3663(R2008), ASTM D-4222-03(R2008), and ASTM D-4641-12(R2008). Samples were degasified at 120°C.

Enzymatic Saccharification

The saccharification was carried out using commercially available Cellic® CTec2 and HTec2 enzyme mixtures of untreated and pretreated AGB samples, which was conducted at 55°C and 150 rpm in 50 mM citrate buffer (pH of 4.8). A 3% biomass loading was used, likewise, untreated AGB were run concurrently with the pretreated samples to eliminate potential differences in temperature history or enzyme loading. The enzyme concentrations of CTec2 and HTec2 were set at 35 FPU/g biomass and 60 CBU/g biomass, respectively. All assays were performed in triplicate.

DNS Assay

The total reducing sugar (TRS) yield of the final hydrolyzate calculated as mg sugar/g biomass was determined by DNS assay (Miller, 1959) on a DTX 880 Multimode Detector (Beckman Coulter, CA, USA) at 550 nm with solutions (0–10 g/L) of D-glucose in water as calibration standards. All assays were performed in triplicate.

Statistical Analysis

Analysis of experimental CCD results was carried out with the software Design-Expert 7.1.5 (Stat-Ease, Minneapolis, MN, USA). Each coefficient in Eq. 1 was calculated and the possible interaction effects of the process variables on the response were obtained. Their significance was checked by variance analysis (ANOVA) of experimental results.

Results and Discussion

Biochemical Composition Analysis of Untreated Agave Bagasse

By following, the National Renewable Energy Laboratory (NREL, Denver, CO, USA) protocols, the composition of untreated AGB in dry basis was 41.5% glucan, 20.3% xylan, 17.0% insoluble lignin, 3.8% soluble lignin, and 5.4% ash, which is consistent with other reported values (Davis et al., 2011; Perez-Pimienta et al., 2013). Glucan and xylan correspond to 61.8% of the total carbohydrates in the AGB.

Model Development

The experimental data were first analyzed, in order to obtain second-order polynomial equations including terms of interaction between the experimental variables using Design-Expert software and the following models for AL and DA pretreatment describes the TRS yield (mg sugar/g biomass) in terms of coded parameters and actual parameters are based on the statistical analysis of the experimental data shown in **Table 2**.

The final equations for AL pretreatment were as follows:

$$TRS\ yield = 513.35 + 21.08 \times A + 3.57 \times B - 16.87 \times C$$
$$- 9.95 \times AB + 4.87 \times AC + 1.67 \times BC - 38.44 \times A^2$$
$$- 2.52 \times B^2 - 18.14 \times C^2 \quad (2)$$

$$TRS\ yield = 277.0937 + 203.1844 * NaOH + 1.0059 * Time$$
$$+ 5.6903 * Solids - 0.4313 * NaOH * Time$$
$$+ 0.7172 * NaOH * Solids + 0.0076 * Time * Solids$$
$$- 53.8367 * NaOH^2 - 0.0034 * Time^2$$
$$- 0.2813 * Solids^2 \quad (3)$$

In the same way, the final equations for DA pretreatment were as follows:

$$TRS\ yield = 427.27 - 5.87 \times A + 0.26 \times B - 12.80 \times C$$
$$- 13.65 \times AB + 2.19 \times AC + 2.92 \times BC$$
$$- 27.48 \times A^2 - 11.45 \times B^2 - 0.62 * C^2 \quad (4)$$

$$TRS\ yield = 305.8687 + 137.0487 \times Acid + 2.1816 \times Time$$
$$- 2.4166 \times Solids - 0.5917 \times Acid \times Time$$
$$+ 0.3231 \times Acid \times Solids$$
$$+ 0.0133 \times Time \times Solids - 384832$$
$$\times Acid^2 - 0.0154 \times Time^2 - 0.0097 \times Solids^2 \quad (5)$$

where A, B, and C are catalyst concentration (NaOH for AL and H_2SO_4 for DA), retention time and solids loading, respectively. An analysis of variance (ANOVA) was performed to test the significance of the developed model and the results are presented for AL and DA pretreatment in **Tables 3** and **4**, respectively. If a p-value (also known as the Prob > D-value) is lower than 0.05 a model in considered significant, indicating only a 5% chance that their respective model could occur due to noise. For both pretreatments, their models effectively describes the response nevertheless the AL pretreatment model have a lower p-value (0.0003) than the DA pretreatment model (0.0247). In addition, the Prob > F values for each model term in AL pretreatment suggest that A, C, and A^2, meanwhile for DA pretreatment suggest that only A^2 are the model terms that have significant effects on the TRS yield. To determine the suitability of the model, the lack of fit test was used, which indicated an insignificant lack of fit with an F-value of 0.1393 and 0.3009 for AL and DA pretreatment, respectively. The coefficient of determination (R^2) of the pretreatment models was 0.9151 for AL and 0.7270 and for DA, implying a good and average correlation between the observed and predicted values of AL and DA respectively, as shown in **Figures 1A,B**. Finally, the quadratic models developed for AL and DA pretreatment are appropriate for predicting TRS yield under different pretreatment conditions within the range used in the present study.

Effect of Pretreatment Conditions on Solids Recovery

The highest solids recovery for AL and DA was obtained in the same run (13) with 87.6 and 86.1%, respectively, with experimental conditions of 0.73% catalyst concentration, 74.8 min and 24.53% solids loading. On the other hand, the lowest solids recovery for AL pretreatment of 60.7% was obtained during run 8

TABLE 3 | ANOVA table for the quadratic model of alkaline pretreatment.

Source	Sum of squares	DF	Mean square	F-value	Prob > F	
Model	34291.73	9	3810.19	11.97	0.0003	Significant
A	5600.42	1	5600.42	17.59	0.0018	
B	97.31	1	97.31	0.31	0.5925	
C	3578.16	1	3578.16	11.24	0.0073	
AB	528.42	1	528.42	1.66	0.2266	
AC	189.46	1	189.46	0.60	0.4583	
BC	14.85	1	14.85	0.047	0.8333	
A^2	21479.83	1	21479.83	67.48	<0.0001	
B^2	40.82	1	40.82	0.13	0.7277	
C^2	4734.21	1	4734.21	14.87	0.0032	
Residual	3183.28	10	318.33			
Lack of fit	2351.53	5	470.31	2.83	0.1393	Not significant
Pure error	831.75	5	166.35			

DF, degree of freedom.

TABLE 4 | ANOVA table for the quadratic model of dilute acid pretreatment.

Source	Sum of squares	DF	Mean square	F-value	Prob > F	
Model	15632.82	9	1736.98	3.79	0.0247	Significant
A	434.69	1	434.69	0.95	0.3529	
B	0.51	1	0.51	0.001	0.9739	
C	2059.98	1	2059.98	4.50	0.0599	
AB	994.66	1	994.66	2.17	0.1713	
AC	38.45	1	38.45	0.084	0.7779	
BC	45.60	1	45.60	0.100	0.7588	
A^2	10975.31	1	10975.31	23.96	0.0006	
B^2	841.41	1	841.41	1.84	0.2051	
C^2	5.61	1	5.61	0.012	0.9140	
Residual	4580.07	10	458.01			
Lack of fit	2843.15	5	568.63	1.64	0.3009	Not significant
Pure error	1736.92	5	347.38			

DF, degree of freedom.

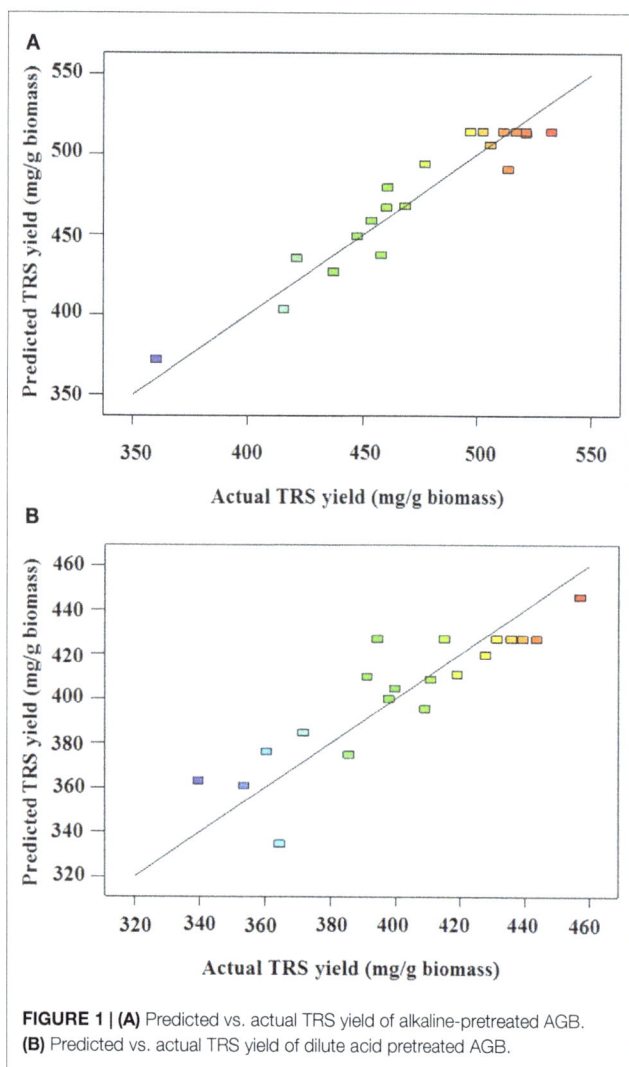

FIGURE 1 | (A) Predicted vs. actual TRS yield of alkaline-pretreated AGB. **(B)** Predicted vs. actual TRS yield of dilute acid pretreated AGB.

(3.00% catalyst concentration, 52.5 min and 16.5% solids loading), while for DA pretreatment was 54.4% with run 10 using 2.42% catalyst concentration, 74.8 min and 8.47% solids loading. The difference between low and high solids recovery, which represents

process severity are 26.9 and 31.7% for AL and DA pretreatment, respectively.

Effect of Pretreatment Catalyst Concentration and Retention Time

The effect of catalyst concentration and retention time in AL and DA pretreatment on TRS yield during enzymatic saccharification using 3% biomass loading of are shown in **Figure 2**. By means of pretreatment shorter retention times and catalyst concentration, the TRS yield became lower and the same applies to longer times and high catalyst concentration for both AL and DA pretreatment.

However, for AL pretreatment from 1.58 to 2.43% NaOH a TRS yield above ~460 mg sugar/g biomass is obtained within the study range of 15–90 min. In the other hand, in DA pretreatment a more distributed region is shown where the highest TRS yields was obtained at the central design points with a relatively shorter differences between the highest yield that occurred in run 7 (457 mg/g biomass) and an average of the central data points (433 mg/g biomass).

Effect of Pretreatment Catalyst Concentration and Solid Loading

The response surface plots presents the effect of catalyst concentration and solid loading on TRS yield of both AL and DA pretreatment is displayed in **Figure 3**. One area for AL pretreatment is clearly defined showing the highest TRS yield region in the middle range of both parameters. A TRS yield above 500 mg/g biomass is obtained in the range of 1.1–2.3% NaOH and solid loading between 4 and 20%. These results are supported with previous reports in AL pretreatment where using the same temperature conditions (121°C), moderate NaOH concentration (1%) and time (30–60 min), which achieved the highest TRS yield (Wang et al., 2010; Xu et al., 2010). During DA pretreatment a clear region where a TRS yield above 430 mg/g biomass was reached within the range of 0.7–2% acid and a solid loading of 3–15%. It is noticeable that such differences between the TRS yields were obtained from the highest experimental runs from both pretreatments at ~533 mg/g biomass from run 15 in AL and ~457 mg/g biomass from run 7 in DA. This differences are encounter from the objective of each pretreatment, which in the case of AL pretreatment is lignin removal whereas for DA pretreatment xylan removal is the main effect, as consequence a lower TRS yield should be obtained as there is lower xylan available as a substrate for the enzymes to be reacted into xylose causing a lower total TRS yield.

Optimization of Pretreatment Conditions

In both of the evaluated pretreatment processes (AL and DA), a lower catalyst concentration, shorter time and high solids loading if preferred to obtain an optimum TRS yield. The optimum catalyst concentration, retention time and solid loading were found to be for AL pretreatment of 1.87% NaOH concentration, 50.3 min and 13.1% solids loading, while DA pretreatment were 2.1% acid concentration, 33.8 min and 8.5% solids loading. For AL pretreatment, an 18% increase in NaOH concentration, 4% reduction in retention time and 20% reduction of solids loading, whereas for DA pretreatment, 33% increase in acid concentration, 35.6% reduction in retention time and 283% increase of solids

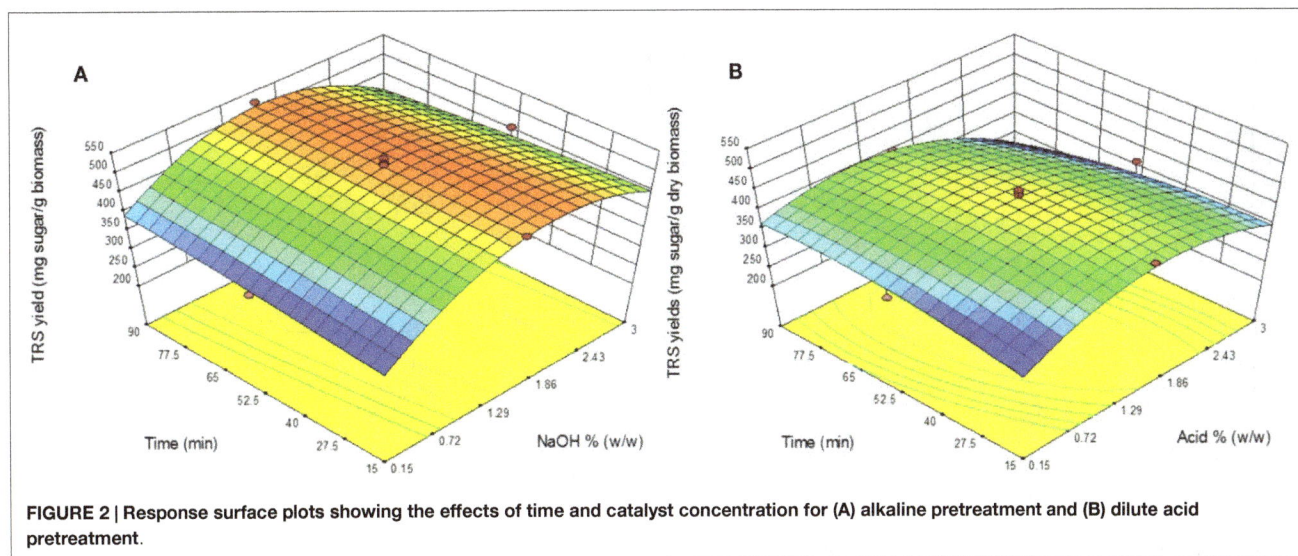

FIGURE 2 | Response surface plots showing the effects of time and catalyst concentration for (A) alkaline pretreatment and (B) dilute acid pretreatment.

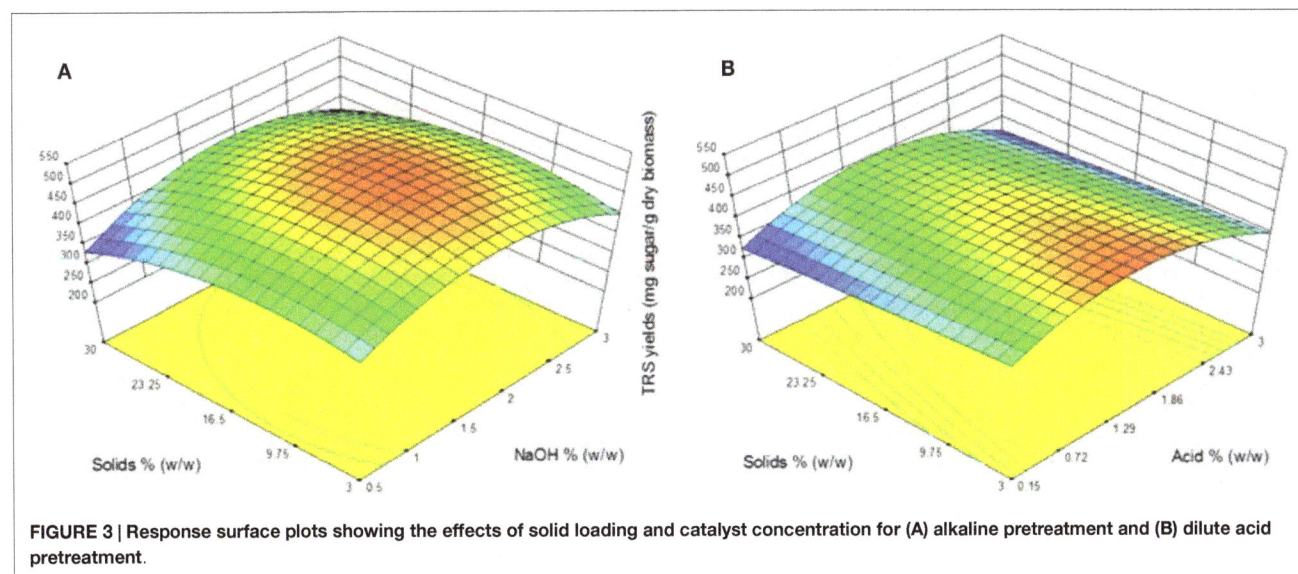

FIGURE 3 | Response surface plots showing the effects of solid loading and catalyst concentration for (A) alkaline pretreatment and (B) dilute acid pretreatment.

loading and when comparing the optimum conditions with the experimental conditions (Run 7, **Table 2**) that gave the highest yields.

Thermogravimetric and Differential Scanning Calorimetry Analysis

Untreated and selected pretreated AGB samples were thermogravimetrically analyzed to compare degradation characteristics in terms of pretreatment. Two samples were selected for TGA analysis for each pretreatment, named AL-1 and DA-1 corresponding to experimental run 8, in addition to AL-2 and DA-2 corresponding to experimental run 16 (one of the CCD points). **Figure 4** shows standards weight loss plots, while in **Figure 5** the differential TGA plots of the untreated and pretreated AGB samples are shown. All samples exhibit three decomposition regions with some initial weight loss from 50 to 125°C (mainly due to moisture evaporation). Up to 200°C, the samples presented thermal stability. The decomposition temperature (T_d) decrease for

both AL and AL pretreated samples as compared to the untreated AGB, shown in **Table 5**. In both of the analyzed pretreatment the lowest values correspond to AL-1 (run 8 sample). These results indicate that AL pretreatment reduced the activation energy that is needed to decompose the AGB in a higher extent than DA pretreatment by deconstructing the tight plant cell wall structures. AL-pretreated AGB samples obtained a lower T_d value when compared to an ionic liquid treated AGB from a recent report (310 vs. 347°C) (Perez-Pimienta et al., 2015). Thermal depolymerization of hemicelluloses and the cleavage of glycosidic linkages of cellulose occurs in the region of 220–300°C, while lignin decomposition extended to the whole temperature range, from 200 until 700°C, due to different activities of the chemical bonds present on its structure and the degradation of cellulose taken place between 275 and 400°C (Deepa et al., 2011). The final decomposition stage for all samples was completed above 400°C, where a weight loss due to thermolysis of carbon containing residues does take place (Fisher et al., 2002). DSC curves of untreated

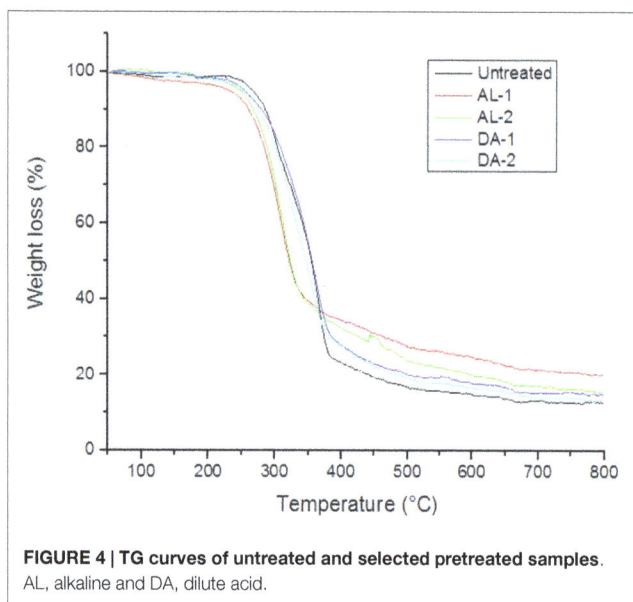

FIGURE 4 | TG curves of untreated and selected pretreated samples. AL, alkaline and DA, dilute acid.

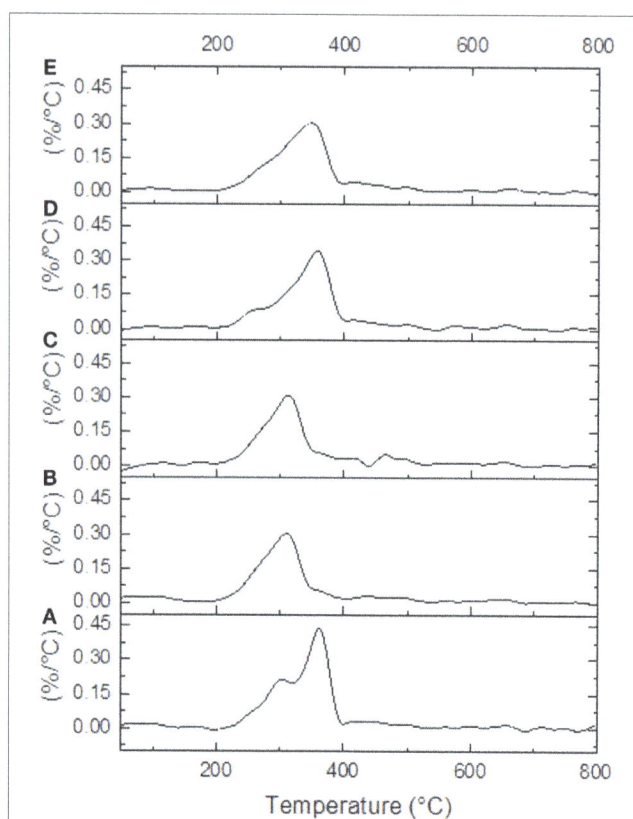

FIGURE 5 | Differential TGA plots are shown for untreated and selected pretreated samples. (A) Untreated AGB, (B) AL-1, (C). AL-2, (D) DA-1, and (E) DA-2.

AGB and selected samples from AL and DA (Figures S1–S3 in Supplementary Material) with two endothermic peaks observed and Table S1 in Supplementary Material summarizes those events. The first thermal is shown below 200°C with low energy between 5.3 and 13.9 J/g°C, where the untreated AGB present the onset temperature at 83°C (8.6 J/g°C), while the AL-4 (run 16 of AL pretreatment) achieved 13.9 J/g°C, whereas for DA the highest energy event was at 12.2 J/g°C with DA-1 (run 8) that employed a 3% acid loading. A similar peak was obtained with an IL-treated AGB sample where the untreated sample showed a dehydration peak at 89°C (Perez-Pimienta et al., 2015). In the other hand, the second thermal event presents a high energy peak for all samples with ∆H in the range of 120–627 J/g°C and temperature above 262 up to 415°C. AL pretreatment achieved its highest energy with run 16 (AL-4) with a peak at 335°C (627 J/g°C), whereas the evaluated DA-pretreated samples was with run 9 (AL-2) at 358°C and 296 J/g, so when compared to the untreated sample it is clear that a pretreated offers a reduction in terms of calorific value turning them into a more digestible biomass.

Scanning Electron and Confocal Fluorescence Microscopy

The SEM images of untreated and pretreated samples (run 16 sample for both AL and DA pretreatment) were taken at 500× (**Figure 6**). Untreated AGB (**Figure 6A**) presents an intact structure without degradation, otherwise AL pretreatments dissolves lignin disrupting the biomass, besides of the increase of pore quantity as can be observed in **Figure 6B**. Finally, DA pretreatment disrupts the lignocellulosic structure by mainly dissolving hemicellulose, hence, major microfibrous cellulose structures remain (**Figure 6C**) and some lignin or lignin–carbohydrate complexes may be condensed on the surface of the cellulose fibers.

Elements content of untreated and pretreated AGB (run 16 from AL and DA pretreatment) are presented in **Table 6**. In the untreated AGB, C and O accounts for a 98.5% of the totals mass fraction remaining only 1.4% of Ca, these attributable to calcium

TABLE 5 | Decomposition T_d temperatures for untreated and pretreated AGB.

Property	Pretreatment				
	Untreated	AL-1	AL-2	DA-1	DA-2
T_d (°C)	366	310	317	360	353

oxalate (CaC_2O_4) crystals in considerable quantities along the surface of the plant cell wall as referred in a previous paper (Perez-Pimienta et al., 2015). In contrast, the DA-treated AGB the available Ca was removed during the process at these conditions (1.58% acid concentration, 52.5 min and 16.5% solids loading). Nonetheless, this Ca removal does not occurred in the AL-treated sample where a small amount of Na (1.2%) was found, possibly, as a result of some of the alkali was converted to irrecoverable salts and/or incorporated into the biomass.

Confocal fluorescence microscopy was used to investigate the surface morphologies of untreated and pretreated AGB (run 16 from AL and DA pretreatment) as presented in **Figures 7A–F**. When compared to the untreated AGB, only the DA-pretreated sample show a significant reduction in the fluorescence signal intensity in cell walls (lignin is represented with a green signal and cellulose with a blue signal), while the AL-treated sample presents only a slight reduction.

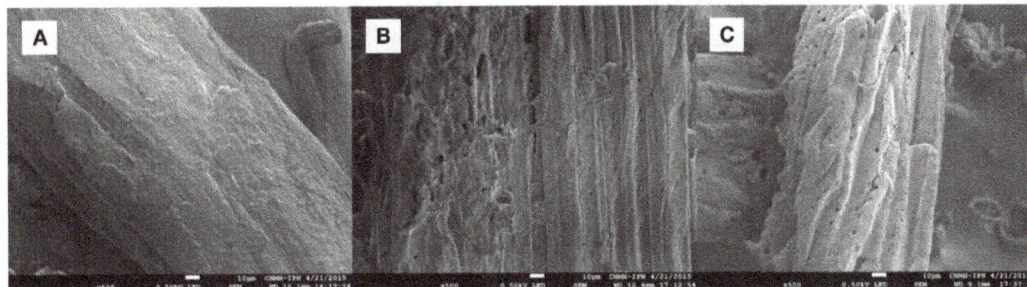

FIGURE 6 | SEM images of AGB samples: (A) untreated, (B) alkaline pretreated, and (C) dilute acid pretreated.

TABLE 6 | Elements content of untreated and pretreated agave bagasse measured by EDS spectroscopy.

Element	Untreated		AL		DA	
	Mass fraction (%)	Atomic mass fraction (%)	Mass fraction (%)	Atomic mass fraction (%)	Mass fraction (%)	Atomic mass fraction (%)
C	51.1 ± 0.9	58.6 ± 0.7	51.7 ± 1.7	59.4 ± 1.7	60.6 ± 5.6	67.2 ± 5.2
O	47.5 ± 0.5	40.9 ± 0.6	45.2 ± 2.2	39.0 ± 2.1	39.4 ± 5.6	32.8 ± 5.2
Ca	1.4 ± 0.4	0.5 ± 0.1	1.9 ± 0.9	1.2 ± 0.5	–	–
Na	–	–	1.2 ± 0.7	0.4 ± 0.3	–	–
Total	100.0	100.0	100.0	100.0	100.0	100.0

FIGURE 7 | Confocal fluorescence images of AGB samples: (A,D) untreated, (B,E) alkaline pretreated, and (C,F) dilute acid pretreated.

Effect of Pretreatment on Biomass Porosimetry

Pretreatment can affect the cellulose accessibility and is often accompanied by variation in the surface area. Surface area, pore volume, and pore average diameter were measured using the Brunauer–Emmett–Teller (BET) method by argon adsorption, which relates the gas pressures to the volume of gas adsorbed, although might not be directly associated to enzyme accessibility since the size differences between argon molecules and enzymes (Li et al., 2013). **Table 7** summarizes surface area, pore volume and pore average diameter of untreated and run 16 (one of the CCD points from both AL and DA-pretreated AGB). When compared to the untreated samples an increment in the surface

TABLE 7 | Comparison of porosimetry parameters in untreated and pre-treated AGB.

	Surface area (m^2/g)	Pore volume (cm^3/g)	Pore average diameter (A)
Untreated	0.6	0.0020	137.7
AL	0.9	0.0023	107.7
DA	1.1	0.0028	106.7

area is noticeable from 0.6 up to $1.1 \, m^2$/g. This is consistent with the changes in the SEM images upon AL and DA pretreatment described above. However, the pore volume of all samples (untreated and pretreated) presents a negligible difference close to $0.0008 \, cm^3$/g, whereas a reduction in the pore average diameter is obtained in the pretreated samples.

Conclusion

The effects of catalyst concentration, retention time and solids loading in terms of TRS yield of AL and DA pretreatment in AGB were investigated. This study demonstrated that AGB is a promising biofuel feedstock that can achieved high sugar yields using both DA and AL pretreatment. For both pretreatments, a model was generated with a high correlation obtained from actual TRS data. Furthermore, the results indicate that TRS yield was enhanced by catalyst concentration and solid loading, but longer retention times does not. Both pretreatment increase porosity and surface area, but AL pretreatment achieved a lower decomposition temperature. Finally, RSM was also used to optimize the pretreatment conditions for maximum TRS yield. The optimum conditions were determined for AL pretreatment: 1.87% NaOH concentration, 50.3 min, and 13.1% solids loading, whereas DA pretreatment: 2.1% acid concentration, 33.8 min, and 8.5% solids loading. Finally, fuel synthesis studies should be performed in the sugars obtained using the best conditions for both pretreatments in order to obtain significant data for a scale-up process.

Acknowledgments

The authors gratefully thank CONACYT through the project 229711 and internal funding from Universidad Autónoma de Nayarit (Autonomous University of Nayarit).

References

Avci, A., Saha, B. C., Dien, B. S., Kennedy, G. J., and Cotta, M. A. (2013). Response surface optimization of corn stover pretreatment using dilute phosphoric acid for enzymatic hydrolysis and ethanol production. *Bioresour. Technol.* 130, 603–612. doi:10.1016/j.biortech.2012.12.104

Avira, P., Tomás-Pejó, E., Ballesteros, M., and Negro, M. J. (2010). Pretreatment technologies for an efficient bioethanol production process based on enzymatic hydrolysis: a review. *Bioresour. Technol.* 101, 4851–4861. doi:10.1016/j.biortech.2009.11.093

Caspeta, L., Caro-Bermúdez, M. A., Ponce-Noyola, T., and Martinez, A. (2014). Enzymatic hydrolysis at high-solids loading for the conversion of agave bagasse to fuel ethanol. *Appl. Energy* 113, 277–286. doi:10.1016/j.apenergy.2013.07.036

Chundawat, S. P. S., Beckham, G. T., Himmel, M. E., and Dale, B. E. (2011). Deconstruction of lignocellulosic biomass to fuels and chemicals. *Annu. Rev. Chem. Biomol. Eng.* 2, 121–145. doi:10.1146/annurev-chembioeng-061010-114205

da Costa Sousa, L., Chundawat, S. P. S., Balan, V., and Dale, B. E. (2009). 'Cradle-to-grave' assessment of existing lignocellulose pretreatment technologies. *Curr. Opin. Biotechnol.* 20, 339–347. doi:10.1016/j.copbio.2009.05.003

Dadi, A. P., Varanasi, S., and Schall, C. A. (2006). Enhancement of cellulose saccharification kinetics using an ionic liquid pretreatment step. *Biotechnol. Bioeng.* 95, 904–910. doi:10.1002/bit.21047

Davis, S. C., Dohleman, F. G., and Long, S. P. (2011). The global potential for agave as a biofuel feedstock. *GCB Bioenergy* 3, 68–78. doi:10.1111/j.1757-1707.2010.01077.x

Deepa, B., Abraham, E., Cherian, B. M., Bismarck, A., Blaker, J. J., Pothan, L. A., et al. (2011). Structure, morphology and thermal characteristics of banana nano fibers obtained by steam explosion. *Bioresour. Technol.* 102, 1988–1997. doi:10.1016/j.biortech.2010.09.030

Fisher, T., Hajaligol, M., Waymack, B., and Kellogg, D. (2002). Pyrolysis behavior and kinetics of biomass derived materials. *J. Anal. Appl. Pyrolysis* 62, 331–349. doi:10.1016/S0165-2370(01)00129-2

Hernández-Salas, J. M., Villa-Ramírez, M. S., Veloz-Rendón, J. S., Rivera-Fernández, K. N., González-César, R. A., Plascencia-Espinosa, M. A., et al. (2009). Comparative hydrolysis and fermentation of sugarcane and agave bagasse. *Bioresour. Technol.* 100, 1238–1245. doi:10.1016/j.biortech.2006.09.062

Kumar, P., Barrett, D., Delwiche, M. J., and Stroeve, P. (2009). Methods for pretreatment of lignocellulosic biomass for efficient hydrolysis and biofuel production. *Ind. Eng. Chem. Res.* 48, 3713–3729. doi:10.1021/ie801542g

Li, C., Tanjore, D., He, W., Wong, J., Gardner, J. L., Sale, K. L., et al. (2013). Scale-up and evaluation of high solid ionic liquid pretreatment and enzymatic hydrolysis of switchgrass. *Biotechnol. Biofuels* 6, 154. doi:10.1186/1754-6834-6-154

Mielenz, J. R., Rodriguez, M., Thompson, O. A., Yang, X., and Yin, H. (2015). Development of *Agave* as a dedicated biomass source: production of biofuels from whole plants. *Biotechnol. Biofuels* 8, 79. doi:10.1186/s13068-015-0261-8

Miller, G. L. (1959). Use of dinitrosalicylic acid reagent for determination of reducing sugar. *Anal. Chem.* 31, 426–428. doi:10.1021/ac60147a030

Modenbach, A. A., and Nokes, S. E. (2012). The use of high-solids loadings in biomass pretreatment – a review. *Biotechnol. Bioeng.* 109, 1430–1442. doi:10.1002/bit.24464

Mosier, N., Wyman, C., Dale, B., Elander, R., Lee, Y. Y., Holtzapple, M., et al. (2005). Features of promising technologies for pretreatment of lignocellulosic biomass. *Bioresour. Technol.* 96, 673–686. doi:10.1016/j.biortech.2004.06.025

Perez-Pimienta, J. A., Lopez-Ortega, M. G., Chavez Carvayar, J. A., Varanasi, P., Stavila, V., Cheng, G., et al. (2015). Characterization of agave bagasse as a function of ionic liquid pretreatment. *Biomass Bioenergy* 75, 180–188. doi:10.1016/j.biombioe.2015.02.026

Perez-Pimienta, J. A., Lopez-Ortega, M. G., Varanasi, P., Stavila, V., Cheng, G., Singh, S., et al. (2013). Comparison of the impact of ionic liquid pretreatment on recalcitrance of agave bagasse and switchgrass. *Bioresour. Technol.* 127, 18–24. doi:10.1016/j.biortech.2012.09.124

Sathitsuksanoh, N., Xu, B., Zhao, B., and Zhang, Y. H. P. (2013). Overcoming biomass recalcitrance by combining genetically modified siwthcgrass and cellulose solvent-based lignocellulose pretreatment. *PLoS ONE* 8:e73523. doi:10.1371/journal.pone.0073523

Saucedo-Luna, J., Castro-Montoya, A. J., Martinez-Pacheco, M. M., Sosa-Aguirre, C. R., and Campos-Garcia, J. (2011). Efficient chemical and enzymatic saccharifcation of the lignocellulosic residue from *Agave tequilana* bagasse to produce ethanol by *Pichia caribbica*. *J. Ind. Microbiol. Biotechnol.* 38, 725–732. doi:10.1007/s10295-010-0853-z

Tan, H. T., Lee, K. T., and Mohamed, A. R. (2011). Pretreatment of lignocellulosic plam biomass using a solvent-ionic liquid [BMIM]Cl for glucose recovery: An optimisation study using response surface methodology. *Carbohyd. Polymers* 83, 1862–1868. doi:10.1016/j.carbpol.2010.10.052

Wang, Z., Keshwani, D. R., Redding, A. P., and Cheng, J. J. (2010). Sodium hydroxide pretreatment and enzymatic hydrolysis of coastal Bermuda grass. *Bioresour. Technol.* 101, 3583–3585. doi:10.1016/j.biortech.2009.12.097

Xu, J., Cheng, J. J., Sharma-Sivappa, R. R., and Burns, J. (2010). Sodium hydroxide pretreatment of switchgrass for ethanol production. *Energy Fuels* 24, 2113–2119. doi:10.1021/ef9014718

Zhang, C., Lei, X., Scott, T., Zhu, J. Y., and Li, K. (2014). Comparison of dilute acid and sulfite pretreatment for enzymatic saccharification of earlywood and latewood of Douglas fir. *Bioenergy Res.* 7, 362–370. doi:10.1007/s12155-013-9376-6

Conflict of Interest Statement: The authors declare that the research was conducted in the absence of any commercial or financial relationships that could be construed as a potential conflict of interest.

17

Immunological approaches to biomass characterization and utilization

Sivakumar Pattathil[1,2]*, Utku Avci[1,2], Tiantian Zhang[1], Claudia L. Cardenas[1†] and Michael G. Hahn[1,2]

[1] Complex Carbohydrate Research Center, University of Georgia, Athens, GA, USA, [2] Oak Ridge National Laboratory, BioEnergy Science Center (BESC), Oak Ridge, TN, USA

Edited by:
Jason Lupoi,
University of Queensland, USA

Reviewed by:
Xu Fang,
Shandong University, China
Arumugam Muthu,
Council of Scientific and Industrial
Research, India

*Correspondence:
Sivakumar Pattathil
siva@ccrc.uga.edu

†Present address:
Claudia L. Cardenas,
Central Piedmont Community
College, Charlotte, NC, USA

Plant biomass is the major renewable feedstock resource for sustainable generation of alternative transportation fuels to replace fossil carbon-derived fuels. Lignocellulosic cell walls are the principal component of plant biomass. Hence, a detailed understanding of plant cell wall structure and biosynthesis is an important aspect of bioenergy research. Cell walls are dynamic in their composition and structure, varying considerably among different organs, cells, and developmental stages of plants. Hence, tools are needed that are highly efficient and broadly applicable at various levels of plant biomass-based bioenergy research. The use of plant cell wall glycan-directed probes has seen increasing use over the past decade as an excellent approach for the detailed characterization of cell walls. Large collections of such probes directed against most major cell wall glycans are currently available worldwide. The largest and most diverse set of such probes consists of cell wall glycan-directed monoclonal antibodies (McAbs). These McAbs can be used as immunological probes to comprehensively monitor the overall presence, extractability, and distribution patterns among cell types of most major cell wall glycan epitopes using two mutually complementary immunological approaches, glycome profiling (an *in vitro* platform) and immunolocalization (an *in situ* platform). Significant progress has been made recently in the overall understanding of plant biomass structure, composition, and modifications with the application of these immunological approaches. This review focuses on such advances made in plant biomass analyses across diverse areas of bioenergy research.

Keywords: glycome profiling, immunolocalization, cell walls, biomass, antibodies

INTRODUCTION

Complexity and Dynamics of Plant Cell Walls Constituting Biomass

Plant biomass, the prime feedstock for lignocellulosic biofuel production, constitutes the principal sustainable resource for renewable bioenergy. Identifying the optimal plant biomass types that are most suitable for biofuel production and optimizing their downstream processing and utilization are at the forefront of modern-day lignocellulosic feedstock research. The focus of much of this research is the examination of diverse classes of plants for their potential as cost-effective and sustainable raw materials for biofuel production. For example, biomass materials originating from classes of plants ranging from herbaceous dicots (e.g., alfalfa), woody dicots (e.g., poplar), perennial monocots (e.g., *Agave* spp.), herbaceous monocots (e.g., grasses such as *Miscanthus*, sugarcane, and switchgrass), and woody gymnosperms (e.g., pines) are regarded as potentially promising resources for biofuel production (Galbe and Zacchi, 2007; Gomez et al., 2008; Somerville et al., 2010).

Cell walls constitute the major part of plant biomass, and physicochemical features of these cell walls vary among biomass materials from diverse plant classes (Pauly and Keegstra, 2008; Popper, 2008; Fangel et al., 2012). For example, cell walls from grass biomass have distinct structural and compositional features [with a higher abundance of glucuronoarabinoxylans and the presence of mixed-linkage glucans (Vogel, 2008)] that are quite different from those of highly lignified woody biomass (Studer et al., 2011) or herbaceous dicot biomass (Burton et al., 2010; Liepman et al., 2010). Even within a plant, the structure and composition of cell walls can vary significantly depending on the cell types, organs, age, developmental stage, and growth environment (Freshour et al., 1996; Knox, 2008). These cell wall variations are the result of differences in the relative proportions and structural dynamics that occur among the major cell wall polymers, which include (but are not limited to) cellulose, hemicelluloses, pectic polysaccharides, and lignin (Pauly and Keegstra, 2008). Several structural models for plant cell walls have been proposed and published (McNeil et al., 1984; McCann and Roberts, 1991; Carpita and Gibeaut, 1993; Carpita, 1996; Cosgrove, 1997; Somerville et al., 2004; Loqué et al., 2015); all of these models focus on the primary wall. To our knowledge, no model has been proposed for secondary plant cell walls, which constitute the bulk of the biomass used for bioenergy production. In vascular plants, non-glycan components such as lignin (especially in secondary cell wall-containing tissues such as sclerenchyma and xylem cells) are important for optimal growth and development of plants by playing important roles in maintaining cell wall integrity to optimally facilitate water transportation, rendering mechanical support and defense against pathogens (Weng and Chapple, 2010; Voxeur et al., 2015). A high abundance of lignin in cell walls is regarded as disadvantageous for biomass utilization for biofuel production as it contributes significantly to recalcitrance. Transgenic plants that are genetically modified for reduced lignin biosynthesis have been shown to exhibit reduced recalcitrance properties (Chen and Dixon, 2007; Pattathil et al., 2012b). The abundance of diverse potential plant biomass feedstocks that are available to be studied and the aforementioned variations among the cell walls constituting them pose a major challenge in lignocellulosic bioenergy research.

Research on the structure, function, and biosynthesis of plant cell walls has received new impetus with advances in genome sequencing that have made available, for the first time, whole genomes from diverse plant families. Thus, complete genomes have been sequenced for plants from diverse phylogenetic classes including both herbaceous [e.g., *Arabidopsis* (The Arabidopsis Genome Initiative, 2000); *Medicago* (Young et al., 2011)] and woody dicots [e.g., *Populus* (Tuskan et al., 2006)] and monocotyledonous grasses [e.g., maize (Schnable et al., 2009), rice (Goff et al., 2002; Yu et al., 2002), and brachypodium (The International Brachypodium Initiative, 2010)]. The availability of these genome sequences has, in turn, dramatically expanded experimental access to genes and gene families involved in plant primary and secondary cell wall biosynthesis and modification. Functional characterization of cell wall-related genes and the proteins that they encode, combined with expanded research on cell wall deconstruction, have dramatically enhanced our understanding of wall features important for biomass utilization.

Genetic Approaches to Studies of Cell Walls with Impacts on Lignocellulosic Bioenergy Research

Cell walls are known for their innate resistance to degradation and specifically to the breakdown of their complex polysaccharides into simpler fermentable sugars that can be utilized for microbial production of biofuels. This property of plant cell walls is referred to as "recalcitrance" (Himmel et al., 2007; Fu et al., 2011). Cell wall recalcitrance has been identified as the most well-documented challenge that limits biomass conversion into sustainable and cost-effective biofuel production (Himmel et al., 2007; Pauly and Keegstra, 2008; Scheller et al., 2010). Hence, identifying cell wall components that affect recalcitrance has been an important target of lignocellulosic bioenergy research (Ferraz et al., 2014). A number of plant cell wall polymers, including lignin, hemicelluloses, and pectic polysaccharides, have been shown to contribute to cell wall recalcitrance (Mohnen et al., 2008; Fu et al., 2011; Studer et al., 2011; Pattathil et al., 2012b).

Most of the studies directed toward overcoming recalcitrance focus on genetically modifying plants by specifically targeting genes involved in the biosynthesis or modification of wall polymers (Chen and Dixon, 2007; Mohnen et al., 2008; Fu et al., 2011; Studer et al., 2011; Pattathil et al., 2012b) with the objective of generating a viable, sustainable biomass crop that synthesizes cell walls with reduced recalcitrance. Identification of target genes for reducing recalcitrance has relied largely on model plant systems, particularly *Arabidopsis*, and then to transfer that information to biofuel crops. This has been particularly successful for genes and pathways that participate directly or indirectly in secondary cell wall biosynthesis and development. Secondary walls constitute the bulk of most biofuel feedstocks and thus become a main target for genetic modification (Chundawat et al., 2011; Yang et al., 2013). Secondary wall synthetic genes that have been investigated in this way include, for example, several genes that are involved in cellulose [such as various *CesA* genes (Joshi et al., 2004, 2011; Taylor et al., 2004; Brown et al., 2005; Ye et al., 2006)] and xylan biosynthesis [*IRX8* (Brown et al., 2005; Ye et al., 2006; Peña et al., 2007; Oikawa et al., 2010; Liang et al., 2013), *IRX9* (Brown et al., 2005; Lee et al., 2007, 2011a; Peña et al., 2007; Oikawa et al., 2010; Liang et al., 2013), *IRX9L* (Oikawa et al., 2010; Wu et al., 2010), *IRX14* (Oikawa et al., 2010; Wu et al., 2010; Lee et al., 2011a), *IRX14L* (Wu et al., 2010; Lee et al., 2011a), *IRX15* (Brown et al., 2011), and *IRX15L* (Brown et al., 2011)] in dicots. In addition, a number of transcription factors including plant-specific NAC-domain transcription factors [*SND1*, *NST1*, *VND6*, and *VND7* in *Arabidopsis* (Kubo et al., 2005; Zhong et al., 2006, 2007b)], WRKY transcription factors [in *Medicago* and *Arabidopsis* (Wang et al., 2010; Wang and Dixon, 2012)], and MYB transcription factors [*MYB83* (McCarthy et al., 2009) and *MYB46* (Zhong et al., 2007a) in *Arabidopsis*] with potential involvement in secondary wall biosynthesis and development have been functionally characterized. Examples of the successful transfer of insights gained in model dicots to studies of orthologous genes in monocots include investigations of rice *IRX* orthologs involved in xylan biosynthesis and secondary wall formation (Oikawa et al., 2010) and experiments on transcription factors controlling secondary wall formation in several grasses (Handakumbura and Hazen, 2012; Shen et al., 2013; Valdivia et al., 2013). These molecular genetic approaches toward understanding and manipulating cell wall-related genes for biofuel feedstock improvement would be assisted by improved methods for rapidly identifying and characterizing the effects of genetic changes on cell wall components.

Need for Efficient Tools for Plant Cell Wall/ Biomass Analyses

The structural complexity of plant cell walls, regardless of their origin, is challenging to analyze, particularly in a high-throughput manner. To date, most of the plant cell wall analytical platforms have been based on the preparation of cell wall materials and/ or extracts that are selectively enriched for particular wall polysaccharides, followed by colorimetric assays (Selvendran and O'Neill, 1987), chemical derivatizations coupled with gas chromatography (Albersheim et al., 1967; Sweet et al., 1974, 1975a,b), mass spectroscopy (Lerouxel et al., 2002), and nuclear magnetic resonance spectroscopy (NMR) (Peña et al., 2008) to gain compositional and structural information about those polysaccharides. Some of these methods have been adapted for biomass analytics [see, for review, Sluiter et al. (2010)]. Overall, these tools have allowed extensive progress in delineating basic structural features of diverse classes of plant cell wall polysaccharides. However, these experimental approaches for plant cell wall/biomass analysis are time-consuming, require specialized and, in some cases, expensive equipment, are low in throughput, and usually provide information only about a single polysaccharide of specific interest. However, given the number of wall components that have already been shown to influence cell wall recalcitrance, and the complex and heterogeneous nature of cell wall components in diverse plants, it is desirable to have additional tools, particularly those with higher throughput and the capability to monitor a broad spectrum of wall polymers. Over the past 10 years, immunological approaches for plant cell wall and biomass analyses have emerged as tools that are broadly applicable to multiple aspects of interests to the biofuel research community, including characterization of genetically altered plant feedstocks, investigations of the effects of diverse biomass pretreatment processes, and the effects of enzymatic or microbial deconstruction of cell walls. In the following sections, we review applications of two immunological tools for studies on plant biomass that employ a comprehensive collection of plant cell wall glycan-directed probes.

PROBES FOR BIOMASS ANALYSES

Currently, well-characterized cell wall-directed probes range from small molecules (Wallace and Anderson, 2012) to larger proteinaceous probes such as carbohydrate-binding modules (CBMs) and monoclonal or polyclonal antibodies (Knox, 2008; Pattathil et al., 2010; Lee et al., 2011b). In this review, we will focus on the latter cell wall-directed probes.

Glycan-Directed Probes
Monoclonal Antibodies

Plant cell wall glycan-directed monoclonal antibodies (McAbs) are among the most commonly used probes for plant cell wall analyses. McAbs, commonly available as hybridoma culture supernatants, are monospecific probes that recognize specific glycan sub-structures (epitopes) present in plant polysaccharides (Knox, 2008; Pattathil et al., 2010). McAbs have several advantages that make them particularly suited for use as glycan-directed probes. First, since each antibody is the product of a single clonal cell line, each McAb is by definition monospecific with regard to the epitope that is recognized. This is important for studies of glycans, whose structures are frequently repetitive and whose substructures can be found in multiple macromolecular contexts (e.g., arabinogalactan epitopes present on glycoproteins and on rhamnogalacturonan I). The monospecific nature of McAbs also means that, in theory, the binding specificity of the antibody can be determined unambiguously, although this is still difficult for glycan-directed antibodies given the complexity of plant cell wall glycan structures. McAbs also typically bind to their epitopes with high affinity ($K_d \sim 10^{-6}$ M), which makes them very sensitive reagents for detecting and quantitating molecules to which they bind. Finally, another significant advantage with McAbs is that their supply is not limited, as cell lines producing them can be cryopreserved indefinitely (some hybridoma lines whose plant glycan-directed antibodies are frequently used today were generated more than 20 years ago) and can be regrown at any time to produce additional McAb, which retains the binding selectivity and affinity of the original McAb, as needed in any quantities required. Currently, a worldwide collection of over 200 McAbs (Pattathil et al., 2010, 2012a) exists (**Figure 1**) that encompasses antibodies recognizing diverse structural features of most major non-cellulosic cell wall glycans, including arabinogalactans, xyloglucans, xylans, mannans, homogalacturonans, and rhamnogalacturonan I. So far, McAbs that bind reliably and specifically to rhamnogalacturonan II have not been reported. The available plant glycan-directed McAbs can be obtained from several stock centers (see **Table 1**) or from the individual research laboratories that generated them. A listing of the McAbs currently available is not practical here. The reader is referred to a plant cell wall McAb database, Wall*Mab*DB,[1] where detailed descriptions of most of the currently available plant glycan-directed McAbs, including immunogen, antibody isotype, and epitope structure (to the extent known), can be obtained.

---

[1] http://www.wallmabdb.net.

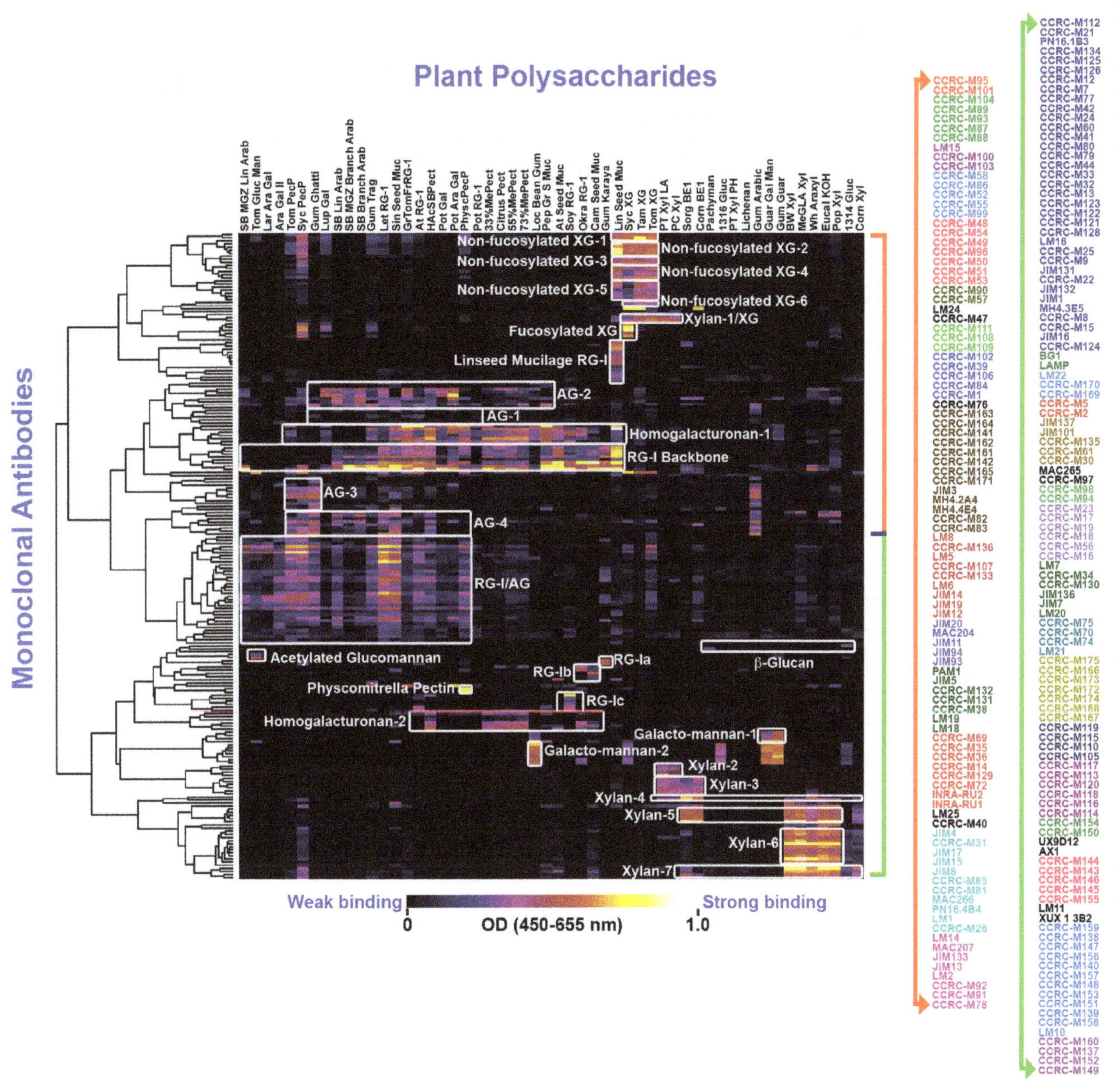

FIGURE 1 | Current worldwide collection of plant cell wall glycan-directed McAbs: the entire collection of ~210 McAbs was ELISA-screened against a panel of 54 structurally known plant cell wall carbohydrate preparations (Pattathil et al., 2010) **and they were clustered to 31 groups (as depicted by the white blocks) based on their binding specificities**. The binding strengths are depicted in a dark blue–red–bright yellow color scheme where maximum and no binding are denoted by bright yellow color and dark blue colors, respectively. The names of individual McAbs are denoted on the right hand panel in different colors denoting 31 groups.

Early studies in our laboratory screened 130 of the plant glycan-directed McAbs available at the time for their binding specificity to 54 structurally characterized polysaccharide preparations from diverse plants (Pattathil et al., 2010). Hierarchical clustering analyses of the resultant binding response data resolved the McAbs into 19 antibody clades based on their binding specificities to the 54 plant glycans tested (Pattathil et al., 2010). A more recent study that included almost all available plant glycan-directed McAbs further resolved the antibody collection into about 31 clades of McAbs (Pattathil et al., 2012a). **Figure 1** shows the data from most recent screening studies employing ~210 plant glycan-directed McAbs. While these broad specificity screens provide considerable information about the binding specificities of the McAbs in the collection, they do not provide complete detailed epitope information for the antibodies. Such detailed epitope characterization studies require the availability

TABLE 1 | List of major CBMs currently used for plant cell wall analyses.

Group	Protein	Enzyme	Organism	Type	Reference
A. Cellulose-binding group	CBM1	Cellulase	*Trichoderma reesei*	Crystalline cellulose	Reinikainen et al. (1992)
	CBM2a	Xylanase 10A	*Cybister japonicus*	Crystalline cellulose	Bolam et al. (1998)
	CBM3a	Scaffoldin	*Clostridium thermocellum*	Crystalline cellulose	Tormo et al. (1996)
	CBM10	Xylanase 10A	*Cybister japonicas*	Crystalline cellulose	Gill et al. (1999)
	CBM4-1	Cellulase 9B	*Cellulomonas fimi*	Amorphous cellulose	Tomme et al. (1996)
	CBM17	Cellulase 5A	*Clostridium cellulovorans*	Amorphous cellulose	Boraston et al. (2000)
	CBM28	Cellulase 5A	*Bacillus* sp. no. 1139	Amorphous cellulose	Boraston et al. (2002)
	CBM9-2	Xylanase 10A	*Thermotoga maritima*	The ends of cellulose chain	Boraston et al. (2001)
B. Xylan-binding group	CBM2b-1-2	Xylanase 11A	*Cellulomonas fimi*	Both decorated and unsubstituted xylan	Bolam et al. (2001)
	CBM4-2	Xylanase 10A	*Rhodothermus marinus*	Both decorated and unsubstituted xylan	Abou Hachem et al. (2000)
	CBM6	Xylanase 11A	*Clostridium thermocellum*	Both decorated and unsubstituted xylan	Czjzek et al. (2001)
	CBM15	Xylanase 10C	*Cybister japonicus*	Both decorated and unsubstituted xylan	Szabó et al. (2001)
	CBM22-2	Xylanse 10B	*Clostridium thermocellum*	both decorated and unsubstituted xylan	Charnock et al. (2000)
	CBM35	Arabino-furano-sidase 62A	*Cybister japonicus*	Unsubstituted xylan	Bolam et al. (2004)
C. Mannan-binding group	CBM27 (TmMan5)	Mannanase 5C	*Thermotoga maritima*	Mannan	Filonova et al. (2007) and Zhang et al. (2014)
	CBM35 (Cjman5C)	Mannanase 5C	*Cybister japonicus*	Mannan	Filonova et al. (2007) and Zhang et al. (2014)
D. Xyloglucan-binding group	CBMXG34	Modified xylanase 10A	*Rhodothermus marinus*	Non-fucosylated xyloglucan	Gunnarsson et al. (2006)
	CBMXG34/1-X	Modified xylanase 10A	*Rhodothermus marinus*	Non-fucosylated xyloglucan	von Schantz et al. (2009)
	CBMXG34/2-VI	Modified xylanase 10A	*Rhodothermus marinus*	Non-fucosylated xyloglucan	von Schantz et al. (2009)
	CBMXG35	Modified xylanase 10A	*Rhodothermus marinus*	Non-fucosylated xyloglucan	Gunnarsson et al. (2006)
E. Pectic galactan-binding group	TmCBM61	GH53 endo-β-1,4-galac-tanase	*Thermotoga maritima*	β-1,4-galactan	Cid et al. (2010)

of purified, structurally characterized oligosaccharide fragments and/or purified and characterized glycosylhydrolases capable of selectively attacking epitope structures. To date, a relatively small number of plant glycan-directed McAbs have had their epitopes characterized in detail using these resources (Meikle et al., 1991, 1994; Puhlmann et al., 1994; Steffan et al., 1995; Willats et al., 2000a; Clausen et al., 2003, 2004; McCartney et al., 2005; Verhertbruggen et al., 2009; Marcus et al., 2010; Ralet et al., 2010; Pedersen et al., 2012; Schmidt et al., 2015). Recent advances in methods for immobilization of oligosaccharides on solid surfaces (Fukui et al., 2002; Wang et al., 2002; Willats et al., 2002; Blixt et al., 2004; Pedersen et al., 2012) is facilitating such epitope characterization studies, but the bottleneck remains the availability of comprehensive sets of purified, well-characterized plant glycan-related oligosaccharides.

Carbohydrate-Binding Modules

Carbohydrate-binding modules are another set of proteinaceous probes that have been used to study plant polysaccharide localization patterns *in vivo* (Knox, 2008). CBMs are amino acid sequences that are contiguous with the catalytic domain in a carbohydrate-active enzyme and are capable of binding to a carbohydrate structural domain (McCartney et al., 2006; Knox, 2008). CBMs have been shown to enhance the efficiency of cell wall hydrolytic enzymes by facilitating sustained and close contact between their associated catalytic modules and targeted substrates (Boraston et al., 2004; Zhang et al., 2014). Although CBMs have been known to occur in several plant enzymes, most CBMs that are used as probes for cell wall glycans are microbial in origin (Boraston et al., 2004; Shoseyov et al., 2006). CBMs, in contrast to the antibody probes described above, are relatively easy to prepare, given that their gene/protein sequences are known (McCann and Knox, 2011). CBMs have been classified into 71 sequence-based families.[2] CBMs from approximately half of these families have been shown to bind to diverse plant cell wall polysaccharides, including cellulose (Blake et al., 2006), mannans (Filonova et al., 2007), xylans (McCartney et al., 2006), and most recently, the galactan side chains of rhamnogalacturonan I (Cid et al., 2010). Protein engineering of a xylan-binding CBM using random mutagenesis, phage-display technology, and affinity maturation has been employed to generate xyloglucan-specific

---

[2]http://www.cazy.org/Carbohydrate-Binding-Modules.html.

CBMs (Gunnarsson et al., 2006; von Schantz et al., 2009, 2012), showing that it is possible to generate CBMs with new and heretofore unseen specificities.

Carbohydrate-binding modules that have been used to detect cellulose, xylan, mannan, xyloglucan, and pectic galactans in plant cells and tissues, together with information about their origins, are listed in **Table 1**. Binding of various CBMs is usually assessed by an indirect triple-labeling immunofluorescence procedure (His-tagged CBM, anti-His mouse-Ig, and anti-mouse Ig fluorescein isothiocyanate) in plant tissue sections (Knox, 2008; Hervé et al., 2010), which is slightly more complicated than the double-labeling procedure used with McAbs (Avci et al., 2012). The binding specificities exhibited by the CBMs enlarge the suite of probes available for biomass analyses, given that at least some of them bind to carbohydrate structures, such as cellulose substructures, for which no McAbs probes have been developed to date. Additional advantages of the CBMs are the availability of their gene and protein sequences and the wealth of structural information, including in many instances X-ray crystal structures, about their binding sites. Potential disadvantages of CBMs are their typically lower affinity for their ligands and the lower selectivity of their binding sites compared with McAb probes. Nonetheless, CBMs are useful probes for analyzing biomass.

Immunological Probes Against Lignin

Lignins are phenylpropanoid polymers comprising 5–30% of biomass weight and have been considered as important sources of renewable aromatics (McKendry, 2002). Lignin composition and structure vary considerably depending on the plant species and on the cell type where lignins are deposited (Ruel et al., 1994; Donaldson, 2001). For example, in gymnosperms, lignins are mainly composed of guaiacyl units, whereas in angiosperms, lignins are formed by guaiacyl and syringyl units (Donaldson, 2001). In angiosperms, the guaiacyl-containing lignins are located mainly in secondary cell walls of vessels while syringyl-containing lignins are found on fibers (Ruel et al., 1994; Joseleau et al., 2004; Patten et al., 2010). Lignin composition and localization are also affected by pretreatment strategies aimed at removing lignin from biomass. For example, potassium permanganate labeling and electron microscopy studies revealed morphological alterations in *Zea mays* lignins subjected to different thermochemical pretreatments (Donohoe et al., 2008).

Lignin is most frequently visualized in plant tissue sections using selectively reactive histochemical stains such as phloroglucinol–HCl and Mäule reaction that can distinguish guaiacyl-enriched from syringyl-enriched cell wall regions (Patten et al., 2010). Although the various histochemical lignin stains provide general information about the localization of different lignin types, they cannot provide detailed information about specific lignin substructures; this would require more highly selective probes.

Given the structural complexity and variability of lignin, several laboratories have undertaken the development of immunological probes for lignins and/or lignin substructures. Much of the early work in this area focused on the production of polyclonal antisera. Thus, polyclonal antisera were raised against synthetic dehydrogenative polymers (DHPs)

prepared from the appropriate p-hydroxycinnamic alcohols [p-hydroxyphenylpropane (H), guaiacyl (G), or syringyl (S), or mixtures of these] (Ruel et al., 1994; Joseleau et al., 2004). These polyclonal sera showed specificity toward the DHPs used to generate them. Other laboratories have generated polyclonal sera against milled wood lignin (Kim and Koh, 1997) or model compounds based on lignin substructures (Kukkola et al., 2003, 2004). The main difficulty with these polyclonal sera is that they are in limited supply, and many of these antisera are no longer available. Thus, new immunizations must be carried out, with uncertain outcomes with regard to the ability to reproduce the specificity of the original antisera; a fundamental problem with polyclonal antisera. In an effort to overcome this limited supply issue, two lignin-related model compounds, dehydrodiconiferyl alcohol and pinoresinol, were used to generate McAbs against these two lignin dimers (Kiyoto et al., 2013); supplies of these antibodies should not be limited. The antibody directed to dehydrodiconiferyl alcohol (KM1) displayed specificity toward a dehydrodiconiferyl alcohol 8-5′ model compound, whereas the antibody directed against pinoresinol (KM2) responded to two 8-8′ model compounds, pinoresinol and syrangaresinol. This recent development suggests that it will be possible, in principle, to generate specific McAbs against diverse lignin substructures. The number and diversity of lignin-directed McAbs will need to be increased in order to fully exploit these probes for greater insights into lignin structural diversity, localization patterns, and integration into the plant cell wall.

TWO MAJOR APPROACHES FOR McAb/CBM-BASED ANALYSES OF PLANT BIOMASS

The use of McAb/CBM probes to define the localization of plant cell wall components has a long history. These probes have been used in basic plant cell wall research to study the effects of mutations in wall-related genes on plant cell wall structure and composition, to study changes in plant cell walls during growth, development, and differentiation, and to study changes in plant cell walls that result from environmental and pathogenic influences. A comprehensive review of this literature is beyond the scope of this minireview and the reader is referred to several recent reviews to gain an overview of this literature (Knox, 1997, 2008; Willats et al., 2000b; Lee et al., 2011b; McCann and Knox, 2011). The use of McAb probes, in particular, is rapidly expanding due to the recent dramatic increase in the number and diversity of plant cell wall-directed antibodies (Pattathil et al., 2010) and the availability of more detailed information about the epitopes recognized by these McAbs (Pedersen et al., 2012; Schmidt et al., 2015).

We will concentrate here on an overview of recent studies that have taken advantage of the availability of the comprehensive collection of cell wall-directed McAb/CBM probes for studying plant biomass of interest as possible lignocellulosic feedstocks for biofuel production. These studies have focused on using these probes to understand the effects of genetic modification on biomass recalcitrance, to study the effects of different pretreatment regimes on biomass digestibility, and to study how microbes being considered for consolidated bioprocessing deconstruct plant biomass. Two complementary experimental approaches have been principally employed in these studies, namely, glycome profiling (Moller et al., 2007, 2008; Pattathil et al., 2012a) and immunolocalization (Avci et al., 2012). The following sections provide an overview of the studies with bioenergy implications done to date using these approaches.

Studies Using Glycome Profiling

Glycome profiling involves the sequential extraction of insoluble cell wall/biomass samples with a series of reagents of increasing harshness and then screening the extracted cell wall materials with McAbs to determine which cell wall polymers are released in which extract. Thus, this experimental method provides two pieces of important information: (1) it provides detailed information about the composition of the biomass/cell walls; and (2) it provides information on how tightly the various wall components that can be detected are linked into the wall structure. The method is limited by the number of probes (McAbs, CBMs, etc.) used in the screen and the extent to which they are able to recognize the full breadth of wall components released by the extractive reagents. The substantial increase in number and diversity of cell wall probes over the past 10 years has dramatically improved the power and versatility of glycome profiling as a technique for rapid screening of cell wall/biomass samples.

The versatility of glycome profiling is also limited by the ability to immobilize the extracted wall components to a solid support. Diverse solid supports have been used, including nitrocellulose (Moller et al., 2007, 2008), glass slides (Pedersen et al., 2012), and multiwell plastic plates (Pattathil et al., 2012a). All of these suffer the limitation that most low-molecular-weight cell wall components that might be released in the wall extracts, especially low-molecular-weight glycans, do not bind to the solid supports without modification and therefore cannot be assayed by glycome profiling. The lower limit of the glycan size that will adhere has not been definitively determined, but is greater than 10 kDa (Pattathil et al., 2010).

The choice of extractive reagents that have been used for glycome profiling analyses has varied, as has their order. However, typically, the extractive reagents are used in order of increasing severity. Thus, relatively mild reagents, such as CDTA (Moller et al., 2007) or oxalate (Pattathil et al., 2012a), are used first, typically extracting primarily arabinogalactans and pectins. Harsher base extractions then follow, in which primarily hemicelluloses (e.g., xylans and xyloglucans) are extracted (Moller et al., 2007; Pattathil et al., 2012a). For samples that contain significant amounts of lignin, which is the case for most biomass samples of interest to the biofuel industry, an acidic chlorite extraction (Ahlgren and Goring, 1971; Selvendran et al., 1975) is used to degrade the lignin and release lignin-associated wall glycans; this chlorite extraction has most frequently been used after the first base extractions (Pattathil et al., 2012a) but has also been used as the first extraction step (de Souza et al., 2013). None of the extraction sequences used to date yield exclusively one kind of polymer in any given extract, an indication that each wall glycan exists as different subclasses that vary in their extent of cross-linking/

interactions within the wall. Ultimately, the choice of extraction reagents and their order depends on the individual investigator and the specific research questions under investigation.

Two approaches for glycome profiling of plant biomass/cell wall samples have been described. The first, termed comprehensive microarray polymer profiling (CoMPP), is a dot blot-based assay system utilizing nitrocellulose as the solid support (Moller et al., 2007, 2008) and typically employs ~20 glycan-directed probes for screening of three sequential extracts [CDTA (50 mM), 4M NaOH, and Cadoxen (33%; v/v)] prepared from plant cell walls. The number of glycan-directed probes that could be used in CoMPP can readily be expanded. An alternative, ELISA-based approach, termed glycome profiling, uses 384-well microtiter plates as the solid support, and uses a broadly diverse toolkit of 155 plant glycan-directed McAbs (Pattathil et al., 2012a) to screen sets of sequentially prepared plant biomass/cell wall extracts [typically, oxalate (50 mM), carbonate (50 mM), 1M KOH, 4M KOH, acidified chlorite, and 4M KOH post-chlorite]. The use of a suite of 155 McAbs ensures a wide-ranging coverage of multiple structural features on most of the major non-cellulosic plant wall glycans (Zhu et al., 2010; Pattathil et al., 2012a). The ELISA-based approach used in glycome profiling lends itself to facile automation and quantitation of antibody binding, hence substantially increasing the throughput of the analyses.

Glycome profiling has seen broad application to diverse experimental approaches in lignocellulosic bioenergy research, including analyzing cell walls from native/genetically modified, variously pretreated, and microbially/enzymatically converted plant biomass (DeMartini et al., 2011; Duceppe et al., 2012; Lee et al., 2012; Tan et al., 2013; Biswal et al., 2015; de Souza et al., 2015; Pattathil et al., 2015; Trajano et al., 2015). Both CoMPP and glycome profiling have been used to undertake comparative glycomics of plant cell wall samples originating from diverse plant phylogenies (Popper et al., 2011; Sørensen et al., 2011; Duceppe et al., 2012; Kulkarni et al., 2012). Examples of such analyses applied to questions related to bioenergy research include a recent study assessing the genetic variability of cell wall degradability of a selected number of *Medicago* cultivars with superior saccharification properties (Duceppe et al., 2012) and an examination of five grass species that revealed commonalities and variations in the overall wall composition and extractability of epitopes among these grasses (Kulkarni et al., 2012). Glycome profiling has also been employed as an effective tool for analyzing cell walls from biomass crops that are genetically modified with the aim of reducing recalcitrance. Examples include examination of the effects on recalcitrance of mutations in lignin biosynthesis in alfalfa [*cad1* (cinnamyl alcohol dehydrogenase 1) (Zhao et al., 2013) and *hct* (hydroxycinnamoyl CoA:shikimate hydroxycinnamoyl transferase) (Pattathil et al., 2012b)] and overexpression of the secondary wall-related transcription factor, PvMYB4 in switchgrass (Shen et al., 2013).

Analyses using cell wall-directed probes have allowed the rapid identification and monitoring of structural and compositional alterations that occur in plant biomass under various regimes of pretreatments (Alonso-Simón et al., 2010; DeMartini et al., 2011; Li et al., 2014; Socha et al., 2014; Pattathil et al., 2015; Trajano et al., 2015). Studies on hydrothermally pretreated wheat straw using CoMPP showed that severe pretreatment regimes induce significant alterations in wheat straw biomass, including reduction in various hemicellulose and mixed-linkage glucan epitopes (Alonso-Simón et al., 2010). In a more recent study, glycome profiling of poplar biomass subjected to low, medium, and severe hydrothermal pretreatment regimes demonstrated that a series of structural and compositional changes occur in poplar cell walls during this pretreatment, including the rapid disruption of lignin–polysaccharide interactions even under mild conditions, with a concomitant loss of pectins and ara-binogalactans, followed by significant removal of hemicellulose (xylans and xyloglucans) (DeMartini et al., 2011). The major inference from this study was that lignin content *per se* does not affect recalcitrance; instead, it is the associations/cross-links between polymers, for example, between lignin and various polysaccharides, within cell walls that play a larger role (DeMartini et al., 2011). Glycome profiling has also been used to examine the effects of other types of pretreatment regimes such as Ammonia Fiber Expansion (AFEX™), alkaline hydrogen peroxide (AHP), and various types of ionic liquids (ILs) on the composition and extractability of wall glycan epitopes in biomass samples from diverse bioenergy crop plants (Li et al., 2014; Socha et al., 2014; Pattathil et al., 2015). These studies demonstrate that, unlike hydrothermal pretreatment, these three types of pretreatment, in general, cause loosening of specific classes of non-cellulosic glycans from plant cell walls, thereby contributing to the reduced recalcitrance exhibited by the pretreated biomasses. Conclusions from these studies contribute significantly to a deeper understanding of pretreatment mechanisms and ultimately will enable optimization of biomass pretreatment regimes and perhaps further downstream utilization processes for biomass from different plant feedstocks.

Glycome profiling has also been used to identify cell wall components that affect biomass recalcitrance. A recent study examined poplar and switchgrass biomass subjected to different pretreatments and correlated pretreatment-induced changes in the biomass with recalcitrance properties of the treated biomass samples (DeMartini et al., 2013). A set of samples with varying composition and structure was generated from native poplar and switchgrass biomass via defined chemical and enzymatic extraction. Subsequently, glycome profiling of the extracts was employed to delineate which wall components were removed and residual solid pretreated biomass samples were analyzed for their recalcitrance features. Major conclusions from this study are that pretreatment regimes affect distinct biomass samples differently and that the most important contributors to recalcitrance vary depending on the biomass. Thus, lignin content appears to play an important role in biomass recalcitrance particularly in woody biomass such as poplar (as they contain higher levels of lignin). However, subclasses of hemicellulose were key recalcitrance-causing factors in grasses such as switchgrass. These results may have important implications for the biofuel industry as they suggest that biomass-processing conditions may have to be tailored to the biomass being used as the feedstock for biofuel generation (DeMartini et al., 2013).

Another bioenergy-related area that has benefited from the use of plant cell wall glycan-directed probes is research into how microbes, particularly those being selected for biomass

deconstruction, degrade plant biomass during culture. Such knowledge will be useful for bioengineering microbes for better biomass conversion. An analysis of biological conversion of unpretreated wild-type sorghum and various *brown midrib* (*bmr*) lines by *Clostridium phytofermentans* examined variations in extractable polysaccharide epitopes of the cell-wall fractions in detail using glycome profiling (Lee et al., 2012). The conclusions were that the loosely integrated xylans and pectins are the primary polysaccharide targets of *C. phytofermentans* and that these are more accessible in the *bmr* mutants than in the wild-type plants (Lee et al., 2012). In another study, an anaerobic thermophilic bacterium, *Caldicellulosiruptor bescii*, was shown to solubilize both lignin and carbohydrates simultaneously in swichgrass biomass at high temperature (Kataeva et al., 2013). Further studies with *C. bescii* demonstrated that deletion of a cluster of genes encoding pectic-degrading enzymes in this organism compromised the ability of *C. bescii* to grow on diverse biomass samples (Chung et al., 2014). A comparative analysis of hemicellulose utilization potentials of *Clostridium clariflavum* and *Clostridium thermocellum* strains demonstrated that *C. clariflavum* strains were better able to grow on untreated switchgrass biomass and degraded easily extractable xylans more readily than do *C. thermocellum* strains (Izquierdo et al., 2014). In all of these studies, glycome profiling proved to be a very effective tool for understanding what was happening to the biomass during culture with the microbes. Studies of this kind provide information about the mode of action of microbial strains on plant biomass, thus identifying wall components that are resistant/recalcitrant to microbial actions.

Studies Using Immunolocalization

Immunolocalization techniques use fixed and embedded (generally in plastic resins) biomass samples (Knox, 1997; Lee et al., 2011b). Primary probes (polyclonal antibodies, McAbs, and CBMs) are applied on semithin sections followed by probing with a fluorescently tagged secondary antibody that allows visualization of glycan epitope localization/distribution under a fluorescent microscope (Avci et al., 2012; Lee and Knox, 2014). This approach for biomass analyses provides information regarding the distribution of cell wall glycans at the cellular and subcellular levels.

A handful of studies thus far have employed this technique in the context of bioenergy research for analyses of cell walls in wall biosynthetic mutants and in pretreated biomass. Examination of *Arabidopsis* and *Medicago* mutants in which a WRKY transcription factor was knocked out revealed secondary cell wall thickening in pith cells caused by ectopic deposition of lignin, xylan, and cellulose. In the *Arabidopsis* mutant, this ectopic secondary wall formation resulted in an approximately 50% increase in biomass density in stem tissue (Yu et al., 2014). The use of three xylan-directed McAbs and a cellulose-directed CBM were instrumental in proving the ectopic deposition of these cell wall glycans in pith cells. In another recent study, the use of two xylan-directed CBMs (CBM2b-1-2 and CBM35 recognizing different degrees of methyl esterification on xylan) on the *Arabidopsis gxmt-1* mutant demonstrated a reduction of 4-*O*-methyl esterification of xylans (up to 75% as detected by chemical analyses) with a concomitant reduction in the recalcitrance of mutant walls (Urbanowicz et al., 2012). Additional studies also implicate the importance of secondary wall xylan for cell wall recalcitrance. Restoration of xylan synthesis in xylan-deficient mutants, as documented using xylan-directed McAbs, could, in some cases, yield plants with reduced xylan deposition compared with wild-type plants, but with normal growth habits and decreased recalcitrance (Petersen et al., 2012). Likewise, reduction of xylan in rice culm cell walls yielded plants with slightly lower stature, but with reduced recalcitrance (Chen et al., 2013).

Plant glycan-directed probes (McAbs and CBMs) can also be used to study the distribution patterns of glycan epitopes in plant biomass after diverse pretreatments used to reduce cell wall recalcitrance. One example of such a study is the demonstration that increasingly harsh hydrothermal pretreatments lead to an increased loss of various hemicellulosic, pectic, and cellulosic epitopes in cell walls of the pretreated tissues (DeMartini et al., 2011). The effects of other pretreatment methods (Alonso-Simón et al., 2010; DeMartini et al., 2013; Li et al., 2014; Socha et al., 2014; Pattathil et al., 2015; Trajano et al., 2015) on glycan epitope distribution patterns have not yet been carried out. Such information could be potentially useful to chemical engineers for the optimization of pretreatment conditions to enable optimal biomass conversion.

Immunolocalization studies have documented lignin distribution patterns in plant cell walls that may be relevant to bioenergy research. For instance, cell wall ultrastructure studies using three polyclonal antisera against DHPs allowed visualization of where these types of lignin-related polymers were located in cells of *Zea mays* L. (Joseleau and Ruel, 1997), *Arabidopsis thaliana*, *Nicotiana tabacum*, and *Populus tremula* (Ruel et al., 2002). These studies showed that H-DHPs were present in cell corners and middle lamella, whereas G-DHPs and G/S-DHPs were mainly present in secondary cell walls. The syringylpropane DHP epitope was visualized mainly in the S2 layer of secondary cell walls of *A. thaliana*, *N. tabacum*, and *P. tremula* (Joseleau et al., 2004). Recently, immunogold labeling analyses using KM1 and KM2 demonstrated the presence of 8-5′ and 8-8′ linked structures, respectively, on either developed xylem or phloem fibers of *Chamaecyparis obtusa* (Kiyoto et al., 2013). It will likely be informative to use these and other lignin-directed probes to monitor lignin distribution patterns in biomass that has been subjected to various pretreatment regimes and/or subjected to microbial degradation in the context of biomass conversion.

Concluding Remarks

The application of high affinity, highly selective molecular probes against plant cell wall polymers clearly has high potential to provide complementary and supplementary data to existing chemical and biochemical analyses for studies on plant biomass structure and conversion. The number and diversity of McAb and CBM probes directed against plant polymers is now sufficiently large that these probes can provide extensive information about cell wall composition and structure in native and pretreated or microbially digested biomass. We have reviewed two main approaches using these probes for biomass characterization and conversion studies. Both glycome profiling/CoMPP and immunolocalization

methods provide distinct but complementary information about the cell walls that constitute the bulk of plant biomass. Glycome profiling and CoMPP provide extensive information about the epitope composition and epitope extractability of polymers present in the biomass. Histochemical approaches using these probes provide valuable information about the spatial distribution of wall epitopes at all levels of organization, ranging from whole plants, to organs, to tissues, to cells, and even to individual cell walls and cell wall domains.

It is important to recognize several attributes of molecular probes directed against cell wall glycan epitopes, in particular, when interpreting the results of experiments. Both McAbs and CBMs are epitope-directed probes, that is, they specifically recognize particular structural motifs. Hence, glycan-directed McAbs and CBMs may not always be polymer-specific, in as much as glycan structures are frequently present in multiple molecular contexts within plant cell walls (e.g., arabinogalactan epitopes present on both polypeptide and polysaccharide backbones). Hence, positive binding of a McAb or CBM probe does not necessarily infer the presence of a particular cell wall glycan polymer. Likewise, the absence of binding of a given McAb or CBM does not unambiguously infer the absence of the glycan detected by this probe; the epitope may be absent or chemically modified (e.g., acetylated or methylated) such that the probe does not bind, but the polymer may still be present (Avci et al., 2012). Furthermore, plant glycans exist as families of polymers, whose epitope composition may not be uniform among all family members. Thus, a single McAb or CBM probe may not bind to all members of a polymer family, and it is therefore advisable to use multiple probes against diverse epitopes on a particular glycan to obtain a comprehensive picture of its abundance either in cell wall extracts or in histochemical localization studies. The size and diversity of the McAb/CBM collections now make such comprehensive studies possible.

Glycome profiling and CoMPP are dependent on the successful immobilization of cell wall-derived molecules to solid supports (e.g., plastic ELISA plates or nitrocellulose). Cell wall glycans with lower molecular masses (less than 20 kDa) have been found not to adhere reliably to the plates (Pattathil et al., 2010, 2012a). Hence, using glycome profiling as a tool to gather information regarding low-molecular-weight cell wall glycans is not advisable unless alternative strategies are employed to ensure adherence of these molecules to a solid support [e.g., covalent attachment directly to the solid support (Schmidt et al., 2015) or to a protein carrier that adheres to the solid support (Pedersen et al., 2012)]. Both glycome profiling and CoMPP also rely on chemical/enzymatic extractions of biomass/cell wall samples. Such extractions are rarely complete or quantitative and thus absolute quantitation of epitope composition in biomass/cell wall samples using these approaches is problematic. Thus, these approaches are best used as initial broad glycome characterization screens, particularly in comparative studies (e.g., mutant vs. wild-type and pretreated vs. untreated) where they provide valuable information regarding changes in the cell wall/biomass samples as a result of a particular experimental manipulation. In histochemical studies, the embedding medium used may influence the results of labeling experiments; in our laboratory, we have found LR White to give the most consistent results with both McAb and CBM probes (Avci et al., 2012).

FUTURE PERSPECTIVES

The molecular probe toolkits (McAb and CBM) currently available provide an invaluable resource for plant biomass analyses of relevance to bioenergy research and biomass conversion process development. In spite of the number and diversity of the probes currently available, there is still a need for additional probes against structural features not encompassed by the binding specificities of the probes currently available. Thus, additional probes against lignin substructures, rhamnogalacturonan II, and cellulose would further enhance the utility of the probe toolkit. In addition, coverage by the current probe collection of the epitope diversity for some cell wall glycans (e.g., mannans, glucomannans, and galactomannans) is limited. Finally, there remains a need to obtain more detailed information regarding the binding

specificities of many of the molecular probes in the toolkit; about one third of the glycan-directed McAbs have had their epitope specificities characterized in detail. Efforts are underway in multiple laboratories to address these needs. Thus, we can look forward to an enhanced toolkit of probes against plant cell wall polymers in the future.

ACKNOWLEDGMENTS

Immunological studies on biomass characterization conducted in our laboratory are supported by the BioEnergy Science Center administered by Oak Ridge National Laboratory and funded by a grant (DE-AC05-00OR22725) from the Office of Biological and Environmental Research, Office of Science, United States, Department of Energy. The generation of the CCRC series of plant cell wall glycan-directed monoclonal antibodies used in this work was supported by the NSF Plant Genome Program (DBI-0421683 and IOS-0923992).

REFERENCES

Abou Hachem, M., Karlsson, E. N., Bartonek-Roxa, E., Raghothama, S., Simpson, P. J., Gilbert, H. J., et al. (2000). Carbohydrate-binding modules from a thermostable *Rhodothermus marinus* xylanase: cloning, expression and binding studies. *Biochem. J.* 345, 53–60. doi:10.1042/0264-6021:3450053

Ahlgren, P. A., and Goring, D. A. I. (1971). Removal of wood components during chlorite delignification of black spruce. *Can. J. Chem.* 49, 1272–1275. doi:10.1139/v71-207

Albersheim, P., Nevins, D. J., English, P. D., and Karr, A. (1967). A method for the analysis of sugars in plant cell-wall polysaccharides by gas-liquid chromatography. *Carbohydr. Res.* 5, 340–345. doi:10.1016/S0008-6215(00)80510-8

Alonso-Simón, A., Kristensen, J. B., Øbro, J., Felby, C., Willats, W. G. T., and Jørgensen, H. (2010). High-throughput microarray profiling of cell wall polymers during hydrothermal pre-treatment of wheat straw. *Biotechnol. Bioeng.* 105, 509–514. doi:10.1002/bit.22546

Avci, U., Pattathil, S., and Hahn, M. G. (2012). Immunological approaches to plant cell wall and biomass characterization: immunolocalization of glycan epitopes. *Methods Mol. Biol.*, 908, 73–82. doi:10.1007/978-1-61779-956-3_7

Biswal, A. K., Hao, Z., Pattathil, S., Yang, X., Winkeler, K., Collins, C., et al. (2015). Downregulation of *GAUT12* in *Populus deltoides* by RNA silencing results in reduced recalcitrance, increased growth and reduced xylan and pectin in a woody biofuel feedstock. *Biotechnol. Biofuels* 8, 41. doi:10.1186/s13068-015-0218-y

Blake, A. W., McCartney, L., Flint, J. E., Bolam, D. N., Boraston, A. B., Gilbert, H. J., et al. (2006). Understanding the biological rationale for the diversity of cellulose-directed carbohydrate-binding modules in prokaryotic enzymes. *J. Biol. Chem.* 281, 29321–29329. doi:10.1074/jbc.M605903200

Blixt, O., Head, S., Mondala, T., Scanlan, C., Huflejt, M. E., Alvarez, R., et al. (2004). Printed covalent glycan array for ligand profiling of diverse glycan binding proteins. *Proc. Natl. Acad. Sci. U.S.A.* 101, 17033–17038. doi:10.1073/pnas.0407902101

Bolam, D. N., Ciruela, A., McQueen-Mason, S., Simpson, P., Williamson, M. P., Rixon, J. E., et al. (1998). *Pseudomonas* cellulose-binding domains mediate their effects by increasing enzyme substrate proximity. *Biochem. J.* 331, 775–781. doi:10.1042/bj3310775

Bolam, D. N., Xie, H. F., Pell, G., Hogg, D., Galbraith, G., Henrissat, B., et al. (2004). X4 modules represent a new family of carbohydrate-binding modules that display novel properties. *J. Biol. Chem.* 279, 22953–22963. doi:10.1074/jbc.M313317200

Bolam, D. N., Xie, H. F., White, P., Simpson, P. J., Hancock, S. M., Williamson, M. P., et al. (2001). Evidence for synergy between family 2b carbohydrate binding

modules in *Cellulomonas fimi* xylanase 11A. *Biochemistry* 40, 2468–2477. doi:10.1021/bi0025641

Boraston, A. B., Bolam, D. N., Gilbert, H. J., and Davies, G. J. (2004). Carbohydrate-binding modules: fine-tuning polysaccharide recognition. *Biochem. J.* 382, 769–781. doi:10.1042/BJ20040892

Boraston, A. B., Chiu, P., Warren, R. A. J., and Kilburn, D. G. (2000). Specificity and affinity of substrate binding by a family 17 carbohydrate-binding module from *Clostridium cellulovorans* cellulase 5A. *Biochemistry* 39, 11129–11136. doi:10.1021/bi0007728

Boraston, A. B., Creagh, A. L., Alam, M. M., Kormos, J. M., Tomme, P., Haynes, C. A., et al. (2001). Binding specificity and thermodynamics of a family 9 carbohydrate-binding module from *Thermotoga maritima* xylanase 10A. *Biochemistry* 40, 6240–6247. doi:10.1021/bi0101695

Boraston, A. B., Ghaffari, M., Warren, R. A. J., and Kilburn, D. G. (2002). Identification and glucan-binding properties of a new carbohydrate-binding module family. *Biochem. J.* 361, 35–40. doi:10.1042/bj3610035

Brown, D., Wightman, R., Zhang, Z. N., Gomez, L. D., Atanassov, I., Bukowski, J. P., et al. (2011). *Arabidopsis* genes *IRREGULAR XYLEM* (*IRX15*) and *IRX15L* encode DUF579-containing proteins that are essential for normal xylan deposition in the secondary cell wall. *Plant J.* 66, 401–413. doi:10.1111/j.1365-313X.2011.04501.x

Brown, D. M., Zeef, L. A. H., Ellis, J., Goodacre, R., and Turner, S. R. (2005). Identification of novel genes in *Arabidopsis* involved in secondary cell wall formation using expression profiling and reverse genetics. *Plant Cell* 17, 2281–2295. doi:10.1105/tpc.105.031542

Burton, R. A., Gidley, M. J., and Fincher, G. B. (2010). Heterogeneity in the chemistry, structure and function of plant cell walls. *Nat. Chem. Biol.* 6, 724–732. doi:10.1038/nchembio.439

Carpita, N. C. (1996). Structure and biogenesis of the cell walls of grasses. *Annu. Rev. Plant Physiol. Plant Mol. Biol.* 47, 445–476. doi:10.1146/annurev.arplant.47.1.445

Carpita, N. C., and Gibeaut, D. M. (1993). Structural models of primary cell walls in flowering plants: consistency of molecular structure with the physical properties of the walls during growth. *Plant J.* 3, 1–30. doi:10.1111/j.1365-313X.1993.tb00007.x

Charnock, S. J., Bolam, D. N., Turkenburg, J. P., Gilbert, H. J., Ferreira, L. M. A., Davies, G. J., et al. (2000). The X6 "thermostabilizing" domains of xylanases are carbohydrate-binding modules: structure and biochemistry of the *Clostridium thermocellum* X6b domain. *Biochemistry* 39, 5013–5021. doi:10.1021/bi992821q

Chen, F., and Dixon, R. A. (2007). Lignin modification improves fermentable sugar yields for biofuel production. *Nat. Biotechnol.* 25, 759–761. doi:10.1038/nbt1316

Chen, X., Vega-Sánchez, M. E., Verhertbruggen, Y., Chiniquy, D., Canlas, P. E., Fagerström, A., et al. (2013). Inactivation of *OsIRX10* leads to decreased xylan content in rice culm cell walls and improved biomass saccharification. *Mol. Plant.* 6, 570–573. doi:10.1093/mp/sss135

Chundawat, S. P. S., Beckham, G. T., Himmel, M. E., and Dale, B. E. (2011). Deconstruction of lignocellulosic biomass to fuels and chemicals. *Annu. Rev. Chem. Biomol. Eng.* 2, 121–145. doi:10.1146/annurev-chembioeng-061010-114205

Chung, D., Pattathil, S., Biswal, A. K., Hahn, M. G., Mohnen, D., and Westpheling, J. (2014). Deletion of a gene cluster encoding pectin degrading enzymes in *Caldicellulosiruptor bescii* reveals an important role for pectin in plant biomass recalcitrance. *Biotechnol. Biofuels* 7, 147. doi:10.1186/s13068-014-0147-1

Cid, M., Pedersen, H. L., Kaneko, S., Coutinho, P. M., Henrissat, B., Willats, W. G. T., et al. (2010). Recognition of the helical structure of b-1,4-galactan by a new family of carbohydrate-binding modules. *J. Biol. Chem.* 285, 35999–36009. doi:10.1074/jbc.M110.166330

Clausen, M. H., Ralet, M. C., Willats, W. G. T., McCartney, L., Marcus, S. E., Thibault, J.-F., et al. (2004). A monoclonal antibody to feruloylated-(1,4)-β-D-galactan. *Planta* 219, 1036–1041. doi:10.1007/s00425-004-1309-3

Clausen, M. H., Willats, W. G. T., and Knox, J. P. (2003). Synthetic methyl hexagalacturonate hapten inhibitors of anti-homogalacturonan monoclonal antibodies LM7, JIM5 and JIM7. *Carbohydr. Res.* 338, 1797–1800. doi:10.1016/S0008-6215(03)00272-6

Cosgrove, D. J. (1997). Assembly and enlargement of the primary cell wall in plants. *Annu. Rev. Cell Dev. Biol.* 13, 171–201. doi:10.1146/annurev.cellbio.13.1.171

Czjzek, M., Bolam, D. N., Mosbah, A., Allouch, J., Fontes, C. M. G. A., Ferreira, L. M. A., et al. (2001). The location of the ligand-binding site of carbohydrate-binding modules that have evolved from a common sequence is not conserved. *J. Biol. Chem.* 276, 48580–48587. doi:10.1074/jbc.M109142200

de Souza, A. P., Kamei, C. L. A., Torres, A. F., Pattathil, S., Hahn, M. G., Trindade, L. M., et al. (2015). How cell wall complexity influences saccharification efficiency in *Miscanthus sinensis*. *J. Exp. Bot.* 66, 4351–4365. doi:10.1093/jxb/erv183

de Souza, A. P., Leite, D. C. C., Pattathil, S., Hahn, M. G., and Buckeridge, M. S. (2013). Composition and structure of sugarcane cell wall polysaccharides: implications for second-generation bioethanol production. *Bioenerg. Res.* 6, 564–579. doi:10.1007/s12155-012-9268-1

DeMartini, J. D., Pattathil, S., Avci, U., Szekalski, K., Mazumder, K., Hahn, M. G., et al. (2011). Application of monoclonal antibodies to investigate plant cell wall deconstruction for biofuels production. *Energy Environ. Sci.* 4, 4332–4339. doi:10.1039/c1ee02112e

DeMartini, J. D., Pattathil, S., Miller, J. S., Li, H., Hahn, M. G., and Wyman, C. E. (2013). Investigating plant cell wall components that affect biomass recalcitrance in poplar and switchgrass. *Energy Environ. Sci.* 6, 898–909. doi:10.1039/c3ee23801f

Donaldson, L. A. (2001). Lignification and lignin topochemistry – an ultrastructural view. *Phytochemistry* 57, 859–873. doi:10.1016/S0031-9422(01)00049-8

Donohoe, B. S., Decker, S. R., Tucker, M. P., Himmel, M. E., and Vinzant, T. B. (2008). Visualizing lignin coalescence and migration through maize cell walls following thermochemical pretreatment. *Biotechnol. Bioeng.* 101, 913–925. doi:10.1002/bit.21959

Duceppe, M.-O., Bertrand, A., Pattathil, S., Miller, J., Castonguay, Y., Hahn, M. G., et al. (2012). Assessment of genetic variability of cell wall degradability for the selection of alfalfa with improved saccharification efficiency. *Bioenerg. Res.* 5, 904–914. doi:10.1007/s12155-012-9204-4

Fangel, J. U., Ulvskov, P., Knox, J. P., Mikkelsen, M. D., Harholt, J., Popper, Z., et al. (2012). Cell wall evolution and diversity. *Front. Plant Sci.* 3:152. doi:10.3389/fpls.2012.00152

Ferraz, A., Costa, T. H. F., Siqueira, G., and Milagres, A. M. F. (2014). "Mapping of cell wall components in lignified biomass as a tool to understand recalcitrance," in *Biofuels in Brazil*, eds Da Silva S. S. and Chandel A. K. (Heidelberg: Springer International), 173–202.

Filonova, L., Kallas, A. M., Greffe, L., Johansson, G., Teeri, T. T., and Daniel, G. (2007). Analysis of the surfaces of wood tissues and pulp fibers using carbohydrate-binding modules specific for crystalline cellulose and mannan. *Biomacromolecules* 8, 91–97. doi:10.1021/bm060632z

Freshour, G., Clay, R. P., Fuller, M. S., Albersheim, P., Darvill, A. G., and Hahn, M. G. (1996). Developmental and tissue-specific structural alterations of the cell-wall polysaccharides of *Arabidopsis thaliana* roots. *Plant Physiol.* 110, 1413–1429.

Fu, C., Mielenz, J. R., Xiao, X., Ge, Y., Hamilton, C. Y., Rodriguez, M., et al. (2011). Genetic manipulation of lignin reduces recalcitrance and improves ethanol production from switchgrass. *Proc. Natl. Acad. Sci. U.S.A.* 108, 3803–3808. doi:10.1073/pnas.1100310108

Fukui, S., Feizi, T., Galustian, C., Lawson, A. M., and Chai, W. (2002). Oligosaccharide microarrays for high-throughput detection and specificity assignments of carbohydrate-protein interactions. *Nat. Biotechnol.* 20, 1011–1017. doi:10.1038/nbt735

Galbe, M., and Zacchi, G. (2007). "Pretreatment of lignocellulosic materials for efficient bioethanol production," in *Biofuels*, Vol. 108, ed. Olsson L. (Berlin: Springer Verlag), 41–65.

Gill, J., Rixon, J. E., Bolam, D. N., McQueen-Mason, S., Simpson, P. J., Williamson, M. P., et al. (1999). The type II and X cellulose-binding domains of *Pseudomonas* xylanase A potentiate catalytic activity against complex substrates by a common mechanism. *Biochem. J.* 342, 473–480. doi:10.1042/0264-6021:3420473

Goff, S. A., Ricke, D., Lan, T.-H., Presting, G., Wang, R., Dunn, M., et al. (2002). A draft sequence of the rice genome (*Oryza sativa* L. ssp *japonica*). *Science* 296, 92–100. doi:10.1126/science.1068275

Gomez, L. D., Steele-King, C. G., and McQueen-Mason, S. J. (2008). Sustainable liquid biofuels from biomass: the writing's on the walls. *New Phytol.* 178, 473–485. doi:10.1111/j.1469-8137.2008.02422.x

Gunnarsson, L. C., Zhou, Q., Montanier, C., Karlsson, E. N., Brumer, H., and Ohlin, M. (2006). Engineered xyloglucan specificity in a carbohydrate-binding module. *Glycobiology* 16, 1171–1180. doi:10.1093/glycob/cwl038

Handakumbura, P. P., and Hazen, S. P. (2012). Transcriptional regulation of grass secondary cell wall biosynthesis: playing catch-up with *Arabidopsis thaliana*. *Front. Plant Sci.* 3:74. doi:10.3389/fpls.2012.00074

Hervé, C., Rogowski, A., Blake, A. W., Marcus, S. E., Gilbert, H. J., and Knox, J. P. (2010). Carbohydrate-binding modules promote the enzymatic deconstruction of intact plant cell walls by targeting and proximity effects. *Proc. Natl. Acad. Sci. U.S.A.* 107, 15293–15298. doi:10.1073/pnas.1005732107

Himmel, M. E., Ding, S.-Y., Johnson, D. K., Adney, W. S., Nimlos, M. R., Brady, J. W., et al. (2007). Biomass recalcitrance: engineering plants and enzymes for biofuels production. *Science* 315, 804–807. doi:10.1126/science.1137016

Izquierdo, J. A., Pattathil, S., Guseva, A., Hahn, M. G., and Lynd, L. R. (2014). Comparative analysis of the ability of *Clostridium clariflavum* strains and *Clostridium thermocellum* to utilize hemicellulose and unpretreated plant material. *Biotechnol. Biofuels* 7, 136. doi:10.1186/s13068-014-0136-4

Joseleau, J.-P., Faix, O., Kuroda, K.-I., and Ruel, K. (2004). A polyclonal antibody directed against syringylpropane epitopes of native lignins. *C. R. Biol.* 327, 809–815. doi:10.1016/j.crvi.2004.06.003

Joseleau, J.-P., and Ruel, K. (1997). Study of lignification by noninvasive techniques in growing maize internodes – an investigation by Fourier transform infrared cross-polarization magic angle spinning ^{13}C-nuclear magnetic resonance spectroscopy and immunocytochemical transmission electron microscopy. *Plant Physiol.* 114, 1123–1133.

Joshi, C. P., Bhandari, S., Ranjan, P., Kalluri, U. C., Liang, X., Fujino, T., et al. (2004). Genomics of cellulose biosynthesis in poplars. *New Phytol.* 164, 53–61. doi:10.1111/j.1469-8137.2004.01155.x

Joshi, C. P., Thammannagowda, S., Fujino, T., Gou, J. Q., Avci, U., Haigler, C. H., et al. (2011). Perturbation of wood cellulose synthesis causes pleiotropic effects in transgenic aspen. *Mol. Plant.* 4, 331–345. doi:10.1093/mp/ssq081

Kataeva, I., Foston, M. B., Yang, S.-J., Pattathil, S., Biswal, A. K., Poole, I. I. F. L., et al. (2013). Carbohydrate and lignin are simultaneously solubilized from unpretreated switchgrass by microbial action at high temperature. *Energy Environ. Sci.* 6, 2186–2195. doi:10.1039/c3ee40932e

Kim, Y. S., and Koh, H. B. (1997). Immuno electron microscopic study on the origin of milled wood lignin. *Holzforschung* 51, 411–413.

Kiyoto, S., Yoshinaga, A., Tanaka, N., Wada, M., Kamitakahara, H., and Takabe, K. (2013). Immunolocalization of 8-5′ and 8-8′ linked structures of lignin in cell walls of *Chamaecyparis obtusa* using monoclonal antibodies. *Planta* 237, 705–715. doi:10.1007/s00425-012-1784-x

Knox, J. P. (1997). The use of antibodies to study the architecture and developmental regulation of plant cell walls. *Int. Rev. Cytol.* 171, 79–120. doi:10.1016/S0074-7696(08)62586-3

Knox, J. P. (2008). Revealing the structural and functional diversity of plant cell walls. *Curr. Opin. Plant Biol.* 11, 308–313. doi:10.1016/j.pbi.2008.03.001

Kubo, M., Udagawa, M., Nishikubo, N., Horiguchi, G., Yamaguchi, M., Ito, J., et al. (2005). Transcription switches for protoxylem and metaxylem vessel formation. *Genes Dev.* 19, 1855–1860. doi:10.1101/gad.1331305

Kukkola, E. M., Koutaniemi, S., Gustafsson, M., Karhunen, P., Ruel, K., Lundell, T. K., et al. (2003). Localization of dibenzodioxocin substructures in lignifying Norway spruce xylem by transmission electron microscopy-immunogold labeling. *Planta* 217, 229–237. doi:10.1007/s00425-003-0983-x

Kukkola, E. M., Koutaniemi, S., Pollanen, E., Gustafsson, M., Karhunen, P., Lundell, T. K., et al. (2004). The dibenzodioxocin lignin substructure is abundant in the inner part of the secondary wall in Norway spruce and silver birch xylem. *Planta* 218, 497–500. doi:10.1007/s00425-003-1107-3

Kulkarni, A. R., Pattathil, S., Hahn, M. G., York, W. S., and O'Neill, M. A. (2012). Comparison of arabinoxylan structure in bioenergy and model grasses. *Ind. Biotechnol.* 8, 222–229. doi:10.1089/ind.2012.0014

Lee, C., O'Neill, M. A., Tsumuraya, Y., Darvill, A. G., and Ye, Z.-H. (2007). The *irregular xylem9* mutant is deficient in xylan xylosyltransferase activity. *Plant Cell Physiol.* 48, 1624–1634. doi:10.1093/pcp/pcm135

Lee, C., Teng, Q., Zhong, R., and Ye, Z.-H. (2011a). Molecular dissection of xylan biosynthesis during wood formation in poplar. *Mol. Plant.* 4, 730–747. doi:10.1093/mp/ssr035

Lee, K. J. D., Marcus, S. E., and Knox, J. P. (2011b). Cell wall biology: perspectives from cell wall imaging. *Mol. Plant.* 4, 212–219. doi:10.1093/mp/ssq075

Lee, K. J., and Knox, J. P. (2014). Resin embedding, sectioning, and immunocyto-chemical analyses of plant cell walls in hard tissues. *Methods Mol. Biol.* 1080, 41–52. doi:10.1007/978-1-62703-643-6_3

Lee, S. J., Warnick, T. A., Pattathil, S., Alvelo-Maurosa, J. G., Serapiglia, M. J., McCormick, H., et al. (2012). Biological conversion assay using *Clostridium phytofermentans* to estimate plant feedstock quality. *Biotechnol. Biofuels* 5, 5. doi:10.1186/1754-6834-5-5

Lerouxel, O., Choo, T. S., Séveno, M., Usadel, B., Faye, L., Lerouge, P., et al. (2002). Rapid structural phenotyping of plant cell wall mutants by enzymatic oligosaccharide fingerprinting. *Plant Physiol.* 130, 1754–1763. doi:10.1104/pp.011965

Li, M., Pattathil, S., Hahn, M. G., and Hodge, D. B. (2014). Identification of features associated with plant cell wall recalcitrance to pretreatment by alkaline hydrogen peroxide in diverse bioenergy feedstocks using glycome profiling. *RSC Adv.* 4, 17282–17292. doi:10.1039/C4RA00824C

Liang, Y., Pattathil, S., Xu, W.-L., Basu, D., Venetos, A., Faik, A., et al. (2013). Biochemical and physiological characterization of *fut4* and *fut6* mutants defective in arabinogalactan-protein fucosylation in *Arabidopsis. J. Exp. Bot.* 64, 5537–5551. doi:10.1093/jxb/ert321

Liepman, A. H., Wightman, R., Geshi, N., Turner, S. R., and Scheller, H. V. (2010). *Arabidopsis* – a powerful model system for plant cell wall research. *Plant J.* 61, 1107–1121. doi:10.1111/j.1365-313X.2010.04161.x

Loqué, D., Scheller, H. V., and Pauly, M. (2015). Engineering of plant cell walls for enhanced biofuel production. *Curr. Opin. Plant Biol.* 25, 151–161. doi:10.1016/j.pbi.2015.05.018

Marcus, S. E., Blake, A. W., Benians, T. A. S., Lee, K. J. D., Poyser, C., Donaldson, L., et al. (2010). Restricted access of proteins to mannan polysaccharides in intact plant cell walls. *Plant J.* 64, 191–203. doi:10.1111/j.1365-313X.2010.04319.x

McCann, M. C., and Knox, J. P. (2011). "Plant cell wall biology: polysaccharides in architectural and developmental contexts," in *Plant Polysaccharides: Biosynthesis and Bioengineering*, Vol. 41, ed. Ulvskov P. (Chichester: Wiley-Blackwell), 343–366.

McCann, M. C., and Roberts, K. (1991). "Architecture of the primary cell wall," in *Cytoskeletal Basis of Plant Growth and Form*, ed. Lloyd C. W. (London: Academic Press Ltd.), 109–129.

McCarthy, R. L., Zhong, R., and Ye, Z.-H. (2009). MYB83 is a direct target of SND1 and acts redundantly with MYB46 in the regulation of secondary cell wall biosynthesis in *Arabidopsis. Plant Cell Physiol.* 50, 1950–1964. doi:10.1093/pcp/pcp139

McCartney, L., Blake, A. W., Flint, J., Bolam, D. N., Boraston, A. B., Gilbert, H. J., et al. (2006). Differential recognition of plant cell walls by microbial xylan-specific carbohydrate-binding modules. *Proc. Natl. Acad. Sci. U.S.A.* 103, 4765–4770. doi:10.1073/pnas.0508887103

McCartney, L., Marcus, S. E., and Knox, J. P. (2005). Monoclonal antibodies to plant cell wall xylans and arabinoxylans. *J. Histochem. Cytochem.* 53, 543–546. doi:10.1369/jhc.4B6578.2005

McKendry, P. (2002). Energy production from biomass (part 1): overview of biomass. *Bioresour. Technol.* 83, 37–46. doi:10.1016/S0960-8524(01)00118-3

McNeil, M., Darvill, A., Fry, S. C., and Albersheim, P. (1984). Structure and function of the primary cell walls of plants. *Annu. Rev. Biochem.* 53, 625–663. doi:10.1146/annurev.bi.53.070184.003205

Meikle, P. J., Bonig, I., Hoogenraad, N. J., Clarke, A. E., and Stone, B. A. (1991). The location of (1,3)-β-glucans in the walls of pollen tubes of *Nicotiana alata* using a (1,3)-β-glucan-specific monoclonal antibody. *Planta* 185, 1–8. doi:10.1007/BF00194507

Meikle, P. J., Hoogenraad, N. J., Bonig, I., Clarke, A. E., and Stone, B. A. (1994). A (1,3;1,4)-β-glucan-specific monoclonal antibody and its use in the quantitation and immunocytochemical location of (1,3;1,4)-β-glucans. *Plant J.* 5, 1–9. doi:10.1046/j.1365-313X.1994.5010001.x

Mohnen, D., Bar-Peled, M., and Somerville, C. (2008). "Cell wall polysaccharide synthesis," in *Biomass Recalcitrance – Deconstructing the Plant Cell Wall for Bioenergy*, ed. Himmel D. E. (Oxford: Blackwell Publishing Ltd.), 94–187.

Moller, I., Marcus, S. E., Haeger, A., Verhertbruggen, Y., Verhoef, R., Schols, H., et al. (2008). High-throughput screening of monoclonal antibodies against plant cell wall glycans by hierarchical clustering of their carbohydrate microarray binding profiles. *Glycoconj. J.* 25, 37–48. doi:10.1007/s10719-007-9059-7

Moller, I., Sorensen, I., Bernal, A. J., Blaukopf, C., Lee, K., Obro, J., et al. (2007). High-throughput mapping of cell-wall polymers within and between plants using novel microarrays. *Plant J.* 50, 1118–1128. doi:10.1111/j.1365-313X.2007.03114.x

Oikawa, A., Joshi, H. J., Rennie, E. A., Ebert, B., Manisseri, C., Heazlewood, J. L., et al. (2010). An integrative approach to the identification of *Arabidopsis* and rice genes involved in xylan and secondary wall development. *PLoS ONE* 5:e15481. doi:10.1371/journal.pone.0015481

Pattathil, S., Avci, U., Baldwin, D., Swennes, A. G., McGill, J. A., Popper, Z., et al. (2010). A comprehensive toolkit of plant cell wall glycan-directed monoclonal antibodies. *Plant Physiol.* 153, 514–525. doi:10.1104/pp.109.151985

Pattathil, S., Avci, U., Miller, J. S., and Hahn, M. G. (2012a). Immunological approaches to plant cell wall and biomass characterization: glycome profiling. *Methods Mol. Biol.*, 908, 61–72. doi:10.1007/978-1-61779-956-3_6

Pattathil, S., Saffold, T., Gallego-Giraldo, L., O'Neill, M., York, W. S., Dixon, R. A., et al. (2012b). Changes in cell wall carbohydrate extractability are correlated with reduced recalcitrance of HCT downregulated alfalfa biomass. *Ind. Biotechnol.* 8, 217–221. doi:10.1089/ind.2012.0013

Pattathil, S., Hahn, M. G., Dale, B. E., and Chundawat, S. P. S. (2015). Insights into plant cell wall structure, architecture, and integrity using glycome profiling of native and AFEX™-pre-treated biomass. *J. Exp. Bot.* 66, 4279–4294. doi:10.1093/jxb/erv107

Patten, A. M., Jourdes, M., Cardenas, C. L., Laskar, D. D., Nakazawa, Y., Chung, B.-Y., et al. (2010). Probing native lignin macromolecular configuration in *Arabidopsis thaliana* in specific cell wall types: further insights into limited substrate degeneracy and assembly of the lignins of *ref8, fah* 1-2 and C4H:F5H lines. *Mol. Biosyst.* 6, 499–515. doi:10.1039/b819206e

Pauly, M., and Keegstra, K. (2008). Cell-wall carbohydrates and their modification as a resource for biofuels. *Plant J.* 54, 559–568. doi:10.1111/j.1365-313X.2008.03463.x

Pedersen, H. L., Fangel, J. U., McCleary, B., Ruzanski, C., Rydahl, M. G., Ralet, M.-C., et al. (2012). Versatile high resolution oligosaccharide microarrays for plant glycobiology and cell wall research. *J. Biol. Chem.* 287, 39429–39438. doi:10.1074/jbc.M112.396598

Peña, M. J., Darvill, A. G., Eberhard, S., York, W. S., and O'Neill, M. A. (2008). Moss and liverwort xyloglucans contain galacturonic acid and are structurally distinct from the xyloglucans synthesized by hornworts and vascular plants. *Glycobiology* 18, 891–904. doi:10.1093/glycob/cwn078

Peña, M. J., Zhong, R., Zhou, G.-K., Richardson, E. A., O'Neill, M. A., Darvill, A. G., et al. (2007). *Arabidopsis irregular xylem8* and *irregular xylem9*: implications for the complexity of glucuronoxylan biosynthesis. *Plant Cell* 19, 549–563. doi:10.1105/tpc.106.049320

Petersen, P. D., Lau, J., Ebert, B., Yang, F., Verhertbruggen, Y., Kim, J. S., et al. (2012). Engineering of plants with improved properties as biofuels feedstocks by vessel-specific complementation of xylan biosynthesis mutants. *Biotechnol. Biofuels* 5, 84. doi:10.1186/1754-6834-5-84

Popper, Z. A. (2008). Evolution and diversity of green plant cell walls. *Curr. Opin. Plant Biol.* 11, 286–292. doi:10.1016/j.pbi.2008.02.012

Popper, Z. A., Michel, G., Hervé, C., Domozych, D. S., Willats, W. G. T., Tuohy, M. G., et al. (2011). Evolution and diversity of plant cell walls: from algae to flowering plants. *Annu. Rev. Plant Biol.* 62, 567–588. doi:10.1146/annurev-arplant-042110-103809

Puhlmann, J., Bucheli, E., Swain, M. J., Dunning, N., Albersheim, P., Darvill, A. G., et al. (1994). Generation of monoclonal antibodies against plant cell wall polysaccharides. I. Characterization of a monoclonal antibody to a terminal a-(1,2)-linked fucosyl-containing epitope. *Plant Physiol.* 104, 699–710. doi:10.1104/pp.104.2.699

Ralet, M.-C., Tranquet, O., Poulain, D., Moïse, A., and Guillon, F. (2010). Monoclonal antibodies to rhamnogalacturonan I backbone. *Planta* 231, 1373–1383. doi:10.1007/s00425-010-1116-y

Reinikainen, T., Ruohonen, L., Nevanen, T., Laaksonen, L., Kraulis, P., Jones, T. A., et al. (1992). Investigation of the function of mutated cellulose-binding domains of *Trichoderma reesei* cellobiohydrolase-I. *Proteins* 14, 475–482. doi:10.1002/prot.340140408

Ruel, K., Faix, O., and Joseleau, J.-P. (1994). New immunogold probes for studying the distribution of the different lignin types during plant cell wall biogenesis. *J. Trace Microprobe Tech.* 12, 247–265.

Ruel, K., Montiel, M.-D., Goujon, T., Jouanin, L., Burlat, V., and Joseleau, J.-P. (2002). Interrelation between lignin deposition and polysaccharide matrices during the assembly of plant cell walls. *Plant Biol.* 4, 2–8. doi:10.1055/s-2002-20429

Scheller, H. V., Singh, S., Blanch, H., and Keasling, J. D. (2010). The Joint BioEnergy Institute (JBEI): developing new biofuels by overcoming biomass recalcitrance. *Bioenerg. Res.* 3, 105–107. doi:10.1007/s12155-010-9086-2

Schmidt, D., Schuhmacher, F., Geissner, A., Seeberger, P. H., and Pfrengle, F. (2015). Automated synthesis of arabinoxylan-oligosaccharides enables characterization of antibodies that recognize plant cell wall glycans. *Chemistry* 21, 5709–5713. doi:10.1002/chem.201500065

Schnable, P. S., Ware, D., Fulton, R. S., Stein, J. C., Wei, F., Pasternak, S., et al. (2009). The B73 maize genome: complexity, diversity, and dynamics. *Science* 326, 1112–1115. doi:10.1126/science.1178534

Selvendran, R. R., Davies, A. M. C., and Tidder, E. (1975). Cell wall glycoproteins and polysaccharides of mature runner beans. *Phytochemistry* 14, 2169–2174. doi:10.1016/S0031-9422(00)91094-X

Selvendran, R. R., and O'Neill, M. A. (1987). "Isolation and analysis of cell walls from plant material," in *Methods of Biochemical Analysis*, 32 Edn, ed. Glick D. (New York, NY: John Wiley & Sons, Inc), 25–153.

Shen, H., Poovaiah, C. R., Ziebell, A., Tschaplinski, T. J., Pattathil, S., Gjersing, E., et al. (2013). Enhanced characteristics of genetically modified switchgrass (*Panicum virgatum* L.) for high biofuel production. *Biotechnol. Biofuels* 6, 71. doi:10.1186/1754-6834-6-71

Shoseyov, O., Shani, Z., and Levy, I. (2006). Carbohydrate binding modules: biochemical properties and novel applications. *Microbiol. Mol. Biol. Rev.* 70, 283–295. doi:10.1128/MMBR.00028-05

Sluiter, J. B., Ruiz, R. O., Scarlata, C. J., Sluiter, A. D., and Templeton, D. W. (2010). Compositional analysis of lignocellulosic feedstocks. 1. Review and description of methods. *J. Agric. Food Chem.* 58, 9043–9053. doi:10.1021/jf1008023

Socha, A. M., Parthasarathi, R., Shi, J., Pattathil, S., Whyte, D., Bergeron, M., et al. (2014). Efficient biomass pretreatment using ionic liquids derived from lignin and hemicellulose. *Proc. Natl. Acad. Sci. U.S.A.* 111, E3587–E3595. doi:10.1073/pnas.1405685111

Somerville, C., Bauer, S., Brininstool, G., Facette, M., Hamann, T., Milne, J., et al. (2004). Toward a systems approach to understanding plant cell walls. *Science* 306, 2206–2211. doi:10.1126/science.1102765

Somerville, C., Youngs, H., Taylor, C., Davis, S. C., and Long, S. P. (2010). Feedstocks for lignocellulosic biofuels. *Science* 329, 790–792. doi:10.1126/science.1189268

Sørensen, I., Pettolino, F. A., Bacic, A., Ralph, J., Lu, F., O'Neill, M. A., et al. (2011). The charophycean green algae provide insights into the early origins of plant cell walls. *Plant J.* 68, 201–211. doi:10.1111/j.1365-313X.2011.04686.x

Steffan, W., Kovác, P., Albersheim, P., Darvill, A. G., and Hahn, M. G. (1995). Characterization of a monoclonal antibody that recognizes an arabinosylated (1,6)-β-D-galactan epitope in plant complex carbohydrates. *Carbohydr. Res.* 275, 295–307. doi:10.1016/0008-6215(95)00174-R

Studer, M. H., DeMartini, J. D., Davis, M. F., Sykes, R. W., Davison, B., Keller, M., et al. (2011). Lignin content in natural *Populus* variants affects sugar release. *Proc. Natl. Acad. Sci. U.S.A.* 108, 6300–6305. doi:10.1073/pnas.1009252108

Sweet, D. P., Albersheim, P., and Shapiro, R. H. (1975a). Partially ethylated alditol acetates as derivatives for elucidation of the glycosyl linkage-composition of polysaccharides. *Carbohydr. Res.* 40, 199–216. doi:10.1016/S0008-6215(00)82603-8

Sweet, D. P., Shapiro, R. H., and Albersheim, P. (1975b). Quantitative analysis by various G.L.C. response-factor theories for partially methylated and partially ethylated alditol acetates. *Carbohydr. Res.* 40, 217–225. doi:10.1016/S0008-6215(00)82604-X

Sweet, D. P., Shapiro, R. H., and Albersheim, P. (1974). The mass spectral fragmentation of partially ethylated alditol acetates, a derivative used in determining the glycosyl linkage composition of polysaccharides. *Biomed. Mass Spectrom.* 1, 263–268. doi:10.1002/bms.1200010410

Szabó, L., Jamal, S., Xie, H., Charnock, S. J., Bolam, D. N., Gilbert, H. J., et al. (2001). Structure of a family 15 carbohydrate-binding module in complex with xylopentaose – evidence that xylan binds in an approximate 3-fold helical conformation. *J. Biol. Chem.* 276, 49061–49065. doi:10.1074/jbc.M109558200

Tan, L., Eberhard, S., Pattathil, S., Warder, C., Glushka, J., Yuan, C., et al. (2013). An *Arabidopsis* cell wall proteoglycan consists of pectin and arabinoxylan covalently linked to an arabinogalactan protein. *Plant Cell* 25, 270–287. doi:10.1105/tpc.112.107334

Taylor, N. G., Gardiner, J. C., Whiteman, R., and Turner, S. R. (2004). Cellulose synthesis in the *Arabidopsis* secondary cell wall. *Cellulose* 11, 329–338. doi:10.1023/B:CELL.0000046405.11326.a8

The Arabidopsis Genome Initiative. (2000). Analysis of the genome sequence of the flowering plant *Arabidopsis thaliana*. *Nature* 408, 796–815. doi:10.1038/35048692

The International Brachypodium Initiative. (2010). Genome sequencing and analysis of the model grass *Brachypodium distachyon*. *Nature* 463, 763–768. doi:10.1038/nature08747

Tomme, P., Creagh, A. L., Kilburn, D. G., and Haynes, C. A. (1996). Interaction of polysaccharides with the N-terminal cellulose-binding domain of *Cellulomonas fimi* CenC. 1. Binding specificity and calorimetric analysis. *Biochemistry* 35, 13885–13894. doi:10.1021/bi961185i

Tormo, J., Lamed, R., Chirino, A. J., Morag, E., Bayer, E. A., Shoham, Y., et al. (1996). Crystal structure of a bacterial family-III cellulose-binding domain: a general mechanism for attachment to cellulose. *EMBO J.* 15, 5739–5751.

Trajano, H. L., Pattathil, S., Tomkins, B. A., Tschaplinski, T. J., Hahn, M. G., Van Berkel, G. J., et al. (2015). Xylan hydrolysis in *Populus trichocarpa* x *P. deltoides* and model substrates during hydrothermal pretreatment. *Bioresour. Technol.* 179, 202–210. doi:10.1016/j.biortech.2014.11.090

Tuskan, G. A., DiFazio, S., Jansson, S., Bohlmann, J., Grigoriev, I., Hellsten, U., et al. (2006). The genome of black cottonwood, *Populus trichocarpa* (Torr. & Gray). *Science* 313, 1596–1604. doi:10.1126/science.1128691

Urbanowicz, B. R., Pena, M. J., Ratnaparkhe, S., Avci, U., Backe, J., Steet, H. F., et al. (2012). 4-O-methylation of glucuronic acid in *Arabidopsis* glucuronoxylan is catalyzed by a domain of unknown function family 579 protein. *Proc. Natl. Acad. Sci. U.S.A.* 109, 14253–14258. doi:10.1073/pnas.1208097109

Valdivia, E. R., Herrera, M. T., Gianzo, C., Fidalgo, J., Revilla, G., Zarra, I., et al. (2013). Regulation of secondary wall synthesis and cell death by NAC transcription factors in the monocot *Brachypodium distachyon*. *J. Exp. Bot.* 64, 1333–1343. doi:10.1093/jxb/ers394

Verhertbruggen, Y., Marcus, S. E., Haeger, A., Ordaz-Ortiz, J. J., and Knox, J. P. (2009). An extended set of monoclonal antibodies to pectic homogalacturonan. *Carbohydr. Res.* 344, 1858–1862. doi:10.1016/j.carres.2008.11.010

Vogel, J. (2008). Unique aspects of the grass cell wall. *Curr. Opin. Plant Biol.* 11, 301–307. doi:10.1016/j.pbi.2008.03.002

von Schantz, L., Gullfot, F., Scheer, S., Filonova, L., Gunnarsson, L. C., Flint, J. E., et al. (2009). Affinity maturation generates greatly improved xyloglucan-specific carbohydrate binding modules. *BMC Biotechnol.* 9:92. doi:10.1186/1472-6750-9-92

von Schantz, L., Håkansson, M., Logan, D. T., Walse, B., Österlin, J., Nordberg-Karlsson, E., et al. (2012). Structural basis for carbohydrate-binding specificity – a comparative assessment of two engineered carbohydrate-binding modules. *Glycobiology* 22, 948–961. doi:10.1093/glycob/cws063

Voxeur, A., Wang, Y., and Sibout, R. (2015). Lignification: different mechanisms for a versatile polymer. *Curr. Opin. Plant Biol.* 23, 83–90. doi:10.1016/j.pbi.2014.11.006

Wallace, I., and Anderson, C. T. (2012). Small molecule probes for plant cell wall polysaccharide imaging. *Front. Plant Sci.* 3:89. doi:10.3389/fpls.2012.00089

Wang, D. N., Liu, S., Trummer, B. J., Deng, C., and Wang, A. L. (2002). Carbohydrate microarrays for the recognition of cross-reactive molecular markers of microbes and host cells. *Nat. Biotechnol.* 20, 275–281. doi:10.1038/nbt0302-275

Wang, H., Avci, U., Nakashima, J., Hahn, M. G., Chen, F., and Dixon, R. A. (2010). Mutation of WRKY transcription factors initiates pith secondary wall formation and increases stem biomass in dicotyledonous plants. *Proc. Natl. Acad. Sci. U.S.A.* 107, 22338–22343. doi:10.1073/pnas.1016436107

Wang, H.-Z., and Dixon, R. A. (2012). On-off switches for secondary cell wall biosynthesis. *Mol. Plant.* 5, 297–303. doi:10.1093/mp/ssr098

Weng, J.-K., and Chapple, C. (2010). The origin and evolution of lignin biosynthesis. *New Phytol.* 187, 273–285. doi:10.1111/j.1469-8137.2010.03327.x

Willats, W. G. T., Limberg, G., Buchholt, H. C., Van Alebeek, G.-J., Benen, J., Christensen, T. M. I. E., et al. (2000a). Analysis of pectic epitopes recognised by hybridoma and phage display monoclonal antibodies using defined oligosaccharides, polysaccharides, and enzymatic degradation. *Carbohydr. Res.* 327, 309–320. doi:10.1016/S0008-6215(00)00039-2

Willats, W. G. T., Steele-King, C. G., McCartney, L., Orfila, C., Marcus, S. E., and Knox, J. P. (2000b). Making and using antibody probes to study plant cell walls. *Plant Physiol. Biochem.* 38, 27–36. doi:10.1016/S0981-9428(00)00170-4

Willats, W. G. T., Rasmussen, S. E., Kristensen, T., Mikkelsen, J. D., and Knox, J. P. (2002). Sugar-coated microarrays: a novel slide surface for the high-throughput analysis of glycans. *Proteomics* 2, 1666–1671. doi:10.1002/1615-9861(200212)2:12<1666::AID-PROT1666>3.0.CO;2-E

Wu, A.-M., Hörnblad, E., Voxeur, A., Gerber, L., Rihouey, C., Lerouge, P., et al. (2010). Analysis of the *Arabidopsis IRX9/IRX9-L* and *IRX14/IRX14-L* pairs of glycosyltransferase genes reveals critical contributions to biosynthesis of the hemicellulose glucuronoxylan. *Plant Physiol.* 153, 542–554. doi:10.1104/pp.110.154971

Yang, F., Mitra, P., Zhang, L., Prak, L., Verhertbruggen, Y., Kim, J.-S., et al. (2013). Engineering secondary cell wall deposition in plants. *Plant Biotechnol. J.* 11, 325–335. doi:10.1111/pbi.12016

Ye, Z.-H., York, W. S., and Darvill, A. G. (2006). Important new players in secondary wall synthesis. *Trends Plant Sci.* 11, 162–164. doi:10.1016/j.tplants.2006.02.001

Young, N. D., Debellé, F., Oldroyd, G. E. D., Geurts, R., Cannon, S. B., Udvardi, M. K., et al. (2011). The Medicago genome provides insight into the evolution of rhizobial symbioses. *Nature* 480, 520–524. doi:10.1038/nature10625

Yu, J., Hu, S. N., Wang, J., Wong, G. K. S., Li, S. G., Liu, B., et al. (2002). A draft sequence of the rice genome (*Oryza sativa* L. ssp. *indica*). *Science* 296, 79–92. doi:10.1126/science.1068037

Yu, L., Shi, D., Li, J., Kong, Y., Yu, Y., Chai, G., et al. (2014). CELLULOSE SYNTHASE-LIKE A2, a glucomannan synthase, is involved in maintaining adherent mucilage structure in *Arabidopsis* seed. *Plant Physiol.* 164, 1842–1856. doi:10.1104/pp.114.236596

Zhang, X., Rogowski, A., Zhao, L., Hahn, M. G., Avci, U., Knox, J. P., et al. (2014). Understanding how the complex molecular architecture of mannan-degrading hydrolases contributes to plant cell wall degradation. *J. Biol. Chem.* 289, 2002–2012. doi:10.1074/jbc.M113.527770

Zhao, Q., Tobimatsu, Y., Zhou, R., Pattathil, S., Gallego-Giraldo, L., Fu, C., et al. (2013). Loss of function of cinnamyl alcohol dehydrogenase 1 leads to unconventional lignin and a temperature-sensitive growth defect in *Medicago truncatula. Proc. Natl. Acad. Sci. U.S.A.* 110, 13660–13665. doi:10.1073/pnas.1312234110

Zhong, R., Demura, T., and Ye, Z.-H. (2006). SND1, a NAC domain transcription factor, is a key regulator of secondary wall synthesis in fibers of *Arabidopsis. Plant Cell* 18, 3158–3170. doi:10.1105/tpc.106.047399

Zhong, R., Richardson, E. A., and Ye, Z.-H. (2007a). The MYB46 transcription factor is a direct target of SND1 and regulates secondary wall biosynthesis in *Arabidopsis. Plant Cell* 19, 2776–2792. doi:10.1105/tpc.107.053678

Zhong, R., Richardson, E. A., and Ye, Z.-H. (2007b). Two NAC domain transcription factors, SND1 and NST1, function redundantly in regulation of secondary wall synthesis in fibers of *Arabidopsis. Planta* 225, 1603–1611. doi:10.1007/s00425-007-0498-y

Zhu, X., Pattathil, S., Mazumder, K., Brehm, A., Hahn, M. G., Dinesh-Kumar, S. P., et al. (2010). Virus-induced gene silencing offers a functional genomics platform for studying plant cell wall formation. *Mol. Plant.* 3, 818–833. doi:10.1093/mp/ssq023

Conflict of Interest Statement: The authors declare that the research was conducted in the absence of any commercial or financial relationships that could be construed as a potential conflict of interest.

18

Evaluation of diverse microalgal species as potential biofuel feedstocks grown using municipal wastewater

Sage R. Hiibel [1,2], Mark S. Lemos [1], Brian P. Kelly [1] and John C. Cushman [1]*

[1] Department of Biochemistry and Molecular Biology, University of Nevada Reno, Reno, NV, USA, [2] Department of Civil and Environmental Engineering, University of Nevada Reno, Reno, NV, USA

Edited by:
Umakanta Jena,
Desert Research Institute, USA

Reviewed by:
Arumugam Muthu,
Council of Scientific and Industrial Research, India
Probir Das,
Qatar University, Qatar
Sivasubramanian Velusamy,
Phycospectrum Environmental Research Centre, India

*Correspondence:
John C. Cushman,
Department of Biochemistry and Molecular Biology, University of Nevada Reno, MS 330, 1664 N. Virginia Street, Reno, NV 89557-0330, USA
jcushman@unr.edu

Microalgae offer great potential as a third-generation biofuel feedstock, especially when grown on wastewater, as they have the dual application for wastewater treatment and as a biomass feedstock for biofuel production. The potential for growth on wastewater centrate was evaluated for forty microalgae strains from fresh (11), brackish (11), or saltwater (18) genera. Generally, freshwater strains were able to grow at high concentrations of centrate, with two strains, *Neochloris pseudostigmata* and *Neochloris conjuncta*, demonstrating growth at up to 40% v/v centrate. Fourteen of 18 salt water *Dunaliella* strains also demonstrated growth in centrate concentrations at or above 40% v/v. Lipid profiles of freshwater strains with high-centrate tolerance were determined using gas chromatography–mass spectrometry and compared against those obtained on cells grown on defined maintenance media. The major lipid compounds were found to be palmitic (16:0), oleic (18:1), and linoleic (18:2) acids for all freshwater strains grown on either centrate or their respective maintenance medium. These results demonstrate the highly concentrated wastewater can be used to grow microalgae, which limits the need to dilute wastewater prior to algal production. In addition, the algae produced generate lipids suitable for biodiesel or green diesel production.

Keywords: microalgae, fresh water, brackish water, salt water, biofuel, municipal wastewater, centrate

Introduction

The world's demand for petroleum fuels continues to grow even as supplies dwindle. In recent years, there has been a strong push to develop alternative energy sources to help supplement or potentially replace fossil fuels. Wind turbines and photovoltaic technologies offer renewable sources of electric power; however, liquid fuels for the transportation sector make up more than 70% of the energy consumed in US (Forsberg, 2009). Biofuels have great potential to help fill this need, and there has been significant research in this area. First- and second-generation liquid biofuels, such as corn ethanol and soy biodiesel, have received considerable interest and resources in the last 20 years, and are considered technically mature. Third-generation biofuels, such as lignocellulosics and microalgae, although not as technologically advanced, hold great promise as sustainable biofuels because they avoid the food *versus* fuel debate that has plagued corn and soy-based biofuels. As part of the Renewable Fuel Standard II, the US government has tapped the third-generation biofuels for 21×10^9 barrels/year of fuel by 2022, with 4×10^9 barrels/year of that coming from non-cellulosic and non-corn-based biodiesel fuels, such as those derived from microalgae (Schnepf and Yacobucci, 2010).

As with many third-generation fuels, microalgae have great potential as a biofuel feedstock source. They are among the most rapidly growing photosynthetic organisms on the planet (Chisti, 2007), and they can be cultured year-round in even cold climates if growth is coupled to a low-cost heat source (i.e., waste or low-grade geothermal). Large-scale production of algae biomass has been demonstrated, but several technical hurdles remain that must be addressed before microalgae can become a viable biofuel feedstock. Currently, biomass harvesting and oil extraction are key processing steps that are energy intensive and cost prohibitive. From the algal growth perspective, a sustainable water source and nutrient supply are paramount to economic microalgae production. Regardless of the strain used (salt or fresh water), a non-saline water supply will be required to make up for evaporative loss in an open-pond cultivation system. Lastly, off-setting part or all of the nutrient supply required to grow microalgae with non-petroleum-derived fertilizer such as municipal wastewater can help improve the carbon budget of microalgal-based biofuel production systems (Chisti, 2007; Wang et al., 2010; Bhatt et al., 2014; Dong et al., 2014; Mu et al., 2014).

Wastewater offers the possibility of serving as both a freshwater and a nutrient source, with most municipalities having a continuous supply. Ideally, the wastewater could be used in its raw state or with minimal treatment so as to reduce the costs of the wastewater treatment plant (WWTP) (Pittman et al., 2011; Bhatt et al., 2014; Mu et al., 2014). Centrate, the liquid fraction after anaerobic digestion, offers a feed stream high in nitrogen and phosphorous, which are two of the main nutrients required for microalgae and also two compounds that cause high removal costs for most municipal WWTPs (Wang et al., 2010; Li et al., 2011a; Bhatt et al., 2014; Dong et al., 2014; Mu et al., 2014). In addition to decreasing the nutrient load to the plant, diverting centrate to algae production also decreases the treatment volume, thereby decreasing treatment costs of other chemical constituents (Li et al., 2011a; Mu et al., 2014). Metal content of wastewaters or centrate are unlikely to inhibit microalgal growth (Wang et al., 2010; Dong et al., 2014).

In this work, centrate was used as a water and nutrient source for the growth of green algae as a biofuel feedstock. A total of 40 microalgae strains comprised of isolates from freshwater, brackish water, and saltwater strains were evaluated for tolerance to and growth in centrate. The growth characteristics, biomass and lipid yields, and lipid profiles were determined for the two most tolerant freshwater strains, *Neochloris conjuncta* and *Neochloris pseudostigmata*, to evaluate the potential of utilizing municipal wastewater centrate to grow microalgae as a biofuel feedstock.

Materials and Methods

Strains and Media

Eleven freshwater *Neochloris* strains, 11 brackish water *Nannochloropsis* strains, and 18 saltwater *Dunaliella* strains were evaluated for their ability to grow on municipal wastewater centrate (**Table 1**). The cultures were obtained from the Culture Collection of Algae at The University of Texas at Austin (UTEX),

the Culture Collection of Algae and Protozoa (CCAP), the Canadian Phycological Culture Centre [CPCC; formerly known as the University of Toronto Culture Collection of Algae and Cyanobacteria (UTCC)], or the Culture Collection of Algae at the University of Göttingen, Germany (SAG). All cultures were grown at room temperature under an 18/6 h light/dark cycle on their respective maintenance media. The freshwater *Neochloris* strains were maintained on Bold's modified Bristol medium (Bold, 1949): 2.94 mM $NaNO_3$, 0.17 mM $CaCl_2$ ($2H_2O$), 0.3 mM $MgSO_4$ ($7H_2O$), 0.43 mM K_2HPO_4, 1.29 mM KH_2PO_4, and 0.43 mM NaCl at pH = 7.7. Saltwater *Dunaliella* strains were maintained on a modified 2ASW (artificial seawater) medium (Gomord et al., 2010); 33.6 g/L of sea salts were used in place of the NaCl. The brackish water *Nannochloropsis* strains were maintained on slightly modified f/2 Medium of Guillard and Ryther (1962) in which the vitamin solution from the 2ASW medium (Gomord et al., 2010), which contained additional vitamin components, was used in place of that described for the original f/2 medium.

Centrate Characteristics

Centrate was collected from the Truckee Meadows Water Authority Reclamation Facility, the local municipal WWTP facility located in Sparks, Nevada. Centrate was filtered through Miracloth (typical pore size 22–25 μm, EMD Millipore), then autoclaved and stored at 4°C until use (<3 days). Average nitrogen, phosphorus, and potassium content of centrate from this facility have been previously reported as 1003.0 ± 174.6 mg/L NH_4-N, 244.5 ± 34.5 mg/L *ortho*-P, and 202 ± 54 mg/L K^+, respectively (Herrera, 2009). The raw centrate had an average pH = 7.8 and contained 1.284 mg/L TDS and 726.2 mg/L bicarbonate, and the major ions by concentration were sodium (77.0 ± 36.5 mg/L) and chloride (189.2 ± 132.4 mg/L). A detailed characterization of the centrate can be found in Herrera (2009).

Preliminary Screening

All strains were screened initially for their tolerance to grow on centrate using a 96-well titer plate format. Plates were loaded with 200 μL of medium and 25 μL of inoculum culture per well and then incubated at 22°C and a 18/6 h light/dark cycle under 100 μmol/m^2 s on an orbital shaking table rotating at 100 rpm. For the *Neochloris* and *Nannochloropsis* strains, the medium consisted of centrate diluted with nanopure water to 0, 2, 4, 6, 8, 10, 12, 14, 15, 20, 25, 30, 40, 50, 60, 70, 80, 90, and 100% v/v. The *Dunaliella* strains were screened with the same centrate concentrations, but dilutions were made with a 33.6 g/L sea-salt solution, as no growth was observed without the addition of sea salts (data not shown). The appropriate maintenance media and sterile water were used as positive and negative growth medium controls, respectively. Each culture was inoculated in duplicate on two plates (four total replicates) to allow for statistical testing power. The 96-well titer plates were monitored visually to monitor growth (Figure S1 in Supplementary Material).

Microalgae Growth

The two freshwater strains with the highest centrate tolerance, *N. pseudostigmata* (UTEX 1249) and *N. conjuncta* (CCAP 254/1),

TABLE 1 | The 40 strains evaluated, their original culture collection source, and the maximum centrate concentration at which growth was observed for each strain in preliminary plate screening tests.

	Species	Source	Max % v/v	Origin and source[a]
Freshwater *Neochloris*	*N. aquatica*	UTEX 138	14	Bloomington, IN (USA); aquarium
	N. minuta	UTEX 776	14	Santa Marta (Cuba); sugar cane field soil
	N. pyrenoidosa	UTEX 777	12	Jovallanos (Cuba); sugar cane field soil
	N. oleoabundans	UTEX 1185	12	Rub al Kali (Saudi Arabia); sand dune
	N. pseudostigmata	UTEX 1249	40	Enchanted Rock, TX (USA); soil
	N. cohaerens	UTEX 1707	0	Bastrop State Park, TX (USA); soil
	N. vigensis	UTEX 1981	20	Travis County, TX (USA); pond
	N. fusispora[b]	UTEX b778	10	Ranchuelo (Cuba); sugar cane field soil (Arce and Bold, 1958)
	N. terrestris[c]	UTEX b947	14	Daniel Town (Jamaica); corn field soil
	N. wimmeri[d]	CCAP 213/4	0	Czechoslovakia; freshwater
	N. conjuncta	CCAP 254/1	40	Travis County, TX (USA); freshwater
Brackish *Nannochloropsis*	*N.* sp.	CCAP 211/78	14	Unknown; marine
	N. sp.	CCAP 849/8	14	Qingdao (China); marine
	N. sp.	CCAP 849/9	14	Japan; marine
	N. gaditana	CCAP 849/5	14	Cadiz Bay (Spain); marine
	N. oceanica	CCAP 849/10	14	Western Norway; marine fish hatchery (Bligh and Dyer, 1959)
	N. oculata	CCAP 849/1	0	Skate Point, Isle of Cumbrae (Scotland); marine
	N. oculata	CCAP 849/7	0	Lake of Tunis (Tunisia); marine
	N. salina	CCAP 849/2	4	Skate Point, Isle of Cumbrae (Scotland); marine
	N. salina	CCAP 849/3	12	Skate Point, Isle of Cumbrae (Scotland); marine
	N. salina	CCAP 849/4	12	Skate Point, Isle of Cumbrae (Scotland); marine
	N. salina[e]	CCAP 849/6	12	Great South Bay, Long Island, NY (USA); marine
Saltwater *Dunaliella*	*D. bardawil*	UTEX LB 2538	50	Bardawil Lagoon, North Sinai (Israel); salt pond
	D. bioculata	UTEX LB 199	50	(Russia); salt lake
	D. salina	UTEX LB 200	50	(Russia); dirty salt lake
	D. tertiolecta	UTEX LB 999	50	Oslo Fjord (Norway); brackish
	D. salina	UTEX LB 1644	30	Baja, CA (USA); marine
	D. primolecta	UTEX LB 1000	40	Plymouth, Devon (England); marine
	D. peircei	UTEX LB 2192	50	Lake Marina, CA (USA); brackish
	D. salina	UTCC 197	0	Unknown
	D. tertiolecta	UTCC 420	50	Unknown; marine
	D. sp.	UTCC 457	40	Sarnia, Ontario (Canada); surface brine storage pond
	D. salina	CCAP 19/18	50	Hutt Lagoon (Western Australia); hypersaline brine
	D. tertiolecta	CCAP 19/6B	50	Oslo Fjord (Norway); brackish
	D. tertiolecta	CCAP 19/24	50	Unknown; possibly marine
	D. tertiolecta	CCAP 19/27	50	Halifax (Canada); unknown
	D. maritima	SAG 42.89	50	Former USSR; marine
	D. sp.	SAG 19-5	14	Wad al Neifur (Egypt); marine
	D. terricola	SAG 43.89	40	Former USSR; marine
	D. granulata	SAG 41.89	14	Former USSR; marine

Max % v/v refers to the maximum volume percent centrate at which growth was observed; the saltwater species were supplemented with 33.6 g/L sea salt. UTEX – Culture Collection of Algae at The University of Texas at Austin; CCAP – the Culture Collection of Algae and Protozoa (CCAP), UTCC – University of Toronto Culture Collection of Algae and Cyanobacteria [now referred to as the Canadian Phycologial Culture Centre (CPCC)]; SAG – the Culture Collection of Algae at the University of Göttingen, Germany.
[a]*Unless noted, origin and source information obtained from the originating culture collection.*
[b]*Currently regarded as a taxonomic synonym of Ettlia fusispora.*
[c]*Currently regarded as a taxonomic synonym of Ettlia terrestris.*
[d]*Currently listed with CCAP as Ettlia carotinosa.*
[e]*Currently listed with CCAP as Nannochloropsis gaditana.*

were grown in 2-L Erlenmeyer flasks to determine growth characteristics. Optical density (measured as absorbance at 600 nm) was used to monitor the cell density of the cultures and to determine the growth phase of the culture. Maximum growth rates were determined by plotting the A_{600} values versus time (Figures S2 and S3 in Supplementary Material) and taking the slope of growth rate plot during the exponential growth phase. Each medium was prepared by mixing the filtered and autoclaved centrate with nanopure water to 10, 25, or 40% v/v centrate, then the pH adjusted to 7.7 with 0.1N HCl or NaOH as required. Bold's modified Bristol medium (Bold, 1949) at pH = 7.7 was also prepared and used as

a baseline. Each batch of medium was divided into 1.5 L aliquots, transferred to the flasks, and then autoclaved. After sterilization, flasks were inoculated with 30 mL of mid-log phase culture and then stoppered with a bubbler apparatus (sterile two-holed stopper with two glass tubes with a small amount of cotton batting). Cultures were grown in a growth chamber (Conviron PGR15) with a 18/6 h light/dark cycle at 26/20°C and 140 μmol/m^2 s light supplied by a mixture of fluorescent and incandescent lamps. Using an aquarium pump, atmospheric air was pumped through a 2-L flask with sterile water, which served to both filter and humidify the air, and then was split into eight flasks using a manifold.

The air bubbles served as both a CO_2 source and agitation for the flasks.

Centrate Acclimation Testing

In a separate experiment, cells were grown to mid-log phase in Bold's modified Bristol medium (Sandnes et al., 2006) or in 10, 25, or 40% v/v centrate. To evaluate the effects of pre-acclimation, 30 mL of each of the liquid cultures were transferred to 2-L Erlenmeyer flasks with their respective fresh media. The cultures were grown and monitored as described above.

Harvesting, Drying, and Lipid Extraction

The algae were harvested via centrifugation ($6000 \times g$ for 5 min) in early stationary phase (3–5 days after log phase) and the algae paste was then stored at $-80°C$. The samples were lyophilized to remove all moisture and dry weights were determined. The dried samples were then re-suspended in water (1 mL/g dry algae) overnight at 4°C. A modified Bligh and Dyer (1959) extraction method was used to extract the lipids (and other chloroform-soluble components). Modifications include centrifugation-assisted phase separation ($2800 \times g$ for 10 min) and a second extraction on the non-lipid phase to ensure complete extraction. Chloroform was evaporated under a stream of nitrogen and dry weights of the lipids were determined.

FAME Preparation and GC/MS

Fatty acid profiles were determined via GC/MS analysis of fatty acid methyl esters (FAMEs). Briefly, the dried lipid extracts were esterified to FAMEs using the rapid BF_3-methanol esterification procedure (Metcalfe et al., 1966). After esterification, samples were dried completely under nitrogen then re-suspended in 1 mL carbon disulfide. FAME profiles were obtained by the Nevada Proteomics Center using GC/MS (Thermo Polaris Q) on an Agilent HP – INNOWAX column (P/N 19091N-136; $60 m \times 0.250 mm$, 0.25 μm film thickness) using helium as the carrier gas at 1.0 mL/min constant flow and a split ratio of 1:10 (split flow 10 mL/min) with 1 μL injection volumes. The GC was operated with an inlet temperature of 225°C and a column temperature starting at 180°C and ramping at 5°C/min to 240°C with a 30 min hold. The transfer line temperature between the GC and the MS was maintained at 250°C. The MS was operated in Full Scan mode with a mass range of 40–450 m/z at 70 eV and an ion source temperature of 200°C. Chromatograms and spectra were analyzed using XCalibur (Thermo Fisher Scientific, v1.3).

Results and Discussion

Preliminary Centrate Screening

Preliminary screening of the various microalgae strains revealed a wide range of tolerance to centrate (**Table 1**). Autoclaved centrate was used to avoid the possible complicating influence of undefined microbial flora on algal growth. Sterile centrate has been used in several other investigations (Wang et al., 2010; Li et al., 2011a; Zhu et al., 2013; Dong et al., 2014). Of the 11 freshwater *Neochloris* strains, two (*N. cohaerens* and *N. wimmeri*) were unable to grow in the presence of any centrate, while two strains (*N. conjuncta* and *N. pseudostigmata*) were tolerant of

concentrations up to 40% v/v. Growth was observed at maximum centrate concentrations between 10 and 20% v/v for the remaining seven strains. The brackish water *Nannochloropsis* strains had a lower centrate tolerance, with growth observed at a maximum of 14% v/v centrate for five strains and two strains unable to grow in the presence of any centrate. This low centrate tolerance might possibly be due to a lack of salts in the growth medium (centrate diluted with nanopure water) rather than as a function of the centrate concentration itself. As expected, saltwater *Dunaliella* strains were unable to grow when the centrate lacked salt and was diluted using nanopure water alone (data not shown). However, when sea salts were added to match the salt concentration of the 2ASW (33.6 g/L) to the diluting water, all but one of the saltwater strains (*D. salina* UTC 197) were able to grow at 14% v/v or greater centrate (**Table 1**). Of the 18 species or strains tested, 11 grew at 50% v/v centrate, and an additional 3 grew at 40% v/v centrate levels. Interestingly, of four different *D. salina* strains (e.g., UTEX LB 200, UTEX LB 1644, UTCC 197, and CCAP 19/18) evaluated, two UTEX strains (LB 200 and LB 1644) and CCAP 19/18 all demonstrated high-centrate tolerance, while UTCC 197 was the only *Dunaliella* strain unable to grow in centrate. Recent work suggests that UTEX 200 should not be designated as *D. salina* based on physiological and molecular markers (Ben-Amotz et al., 2009).

These tolerances to centrate were lower than those reported previously for several other algal species and strains (Wang et al., 2010; Li et al., 2011a; Zhu et al., 2013). However, such differences might be due to differences in centrate characteristics arising from the unique inputs and processing steps of a particular WWTP. Wang and Lan (2011) reported growth with a wild-type *Chlorella* sp. on a 100% centrate medium, which contained 72 mg/L total N, and would correspond to ~7% v/v centrate used in this work on a total N basis. Further evaluation with this same centrate source found a total of fourteen strains from genera *Chlorella*, *Haematococcus*, *Scenedesmus*, *Chlamydomonas*, and *Chloroccum* that were able to grow on the pure centrate (Li et al., 2011b). From the present work, the two freshwater strains (*N. conjuncta* and *N. pseudostigmata*) were considered for further evaluation because of their high-centrate tolerance and no need for additional salt supplementation.

Microalgal Growth

The *N. conjuncta* and *N. pseudostigmata* were grown in 2-L flasks with modified Bristol medium and various concentrations of centrate. No significant differences ($p > 0.05$) were observed in the maximum growth rates for either species (**Table 2**). For both species, no significant difference in the acclimation period was observed between modified Bristol medium and 10% v/v centrate. However, significant time lag periods in growth were observed for both species when the centrate concentration was increased to 25% v/v ($p < 0.01$), and additional lags were observed when the concentration was increased to 40% ($p < 0.01$ for *N. conjuncta*; $p < 0.05$ for *N. pseudostigmata*). At the highest centrate concentrations, a delay of more than 40 days was observed for both species. This increase in the lag time with increasing centrate concentration suggested the presence of a compound(s) that inhibit microalgal growth in the centrate, although the cells were able to

TABLE 2 | Maximum growth rate (μ_{max}) and observed lag time of *N. conjuncta* and *N. pseudostigmata* grown in 10, 25, and 40% v/v centrate and in Bold's modified Bristol medium (Bold, 1949).

	N. conjuncta		N. pseudostigmata	
	μ_{max} **(1/days)**	**Lag (days)**	μ_{max} **(1/days)**	**Lag (days)**
NON-ACCLIMATED CELLS				
Bristol medium	0.004 ± 0.001	2.3 ± 1.5	0.0072 ± 0.0004	0.8 ± 1.5
10% Centrate	0.006 ± 0.002	2.3 ± 1.5	0.007 ± 0.001	1.5 ± 1.7
25% Centrate	0.009 ± 0.003	16.0 ± 7.7	0.008 ± 0.002	17.8 ± 3.8
40% Centrate	0.007 ± 0.002	46.3 ± 3.4	0.009 ± 0.002	41.5 ± 20
PRE-ACCLIMATED CELLS				
Bristol medium	0.009 ± 0.002	0.0 ± 0.0	0.0099 ± 0.0006	0.0 ± 0.0
10% Centrate	0.011 ± 0.001	2.0 ± 0.0	0.012 ± 0.004	1.5 ± 1.2
25% Centrate	0.006 ± 0.002	7.0 ± 0.0	0.007 ± 0.001	10.0 ± 6.5
40% Centrate	0.015 ± 0.005	33.8 ± 9.8	0.016 ± 0.003	44.3 ± 4.0

Error ranges represent the SD of four biological replicates.

grow given enough time. The identity of the putative inhibitor(s) is unknown, but likely candidates might include ammonia or urea (Moazeni, 2013). Additional studies are needed to verify the presence of ammonia or urea in centrate and to test this possibility in future studies. However, neither metals nor ammonia ions were reported to be likely candidates for toxicity responses to centrate (Dong et al., 2014).

The long lag period in growth at elevated centrate concentrations would be problematic for large-scale cultivation, especially in open-pond systems, due to the prolonged opportunities for contamination of the cultures by other species (Chisti, 2007; Bhatt et al., 2014). To evaluate whether the cells could be acclimating to the inhibiting compound(s), and thereby decrease the observed lag time, cells exposed to centrate were used as inoculum for serial growth trials. For these tests, *N. conjuncta* and *N. pseudostigmata* grown in either Bold's modified Bristol medium (Bold, 1949) or in 10, 25, and 40% v/v centrate were inoculated into fresh batches of their respective medium and monitored (**Table 2**; Figures S2 and S3 in Supplementary Material). In general, the maximum growth rates of the pre-acclimated cells increased relative to the non-acclimated cells. As expected, no difference in the lag time was observed with the cells grown in Bristol medium and the 10% v/v centrate for either species. However, the average lag times for the pre-acclimated *N. conjuncta* decreased by 9 and 12.5 days when grown on 25 and 40% v/v centrate, respectively. The trend was not as strong with the pre-acclimated *N. pseudostigmata* cells, although the average lag time of those grown on 25% v/v centrate decreased 7.8 days (**Table 2**; Figures S2 and S3 in Supplementary Material). These results suggest that the inhibitory compound found in centrate is one that the freshwater *Neochloris* species may be able to acclimate to, and a further reduction in the lag time might be achieved with additional acclimation cycles.

Biomass Production and Productivity

The *N. conjuncta* and *N. pseudostigmata* cultures were harvested via centrifugation ~5 days after reaching stationary phase, as determined by A_{600} measurements. After harvesting, the cell pellets were frozen and then water removed via lyophilization. For both species, the dry biomass obtained (normalized to the culture volume harvested) increased with increasing centrate concentrations (**Figure 1A**). Biomass concentrations ranged over nearly

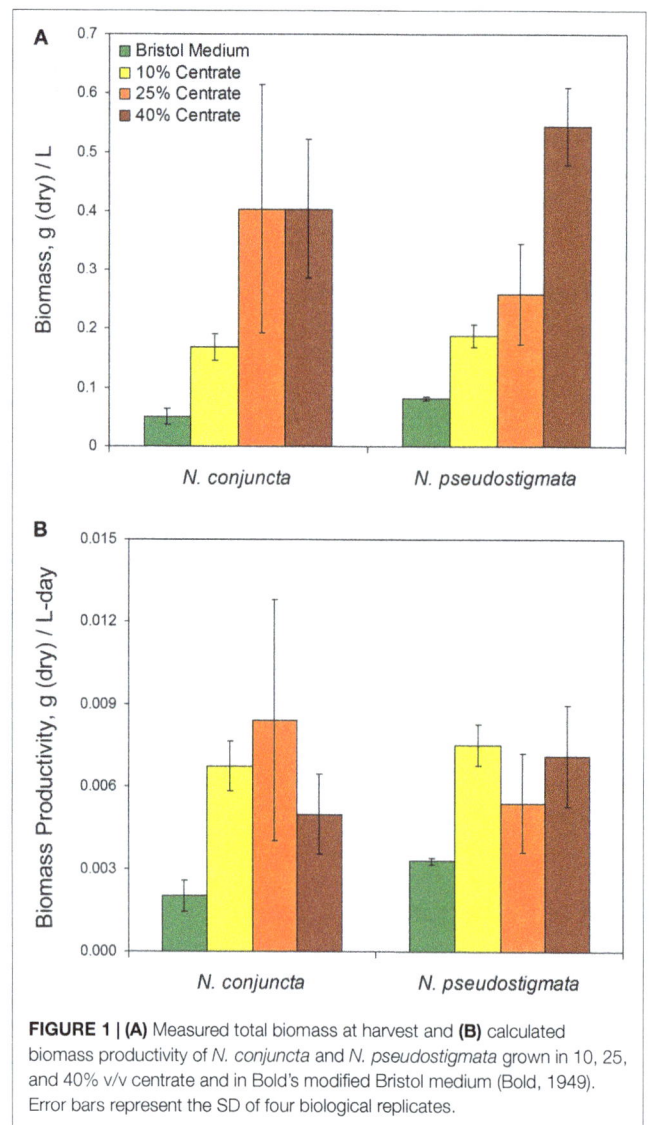

FIGURE 1 | (A) Measured total biomass at harvest and **(B)** calculated biomass productivity of *N. conjuncta* and *N. pseudostigmata* grown in 10, 25, and 40% v/v centrate and in Bold's modified Bristol medium (Bold, 1949). Error bars represent the SD of four biological replicates.

an order of magnitude, from 0.045 to 0.404 g dry biomass/L for *N. conjuncta* and from 0.107 to 0.544 g dry biomass/L for *N. pseudostigmata*. The *N. conjuncta* cultures grown on centrate

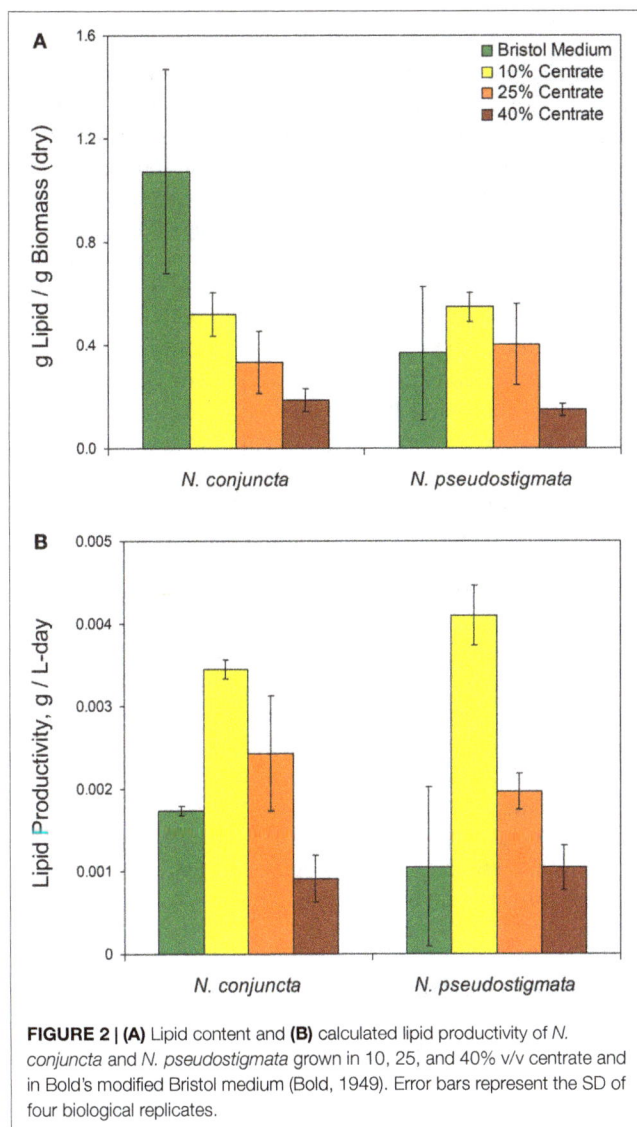

FIGURE 2 | (A) Lipid content and (B) calculated lipid productivity of N. conjuncta and N. pseudostigmata grown in 10, 25, and 40% v/v centrate and in Bold's modified Bristol medium (Bold, 1949). Error bars represent the SD of four biological replicates.

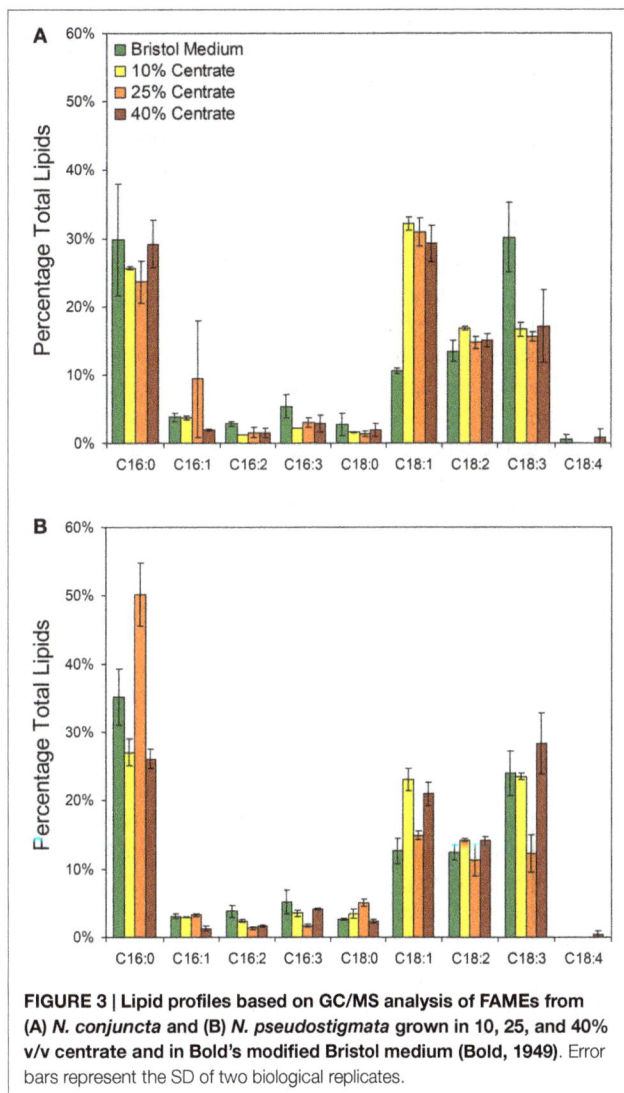

FIGURE 3 | Lipid profiles based on GC/MS analysis of FAMEs from (A) N. conjuncta and (B) N. pseudostigmata grown in 10, 25, and 40% v/v centrate and in Bold's modified Bristol medium (Bold, 1949). Error bars represent the SD of two biological replicates.

all had significantly ($p = 0.0026$) more biomass than the Bristol culture, whereas only the N. pseudostigmata cultures grown on the two higher centrate concentrations had significantly ($p < 0.05$ for 25% v/v; $p < 0.01$ for 40% v/v) more biomass. Because centrate has a very high nutrient content (e.g., ~1000 mg/L N and ~200 mg/L P), the greater biomass amounts were expected.

Despite the greater biomass amounts with increasing centrate concentrations, the amount of dry biomass obtained per day did not appear to be affected by the centrate concentration (**Figure 1B**). There was no significant difference ($p > 0.05$) for either species across the various centrate concentrations, with productivity values ranging from 0.0050 to 0.0084 g dry biomass/L-day. However, biomass productivity of N. conjuncta was significantly ($p = 0.012$) lower at 0.0018 g dry biomass/L-day when grown on Bristol medium. These values were lower than those obtained by Wang and Lan (2011) who reported a biomass productivity of 0.233 g dry biomass/L-day with N. oleoabundans grown on secondary municipal wastewater effluents enriched with nitrogen. Sun et al. (2014) also reported higher biomass

productivities of 0.113 g dry biomass/L-day for N. oleoabundans grown in basal SE medium. Although biomass was not measured for the cultures inoculated with pre-acclimated cells, the reduction in lag times with the higher centrate concentrations would be expected to result in higher biomass productivities due to the shorter cultivation times. However, biomass productivities can vary as a result of other influences, such as mixing of the cultures, CO_2 supplementation, light intensity, and carbon source.

Lipid Production and Productivity

Unlike the biomass, the mass of lipids produced (per mass of dry biomass) was found to decrease with increasing centrate concentrations for both Neochloris species (**Figure 2A**). The N. conjuncta grown in Bristol medium was found to have a lipid content ($1.07 +/- 0.4$ g lipid/g dry biomass) significantly ($p = 0.0004$) higher than any other species/medium combination, which ranged from 0.5 to 50 g lipid/g dry biomass. These values agree with the 0.27 (Sun et al., 2014), 0.50 (Griffiths et al., 2012), and 0.52 (Gouveia et al., 2009) g lipid/g dry biomass reported for N. oleoabundans.

The lipid productivity also decreased with increasing centrate concentrations for both species (**Figure 2B**). Interestingly, the highest lipid productivity (0.0041 g lipid/L-day, *N. pseudostigmata*) was observed with 10% v/v centrate rather than on the Bristol medium with both species. The lowest productivities (0.001 g lipid/g dry biomass) were observed with 40% v/v centrate. This range is similar to the 0.003 g lipid/L-day (Sun et al., 2014) and 0.0029 g lipid/L-day (Griffiths et al., 2012) reported for *N. oleoabundans* under nitrogen-limited conditions. As mentioned previously, the lipid productivities would be expected to increase with the higher centrate concentrations when pre-acclimated cells are used. Therefore, centrate-grown cultures all would likely have greater lipid productivities than those grown in Bristol medium.

In addition to the total lipid measurements, fatty acid profiles were obtained for the two species grown with Bristol medium and various centrate concentrations with the most prevalent fatty acids presented in **Figure 3**. The dominant fatty acids for both *Neochloris* species were palmitic (C16:0), oleic (C18:1), linoleic (C18:2), and linolenic (C18:3). Although the lipid profiles varied by species, the profiles in **Figure 3** agree generally with what others have observed for various *Neochloris* species (Griffiths et al., 2012; Sun et al., 2014) and other green algae (Islam et al., 2013). The most notable differences between the lipid profiles of *N. conjuncta* was the significant increase in C18:1 ($p < 0.01$) and decrease in C18:3 ($p < 0.05$) when changing from Bristol medium to centrate (**Figure 3A**). There were no significant differences in any of the other fatty acids with any growth media for the *N. conjuncta*. Interestingly, very similar changes (e.g., C18:1 increase, C18:3 decrease) were observed when going from nitrogen-replete to nitrogen-limited conditions using *N. oleoabundans* (Griffiths et al., 2012). Lipid accumulation in microalgae is known to occur as a result of stress events, including nutrient starvation, temperature or pH shock, or light limitation (Wang et al., 2010; Li et al., 2011a; Sharma et al., 2012). With these experiments, the cultures were grown until early stationary phase (based on A_{600} before harvesting, which corresponds to ~5 days of growth when at least one nutrient was limiting. The differences in lipid productivity observed between the cells grown in Bristol medium and the centrate solutions are possibly due to a different nutrient becoming limited. Xin et al. (2010) demonstrated that *Scenedesmus* subjected to nitrogen starvation had a 30% increase in lipids, whereas the same cells subjected to phosphorus starvation had a 53% increase, suggesting that nutrient stress affects the degree of lipid accumulation (Xin et al., 2010).

The *N. pseudostigmata* profiles had more variability (**Figure 3B**), but a significant decrease in C18:1 ($p < 0.05$) was observed in the Bristol medium compared to the 10 and 40% v/v centrate cultures. The increase in C18:3 in Bristol was not observed, as was the case with *N. conjuncta*. Instead, a decrease in C18:3 with corresponding increases in C16:0 and C16:3 were observed for the *N. pseudostigmata* culture grown in 25% v/v centrate. Sun et al. (2014) also observed increases in C16:0 accumulation with increasing nutrient stress duration for *N. oleoabundans*; however, they also found an increase in C18:1, which is in contradiction of what was observed here for *N. conjuncta*.

Conclusion

Wastewater centrate offers a promising alternative water source for the cultivation of microalgae for biofuels production. Forty microalgae species were screened for tolerance; two freshwater species were found to grow in up to 40% v/v centrate and multiple saltwater species could grow in up to 50% v/v centrate when supplemented with sea salt. Despite these promising results, many of the strains evaluated were very sensitive to centrate and failed to grow at even low concentrations, and even those strains with high tolerance to centrate had increased lag times with increasing centrate concentration. This lag time could be partially reduced by pre-acclimating the cells to the centrate. Using the two freshwater *Neochloris* strains as models, improvements in both biomass productivity and lipid productivity were observed when the cells were grown on centrate, relative to defined maintenance medium. The lipid profiles of the microalgae grown with centrate and with the maintenance medium were similar, although a significant increase in C18:1 and significant decrease in C18:3 were observed in *N. conjuncta*. Screening of additional microalgae strains should be continued, especially of environmental strains with pre-exposure to the centrate, to identify those strains that can be grown with higher centrate concentrations and also those strains with high lipid content to be used for biofuel feedstock production. The identification of microalgal species or strains adapted to the nutrient profile of a particular waste stream would likely be less expensive than the potential costs associated with supplementing the nutrient profile of a particular waste stream in order to attain optimal growth. Lastly, high performing strains should be evaluated in larger volumes and in raceway ponds or photobioreactors to further evaluate their applicability for full-scale use.

Author Contributions

SRH conducted the research, designed experiments, and wrote the bulk of the manuscript. MSL assisted with experimental design and experimentation and assisted with the preparation of the manuscript. BPK assisted with centrate testing studies, contributed to interpretation of results, and assisted with the preparation of the manuscript. JCC initiated research on wastewater algae, conceived of the study, helped design experiments, and revised the manuscript.

Acknowledgments

Financial support was provided by Department of Energy's Nevada Renewable Energy Consortium (Grant DE-EE0000272). The Nevada Proteomics Center receives financial support from Nevada INBRE, which is funded by NIH Grant number P20 RR-016464 from the INBRE Program of the National Center for Research Resources. The authors would also like to thank Rebecca Albion for all of her assistance in the lab and Rebekah Woolsey for performing the GC-MS analyses.

References

Arce, G., and Bold, H. (1958). Some Chlorophyceae from Cuban soils. *Am. J. Bot.* 45, 492–503. doi:10.2307/2439186

Ben-Amotz, A., Polle, J., and Rao, D. (2009). *The Alga Dunaliella: Biodiversity, Physiology, Genomics, and Biotechnology.* Enfield, NH: Science Publishers, Inc.

Bhatt, N., Panwar, A., Bisht, T., and Tamta, S. (2014). Coupling of algal biofuel production with wastewater. *ScientificWorldJournal* Article ID 210504, 10. doi:10.1155/2014/210504

Bligh, E., and Dyer, W. (1959). A rapid method of total lipid extraction and purification. *Can. J. Physiol. Pharmacol.* 37, 911–917. doi:10.1139/y59-099

Bold, H. (1949). The morphology of *Chlamydomonas chlamydogama* sp. nov. *Bull. Torrey Botanical Club* 76, 101–108. doi:10.2307/2482218

Chisti, Y. (2007). Biodiesel from microalgae. *Biotechnol. Adv.* 25, 294–306. doi:10.1016/j.biotechadv.2007.02.001

Dong, B., Ho, N., Ogden, K., and Arnold, R. (2014). Cultivation of *Nannochloropsis salina* in municipal wastewater or digester centrate. *Ecotoxicol. Environ. Saf.* 103, 45–53. doi:10.1016/j.ecoenv.2014.02.001

Forsberg, C. (2009). Sustainability by combining nuclear, fossil, and renewable energy sources. *Prog. Nucl. Energy* 51, 192–200. doi:10.1016/j.pnucene.2008.04.002

Gomord, V., Fitchette, A., Menu-Bouaouiche, L., Saint-Jore-Dupas, C., Plasson, C., Michaud, D., et al. (2010). Plant-specific glycosylation patterns in the context of therapeutic protein production. *Plant Biotechnol. J.* 8, 564–587. doi:10.1111/J.1467-7652.2009.00497.X

Gouveia, L., Marques, A., Da Silva, T., and Reis, A. (2009). *Neochloris oleoabundans* UTEX #1185: a suitable renewable lipid source for biofuel production. *J. Ind. Microbiol. Biotechnol.* 36, 821–826. doi:10.1007/s10295-009-0559-2

Griffiths, M., Van Hille, R., and Harrison, T. (2012). Lipid productivity, settling potential and fatty acid profile of 11 microalgal species grown under nitrogen replete and limited conditions. *J. Appl. Phycol.* 24, 989–1001. doi:10.1007/s10811-011-9723-y

Guillard, R., and Ryther, J. (1962). Studies of marine planktonic diatoms. *Cyclotella nana* Hustedt and *Detonula confervaceae* (Cleve) Gran. *Can. J. Microbiol.* 8, 229–239. doi:10.1139/m62-029

Herrera, N. (2009). *Analysis of Centrate Composition and Evaluation of its Applicability as a Nutrient Supplement to Irrigation Water.* M.S. Reno: University of Nevada.

Islam, M., Magnusson, M., Brown, R., Ayoko, G., Nabi, M., and Heimann, K. (2013). Microalgal species selection for biodiesel production based on fuel properties derived from fatty acid profiles. *Energies* 6, 5676–5702. doi:10.3390/en6115676

Xin, L., Hong-ying, H., Ke, G., and Ying-xue, S. (2010). Effects of different nitrogen and phosphorus concentrations on the growth, nutrient uptake, and lipid accumulation of a freshwater microalga *Scenedesmus* sp. *Bioresour. Technol.* 101, 5494–5500. doi:10.1016/j.biortech.2010.02.016

Li, Y., Chen, Y.-F., Chen, P., Min, M., Zhou, W., Martinez, B., et al. (2011a). Characterization of a microalga *Chlorella* sp. well adapted to highly concentrated municipal wastewater for nutrient removal and biodiesel production. *Bioresour. Technol.* 102, 5138–5144. doi:10.1016/j.biortech.2011.01.091

Li, Y., Zhou, W., Hu, B., Min, M., Chen, P., and Ruan, R. (2011b). Integration of algae cultivation as biodiesel production feedstock with municipal wastewater

treatment: strains screening and significance evaluation of environmental factors. *Bioresour. Technol.* 102, 10861–10867. doi:10.1016/j.biortech.2011.09.064

Metcalfe, L., Schmitz, A., and Pelka, J. (1966). Rapid preparation of fatty acid esters from lipids for gas chromatographic analysis. *Anal. Chem.* 38, 514–515. doi:10.1021/ac60235a044

Moazeni, F. (2013). *Investigating the Feasibility of Growing Algae for Fuel in Southern Nevada.* Ph.D. Dissertation Las Vegas: University of Nevada.

Mu, D., Min, M., Krohn, B., Mullins, K., Ruan, R., and Hill, J. (2014). Life cycle environmental impacts of wastewater-based algal biofuels. *Environ. Sci. Technol.* 48, 11696–11704. doi:10.1021/es5027689

Pittman, J., Dean, A., and Osundeko, O. (2011). The potential of sustainable algal biofuel production using wastewater resources. *Bioresour. Technol.* 102, 17–25. doi:10.1016/j.biortech.2010.06.035

Sandnes, J., Ringstad, T., Wenner, D., Heyerdahl, P., Kallqvist, I., and Gislerod, H. (2006). Real-time monitoring and automatic density control of large-scale microalgal cultures using near infrared (NIR) optical density sensors. *J. Biotechnol.* 122, 209–215. doi:10.1016/j.jbiotec.2005.08.034

Schnepf, R., and Yacobucci, B. (2010). *Renewable Fuel Standard (RFS): Overview and Issues.* Washington, DC: Congressional Research Service.

Sharma, K., Schuhmann, H., and Schenk, P. (2012). High lipid induction in microalgae for biodiesel production. *Energies* 5, 1532–1553. doi:10.3390/en5051532

Sun, X., Cao, Y., Xu, H., Liu, Y., Sun, J., Qiao, D., et al. (2014). Effect of nitrogen-starvation, light intensity and iron on triacylglyceride/carbohydrate production and fatty acid profile of *Neochloris oleoabundans* HK-129 by a two-stage process. *Bioresour. Technol.* 155, 204–212. doi:10.1016/j.biortech.2013.12.109

Wang, B., and Lan, C. (2011). Biomass production and nitrogen and phosphorus removal by the green alga *Neochloris oleoabundans* in simulated wastewater and secondary municipal wastewater effluent. *Bioresour. Technol.* 102, 5639–5644. doi:10.1016/j.biortech.2011.02.054

Wang, L., Min, M., Li, Y., Chen, P., Chen, Y., Liu, Y., et al. (2010). Cultivation of green algae *Chlorella* sp. in different wastewaters from municipal wastewater treatment plant. *Appl. Biochem. Biotechnol.* 162, 1174–1186. doi:10.1007/s12010-009-8866-7

Zhu, L., Wang, Z., Shu, Q., Takala, J., Hiltunen, E., Feng, P., et al. (2013). Nutrient removal and biodiesel production by integration of freshwater algae cultivation with piggery wastewater treatment. *Water Res.* 47, 4294–4302. doi:10.1016/j.watres.2013.05.004

Conflict of Interest Statement: The authors declare that the research was conducted in the absence of any commercial or financial relationships that could be construed as a potential conflict of interest.

Permissions

All chapters in this book were first published by Frontiers; hereby published with permission under the Creative Commons Attribution License or equivalent. Every chapter published in this book has been scrutinized by our experts. Their significance has been extensively debated. The topics covered herein carry significant findings which will fuel the growth of the discipline. They may even be implemented as practical applications or may be referred to as a beginning point for another development.

The contributors of this book come from diverse backgrounds, making this book a truly international effort. This book will bring forth new frontiers with its revolutionizing research information and detailed analysis of the nascent developments around the world.

We would like to thank all the contributing authors for lending their expertise to make the book truly unique. They have played a crucial role in the development of this book. Without their invaluable contributions this book wouldn't have been possible. They have made vital efforts to compile up to date information on the varied aspects of this subject to make this book a valuable addition to the collection of many professionals and students.

This book was conceptualized with the vision of imparting up-to-date information and advanced data in this field. To ensure the same, a matchless editorial board was set up. Every individual on the board went through rigorous rounds of assessment to prove their worth. After which they invested a large part of their time researching and compiling the most relevant data for our readers.

The editorial board has been involved in producing this book since its inception. They have spent rigorous hours researching and exploring the diverse topics which have resulted in the successful publishing of this book. They have passed on their knowledge of decades through this book. To expedite this challenging task, the publisher supported the team at every step. A small team of assistant editors was also appointed to further simplify the editing procedure and attain best results for the readers.

Apart from the editorial board, the designing team has also invested a significant amount of their time in understanding the subject and creating the most relevant covers. They scrutinized every image to scout for the most suitable representation of the subject and create an appropriate cover for the book.

The publishing team has been an ardent support to the editorial, designing and production team. Their endless efforts to recruit the best for this project, has resulted in the accomplishment of this book. They are a veteran in the field of academics and their pool of knowledge is as vast as their experience in printing. Their expertise and guidance has proved useful at every step. Their uncompromising quality standards have made this book an exceptional effort. Their encouragement from time to time has been an inspiration for everyone.

The publisher and the editorial board hope that this book will prove to be a valuable piece of knowledge for researchers, students, practitioners and scholars across the globe.

List of Contributors

Wegi A. Wuddineh, Mitra Mazarei and C. Neal Stewart Jr.
Department of Plant Sciences, University of Tennessee, Knoxville, TN, USA
Bioenergy Science Center, Oak Ridge National Laboratory, Oak Ridge, TN, USA

Geoffrey B. Turner, Robert W. Sykes, Stephen R. Decker and Mark F. Davis
Bioenergy Science Center, Oak Ridge National Laboratory, Oak Ridge, TN, USA
National Renewable Energy Laboratory, Golden, CO, USA

Kai Deng
US Department of Energy Joint BioEnergy Institute, Emeryville, CA, USA
Sandia National Laboratories, Livermore, CA, USA

Taichi E. Takasuka and Lai F. Bergeman
US Department of Energy Great Lakes Bioenergy Research Center, Madison, WI, USA

Christopher M. Bianchetti
US Department of Energy Great Lakes Bioenergy Research Center, Madison, WI, USA
Department of Chemistry, University of Wisconsin-Oshkosh, Oshkosh, WI, USA

Paul D. Adams
US Department of Energy Joint Bio Energy Institute, Emeryville, CA, USA
Lawrence Berkeley National Laboratory, Berkeley, CA, USA
Department of Bioengineering, University of California Berkeley, Berkeley, CA, USA

Trent R. Northen
US Department of Energy Joint Bio Energy Institute, Emeryville, CA, USA
Lawrence Berkeley National Laboratory, Berkeley, CA, USA

Brian G. Fox
US Department of Energy Great Lakes Bioenergy Research Center, Madison, WI, USA
Department of Biochemistry, University of Wisconsin-Madison, Madison, WI, USA

Nam V. Hoang
Queensland Alliance for Agriculture and Food Innovation, The University of Queensland, St. Lucia, QLD, Australia
College of Agriculture and Forestry, Hue University, Hue, Vietnam

Agnelo Furtado and Robert J. Henry
Queensland Alliance for Agriculture and Food Innovation, The University of Queensland, St. Lucia, QLD, Australia

Frederik C. Botha
Queensland Alliance for Agriculture and Food Innovation, The University of Queensland, St. Lucia, QLD, Australia
Sugar Research Australia, Indooroopilly, QLD, Australia

Blake A. Simmons
Queensland Alliance for Agriculture and Food Innovation, The University of Queensland, St. Lucia, QLD, Australia
Joint Bio Energy Institute, Emeryville, CA, USA

Kai Deng, Joel M. Guenther, Richard Heins, Kenneth L. Sale, Blake A. Simmons, Huu Tran and Anup K. Singh
US Department of Energy Joint Bio Energy Institute, Emeryville, CA, USA
Sandia National Laboratories, Livermore, CA, USA

Jian Gao and Benjamin P. Bowen
Lawrence Berkeley National Laboratory, Berkeley, CA, USA

Vimalier Reyes-Ortiz, Xiaoliang Cheng, Noppadon Sathitsuksanoh, Dominique Loqué and Trent R. Northen
US Department of Energy Joint BioEnergy Institute, Emeryville, CA, USA
Lawrence Berkeley National Laboratory, Berkeley, CA, USA

Taichi E. Takasuka and Lai F. Bergeman
US Department of Energy Great Lakes Bioenergy Research Center, University of Wisconsin, Madison, WI, USA

Henrik Geertz-Hansen
US Department of Energy Joint BioEnergy Institute, Emeryville, CA, USA

Samuel Deutsch
Lawrence Berkeley National Laboratory, Berkeley, CA, USA
Joint Genome Institute, Walnut Creek, CA, USA

Paul D. Adams
US Department of Energy Joint Bio Energy Institute, Emeryville, CA, USA
Lawrence Berkeley National Laboratory, Berkeley, CA, USA
Department of Bioengineering, University of California Berkeley, Berkeley, CA, USA

Brian G. Fox
US Department of Energy Great Lakes Bioenergy Research Center, University of Wisconsin, Madison, WI, USA
Department of Biochemistry, University of Wisconsin, Madison, WI, USA

Fernanda Vargas e Silva and Luiz Olinto Monteggia
Institute of Hydraulic Research, Federal University of Rio Grande do Sul, Porto Alegre, Brazil

Gaojin Lyu
Key Lab of Pulp and Paper Science and Technology of Ministry of Education, Qilu University of Technology, Jinan, China
State Key Lab of Pulp and Paper Engineering, South China University of Technology, Guangzhou, China

Shubin Wu and Hongdan Zhang
State Key Lab of Pulp and Paper Engineering, South China University of Technology, Guangzhou, China

Jennifer J. Stewart and Kathryn J. Coyne
College of Earth, Ocean, and Environment, University of Delaware, Lewes, DE, USA

Colleen M. Bianco
Department of Microbiology, University of Illinoisat Urbana-Champaign, Urbana, IL, USA

Katherine R. Miller
Department of Chemistry, Salisbury University, Salisbury, MD, USA

Adam L. Healey, Agnelo Furtado and Robert J. Henry
Queensland Alliance for Agriculture and Food Innovation, University of Queensland, St. Lucia, QLD, Australia

David J. Lee
Forest Industries Research Centre, University of the Sunshine Coast, Maroochydore, QLD, Australia
Department of Agriculture and Fisheries, Forestry and Biosciences, Agri-Science Queensland, Gympie, QLD, Australia

Blake A. Simmons
Queensland Alliance for Agriculture and Food Innovation, University of Queensland, St. Lucia, QLD, Australia
Joint BioEnergy Institute, Lawrence Berkeley National Laboratory, Emeryville, CA, USA
Biological and Engineering Sciences Center, Sandia National Laboratories, Livermore, CA, USA

Matteo Marsullo, Giovanni Manente and Andrea Lazzaretto
Department of Industrial Engineering, University of Padova, Padova, Italy

Alberto Mian and François Marechal
Industrial Process and Energy System Engineering Group (IPESE), École Polytechnique Fédérale de Lausanne, Lausanne, Switzerland

Adriano Viana Ensinas
Industrial Process and Energy System Engineering Group (IPESE), École Polytechnique Fédérale de Lausanne, Lausanne, Switzerland
Universidade Federal do ABC, Santo Andre, Brazil

Naga Sirisha Parimi, Manjinder Singh, James R. Kastner and Keshav C. Das
College of Engineering, The University of Georgia, Athens, GA, USA

Lennart S. Forsberg and Parastoo Azadi
Complex Carbohydrate Research Center, The University of Georgia, Athens, GA, USA

Quynh Anh Nguyen and Jakyun Jung
Department of Bioenergy Science and Technology, Chonnam National University, Gwangju, South Korea

Dae-Seok Lee
Bio-Energy Research Center, Chonnam National University, Gwangju, South Korea

Hyeun-Jong Bae
Department of Bioenergy Science and Technology, Chonnam National University, Gwangju, South Korea
Bio-Energy Research Center, Chonnam National University, Gwangju, South Korea

Jaya Shankar Tumuluru
Idaho National Laboratory, Idaho Falls, ID, USA

Fan Lin and Laura E. Bartley
Department of Microbiology and Plant Biology, University of Oklahoma, Norman, OK, USA

Christopher L. Waters, Richard G. Mallinson and Lance L. Lobban
School of Chemical, Biological, and Materials Engineering, University of Oklahoma, Norman, OK, USA

Yaqin Qiao
Key Laboratory of Algal Biology, Institute of Hydrobiology, Chinese Academy of Sciences, Wuhan, China
University of Chinese Academy of Sciences, Beijing, China

Junfeng Rong
SINOPEC Research Institute of Petroleum Processing, Beijing, China

Hui Chen, Chenliu He and Qiang Wang
Key Laboratory of Algal Biology, Institute of Hydrobiology, Chinese Academy of Sciences, Wuhan, China

Thomas François Robin, Andrew B. Ross , Amanda R. Lea-Langton and Jenny M. Jones
School of Chemical and Process Engineering, University of Leeds, Leeds, UK

Abimael I. Ávila-Lara, Jesus N. Camberos-Flores and Jose A. Pérez-Pimienta
Department of Chemical Engineering, Universidad Autónoma de Nayarit, Tepic, Mexico,

Jorge A. Mendoza-Pérez
Department of Engineering in Environmental Systems, Instituto Politécnico Nacional, Mexico City, Mexico

Sarah R. Messina-Fernández and Claudia E. Saldaña-Duran
Cuerpo Académico de Sustentabilidad Energética, Universidad Autónoma de Nayarit, Tepic, Mexico,

Edgar I. Jimenez-Ruiz and Leticia M. Sánchez-Herrera
Food Technology Unit, Universidad Autónoma de Nayarit, Tepic, Mexico

Siva kumar Pattathil, Michael G. Hahn and Utku Avci
Complex Carbohydrate Research Center, University of Georgia, Athens, GA, USA
Oak Ridge National Laboratory, BioEnergy Science Center (BESC), Oak Ridge, TN, USA

Tiantian Zhang and Claudia L. Cardenas
Complex Carbohydrate Research Center, University of Georgia, Athens, GA, USA

Sheikh Adil Edrisi and P. C. Abhilash
Institute of Environment and Sustainable Development, Banaras Hindu University, Varanasi, India

Sage R. Hiibel
Department of Biochemistry and Molecular Biology, University of Nevada Reno, Reno, NV, USA
Department of Civil and Environmental Engineering, University of Nevada Reno, Reno, NV, USA

Mark S. Lemos, Brian P. Kelly and John C. Cushman
Department of Biochemistry and Molecular Biology, University of Nevada Reno, Reno, NV, USA